Molecular genesis

First cellular life, prokaryotes, cellular evolution

Complex cellular life, inception of eukaryotes

Eukaryotes organize, metazoans arise

Snowball earth

Multicellular animals evolve, first evidence of multicellular life

Evolution of complex life forms

Initial flourish of Animalia

Explosion of animalian body designs, first known eye

Trilobites and Mollusca expand

Bony fish appear, plants begin to populate land

Age of fish, terrestrial insects arise, tetrapods come ashore

Oxygen levels rise, trees predominate, insects flourish

Life expands, pelycosaurs and amphibians predominate

Age of dinosaurs

Turtles and crocodilians appear

Pangaea breaks apart, ichthyosaurs flourish

Dinosaurs predominate, birds arise, angiosperms begin

Mammals diverge into many different environments

Birds predominate initially, mammals thrive

Humans appear, life flourishes

Evolution's Witness

Odontodactylus scyallurus

Mantis Shrimp

Image © James Brandt, MD

Evolution's Witness

How Eyes Evolved

Ivan R. Schwab University of California, Davis

Histology by Richard Dubielzig, DVM, and Charles Schobert, DVM
Comparative Ocular Pathology Laboratory of Wisconsin (COPLOW)

OXFORD
UNIVERSITY PRESS

OXFORD
UNIVERSITY PRESS

Oxford University Press, Inc., publishes works that further
Oxford University's objective of excellence
in research, scholarship, and education.

Oxford New York
Auckland   Cape Town   Dar es Salaam   Hong Kong   Karachi
Kuala Lumpur   Madrid   Melbourne   Mexico City   Nairobi
New Delhi   Shanghai   Taipei   Toronto

With offices in
Argentina   Austria   Brazil   Chile   Czech Republic   France   Greece
Guatemala   Hungary   Italy   Japan   Poland   Portugal   Singapore
South Korea   Switzerland   Thailand   Turkey   Ukraine   Vietnam

Published by Oxford University Press, Inc.
198 Madison Avenue, New York, New York 10016
www.oup.com

Library of Congress Cataloging-in-Publication Data

Schwab, Ivan R.
Evolution's witness : how eyes evolved / Ivan R. Schwab ; histology by
Richard Dubielzig, Charles Schobert.
    p. cm.
Includes bibliographical references and index.
 ISBN 978-0-19-536974-8 (hardback)
1. Eye—Evolution. 2. Adaptation (Biology). 3. Anatomy, Comparative.
4. Evolution (Biology) I. Title.
[DNLM: 1. Eye—anatomy & histology. 2. Adaptation, Biological.
3. Anatomy, Comparative. 4. Biological Evolution. 5. Vision,
Ocular--physiology. WW 101]
QP475.S374 2012
612.8'4—dc23   2011016413

3 5 7 9 8 6 4 2
Printed in China
 on acid-free paper

To all my students, who have taught me much

CONTENTS

Evolution's Witness is a wonderful contribution to all the celebrations of the 200th anniversary of Darwin's birth and the 150th anniversary of publication of the *Origin of Species*. We know that the eye posed a challenge for Darwin that he rightly acknowledged saying:

> "To suppose that the eye, with all its inimitable contrivances for adjusting the focus to different distances, for admitting different amounts of light, and for the correction of spherical and chromatic aberration, could have been formed by natural selection, seems, I freely confess, absurd in the highest possible degree. . . . "

Over the years, his thoughtful comments have often been taken out of context by anti-evolutionary activists who purposely ignored the rest of his statement that:

> . . . yet reason tells me, that if numerous gradations from a perfect and complex eye to one very imperfect and simple, each grade being useful to its possessor, can be shown to exist; if further, the eye does vary ever so slightly, and the variations be inherited, which is certainly the case; and if any variation or modification in the organ be ever useful to an animal under changing conditions of life, then the difficulty of believing that a perfect and complex eye could be formed by natural selection, though insuperable by our imagination, can hardly be considered real. (*On the Origin of Species, 1859*)

So now, in your hands are the current versions of those "variations" in eyes Darwin could not have known. Professor Schwab's remarkable collection of "witnesses" to the evolutionary history of the eye offers a clear, comprehensive historical look at eyes as they evolved and their remaining versions. By providing a glimpse of eyes and eye evolution within each era of time from the beginnings of life to the present, he provides exceptional and impressive scholarship, likely to be consulted by everyone interested in evolution and eyes. The vivid descriptions beginning with the chemistry of early life on earth and continuing through to the present reveals the consequences for organisms at every stage and evokes a wonderful sense of place and time. This is a book that will be hard to put down.

Why is it that eye evolution seems hard to understand? Partly because although some eyes left a wonderful fossil record, for vertebrate eyes, there was a fossil trace only of where they were on the body but not much else! Soft tissue, like behavior, has had to be reconstructed by clever use of the geography of the fossil record, slender glimpses of likely scenarios, and now, by DNA sequence comparisons. But perhaps more important, eyes are not one simple "organ" that might be traced with some ease starting now and reaching back through time, using DNA or other methods. Instead, eyes are actually an amalgam of structures, some of independent origin with completely separate evolutionary origins and trajectories. This has been a fundamental puzzle for sleuths of eye evolution.

Quite early on, molecular tools told us that all eyes shared a protein, opsin, for collecting light, and that led to a burst of DNA sequences of opsin, revealing just how it had been modified to match the light it had evolved to catch. And families of opsins matched the family trees of their carriers yielding a satisfying window into opsin evolution tracking morphological evolution. But when the same techniques were applied to lenses, the story seemed murky at best. Instead of a common lineage, open to easy understanding, lenses were clearly different. Lenses are made of many proteins, and a few of these seemed to have a long history, but many, in the same lenses, did not. Instead, the lens proteins seemed to be a ragtag collection of molecules from various different sources, depending on the animal. Enzymes normally found in liver turned up in some bird lenses and the gene encoding these proteins was shared between these organs, an evolutionary first. Not only were the lens proteins gathered from many sources, making them hard to trace phylogenetically, but lenses were much more structurally diverse than opsins. Some were simple, like ours, but others had versions of telescope optics and other arrangements, often requiring multiple components. So, rather than being in lockstep with opsin evolution, lens evolution was much harder to understand. Possibly this was because converting photons into electrons, the way opsins do, solved a problem in a unique way while nearly any protein could serve as a lens if it were shaped properly.

This puzzling turn of events has been slowly unraveled to show that lenses are an add-on to many eyes, added from many sources at different times and yet in all instances solving the same problem of collecting and focusing light to make eyes work.

All of this evolutionary history is woven seamlessly into this entertaining prose and laced with useful metaphors and sly humor. Prof. Schwab has written a book that will clearly be a classic reference much like *The Vertebrate Eye* (Gordon Walls, 1942) and *The Vertebrate Retina* (Bob Rodieck, 1973). These two tomes are in the library of everyone who studies eyes, and now they will be joined by *Evolution's Witness*. I imagine with delight that students in future classes will be able to see the vast diversity of eyes in all their glory instead of being subjected to dry descriptions and line diagrams. The volume may also inspire scientists to study eyes that lie outside the normal range, giving us greater insights into their functions in the lives of organisms.

Vision as a sense and eyes as the organ allowing us to use light information will be opened to a new range of experimentation that will have effects over many fields. Scientists are fortunate that Professor Schwab has undertaken this labor of love.

—Russell Fernald, PhD

This book could not have been accomplished without the help and support of many people, and to these individuals I owe a debt of gratitude. I am truly amazed, honored, and humbled that all of these people helped with a genuine sense of stewardship. Some helped in spite of knowing that, at times, I was way out of my field. All provided invaluable assistance.

I wish to thank Joanna Aizenberg PhD, Catherine Arrese PhD, Ron Atkinson, Foster Bam, Karlheinz Baumann, Giff Beaton, Roy W. Bellhorn DMV MSc, John Binns, Thomas Blankenship PhD, Liz Borda PhD, Brian Boyle, James Brandt MD, Dennis Brooks DVM PhD, Denise Brundenall MA,VetMB, Joe Burgess, Marie Burns PhD, Nedim Buyukmihci VMD, Michelle Campbell MS, William Capman PhD, Catharine Carlin, Tammy Chi, Melissa Coates PhD, Bernie L. Cohen PhD, Carmen MH Colitz DVM PhD DACVO, Shaun Collin BSc, MSc, Kristaan D'Aoüt PhD, William Debello PhD, Guido Dehnhardt PhD, Steve Dolberg, Richard Dubielzig DVM, Jack Dumbacher PhD, Casey Dunn PhD, Ralph C. Eagle Jr. MD, Doug Erwin PhD, Mark Feldman MD, Dante Fenolio, Brock Fenton PhD, Russell D. Fernald PhD, Frank D. Ferrari PhD, Ron Fishman MD, Erika Fitzpatrick, Krisztina Forward, Kerstin Fritsches PhD, Jim Gehling BSc, MSc, PhD, Judy Gire, Tom Gire, Penelope Gullan BSc, Bruce Hallett, Frederike Hanke PhD, Duane P. Harland PhD, Ed Harper, Nathan Hart PhD, Phil Hastings PhD, S. Blair Hedges PhD, Ron Hedrick PhD, Gordon Hendler PhD, Tim Hengst, Lawrie Hirst MD, Viet Ho MD, Nickolas Holland BA PhD, Creig Hoyt MD, Denise Imai DVM PhD, Andrew Ishida PhD, Robert R. Jackson BSc PhD, Darren Jew, Sönke Johnsen PhD, Tom Jorstad, Stephen Kajiura PhD, Michael D. Kern, Lynn Kimsey PhD, William Kirsten Jr. PhD, Cynthia Klepaldo, Mikki Kobza-McComb PhD, Dan Kramer PhD Cyano Biotech GMBH, D'Aout Kristiaan PhD, William Kristan PhD, Pat Kysar, Trevor Lamb ScD, Michael Land PhD, Carlito Lebrilla PhD, Lena Linck PhD, Ronnie Lipton, William Lloyd MD, Martin Mach, Aynsley MacNab, Larry P. Madin PhD, David Maggs PhD DVM, James Major Jr. MD, PhD, Justin Marshall PhD, Georg Mayer PhD, John McCosker DSc PhD, Robert McDowall MSc PhD, Margaret McFall-Ngai PhD, Paul McMenamin BSc, MSc, DSc, PhD, Beno Meyer-Rochow MSc, BSc, PhD, David Miller MD, Anthony C. B. Molteno MD, Robert Munn PhD, Chris Murphy DVM, Mike Murray DVM, Dan Eric Nilsson PhD, Ichiro Nishii PhD, Dieter Oesterhelt PhD, Rebecca Papendick DVM, David Pearson PhD, Simon Petersen-Jones DVM PhD, Jack Pettigrew BSc (Med), MSc, MB, BS, FRS. Michael Pfaff, Joram Piatigorsky PhD, Robert L. Pitman PhD, Edward Pugh PhD, Pat Randolph PhD, Ellen Redenbo, Dennis A. Redfern, Christopher M. Reilly DVM, Greg Rouse PhD, Jes Rust PhD, William Saidel PhD, Mark Schneegurt PhD, Lars Schmitz PhD, Charles S. Schobert MS, DVM, J. Anthony Seibert PhD, Ricardo Setti-Levi PhD, Thomas Shahan, Brad Shaffer PhD, Amber Shawl, Myron Shekelle PhD, Brad Shibata, Neil Shubin PhD, Mark Siddall BSc, MSc, PhD, Jacob Sivak PhD, David B. Snyder PhD, Dan Speiser PhD, Jan Storey PhD, Ken Storey PhD, F.J.R. "Max" Taylor PhD, Annie Townsend, Mick Turner, Casey Y.-J. Ung MD, MBBS (Qld), John Utterback, Utterback Farms, Jose Valadez, Michael Vecchione PhD, Peter Wainwright PhD, WJ Walker, Eric Warrant PhD, Richard Young PhD, and Gavin Young BSc, PhD.

Throughout the book there are numerous images of ocular histology, and I wish to thank Janice Lokkem at the Eye Pathology Laboratory of Wisconsin for preparation of this histology.

I wish to thank institutions that have helped with this work including the California Academy of Sciences and the Steinhart Aquarium San Francisco; Dolphin Quest; the Entomology Department of the University of California, Davis; the International Reptile Conservation Foundation; the Monterey Bay Aquarium; the Natural History Museum of Los Angeles County; the National Institute of Water and Atmospheric Research, New Zealand; the National Museum of Natural History, Washington, DC; the Natural Museum London; Oxford University Press; the Perth Zoo; the Royal Ontario Museum and Parks, Canada; the School of Biological Sciences, and the Vision Touch and Hearing Research Centre at the University of Queensland, Brisbane, Australia; the San Diego Zoo; the Smithsonian Institution, the South

Australia Museum, the West Virginia Raptor Rehabilitation Center, and the Woods Hole Oceanographic Institution.

My reviewers have been a godsend to me in advice and counsel. These include Shaun Collin BSc, MSc, PhD, Mark Feldman MD, Ron Fishman MD, Judy Gire, Tom Gire, Creig Hoyt MD, John L. Keltner MD, Lena Linck PhD, Mark Mannis MD, Joram Piatigorsky PhD, Greg Rouse PhD, William Saidel PhD, my wife Nora Schwab, and Peter Wainwright PhD.

The editor Ronnie Lipton and publisher Catherine Barnes as well as Oxford University Press have been extraordinary in their assistance and hard work to see this published and have provided invaluable assistance.

Life's 3.75 billion years of history so staggers our thinking as to distort our understanding of evolution. Yet the time-frame is not the only reason for distortion. Along the evolutionary trail, much of life on earth came and went with little record, because so many of these initial details were lost and not preserved. Still, there is one enduring record: evolution's witness is photoreception. Following photoreception's thread will help us to understand evolution in general and the evolution of the eye in particular because life evolved with light and its perception. That light stimulus would eventually lead to the eye.

This is the story of the evolution of the eye. Like evolution of life in general, the eye's early trail is obscure. Most eyes are composed of soft tissue so they do not fossilize, and so much detail has been lost in mud, volcanic debris, and deep time. Nevertheless, we know much about how the eye evolved, pieced together from available fossil records, comparative anatomy, physiology, optics, and genomics. In this volume we will review these bits of evidence in the context of time and place. For instance, available fossil records illustrate that the Cambrian explosion spawned the simultaneous birth of the principal invertebrate compound eye and the vertebrate camera-style eye. Evidence suggests that these basic morphologic forms evolved and expanded, all within 10 million years or so. But, we cannot know exactly how much organization in cellular function or morphology occurred in the Cambrian or in the preceding periods. For that matter, eyes probably began in some form much earlier in the pre-Cambrian. More than likely, much was accomplished very quickly in the Cambrian as basic morphology, and visual processing was likely well established in the pre-Cambrian.

In this text chronology dictates the initial narrative, but the text then follows each lineage in most phyla or classes to current time to illustrate the evolution within those related species and within the lineage. Hence, a particular animal and its eye may be used as a model for discussion and illustration even if that animal described may not have lived in that period. For example, the avian eye is discussed in the Cretaceous time frame even though most of the bird species discussed never lived in that period. We do not know precisely when the theropods (a clade of dinosaurs) evolved into birds, but our best evidence suggests the Cretaceous. We also do not know the eye's form during that transition, and certainly the avian eye has evolved since the Cretaceous. But we *do* know that the theropods lived during the Cretaceous. And we know the reptilian and avian eyes of today, so we can compare the similarities and differences of those eyes. With that knowledge, we can predict, or at least better understand, the eye of the last common ancestor that would radiate into reptiles and birds. In particular, we have some understanding of the state of the eye of the theropods, the dinosaur lineage that would sire birds.

Individual eyes are of interest in their position within an era, to be sure. But, the eyes are even more interesting in the animals that possess them. On a geologic timeframe, each eye is evolving with its host, so although the description is a moving target, it moves slowly. This relentless pace leads to rough estimates of the exact timing of any eye's description. For example, depending on classification, reptiles have various ancestral beginnings ("polyphyletic"), and that lineage includes various groups such as true lizards, snakes, tuataras, crocodiles, and turtles, not to mention the extinct lineages. So, although the "reptilian" eye was probably advanced well before the Triassic, we cannot consider that all reptiles harbor the same eye. Nevertheless, because of the common features of each reptilian eye, we can assume a common ancestor, at some point. The eye of that last common ancestor can be illustrated by the commonality exhibited in its progeny.

For example, the Tuatara, a basal lizard, possesses one of the oldest living species of the "reptilian" eye, even presaging the dinosaurs. Although the "reptilian" eye is not homogeneous, yet, at some point, all of the reptiles and their associated eyes radiated from a common ancestor, and the Tuatara was close to that ancestor phylogenetically. Although this ancient species does not have the same eye it did when it first evolved, the commonality with other reptiles permits us to understand the evolution of the reptilian eye. Hence, you find the entire description of the reptilian eye in the context of its close relatives and in one position in the book, although the different reptilian eyes developed over hundreds of millions of years and sometimes in different directions.

We review the eye of a class within a phylum already discussed when that class appears later in the story if these animals have experienced significant additions or changes in the eye's morphology, physiology, or neurology. For example, the invertebrate and vertebrate eyes are revisited when each takes flight, as the aerial challenge required different eyes and different brains. The contextual tapestry of time and surrounding events is important to understanding the divergence of eyes even within the phylum at a class level. Hence, some phyla will reappear in the text and be parsed to the class level when such changes exist.

Some of the book's stories about comparative anatomy and physiology are derived from a monthly series of essays I wrote for the *British Journal of Ophthalmology*. Some accounts are waypoints for the story to illustrate a biologic principle or describe a particular eye. These and other included essays illustrate the visual mechanisms, but not necessarily the organism, described at that point in the evolutionary history. For example, although we cannot know how the first worm's eye formed, we can illustrate the process with an extant worm species.

Whereas evolution occurred on a grand scale, animals go about their individual lives on a much more limited, parochial, and more interesting scale. It is these individual quotidian lives that we will discuss. For example, it is equally interesting to know what the mantis shrimp does with its sixteen or so visual pigments (compared to our three) as it is to know the abstract evolution of its visual pigments.

Yet we must also examine a few abstract evolutionary principles, such as the evolutionary concept of homologous versus analogous. Homologous means that a molecular combination or morphologic structure has shared ancestry; analogous means that the combination or structure has the same or similar function but no shared ancestry. For example, a monkey's hands and a dolphin's flippers are homologous as forelimbs, as these structures have shared ancestry. But, the wings of bats and bird's are analogous, as they share a purpose but not an ancestry as wings.

Another, and more important, abstract principle we must address here is a definition of an eye since we are chronicling its evolution from first photoreception to current adaptations. What, then, defines an eye? A common

definition: an eye is a structure that can compare the amount of light coming from different directions. That definition is not precise enough because although the eye started with photoreception, it is more than that. For a more precise definition, and for this text, we will consider the eye as an optical device that receives and recognizes light and has the ability to define spatial detail. Optically, that includes the comparison of light coming from different directions as described in the first definition.

Such a complex sensory end organ as the eye had a gradient of changes that permitted its establishment and growth, and we cannot know all of the details of these changes. Some molecular similarities across phyla suggest that much of that ocular evolution occurred earlier and more quickly than has been generally believed, and some differences are so profound between phyla to suggest multiple origins of eyes. Importantly, ocular evolution has been influenced by random external events such as comet strikes, and must be viewed in that context.

The general conditions of each era will be discussed, giving you some perspective on the timeframe in question. The overleaf and inside cover contain a color-coded timeline, and each page is edged in the color that corresponds to the time period being discussed.

The prologue and the first three chapters prepare the ground for the evolution. Subsequent chapters follow different eyes and phyla against the backdrop of the timeframe. Boxed highlighted texts in blue are supplemental information that provide more complex details for the interested reader but are not necessary for the flow of the general story. A glossary can be found at the end of the book for words that might initially be unfamiliar.

The bibliography will help the interested reader to discover more of this history of sensory perception. Citations appear according to chapters at the end of the text, although the general references (key books as suggested reading) provide basic and expanded degrees of information if desired. These are grouped at the end of the bibliography section.

Appendix A shows basic eye anatomy with a labeled illustration of a human camera-style eye to assist understanding terminology for those who do not work with the eye on a regular basis. This illustration with its accompanying explanation will help you understand the morphology of the eye. Appendices B through H add detail for those with special knowledge or deeper interest in a specific topic.

Evolution's Witness

PROLOGUE

MOLECULAR GENESIS

HADEAN EON

4600–3750 MILLION YEARS AGO

Hadean Birth

It must have been the most hellacious of births. Approximately 4600 million years ago (mya), in a condensing, noxious, hissing steam of raw materials, early earth formed from amorphous chunks of debris swirling together and coalescing into a mass large enough to generate its own gravity. That unspeakable violence and energy associated with bombardment and bleak space would have permitted, encouraged, and even forged chemical change. The prebiotic bouillabaisse of amino acids, reactive molecules, unstable metals, and raw energy of the planet was certainly pummelled and pounded time and again. Asteroids, small planets, and water-laden comets hammered the new planet. Many of these objects contained organic-like molecules and possibly the seed of first life. Most of the Hadean eon had little to no possibility of life, as these sulfurous times were beyond harsh. This eon set the stage, though, with prebiotic molecules capable of assembly into life. Eventually, and by chance, chemical change would have hit upon a molecule that was self-replicating.

THE AGE OF FIRST CELLULAR LIFE

ARCHEAN EON

3750–2500 MILLION YEARS AGO

1

Only a tiny fraction of the sun's light and energy bathes the earth, but it is all we need. Virtually all life on earth depends directly or indirectly on the sun's radiant energy. Thus, the beginning of photoreception on earth is close, at least geologically, to the beginning of life.

Our story begins when life forms first recognized light. Although we cannot know exactly when that occurred, it had to be early in life's history—probably within the first few million years if not earlier.

Prokaryotes

Early Cells—Protobionts

No matter when life began, the first few hundred million years of cellular life differed markedly from anything we know today. The sun's energy was more intense, with almost no atmosphere to protect the surface and nascent life. Temperatures were probably 60–71°C (140–160°F); they may have cooled somewhat by the end of that period, but not to the temperatures of today. The days were shorter as the earth spun faster. Volcanoes were exploding, and noxious chemicals were swirling about, indiscriminately snuffing out all but the hardiest of new life forms. These early torturous conditions must have challenged, shaped, and extinguished early life—perhaps many times.

The constant bombardment of meteors and comets energized and guided the formation of the chemical milieu on the surface. There was little atmosphere and no free oxygen. Carbon dioxide, nitrogen, and methane were what passed for an atmosphere. Scattered landmasses were present. So too were global oceans, with the water coming mostly from comets. The water likely contained dilute organic compounds coated or combined with oily films.

This random development led to the formation of molecules that combined and recombined, beginning a crucial process. Amino acids appeared and combined, creating peptides, followed by proteins. Then thin bubbles, like those that form on a stagnant pond, appeared. Next came lipids, perhaps on a rocky surface or in a shallow aquatic environment, formed by more random combinations, with a reactive metal or clay as a catalyst. Eventually, the combined proteins and membranes created the first cell.

Although life's origins remain somewhat a mystery, several possibilities exist. Perhaps life began beside the fumaroles on the ocean's floor with access to sunlight and nutrients, as these vents were far shallower than they are now. Perhaps life began in shallow pools and radiated downward toward those vents in the ocean floor. And perhaps most of early life was wiped out many times only to spring again from the vents or from small puddles on rock surfaces.

Wherever cellular life began, it did so as no more than self-replicating molecules requiring energy input and a constant supply of chemicals to continue. The enormity of time combined with molecular trial and error randomly spawned successful replicating molecules. But once replication began, the course was set. These molecules succeeded because they represented improvements in replication and energy conservation, hence, survival. These precursors—called protobionts—may not have been living in all senses of the word, but they did share certain traits with living cells.

The prebiotic soup would have become very rich, especially with replicating molecules. Having no nucleus, the early cells would have had less internal organization and regulation than later generations with nuclei but still enough to winnow less competitive cells. Most of the early cells' activities would have been strictly biochemical, perhaps guided by early ribonucleic acid (RNA).

One domain of single-celled organisms, the *Archaea*, arose and continued development toward multicellular organisms. We can only speculate about the timing and hierarchy.

First Life

The first truly living cells are called *prokaryotes*. These single-celled organisms play a key role in providing the raw

BOX 1.1 CELLULAR EVOLUTION

A favorite and entirely plausible theory of first cells' origin is called "RNA world." Biomolecules eventually developed the ability to replicate as RNA, which probably preceded life or at least appeared in very early life. Deoxyribonucleic acid (DNA), a chemically more stable repository of genetic rules than RNA, probably came later. (But DNA and its proteins did appear in these early cells before internal structures, such as a nucleus, did.)

materials of photobiology to the Metazoa. The relationship of prokaryotes to eukaryotes (more complicated cells with a nucleus and membrane-enclosed internal structures, called organelles or "little organs") is not clear, but here is a possible scenario within the context of the visual witness.

A last common ancestor—descendants of the only surviving line of the first cellular life—gave rise to two prokaryotic domains and one eukaryotic domain. One prokaryotic domain consists of the bacteria (Figure 1.1), which many believe came first. The other consists of the Archaea (Greek for ancient and the name of the eon) (Figure 1.2). Archaea differ biochemically and in other significant manners from bacteria and include the extremophiles that populate the geysers and hot springs of Yellowstone, among other locations.

The third domain includes all eukaryotes—*Eukaryota*—which are aerobic organisms (at least all living eukaryotes are aerobic) with cells that contain a nucleus and organelles. In addition to protists, which are organisms of a single cell with a nucleus, this group also includes all multicellular organisms. Virtually all multicellular organisms are Metazoa, meaning that they possess tissues such as muscles or nerves. The exceptions, including sponges, are known as Parazoa. It is not completely clear how closely related single-cell eukaryotes (protists) are to multicellular eukaryotes (Metazoa), but protists probably radiated into Metazoa, perhaps through some extinct Parazoa but probably not sponges. Importantly, the Archaea are more closely related to eukaryotes than are bacteria, and hence Archaea is more closely related to Metazoa than are bacteria (Figure 1.1).

Eventually, one of these Archaea added endosymbiosis to its repertoire of survival mechanisms. Endosymbiosis means that one organism will engulf or consume another organism without digesting it. The engulfed or consumed endosymbiont will create a cooperative network with the consuming cell, causing the relationship to prosper and become more intricate. The cellular organization and atmospheric conditions required for this acquisition of an endosymbiont—and much development of photoreception and the eye—took approximately 1.5 billion years.

First Witness

The first witness remains unknown, but examples of what that first prokaryote might have been are still alive. These can help us understand how early cellular life harnessed the power of the sun using chlorophyll.

Phylum Cyanobacteria

Many direct descendants of these hardy pioneers believed to be among the first single-celled organisms (prokaryotes) are still

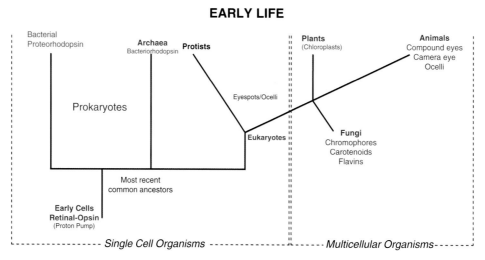

EARLY LIFE

Figure 1.1 Cladogram of early life.

Figure 1.2 Grand Prismatic Spring. Large colorful hot spring in Yellowstone National Park. Hot springs are home to many species of thermophiles—organisms that thrive at high temperatures. The spectacular bright colors in these hot springs are produced by the organisms themselves. This hot spring maintains a temperature of approximately 22° C (72°F) the year round, but the water cools as it moves toward the edges of the pool. Specific Archaea and algae thrive at the different temperatures, accounting for the different colors in the pool. *Images © Steve Dolberg.*

with us, although evolved, and can help us understand photoreception. The ubiquitous cyanobacteria are perhaps one of the oldest living organisms, having originated between 3.75 and 2.6 bya (Figure 1.3). Although cyanobacteria are commonly called blue-green algae, they are a species of bacterium, not an alga or a plant.

Cyanobacteria are photoreceptive through a compound they possess—chlorophyll, a photopigment. Chlorophyll may well have been the first photoreceptive witness, although there are other candidates. Using chlorophyll, cyanobacteria employ light to produce energy in the form of a carbohydrate. Cyanobacteria are our first known proof, or witness, of photoreception with a response to the light, shaped by the early sun's energy and the climate of that day. Cyanobacteria exhibit one key feature of early life on earth that would be essential to later evolution. That is, they are capable of photosynthesis in which organisms

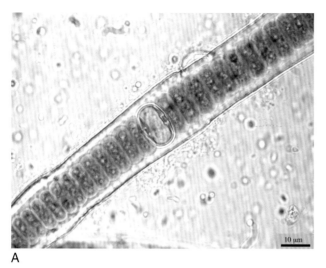

A

B

Figure 1.3 (A) *Scytonema*, a common, highly motile, aquatic photosynthesizing cyanobacterium. This organism contains *phycoerythrin*, a red protein, and chlorophyll, a green pigment. Both proteins combine in the process of photosynthesis. Image by Dan Kramer (Cyano Biotech GMBH), Mark Schneegurt (Wichita State University), and Cyanosite (www-cyanosite.bio.purdue.edu).
(B) *Cyanobacteria*. A stromatolite formation that produces oxygen as a waste product. The bubbles are oxygen.

make nutritional substances internally with the use of sunlight (also called photo-autotrophs). Through endosymbiosis, a common agent of change in life's early history, protists engulfed the cyanobacteria, incorporating them to form what we now know as a chloroplast (an organelle). Chloroplasts convert sunlight to energy and create oxygen as a waste product. Over hundreds of millions of years that oxygen accumulated to create an atmosphere that would support aerobic eukaryotes and eventually the broad range of Animalia (Figure 1.3B).

Other prokaryotes would incorporate another compound (bacteriorhodopsin or proteorhodopsin) into their outer cellular membrane, also for the transfer or capture of energy. But this was not endosymbiosis. Prokaryotes (and even eukaryotes) have other common and interesting ways, including lateral gene transfer, to transfer genetic material.

Bacteriorhodopsin and proteorhodopsin or, more likely, their unknown predecessors, are other candidates for the first photoreceptive witness. These compounds appeared for purposes other than vision, although they are photoreceptive. That is, they receive light, and light provokes a response.

The Road to Cellular Success

Retinal (or Retinaldehyde, a Form of Vitamin A)
When the first cell arose, it was likely sedentary and moved only with the substrate in which it lived. Motility would have represented a major innovative step, providing access to additional food sources. But motility would require energy. A successful cell, then, would need an energy source.

The rhodopsin family in single-celled organisms, which includes bacteriorhodopsin, proteorhodopsin, and more than 800 other members, can capture the energy in light and transfer that energy into a cell by moving protons. That ability, provided by vitamin A–derived retinal, is known as a proton pump. Retinal belongs to a class of compounds known as chromophores, and almost all of them, including retinal, are colorful, which as we will see is useful to the cell. Retinal combines with a protein, opsin, to create rhodopsin. The opsin assists in selecting the peak wavelength of light that will stimulate the rhodopsin molecule.

All rhodopsin family members in single cells are proton pumps because of retinal. The first early cell to incorporate this molecule into its membrane by simple chance would have found a powerful energy-producing compound. Although a proton pump does not *create* energy, it does create the electrochemical gradient across a membrane that *stores* energy. This mechanism would give one cell a competitive advantage over other early cells.

A photosynthetic pigment, bacteriorhodopsin, is a proton pump for many Archaea. Proteorhodopsin works similarly in many bacteria. Both permit their hosts to function and grow with only light as an energy source, a significant advantage, and yet in these organisms, neither compound has a role in vision.

Halorhodopsin, another rhodopsin found in a salt-tolerant Archaea, Halobacteria, is both an energy source *and* a light-detection device. It may have been our first photoreceptive molecule or visual witness engaged for "sensory" purposes and illustrates the ability of evolution to co-opt molecules for other purposes, even at a basic level.

All rhodopsins would potentially provide an energy source for these single cells, perhaps even light detection. An energy source would permit early phototaxis, which is the ability of the organism either to align to or avoid light—key advantages in the evolution of life.

This incorporation of at least one of the rhodopsin compounds occurred with the Archaea; evidence also suggests that incorporation of proteorhodopsin with the first bacteria occurred separately. Possibly that last common ancestor had a more basal compound and transferred the basic genetic code for each compound to its successors. Today, Archaea are commonplace and found in many environments including soils and in the boiling hot springs, fumaroles, and geysers of Yellowstone National Park and even the oceanic vents with their intense pressures and temperatures. These unusual organisms are quite tolerant of very high temperatures, high pressure, and even high salt concentrations (Figure 1.2).

So bacteriorhodopsin or a related compound initially was incorporated into the cell membrane for more than one reason, but vision is probably not one of them. Although bacteriorhodopsin may have been our first visual witness, it was not an *eye* in any sense. But this witness *was* winnowed out of the early soup because of its ability to respond to light.

The innovation had powerful consequences. Photoreception and transduction (the transfer of energy to the cell, in this case) brought significant evolutionary advantages. After all, light provides energy. And, as we have seen before, photoreception was an early and evolving witness to evolution. Retinal, the photosensitive vitamin A derivative, and perhaps the opsins, were present in that last common ancestor of Archaea and bacteria. Hence, they originated once and were inherited by all descendants. These compounds could be easily transferred from one cell to another. So-called lateral gene transfer between prokaryotic cells permits distantly related, or even unrelated, organisms to exchange information. Information may transfer in one direction from one cell to another (through a virus-like organism called a phage), in a mutual exchange between two single-celled organisms (in a process called conjugation), or by other methods. Lateral gene transfer represents an important channel of genetic exchange in addition to cellular replication and reproduction.

Photolyases and Cryptochromes

Early in the Archean Eon, ultraviolet (UV) light was plentiful as there was no oxygen to produce ozone to block it.

Although UV light is toxic to cells, it is also a powerful energy source that could activate or hasten biochemical reactions. The first cellular life, then, would have faced the challenge of turning light (permeated with UV) into energy while preventing cellular damage.

In addition to chlorophyll and the rhodopsin family, other candidates for first visual witness include the photolyases and their chemical descendants—cryptochromes. These proteins appeared as reparative mechanisms soon after the initial cells arose and may have been one of the first light-sensitive and photoreceptive compounds. They use blue light to repair RNA and DNA and are found in living prokaryotes.

Cryptochromes help to regulate photoperiodicity (an organism's behavior in response to a daily cycle of light and dark) such as the circadian rhythm. As a result, they are found throughout the animal kingdom—from single-celled bacteria to invertebrates and vertebrates—and in plants. But although they are part of the photoreception system, cryptochromes are not rhodopsins. As a chromophore, they use flavins—such as riboflavin (vitamin B_2)—rather than any vitamin A derivative.

Photoreceptive compounds, including the rhodopsin family, photolyases, and cryptochromes, and chlorophyll all would be helpful to early cells in several ways: (1) shielding them from the molecule-disruptive effect of UV light, (2) mediating negative phototaxis to move away from and avoid intense UV light, (3) repairing RNA and other molecules, and (4) capturing and storing energy. In all cases, blue light, but perhaps UV light, would have been the most influential.

Sunlight and Blue Light

Of the entire electromagnetic spectrum available in the energy reaching the earth, why did life select that very narrow window of visible and ultraviolet wavelengths?

Life began in water. Availability of energy to cells was key, and the shorter the wavelength is, the higher the energy.

The wavelength will control what a cell can receive effectively and use in some manner. For example, radio waves can be so large (longer, and hence wider, wavelengths) that they require a large "dish" to focus the signal. Such a wave would not "fit" into a cell to produce a signal or provide enough energy to provide much survival advantage.

On the other hand, wavelengths shorter than the blue-violet range tend to damage cells—even ultraviolet light will damage cells to some extent. Wavelengths shorter than ultraviolet are worse yet—x-rays and gamma rays can kill cells or, given more time, even animals. Generally, these shorter wavelengths bring energies that are simply too high for cellular life. Furthermore, electromagnetic waves of visible and ultraviolet light penetrate water approximately six orders of magnitude more than shorter wavelengths. The sweet spot of penetration of the electromagnetic spectrum was (and is) in the blue range. In addition, the shorter visible wavelengths we perceive as violet carry nearly twice as much energy as the longer wavelengths that we would perceive as red. This higher energy level of the shorter wavelength makes it easier for a receptive cell to discharge. Hence, the higher energy means that the visual pigments that perceive the shorter wavelengths are more stable and less subject to degradation or even mutation. As it turns out, the bluer wavelengths are at the peak of the combination of energy delivery, safety to cells, transmission in media, and penetration in media such as water.

Beginning to Organize

The most recent common ancestor to the Archaean organisms and to Eukaryota shepherded much of the biochemical necessities for vision to the eukaryotes. Whatever that ancestor was, it brought the transmembrane retinal and opsin combinations to the Archaean domain.

One class of highly salt-tolerant Archaea has so much bacteriorhodopsin (and halorhodopsin) studding its outer membrane that in a colony it has a purple or reddish bloom from the chromophore—retinal (Figure 1.4). The retinal, and hence, the bacteriorhodopsins, are inherently colorful, and depending on pH, concentration, and other molecules around these compunds, they may be purple, blue or red (Figure 1.4). These Archaea react strongly to light, but their visual pigments—bacteriorhodopsin and the three other photosensitive compounds, including halorhodopsin—are biochemically distinct from those found in metazoan organisms and even single-celled eukaryotes. All metazoan opsins are similar enough for us to conclude that they had a common ancestor. We do not know how Archaea transferred the necessary opsins to the protists or how protists, in turn, transferred them to Metazoa. But there is evidence that it did happen.

Genetic Machinery—The Toolkit

Some basic genes and gene products had to have been produced within the first few hundred million years after life's emergence. These basic genes included a primitive form of what are known as Pax genes and their proteins. The Pax genes are major players in the formation of the head, nose, and significantly, the eye. A Pax gene, *Pax6* (or its closely related predecessors), is highly conserved across many metazoan lineages and represents an important portion of the necessary genes for the production of the eye. This fragment of DNA (or RNA), a primitive Pax gene, created the necessary proteins that would span hundreds of millions, or even billions, of years of ocular development and would itself evolve and mature by a variety of mechanisms.

Prokaryotic Gifts

The single-celled Archaea led to eukaryotes and, following further modifications, probably led to Metazoa. An organism called *Volvox*—in the murky area between a single-celled organism and a metazoan—might help us to understand this process. This organism has a rhodopsin that is surprisingly similar to that of all Metazoa, as we discuss in the next chapter, and the evidence suggests that all metazoan opsins are related or homologous.

Walter Gehring, a noted evolutionary biologist, has proposed a "Russian doll" hypothesis, which is attractive and may be true at least in part. To understand this proposal it will help to note that, somewhere in evolution, the cells that would become metazoan organisms acquired mitochondria (energy organelles in cells) as endosymbionts. Gehring proposes that endosymbiosis occurred many times in other ways: first, red algae absorbed and essentially used a photosynthetic organism (cyanobacteria) as a chloroplast or energy source. At some later time a certain phylum of protists called dinoflagellates absorbed these red algae (now containing cyanobacteria). These dinoflagellates modulated the red algae and their components into a single subcellular eye before being themselves absorbed into certain jellyfish. The acquiring jellyfish may have further modulated the dinoflagellates into a true eye to acquire "sight." This would not be true sight as we think of it because there is no brain to interpret the signals. These jellyfish could have incorporated the protist genes directly into their genome. That process could have occurred for bacteria or Archaea with the right rhodopsin, leading to an organism like *Volvox* and eventually true Metazoa. This is quite possible because other organisms alive today, such as coral and certain clams we will meet later, incorporate dinoflagellates into their body structure for mutual benefit.

As single-celled organisms grew more complex, they began to congregate into larger, more complex organisms (much like the condominium complexes formed when homeowners move into them from single-family homes). Also like the condominium residents, the cells of this larger organism would remain relatively autonomous. These cellular collections were probably similar to early Parazoa. Eventually, cellular specialization, including specialization in photoreception, would occur.

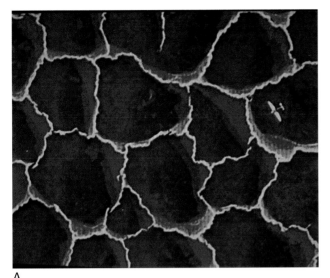

A

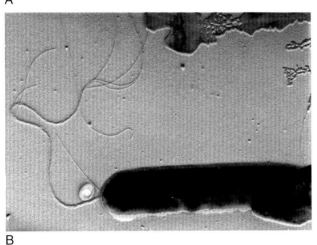

B

Figure 1.4 (A) *Halobacterium* salinarum. A model organism for understanding Archaea, *H. salinarum* lives in environments of high salinity that is completed saturated or nearly so (4 M salt and higher). This halophile can grow aerobically or anaerobically and is capable of photosynthesis. The image shows huge mass cultures (note the airplane and its shadow in the upper right hand corner for size estimation) of the organisms. The color originates from bacterioruberins (carotenoid pigments found in some halophilic aerobic archaebacteria giving them a red coloration). This organism contains bacteriorhodopsin, which acts as a proton pump for its energy source. The pigments are used to provide energy to create and use ATP. Using only light as an energy source, this organism can colonize salt solutions with very high concentrations of salt. Although the bacteriorhodopsin is purple, the other retinal proteins including halorhodopsin and two sensory rhodopsins give a more reddish coloration. These compounds combine to give large blooms like these. The halorhodopsin acts as a chloride pump and assists in the production of energy from light. Image by Professor Dieter Oesterhelt. Further information can be found at http://www.biochem.mpg.de/oesterhelt/web_page_list/Org_Hasal/index.html. **(B)** *H. salinarum*. Scanning electron microscopic image of the organism of approximately 13,500× magnification. This organism is a rod-shaped motile organism shown here with its flagellar bundle. The organism is strongly phototaxic, that is, attracted to light and uses the light as its energy source. Image by Professor Dieter Oesterhelt. Further information can be found at http://www.biochem.mpg.de/oesterhelt/web_page_list/Org_Hasal/index.html.

The gradual evolution of cellular components and combinations into tissue and organs, and hence Metazoa, has a darker side. Natural selection favors whatever variant is successful *for that time and place* and will outcompete and supplant other cells and other single- or multicellular creatures. Perhaps a better cellular or even metazoan design is lost because it does not fit that particular niche at that particular time.

Further Organization

The last common ancestor of all life on earth began in an anaerobic environment. The two lineages of the prokaryotes (bacteria and Archaea) ruled the earth for at least one and possibly two billion years as anaerobes until the development of protists.

But there is substantial evidence that the domain of Eukaryota began as aerobes. This implies that the ancestor of all eukaryotes had the subcellular energy factories mentioned above—mitochondria. Mitochondria had been incorporated as endosymbiont organelles within the period after oxygen began to accumulate, approximately 2.7 to 0.9 billion years ago.

Photobiology had begun, and the eye would be the beneficiary.

THE AGE OF COMPLEX CELLULAR LIFE

PROTEROZOIC EON

2500–543 MILLION YEARS AGO

CRYOGENIAN PERIOD

850–650 MILLION YEARS AGO

EDIACARAN PERIOD

650–543 MILLION YEARS AGO

2

Inception of Eukaryotes

A yawning gap exists between prokaryotes and eukaryotes. The introduction of a nucleus represents a biological Rubicon and is one of the key steps in the history of life on earth. The first eukaryotic cells appeared on earth approximately 1500–1000 million years ago (mya) according to most investigators.

Beyond the divide between individual cells and their organization, at least one more great divide occurs in the organization of life: the one between single-celled eukaryotic organisms (that remain single-celled throughout their lives) and multicellular organisms. The single-celled eukaryotes are protists; multicellular organisms are metazoans. All Metazoa are composed entirely of eukaryotic cells.

The visual witness and its necessary genetic information had to leap each of these two great divides: first, from cells without a nucleus to cells with a membrane-bound nucleus, and second, from single nucleated cells to organisms with many nucleated cells. The necessary equipment to produce an eye had to leap those divides or appear anew more than once. How this was accomplished is not entirely clear, although lateral gene transfer is one contributing mechanism. Although lateral gene transfer is insufficient to explain completely how single-celled eukaryotes evolved into Metazoa, this process would have had a profound effect on our visual witness.

Nucleated Kleptomaniacs

Eukaryotes are defined by having a nucleus, but the differences between them and prokaryotes are actually much more extensive. In general eukaryotes are larger cells, divide differently, and have more subcellular organelles (specialized internal parts) than prokaryotes. Despite being single-celled, these organisms occupy much of the earth. The protists are far more diverse in size, shape, and motility (but less versatile and basal) than are the prokaryotes. By the time the eukaryotes appeared approximately 1500–1000 mya, oxygen had been present for perhaps 500 million years and was gradually increasing as an excretory product of cyanobacteria. Oxygen would facilitate the expansion of eukaryotes.

By the process of endosymbiosis, eukaryotic cells gained mitochondria—those energy factories in the cell—to use the oxygen in the forming atmosphere. This mitochondrial metabolism would have greatly enhanced the cell's ability to survive and prosper. As one well-accepted theory goes, the eukaryotes began with endosymbiosis when a single cell (a prokaryote) subsumed another single cell (another prokaryote) with a different function. It is not clear whether all of the organelles of the eukaryote came about in this manner, but the evidence is strong that some organelles, such as chloroplasts and mitochondria, did.

Chloroplasts arose from assimilated blue-green algae (cyanobacteria) and, like the mitochondria, represent a major step in cellular evolution. This plastid continued on in those cells that would radiate into plants and fungi, but the eukaryotes that headed toward Animalia would discard it.

Protists are plentiful, yet surprisingly unknown. Few investigators would hazard a guess as to how many nucleated algae, water molds, slime nets, diatoms, and other such species exist. These diverse and adaptable species often have specialized features including the aforementioned chloroplasts, toxins (responsible for the red tide), bioluminescence, and flagella, allowing them to occupy otherwise challenging niches. Many are commensals, saprophytes, or parasites—sometimes they parasitize each other. Protists are mainly aquatic, but some terrestrial species exist in moist soils.

Protists, then, developed complexity and continued to evolve, even to the point of developing methods of adhering to one another, but the protist story does not end there. These somewhat obscure but very common creatures remain with us and hold clues to the development of the visual witness.

As discussed in Chapter 1, an unknown protist probably consolidated the rhodopsin that studded the cellular membrane of an Archaea, and an eyespot was born (Chapter 1 and Figure 1.4B). *Euglena* represents just such an organism.

Euglena gracilis

Like a flag waving in the breeze, Euglena *waves its flagellum, achieving a form of swimming common to the Protista kingdom to which it belongs (Figure 2.1). We do not think of a green alga as being highly motile or even graceful, but* Euglena *is both. Commonly found as the scumlike film on the surface of stagnant water or washing in from oceanic blooms,* Euglena *appears entirely different when observed at its own scale.*

The eyespot near the base of the flagellum was originally believed to contain a photopigment based on a flavin. Although flavins are present, the eyespot turns out to contain a rhodopsin. The retinal-opsin combination is not quite the same as we find in metazoans but is a rhodopsin. The eyespot, at the base of the

flagellum, works in concert with the flagellum, permitting the organism to find and move to the highest light concentration for photosynthesis. The flagellum works like a propeller, not a tail. Both the flagellum and the eyespot are at the "head" of the organism. Hence, the organism travels "forward" in the direction of the flagellum.

Euglena *is a strongly phototaxic organism, but its photoresponse comes from more than just the eyespot. This lovely creature has other chromophores within its cytoplasm. These chromophores permit* Euglena *to respond both positively and negatively to light at different times depending on its needs and on available light.*

In many ways, Euglena *acts like a plant.* Euglena *is green, harvests sunlight, contains chloroplasts, and is photosynthetic—just like a plant. But this protist acts as though it belongs in the zoological kingdom because it has an eyespot and can use its propeller or flagellum to move it toward light. It represents a time before the metazoans and before the split into plant and animal.* Euglena *illustrates a pathway to both. When* Euglena *has enough sunlight, photosynthesis produces ample energy for the organism, but when sunlight is inadequate,* Euglena *will act more like a predator, engulfing other protists and plankton.*

The appearance of the organism and its organelles fosters at least one intriguing speculation on the evolution of plants and animals. Perhaps a protist resembling Euglena *with a retinal-opsin combination developed a cilium. Suppose that, instead of the visual pigment spanning the eyespot, it concentrated only within the flagellum. Perhaps the proton-pump effort of the rhodopsin provided an energy source for the flagellum. As the cell evolved further, the rhodopsin was concentrated into the cilium, and this single-celled protist became the first ciliated photoreceptor. As simply another form of endosymbiosis, the newly formed "photoreceptor" may have taken up residence in a jellyfish only to become the ciliated photoreceptor of the eye of* Tripedelia cystophora, *a jellyfish we will meet later.*

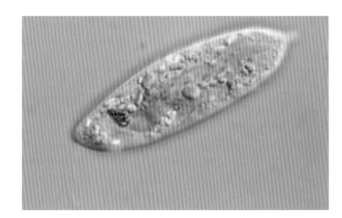

Figure 2.1 *Euglena gracilis,* a unicellular eukaryote, or protist. Note the eyespot, or stigma, in the anterior one-third of the specimen. The "head" of the organism contains the flagella, and both twirl like propellers. The eyespot contains a rhodopsin as well as flavins for photoreception. The organism can ingest food in a process called phagocytosis, making this organism a heterotroph, and the animal can photosynthesize making it an autotroph. Hence, it has elements of animal and plant. *Image © Greg Rouse, PhD.*

Protists have continued to evolve from their beginning 1.5 billion years ago and have developed "eyes" with minimal, if any, neurologic organization that we would call a brain. So, those early protists had to have competed using various modes of sensory and motor output, and this likely means the visual witness was already present with a strong foothold in life.

Evolution illustrates the importance of such sensory input, and this emphasis on sensory input can best be illustrated by a living protist with a subcellular eye, an eye whose elements are all smaller than the cell itself. The subcellular eye is discussed here because it epitomizes protists' capability for evolving an eye if the tools and the ecology are appropriate.

Erythropsidium

Some protists can sense direction and perhaps even intensity of light with the grouped pigment granules in the eyespot. That is a visual witness, true enough. By any definition, though, some members of one group of protists—the dinoflagellates—actually have an eye. Although dinoflagellates have been traced at least to the early Cambrian, it is unknown when this lineage of organisms developed an eye.

The planktonic dinoflagellate Erythropsidium qualifies as a small and important step beyond Euglena (Figure 2.2). Importantly, the reader should understand that Erythropsidium is not a descendent of Euglena but only serves to illustrate how an eye might have formed from within the lineage of protists.

This eye must be among the smallest of eyes because the entire creature is only 50–60 μm in diameter. Erythropsidium cannot interpret images, at least as we understand interpretation, because it has no brain or even neurological elements. Next to nothing is known about Erythropsidium's visual mechanisms or genetics, but this remarkable eukaryote speaks volumes about evolution's creativity and emphasis on vision.

Erythropsidium is one of several dinoflagellates that are known to have an ocelloid. An ocelloid is an eyespot with a lens. In this ocelloid, a clear space overlies a cup that is filled with a folded membrane that is studded with an unknown visual pigment. It is, in essence, a tiny eye.

Although Erythropsidium's visual pigment is not known, a related dinoflagellate, Peridinium foliaceum, has a visual pigment that absorbs and responds to blue-green light. This emphasizes the selection of blue light by many of these single-celled creatures.

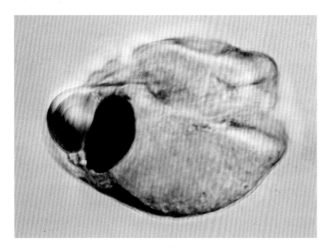

Figure 2.2 Erythropsidium, a protist with a complicated eyespot—an ocellus. This organism contains perhaps the smallest camera-style eye, although, because of its simplicity, it is called an ocellus or better termed an ocelloid. Note the pigment cup and lens to the left of a groove in the organism. This organism is approximately 50 μm in diameter. Image © F. J. R. "Max" Taylor, PhD.

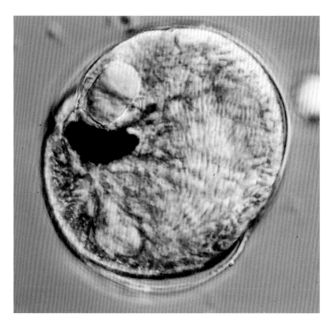

Figure 2.3 Warnowia, a dinoflagellate with an ocelloid complete with a lens, a pigment cup, and a "retina" with photoreceptive compounds. Note the diffuse chromosomes in the cytoplasm that appear to be faint streaming ribbons of cytoplasm. This protist also has chloroplasts, meaning that the organism has elements of the plant kingdom. Image © F. J. R. "Max" Taylor, PhD.

These rather perplexing creatures—dinoflagellates and in particular Erythropsidium—are neither plant nor animal but have components of each. Erythropsidium has been seen to chase prey. Yet they also boast chloroplasts. Other protists, such as Euglena, are known to be so opportunistic as to be autotrophic (photosynthetic and capable of making all of the energy and molecular components necessary for life from surrounding inorganic materials) when light is plentiful and heterotrophic (capable of eating other organisms to acquire necessary molecules) when it is dark.

Lynn Margulis, in her recent book, Acquiring Genomes, asks a rather profound question with knowledge of these dinoflagellates and their miniature eye: Could these creatures have been the source of the first metazoan eye? Perhaps, rather than evolving an eye, another protist or metazoan co-opted one from a dinoflagellate by ingesting the dinoflagellate. In that scenario, evolutionary refinements and improvements eventually led to other forms of eyes, including both compound and camera styles. In reality, the dinoflagellates probably did not evolve an ocellus at such an early stage in life's history. But dinoflagellates must have had the genetic machinery to do so and could have provided those genes to any multicellular animal that ingested them, so the acquisition of an eye or its genetic blueprint by endosymbiosis is not so far fetched (Figure 2.3).

So, by approximately 1500–1000 mya, eukaryotes had become acquisitive and had begun building their complexity by subsuming, or endosymbiosis, different organisms such as prokaryotes or even other eukaryotes. Protists were acquiring organelles, expanding, and diverging. These nucleated kleptomaniacs carried the retinal-opsin combinations even though these combinations may have diverged, laterally transferred, or mutated from the original forms. Nevertheless, still more tools would be needed to make or modify a proper eye in a metazoan.

Bridging the Gap to Metazoa

For protists to cross that second great divide—from eukaryote to metazoan—single-celled eukaryotes would be required to congregate, aggregate, and divide cellular tasks. That step remains a bit of a mystery. Nevertheless, we know that some eukaryotes did carry on in just that direction with aggregation and task division. Some of these ubiquitous organisms tend to congregate, and many of protists' successful forms have gained and used visual perception, as discussed with *Erythropsidium*. But the dinoflagellates did not have the first eye because they probably have evolved it rather recently. A candidate for that first eye, or proto-eye, does exist, though, and it is in a multicellular organism.

An earlier, if not the first, "eye" can be found in an extant organism—colonial green algae. Although few people probably would call this an eye, it qualifies if we use our definition in the Introduction: an "eye" is an organ that receives and recognizes light and has the ability to define spatial detail.

With at least 6000 species, green algae are a surprisingly large group of organisms. As protists, green algae's chloroplasts have translated light into energy for many species. Surprisingly, as we have seen, some of these green algae also have photoreceptor cells that contain rhodopsin. Sometimes these green algae will form long filaments or colonies, even primitively cooperative ones. Although these colonies resemble metazoans, they are not true metazoans. Nevertheless, they suggest methods by which single eukaryotic cells may congregate to become an organized multicellular animal.

One humble colonial alga may be an example to help explain how single-celled eukaryotes transitioned into multicellular organisms and brought opsins along with them.

Pre-Metazoa Volvox carteri

Volvox carteri is a spheroidal, colonial alga with at least two distinctly different cells revealing primitive division of labor (Figure 2.4). The colony may have up to 4000 individual cells, known as somatic cells, surrounding up to sixteen select larger,

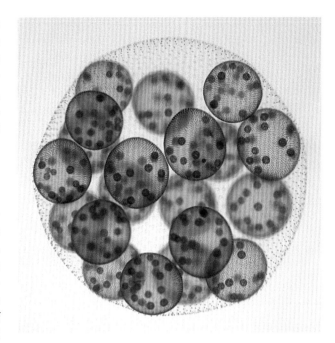

Figure 2.4 *Volvox carteri*. This green alga usually occurs as a spherical colony. In this image, the individual green "dots" lining the walls of all level of spheres are the individual cells, and the multiple green spheres (about twenty) within the single large sphere are daughter colonies. Within each daughter colony, there are beginning new daughter colonies. Each small green algal cell within all level of spheres has two flagella, which can undulate in unison. Each individual alga is connected to others within the gelatinous extracellular matrix (mucilage) that surrounds them. In this way, the colony is coordinated and can move in a synchronized fashion. The cells coordinate their flagella to beat to propel the colony toward light, so only some of the cells will move their flagella. Eyespots occur in each individual alga cell, visible as red spots, although they are not easily visible at this magnification. *Image © Ichiro Nishii, Nishii Initiative Research Unit.*

replicative cells, known as gonidia, in the center. Each somatic cell has an eyespot (giving the organism thousands of eyes) and two flagella, which provide movement.

This combination of flagella and eyespots leads to phototaxic responses, although not all wavelengths of light will stimulate these eyespots. Peak responses from its photosensitive pigment occur between 490 and 520 nm—not surprisingly in the blue-green range. This visual pigment, a rhodopsin, will generate a signal when stimulated at this wavelength. These responses are purely reflex because even minimal neurologic tissue has not developed in this organism. This volvoxrhodopsin of V. carteri functions much like the metazoan opsins, which suggests that there can be an evolutionary bridge for such needed molecules between the unicellular eukaryotes and metazoans. This rhodopsin even may be the ancestral photopigment to all metazoan opsins, as it appears to be related to both vertebrate and invertebrate opsins.

As single-celled eukaryotes began to tack the upwind course to Metazoa, several course changes had to have been made. First, protists had to congregate and organize, perhaps by a process known as the collective mode. Much like flocking birds, schooling fish, or even the "wave" at a sporting event, the collective mode overtakes the individual organisms. They begin to function as a unit and, with the necessary molecular components, may unite cell to cell. After union came specialization, as certain cells were assigned tasks. Such a unit would outcompete individuals to form metazoan organisms. Volvox is an example of how unicellular eukaryotes in a collective mode may have led to a metazoan. Here is a possible scenario.

A protist, such as Euglena, became a colonial organism, resembling Volvox (a result of collective modes) with gradual cell specializations such as reproductive cells and visual cells. Once a colonial organism is successful, it becomes more organized, neurologic tissues develop for communication, and eventually an organism such as a ctenophore or a sponge evolves. That protist in question carried an archaeal rhodopsin, which is gradually or rapidly mutated into metazoan opsins, which then populate the metazoan phyla.

This suggests that a protist can become a colonial organism, leading to Metazoa, and can bring the opsins across these boundaries. V. carteri represents an example of what could be a missing physiologic link—a humble beginning to multicellularity.

But as we have seen, we shall need to go further than colonial algae to meet the first true metazoan. Other components besides a photopigment would be needed. Perhaps the simplest multicellular organism can help describe the course of this evolution.

EUKARYOTES ORGANIZE AND

METAZOANS ARISE

NEOPROTEROZOIC ERA

1000–543 MILLION YEARS AGO

CRYOGENIAN PERIOD

850–650 MILLION YEARS AGO

EDIACARAN PERIOD

650–543 MILLION YEARS AGO

Multicellular Animals

The transition from single cells to organized multicellularity represents a major leap for life on earth and is still a bit of a mystery. A colonial alga such as *Volvox* (discussed in Chapter 2) illustrates that individual eukaryotic cells can assemble and begin to divide responsibilities, much like a metazoan, but when does it qualify as a metazoan? One of the simplest known organisms can help answer that question for Metazoa and for the visual witness.

Trichoplax adhaerens

One very basal (some would say primitive) extant metazoan has no organs, lacks most tissues, and does not have a brain. It is likely related to jellyfish and corals, although it is less derived and in its own phylum. Nevertheless, Trichoplax adhaerens *can teach us about the acquisition of the genetic toolkit for the production of eyes.*

Believed to be the simplest multicellular organism, with only three layers and four types of cells, Trichoplax adhaerens *has the smallest amount of DNA of any metazoan. Included in that DNA, though, are some of the genes needed for the formation of an eye, even though those genes do not express the proteins necessary to form an eye. (Genes generally act by making proteins, and the proteins define the result, such as an enzyme, tissue, or organ.) In this organism, the* Pax6 *gene, so important in the formation of an eye, is incomplete and has not been activated is if it is in a formative stage.*

The pioneer metazoans of 1.2–1.0 bya, likely now extinct, radiated into creatures we recognize—the ctenophores (comb jellies), cnidarians (corals and jellyfish), and porifera (the sponges). Although they do not have eyes, some of these descendents of early organisms have distinct, although less derived, predecessors to the Pax gene family. The genetic toolkit was in place, perhaps serving other functions, in these early predecessors. So these first metazoans exhibit the genetic *anlage* (early rudimentary blueprint) that eventually would lead to the production of the eye.

Why is the Pax family of genes, and especially *Pax6*, so important to our first visual witness? As mentioned in Chapter 1, the Pax family, or their orthologues (the genetic predecessors or sister genes that are seen in other phyla such as arthropods), are very old and play a significant role in olfaction and the development of the front or head of a metazoan organism as well as the eye.

The *Pax6* gene is important for the development of the eye but has gene partners to help. Corals and jellyfish have a gene that closely resembles *Pax6*. They are so-called outgroups to the bilaterally symmetrical animals, which means that these organisms did not radiate directly into more complex animals, such as bilaterians or even other phyla.

Think of outgroups this way: your great-grandfather has three sons, only one of whom has offspring. Your grandfather, in turn, also has three sons, and again only one, your father, has offspring. Before your childless great-uncles die, leaving no genetic lines, they are on earth at the same time as your father. Your two childless uncles are on earth with you and your father and represent different outgroups.

With their Pax gene family character, then, the corals seem to be at an early pivotal point in photoreception and transduction of the signal. One such coral, *Goniopora*, illustrates just how photoreception, even without eyes, guides these species in what may be the largest mass sex act on earth.

Corals: Diploastrea Heliopora *and* Goniopora *Species*

The corals are very light sensitive and light responsive—phototaxic—although they have neither an obvious visual instrument nor a light receptacle (Figure 3.1 and 3.2). Nevertheless, corals, like almost every other metazoan, have a rather complex interaction with light. In some species and some locations, this interaction is incredibly precise.

In the early spring light of a full moon, coral species in the central Indo-Pacific perform one of the earth's greatest natural events: They broadcast spawn. Occurring on a single night, once a year, this curious event represents the largest spawning event on earth. It requires the animals to recognize a rather subtle variance in the moon's blue light on that night.

Coral, and many other Metazoa including humans, are very responsive to blue light in their circadian rhythms. Blue light was omnipresent in the early oceans when life began, and organisms that were entrained by it (synchronized with its daily rhythm), responded more vigorously and made an easy evolutionary choice.

Coral recognizes blue light by using cryptochromes, which are ancient photoreceptor proteins (discussed in Chapter 1). Corals are among the first metazoans, especially the first organized metazoans, that had specialized cells and proteins for specific purposes such as photoreception and vision. This emphasis on the visual witness in the most ancient Metazoa confirms the imperative and success of such a witness. Coordinating such complex sensory mechanisms requires a conductor, and cryptochromes fill that role.

Although corals do not have eyes, they are phototaxic. They respond to the periodicity of light through at least two genes, cry-1 and cry-2, which become active and produce increased cryptochromes. These cryptochromes, in turn, induce the mass spawning.

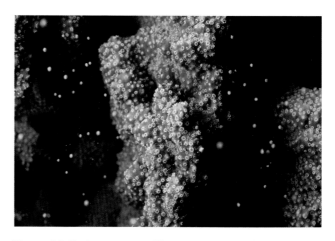

Figure 3.2 *Goniopora species.* The synchronization of the coral spawn will frustrate predators by overwhelming them. When released en masse, all eggs and sperm cannot be consumed, and some will survive to settle to another site and begin growing as a new coral. *Image © Darren Jew.*

So these early cryptochromes have echoed through metazoan phyla and predate eyes, as we know them. Even primates have cryptochromes, but without light-sensing abilities, at least that we are conscious of. They are essential to our circadian rhythm.

A Pax6-like gene is another molecular candidate for the conductor of light and vision in coral. Eyes eventually were needed, and the genetic framework to code for eyes, or at least to begin the cascade, would be essential. Coral's ancestral Pax6-like gene laid the groundwork for eyes.

Coral exhibits other aspects of the visual witness that illustrate evolution's emphasis on photoreception. These aspects provide clues to molecular evolution.

In another example of endosymbiosis, most hard corals rely on a photosynthetic dinoflagellate (a single-celled organism, or protist, that often has two flagella) to provide the energy necessary to power metabolism and reef-building. This reliance means that corals must have adequate access to sunlight, so they usually live in relatively shallow water. But corals also must protect themselves from light, which can be toxic. Evolution had to solve this problem.

Most corals are colorful, and many of these colors are fluorescent. The color comes from proteins and chromophores, much like the retinal and flavins discussed in Chapter 1. The chromophores protect the corals by reflecting or absorbing harmful light waves and permit those necessary for the photosynthetic dinoflagellate that powers photosynthesis.

These proteins degrade and require energy to reproduce and maintain, but they are worth the expense: with the fluorescent proteins and the cryptochromes, evolution has whittled the visible spectrum to a few selected wavelengths necessary for reproduction and life.

Figure 3.1 *Diploastrea heliopora,* a coral species, with *Phyllidia carlsonhoffi,* a nudibranch, on the surface of the coral. *Image © Foster Bam.*

A single gene, probably present very early in evolution, directs the production of these proteins. Arthropods have a related protein, suggesting that the last common ancestor to corals and arthropods had a similar gene and protein. These compounds were likely present in the oldest of metazoans.

With the help of these endosymbionts, hard corals manufacture a mineral skeleton composed of a rather pedestrian compound, calcium carbonate, which builds into limestone. Corals are colonial and highly cooperative individual organisms that have built reefs and islands, and they have contributed mightily to building the continental shelf.

Another ancient multicellular animal related to corals, jellyfish provide further insight into the development of an eye. Although these entrancing and rather basal creatures represent an evolutionary cul-de sac, not a vast new lineage, they are successful and, much like the protists, have pursued an evolutionary path that helps tell the story of our witness. Most jellyfish are light sensitive; some have rudimentary eyes. A few in a small class, Cubozoa, have evolved rather sophisticated eyes. This class illustrates that the march toward a better eye has surprising converging tendencies and implications, and it continues even in outgroups.

Cubozoan Jellyfish: Tripedalia cystophora

Jellyfish go back at least 600 million years, preceding the Cambrian explosion by tens of millions of years. They were the first multicellular organisms to become truly motile. Jellyfish are a successful lineage illustrating the morphology of early metazoans. They are radially symmetric as opposed to being bilaterally symmetric as are most other animals. These gelatinous creatures have but two layers of tissue instead of three, as do most other animals. This middle, third layer appears later in evolution and is responsible for organs and muscles within the body. And, instead of a brain, in a classic sense, jellyfish have a neural network organized into nerve rings and ganglia, suggesting that the eye does not necessarily need a brain. Bilateral symmetry, the creation of muscle, and the organization of the brain will be important to the direction of the developing eye, as we will see. And yet, jellyfish can teach us much about the state of the developing eye of 600 million years ago.

Most jellyfish have rather simple visual mechanisms that suggest their humble beginnings, but in vision, one jellyfish class stands out.

The cubozoan jellyfish have surprisingly sophisticated camera-style eyes, much like squid or even vertebrates. The members of this jellyfish class vary greatly in size, and Tripedalia cystophora is one of the smallest (Figures 3.3–3.5). This Caribbean species must rely on its vision to stay in the light shafts that filter between the roots where its prey–tiny

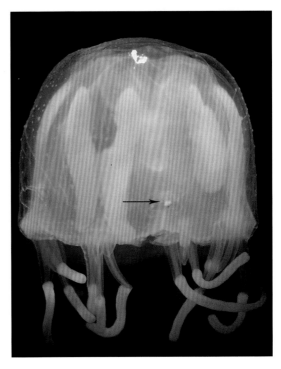

Figure 3.3 *Tripedalia cystophora,* one of the members of the family of box jellies, and this one has eyes. Surprisingly, this beautiful animal has four "arms" that hang down on the four sides of the bell, and at the end of these rhopalia is a small knot of tissue that contains six eyes. Two are camera-style eyes as detailed in the text. The black arrow points to one of these sets of eyes.

shrimp-like zooplankton––swarm. The intricate anatomy of its eye reveals the evolutionary potential present when this class diverged from the early metazoan lineage.

Near the lower edge of the bell's exterior are four flower-like structures known as rhopalia. These dangle from a sinuous stalk, like a watch on a chain, and represent the organism's sensory elements. Each golf-club-shaped rhopalium has petal-like leaves that surround a center knob with six eyes. Four of the eyes are ocelli, which can probably see only light and dark, consisting of a cup, a pigment layer, a few sensory cells, and a lens.

The other two, however, are surprisingly complex camera-style eyes. (Perhaps they evolved from ocelli in the same rhopalium.) Each has a cornea, lens, and a multilayered retina. The cornea consists of a flattened layer of epithelium directly in contact with the lens without an intervening chamber. The spherical lens is well organized, having a novel family of proteins with a decreasing index of refraction (that is, with varying indices). This lens design will be seen in later aquatic animals, as it permits improved focusing and minimizes spherical aberration.

The retina, or at least the tissue immediately behind the lens, consists of a vitreous layer followed by ciliated photoreceptors

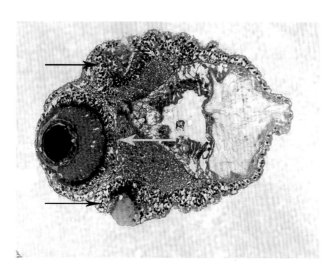

Figure 3.4 *Tripedalia cystophora* eye. Note the section through the principal eye (yellow arrow). There is a lens, and photoreceptors are present behind the lens. Two ocelli are illustrated by the black arrows.

without intervening neural tissue, as is found in our own eye and that of other vertebrates. Unlike those found in vertebrates though, these ciliated photoreceptors have fine microvilli extending from the cilia, as if they bridge the cells between invertebrates and vertebrates. This hints that the last common ancestor had a cell that would give rise to both types of photoreceptors in the metazoan world.

The investigated cubozoan species use the larger of the two complex eyes to view the depths of the water column below and to view the interior of the bell. They use the smaller complex eye to view the water surface and the apex of the bell above. The

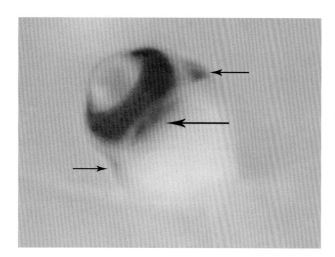

Figure 3.5 *Tripedalia cystophora* rhopalia. Magnified image showing lens protruding through pupil. The pupil of this principal eye actually constricts to stimulation by light. Directly beneath the principal eye is a slit eye (with blue arrow), and just to the right and left of the principal eye are two simple eyes (black arrows).

organisms are positively phototactic (they travel toward light), and their photoreceptors contain rhodopsin but probably only a single functioning visual pigment. That means that these creatures do not see color. But the eyes are able to resolve images and perceive motion, at least to some extent, and they even contain an iris and a pupil that responds to light.

Carybdea marsupialis, a related jellyfish with similar eyes dangling on its rhopalia, has been shown to have photoreceptors containing other visual pigments, including opsins for blue, green, and ultraviolet. This work, however, is controversial and unconfirmed. Other investigators have found only one visual pigment with its peak in the ultraviolet range. Still, this illustrates that the cnidarians, which represent an outgroup to the bilaterally symmetrical animals, had many of the necessary tools to assemble an eye—and a sophisticated one at that— even after the divergence from the principal line that would lead to the bilaterians.

Although of course we do not know when these jellyfish eyes appeared, if it were during the Ediacaran period, eyes of this sophistication did not appear again for tens of millions of years or (more likely) much later. Even if the jellyfish did not evolve eyes until tens or hundreds of millions of years after those eyes of the bilaterians, the similarities indicate that the genetic and cellular anlagen were present in the cnidarians at the time of the split.

Mysteries abound with this cubozoan jellyfish, and the answers to these mysteries will lead to a better understanding of early ocular evolution. Significantly, the ciliary cells that later compose the vertebrate retina were already in place in that last common ancestor before diverging into the cnidarians.

Furthermore, this animal's larval form has a cell that is midway between the microvillus and ciliary cells. These cells may be the precursors of the rhabdomeric cells of the bilaterally symmetrical invertebrates. Cryptochromes, mentioned in the discussion on coral, have been present in the jellyfish too. All the elements for our visual witness were present in Animalia when the jellyfish diverged from the metazoan line.

Although the complex camera-style eyes in this species of jellyfish are surprisingly well organized, image processing is another matter. No central processing of the image exists, but the second-order neurons (first-order neurons are the photo-receptors themselves) that progress from the photoreceptors do connect to the decentralized neural ring that connects the rhopalia. The bell contains small "sensory miniganglia" that probably perform some form of neural processing and coordination. As predators, these species must be able to coordinate their eyes, bell, and tentacles. The multiple eye types are probably specialized to collect different visual information leading to a form of "hard-wiring" of the visual system and reliance on reflex rather than input processing.

Some cubozoans may actually "chase" fish with tentacles that must resemble the long arm of the law. C. fleckeri, another *cubozoan of northeastern Australia waters, can swim at up to 3–4 knots per hour. Although most jellies, including T. cystophora, feed on plankton or small copepods, some of the larger box jellies actually kill and eat fish using their extremely potent venom. Presumably the venom provides the ability to kill quickly and avoid harm from the thrashing of the prey.*

Ancient animals such as jellyfish possess little or no central processing system even after hundreds of millions of years, although the ganglia may be more sophisticated than we realize. Nevertheless, sensory inputs from the rhopalia or other sensory organs probably stimulated the organization of a true brain.

BOX 3.1 CILIARY AND RHABDOMERIC PHOTORECEPTOR CELLS

Metazoans have at least two major classes of photoreceptive cells: ciliary and rhabdomeric (the latter also called microvillus). Generally, ciliary cells have a single long projection called a cilium, and rhabdomeric cells have multiple shorter projections called microvilli. To maximize surface area, each cilium is modified with many folds lined with photoreceptive transmembrane protein creating a stacked appearance. The multiple microvilli serve the same purpose in rhabdomeric cells. In general, ciliary photoreceptors are found in vertebrates, and rhabdomeric photoreceptors are found in invertebrates, but some animals have both cells acting as photoreceptors. The difference extends beyond the cell type and distribution of visual pigment, however, because the opsins and, in particular, the proteins in the cascade to create a signal are different, too. Additionally, many animals have both types of cells involved in the photoreception and transmission of light even if they are not exactly photoreceptors.

For example, cubozoan jellyfish have both types of cells, although the rhabdomeric cells are not directly involved in photoreception. Importantly, this suggests that the cells that function as photoreceptors diverged from a common cellular ancestor *before* jellyfish diverged from the principal line of Metazoa. Not surprisingly, single cells evolve too, and the photoreceptor is no exception. The genetic machinery needed to assemble these two types of cell had to be present very early, probably in that single-celled protist that led to a colonial organism that led to Metazoa. (Please see additional information in Appendix G.)

Sensory Input

Identification of potentially toxic or nutritional stimuli in the surrounding environment was essential for early prokaryotes. The sensory input, though, might have been very different from any we encounter today. Conceptually, we consider olfaction and vision as intricate and even elegant senses, but this was not always true. The first sensory systems were surely crude and likely reflexive in action. Similarly, early cellular reaction to sensory input was crude, not requiring complex analysis. The categorization, analysis, and storage of these sensory inputs came later, as the sensory input increased. After all, you need a file cabinet only when your desk has too many papers to maintain organization.

The Eye and the Brain

WHICH CAME FIRST? Sensory input, with vision as the most prominent proxy, probably pushed the evolution of the brain, at least as one of the forces. A repository for information is unnecessary without the input of complex and rapidly changing sensory information, particularly if that information requires categorization, storage, and executive decisions.

Photoreception, in its most rudimentary form, probably began very early in life's history. In the long course of evolution, sensory mechanisms began slowly, and many components probably were co-opted from other functions. The sun provided the principal source of the energy that permitted life to begin and flourish. Incorporation of rhodopsin, as a proton pump and simple energy source, conferred a distinct advantage over those single cells that did not have such an energy source. Did such concentrations of rhodopsin in membranes of single cells constitute the first eye? After all, the rhodopsin was not present to serve sight.

Neither negative nor even positive trophic responses toward light seems to qualify for "real" vision as spatial resolution is the core concept of any true visual system. For such spatial capabilities, an organized system of collection or an inherent difference at the stimulus level is needed.

Storage and analysis are not required initially—just a simple reflex reaction to the stimulus. Collection and storage require additional cells, so the early brain could not appear until the appearance of Metazoa. Such collection and storage probably provided the basis for the stimulus and beginning of a brain.

The protists progressed in diverse ways, and one radiation of single-celled diaspora eventually led to the dinoflagellates (Figures 2.2 and 2.3). Each of these creatures has a rudimentary but entirely subcellular eye. Yet none has a brain. No one knows exactly how this creature transmits the signals it receives from its eye to its flagellum, but it must do so, because it preys on other dinoflagellates. Clearly, a "brain" is unnecessary for this organism.

Jellyfish have eyes with little or no neurologic organization or, at most, neural nets with only reflexive output and no central control. Minimal processing of the neuronal input occurs and is limited only to spatial relaying. And yet, the eyes can be surprisingly sophisticated (Figures 3.4 and 3.5).

The larvae of *Tripedalia cystophora*, discussed above, have primitive eyes—ocelli. These are very simple, each with a single photoreceptor cell that contains pigment granules in a cup. Within the pigment cup are microvillus photoreceptive cells. These, then, are rhabdomeric photoreceptors, which most invertebrates have.

Furthermore, this single-cell larva has a cilia believed to be used only for motility. Because this single-cell ocellus is driven by light and has all of the eye components, it resembles the adult dinoflagellate in its ability to move toward a visual stimulus without the use of neuronal tissue. A brain is not necessary in these jellyfish larvae, either.

Once the sensory mechanisms became organized, storage and executive function had to follow, especially if muscular elements–as appeared in the early worms–required an organized output. This still does not guarantee a brain, however.

Andrew Parker believes that vision drove evolution, according to his book *In the Blink of an Eye*. Of course it cannot be that simple, as many forces, including climate, meteor strikes, volcanoes, new competitive organisms, and serendipity, influence the direction of evolution. Nevertheless, several forms of sensory input must have had and continue to have a critical role in evolution. Sensory input, particularly vision, became more sophisticated in coevolution with neurologic development because the sensory input could be useful only if the organism had the necessary machinery to decode, translate, organize, remember, and integrate that input. Conversely, no such machinery is needed for more pedestrian sensory input. Hence, sensory input, with our visual witness of photoreception leading the way, stimulated the formation of the necessary neurologic machinery.

Comb jellies, sponges, and jellyfish were among the first groups of multicellular animals in the watery world of a billion years ago, and these have persisted. Conditions were changing, though, and a new order was coming. Innumerable prokaryotes (such as cyanobacteria) that produced the waste product oxygen were signing their own death warrant, and oxygen levels were rising. But the earth was unstable, and the climate was about to change. This climate change certainly narrowed the living organisms and cleared the stage for the Cambrian explosion.

The Cryogenian Period (850–635 mya) or "snowball earth" was a time of winnowing. It is not universally accepted that this severe glaciation actually took place, but the period is known as a time of severe climate change. Fossil evidence suggests that the climate change led to neither the decimation nor expansion of marine fauna. But the eventual warming at the end of the Cryogenian led to a period of fauna expansion, the Ediacaran, and probably the beginning of real development of the eye. All the tools were in place.

THE MAKING OF AN EYE But how does evolution make an eye? We know a surprising amount: We have learned that the *Pax6* gene, or at least one of its related predecessors, was present very early in the history of Animalia. In fact, almost all metazoans have the Pax genes or similar genes. The *Pax6* gene is one of the important genes coding for proteins to initiate the formation of an eye.

The necessary molecular products, biochemical processes, and appropriate genetic fragments were assembled and were ready to form an eye somewhere deep in the preneoproterozoic waters a billion or more years ago. Biology can form an eye in several different ways, and all of these would be used in the coming assemblage of Metazoa.

The process of building an eye begins with a photoreceptive cell, as we have seen in the preceding chapter. The photoreceptive molecule may have been studded throughout the membrane of the cell as in H*alobacterium salinarium* or isolated in a single area as in some protists. Photoreceptive molecules in individual cells are more effective, especially to define direction, if they have a way to block any stray photon stream. These stray photons can stimulate adjacent photoreceptive cells and eliminate any spatial definition or light direction. The early developing eye improved as evolution stumbled on the combination of dark pigment granules and photoreceptive cells. Pigment granules are helpful in limiting the direction of the light. Biological pigments, such as melanin, were readily available. A cell with pigment granules in it is entirely self-contained, but eventually cells evolved to contain pigment exclusively as cellular functions were distributed. These dark pigment granules are distinct and different from the photopigment. These pigment granules are used to stop light as an opaque barrier, not respond to it.

Perhaps several of these cells congregated, or perhaps the same cell duplicated itself. In any case, they combined to create an eyespot, which by some definitions, but not quite our definition, could be an eye. It contained some limited directional information but no form or much directionality accompanying that stimulus.

Once an eyespot is established, the ability to recognize spatial characteristics—our eye definition—takes one of two mechanisms: invagination (a pit) or evagination (a bulge). In invagination, a pit that deepens and begins to form a cystlike structure is a camera-style eye. A simple collection of transparent or translucent concentrated cells or proteins within the pit adds a focusing lens to the system. This combination becomes a simple eye, an ocellus.

Invagination is used to form a camera-style eye for invertebrates or vertebrates. As a result, some invertebrate (octopus) and vertebrate (fish) eyes appear and are very much alike, at least superficially.

In evagination, the second method, the protrusion of the eyespot to form a rounded bulge creates directional information even with no additional morphologic changes to the "eye." Individual eyespots can have single isolated cells with transparent cells overlying them and pigment cells backing them. These combinations can duplicate many times, and these cells can individually or collectively evaginate to form a protrusion, then a compound eye.

Several iterations on this system can combine with a variety of focusing mechanisms. Invagination and evagination, then, have led to the many different types of eyes and even a few novel systems as well. At least twelve evolution-created ocular designs or "bauplans" remain with us today (Table 4.1). Evolution has randomly settled on these methods because chance, error, time, and competition winnow out the less favorable mechanisms *for that particular niche*.

TABLE 3.1 *Ocular bauplans (eye plan or design)*

Ocelloid of Protists
Camera-style eyes
        Ocellus
        Pinhole camera
        Lenticular camera
Compound eye (see Table 7.1)
        Apposition eye
                Focal apposition eye (several variations)
                Afocal apposition eye
                Neural superposition eye
        Superposition eye
                Refracting superposition eye
                Reflecting superposition eye
                Parabolic superposition eye
Miscellaneous eyes
        Telephoto scanning lensed eye of *Copilia*
        Scanning lensed eye of jumping spiders (ocellus-like)
        Mirror optics of scallop eye (camera style)

PHOTORECEPTORS Evidence suggests that both the rhabdomeric photoreceptor cells (type seen in most invertebrates) and ciliary photoreceptor cells (type seen in most vertebrates) began as one single cell type and diverged into two. If so, this happened very early in the differentiation of Metazoa because these two cell types have very different methods of communication within the cell and between cells. This key point illustrates that although the opsins and other photoreceptive compounds are related, the photoreceptive cells were far from mature when they diverged—well before the existence of bilateral organisms and probably about the time of the first metazoan organisms. The opsin-retinal molecules were probably present, but the rest of the system for the communication of the photoreception was not. This molecule was likely co-opted from its function as a proton pump—a relatively unsophisticated agent of photoreception. Nevertheless, all metazoan opsins are related, indicating that the retinal-opsin combination is ancient, even preceding Metazoa.

THE CRYSTALLINE LENS Different animals have synthesized focusing lenses using different methods and compounds. The compounds that make up lenses and lens systems are varied, ranging from minerals to carbohydrates in invertebrates to proteins in vertebrates. Evolution cobbled together whatever transparent material it could find to create focusing lens systems that evolved into surprisingly sophisticated devices. The material is frequently specific with the taxon, and generalities are more difficult.

Crystallin protein lenses offer a mini laboratory in evolution as the crystallins that compose the lenses, especially in vertebrates, vary according to the animals' needs. These proteins were chosen by evolution to fit the niche. (See Appendix F.)

EXTRAOCULAR MUSCLES AND SUPPORTING STRUCTURES (ADNEXA) The extraocular muscles and other surrounding structures needed to move and stabilize the eyes came along as the body developed. Muscles appear in the Ediacaran, but the organization of the extraocular muscles awaited the appearance of early vertebrates, such as lampreys' ancestors. The extraocular muscles likely arose to stabilize the gaze as the animal moved rather than to follow the movement of other animals.

These muscles appear rather suddenly in the vertebrates probably because of the requirements of the ecological niche and the appearance of the neural crest, an area of specialized cells in the developing embryo. These cells are important to the development of the vertebrate brain and eye. Muscles appear suddenly because muscles depend on, and arise from, the neural crest cells.

The adnexa (supporting structures around the eye), such as the tear glands and ducts, also appeared, almost on cue, when they were essential to the functioning of any critical sensory element. For example, tears and tear glands would not be needed in a watery environment, so they did not appear until animals began a terrestrial life.

Metazoans and Their Eyes

Evolutionary controversy surrounds the eye. One asks "did the eye evolve just once, or many times?" As with other scientific questions, the answer requires some definition.

Some eye components evolved only once—the metazoan opsins and the master genes are prime examples, although each has continued to evolve and change during the ensuing hundreds of millions of years. But most of the components—such as the lenses, which are enzymes and proteins—have evolved individually many times and from different stimuli.

Each of these components and their evolution gives clues to the timing, progress, and success of the evolutionary process. If the opsin-retinal complex alone could be considered an eye, perhaps the eye evolved only once. But even the fundamental differences between the basic photoreceptive cells of invertebrates and vertebrates confirm that the eye evolved more than once.

So the "mature" eyes in living species evolved many times, but some basic and important components, such as the retinal, opsins, and basic gene proteins, evolved only once and are highly conserved. These components may have been the problem's best or only solution. More likely, it was the most expedient solution and retained a predominant position as much by chance as by effectiveness.

Some of the components in the more mature eyes are homologous, meaning that the cells, tissues, organs, or other biological structures have the same evolutionary origins and usually a similar structure and position, but perhaps different functions. Many other components are analogous, which means that the biological structure has similar function but a different evolutionary origin. For example, the ocular lenses of a trilobite and a tapir are analogous but not homologous. On the other hand, the retinular cells (photoreceptors) of the Humboldt squid and Union Jack butterfly are homologous to one another but analogous to our own photoreceptors.

These first three chapters illustrate that although some of the eye's components were present very early and the visual witness, at least in the form of photoreception, had begun, the eye, as we think of it, had not yet evolved. Still, it was on its way, probably within the next era.

The fantastic array of variability in ocular morphology and physiology becomes apparent with the study of the visual witness in various taxa. The evolutionary process to reach these many varieties of eyes began with a few basic tools. Over time, using trial and error, this indiscriminately random process of evolution has tinkered with many different eyes from those few basic components. When additional components, such as a crystalline lens, were needed, evolution simply assembled available biological components commandeered from other uses. In this sense, the eye has evolved many times over the last 3.75 billion years.

The Proterozoic was nearing its end as the Ediacaran began. The pump was primed at the beginning of the Ediacaran, but more development was required before the Cambrian could explode with new phyla and new eyes.

EARLY ANIMALS PREPARE
THE GROUND
EDIACARAN PERIOD
650–543 MILLION YEARS AGO

4

Metazoan life was in full swing by the early phases of the Ediacaran, approximately 650 million years ago, but these first steps were small aggregations of cells with little more than eyespots or early cups of pigment lined with photoreceptive cells. Sponges, ctenophores, and jellyfish had an early foothold. The toolkit with the genetic components for ocular assembly was combined with all the molecular components for photoreception and signal transduction—the molecules needed to transform and transmit a signal after the photoreceptor had fired. The visual witness was still relatively primitive, but organized photoreception and visual processing had begun. This time period would blaze the trail with all manner of new animals as the world came alive. Most of them would need eyes, and just as these animals would be experiments in nature, so would their eyes.

The Ediacaran began with a whimper, not a bang, as no major single event such as a meteor strike appears to have initiated it. The earth had emerged from the Cryogenian period (snowball earth), ready for the expansion of metazoan organisms.

The Ediacaran saw the climate change and warm, causing the glaciers to retreat. Oxygen levels increased, and within 20 million years of the beginning of the Ediacaran, the first metazoan glimmer would appear on earth. These warmer, shallow, oxygenated seas provided the necessary algal and bacterial nutrients that encouraged the growth of animals. When animals began to enlarge, they became a food source themselves. Larger predatory animals would have required energy, which would necessitate more of an oxidative metabolism as well as the temperature to allow that metabolic engine to flourish. The toolkit for this expansion was present, and the tinkering with the various potential DNA switches had begun. Within the next few million years, the first recognizable animals appeared, although we know little about them because they had no hard body parts to fossilize. We do have some precious evidence of the time, though, and we know soft-bodied creatures were abundant. The earliest known multicellular animal fossils were present as imprints of embryos, eggs,

and even skeletal tubes. Soon thereafter, the protists, ctenophores, primitive sponges and corals, lichens, and seaweeds were all present.

Sea pen–like creatures existed, some with fronds 2–3 meters high gently swaying in the warm seas (Figure 4.1). The flora and other fauna of the Ediacaran were probably more vigorous than we appreciate, but the fossil record is comparatively meager. Almost certainly there were primitive eyes, the specialized areas for photoreception called eyespots or stigmata.

In these early stigmata, primitive photoreceptive cells probably resembled those found today in echinoderms (sea stars or urchins), sea squirts (Figure 4.2), or salps (Figure 4.3), if you allow for cellular evolution. Today's animals surely have more highly derived photoreceptive cells than would have been found in Ediacaran organisms. At that time, these organisms may have seemed more like plants or jellyfish, but they were predecessors of the vertebrates.

A few pigment-backed photoreceptive cells combining in a shallow depression and a condensation of cells or proteins creating a lens bring about the simple eye or ocellus (Figure 4.4). Simple eyes were likely present in early worms (predecessors of flatworms, annelids, and others) in the Ediacaran, as these creatures first appeared during this time. The condensation of properly shaped cells or other proteins provided a lens for focusing. This ability to focus, however primitive and coarse, added spatial information that was amplified beyond what the simple cup would have interpreted. This condensation occurred early, evolutionarily speaking, and quickly, as it would have been a distinct advantage.

The Ediacaran was a key period, even if we understand very little about the visual witness of the time. The ctenophores (comb jellies) and cnidarians (corals and jellyfish) appeared. Both of these phyla are phototaxic (react to light—including positive and negative phototropism), although only some jellyfish have eyes (Figures 3.3–3.5). Those species without eyes (ctenophores and most cnidarians) do, however, have cryptochromes.

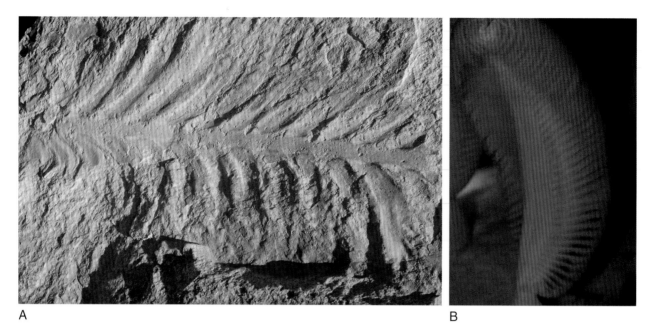

A

B

Figure 4.1 (A) *Charniodiscus*. Imprint of frond of sea pen–like creature believed to be related to the corals or jellyfish (cnidarian). Little is known of these organisms, but they were probably light sensitive and phototrophic. **(B)** *Ptilosarcus gurneyi*, giant sea pen. Extant sea pen as comparison to *Charniodiscus*.

Swaying to their own rhythm, jellyfish continued on a separate path and did not radiate into other extant phyla. These jellyfish species competed, matured, and became more sophisticated over the next 600 million years. The ancient split of the jellyfish and other cnidarians from the original line of multicellular organisms occurred before the formation of known and recognizable eyes. But

Figure 4.2 *Didemnum vexillum*, purple and yellow sea squirts (subphylum Urochorodata, also called tunicates), and *Chromodoris magnifica*, a nudibranch in this image. Although these sea squirts are invertebrates, they are deuterostomes and more closely related to the vertebrate lineage. They are considered to be chordates and have a notochord in their larval stage. *Image © Foster Bam.*

despite the split, and despite having predecessors with no eyes, the jellyfish developed eyes with strong similarities to our own.

That last common ancestor that radiated into jellyfish would also radiate toward vertebrates and other invertebrates (besides cnidarians). Each of these sister groups (sister groups to jellyfish) would diverge and develop eyes as well—often very similar to those the jellyfish would develop.

The visual apparatus that is common to jellyfish, other invertebrates, and vertebrates suggests that the visual system, or at least its potential, must have been surprisingly mature when the comb jellies (ctenophores) and the jellyfish (cnidarians) diverged. The photoreceptor also was relatively mature, as it is quite similar to those of other animals that diverged in that dim past.

So, well before the beginning of the Cambrian explosion, most of the necessary cellular and molecular components were present, assembled, and prepared to manufacture eyes, even if in a more basal form. It is likely that eyes were already being assembled in the Ediacaran in precursors to the trilobites, other arthropods, mollusks, and even precursors to the vertebrates.

Likely, the metazoan eye began more than once, and both mechanisms (pit or bulge) were at work at different times. But at that time, there would have been almost no visual processing to manage the information, just a reflex response, as in some jellyfish eyes even now.

Figure 4.3 *Salpa aspera.* A long stringlike colony of salps with an amphipod in the center of the image. The phylum Chordata includes the urochordates such as the salps and sea squirts (illustrated in Figure 4.2); the cephalochordates, such as amphioxus; and the craniates, including the hagfish and all vertebrates. An early ancestral animal diverged into the urochordates and separately into the two more closely related groups—the cephalochordates and the vertebrates. That original ancestral animal may have resembled the sea squirts or the salps or something else entirely, as these living organisms are now, themselves, more derived and evolved. It is likely we will never know. Nevertheless, there is an important clue to the development of the eye in the salps and other tunicates as compared to the cephalochordates and the craniates. Salps have a simple eye resembling an ocellus that is dorsal but not frontal in its location on the "head." Furthermore, that eye contains microvillus and not ciliary cells for photoreception. The cephalochordates and craniates, on the other hand, have ciliary cells for photoreception. That most recent common ancestor may have had both types of cells, or the ciliary cell may have arisen after the divergence of the urochordates. Currently, this is poorly understood but may represent a key step in the development of the vertebrate eye. The single eye is not visible. The orange mass in each salp is plankton in its intestine near its anal opening. The single round white structure is an egg. *Image © Casey Dunn.*

Other changes would stimulate the need for better sensory mechanisms. At least two important morphologic events during this period would relate directly to the visual witness.

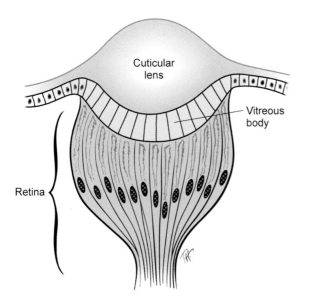

Figure 4.4 Ocellus or simple eye. The eye begins as a pit with photoreceptive cells at the base. The condensation of tissue above the pit creates the simple eye or ocellus. *Art by Tim Hengst.*

A Major Genetic Step

Approximately 600 mya, a monumental genetic change—the framework for bilaterality—was established. This would lead to bilateral symmetry, contributing directly to the presence of paired eyes in most species of invertebrates and vertebrates. This stroke of genetic vigor would eventually produce the majority of phyla diverse enough to include flies, crabs, clams, cephalopods, birds, reptiles and most other creatures that inhabit the earth. Even spiders, which have multiples of eye pairs, are bilaterally symmetrical.

Urbilateria

The clade of that first bilaterian (bilaterally symmetrical) creature is immense. Urbilateria radiated into worms, the ancient ancestors of the other invertebrates, and eventually the vertebrates. Of course, many of the Ediacaran animals were not bilaterally symmetrical. Creatures such as corals and adult jellyfish, for example, are instead radially symmetrical. But bilateral symmetry was to be a very successful design, especially for larger creatures and for the evolution of the visual witness.

Symmetry may have been so successful in Animalia because it permits repeated segmentation. This permits easier specialization of a part along the axis of symmetry, especially after the necessary genes—the Hox gene kit—become established. A body plan can be built with these individual building blocks in an organized, relatively simple, energy-saving manner. Spatial dimensions—up-down, front-back, and right and left—are more easily established.

We know little about the forces that propelled the birth of bilaterality, but it occurred between 650 and 600 mya. We know this because early imprints of the Ediacaran's soft-bodied creatures are bilaterally symmetrical, and that tendency has proven to be a win-ning trend. The Ediacaran period, then, established bilaterality.

Another Major Genetic Step

Another important step occurred within the Ediacaran's almost 100-million-year span. Some of the Ediacaran animals moved of their own accord—that is, by their muscles, not by drifting.

Wormy Beginnings

Worms of some variety appeared in the Ediacaran, and they had a critical third cell layer. The critical element in this layer was muscle, permitting purposeful locomotion. This middle layer made them triploblasts (three layers). Because jellyfish have less well-organized muscle, they are considered diploblasts (two layers) by many, although recent evidence suggests otherwise. Nevertheless, some members of the same phylum, Cnidaria, are diploblasts.

During embryological development, all animals have either two or three major cellular developmental layers. The diploblasts' two major layers of cells (ectoderm and endoderm) expand, proliferate, differentiate, and eventually produce the animal. Mesoderm, the third layer of triploblasts (which also include mammals), differentiates into more organized, controllable, muscles as well as into internal organs.

Until recently jellyfish were thought to be diploblasts, but they can move in a predetermined direction to some extent. Contractile muscular elements, including muscle much like our own, and energy-conserving passive elements (called mesogleal matrix) in the bell of some jellyfish can cause seemingly purposeful swimming movement, but it is not quite the same vigorous muscular-based movement as our own. And despite this ability, jellyfish generally hitch a ride with currents and tides. Jellyfish do not have the classic triloblastic arrangements but may be more intermediate in form.

Traces in the sands of the Ediacaran hills of Australia (namesake of the period) indicate that worms were "on the move" (Figure 4.5). They were the first truly and freely mobile animals. This purposeful movement must have been motivated by one of three reasons: (1) food (seeking vegetation or active predation), (2) protection or shelter (avoiding predation), or (3) reproduction. Movement would prove to be very important for sensory mechanisms, especially for the eye, because it would have provided a high degree of sophisticated evolutionary advantage for feeding and protection.

These early bilaterally symmetrical worms, appearing at least by 560 mya, probably were much like the key species Urbilateria, last common ancestor of the all bilaterians.

These newly gifted triploblasts had to be among the most complex of the new denizens of the Ediacaran. These animals had made one of life's key steps toward larger and more complex metazoan animals. On the other hand, jellyfish probably radiated off on their own without developing other lineages but may have developed a form of triploblasty. Remember, evolution is a branching process with a common ancestor that is no longer living, not a linear process with one living animal morphing into another.

During the Ediacaran the last common ancestor must have branched in at least two directions, one towards diploblasts and eventually modern jellyfish, and another more robust branch toward the triploblasts.

Animalia would progress in different directions. Although the basic proteins and genetic necessities for the molecular development of eyes were present by the late Ediacaran, there would still be many twists and turns. The visual witness would be reinvented multiple times from these early beginnings.

That last common ancestor had the sensory toolkit to develop eyes. Moreover, that ancestor must have bequeathed

Figure 4.5 Traces of worms in the Ediacaran sand.

the toolkit to the first triploblast, which exhibited many of the same molecular and genetic mechanisms. This creature may have had surprisingly mature eyes of its own, even at that divergence in the Ediacaran.

The early urbilaterian that gave us those first wormlike animals radiated into at least two distinct directions—the protostomes and the deuterostomes (Figure 4.6)—that would determine the course of animal life on earth. Protostome means first mouth, or one opening, and deuterostome means two mouths. In both definitions, "mouths" refers to the openings in the early embryo. In protostomes, the mouth develops from the first opening, and in deuterostomes the mouth develops from the second opening.

These openings are relevant to the visual witness in that deuterostomes would eventually lead to the vertebrates and the protostomes to the invertebrates. We will return to this in a moment.

Genetic Development of Eyes

Among other duties, Pax genes direct or participate in and regulate the development of the head and eyes. This DNA molecule must have appeared very early in metazoan evolution because it seems to be part of that early toolkit of genes. Although corals and jellyfish do not have the same Pax genes as the bilaterians, they do have a *PaxB* gene that is similar enough to suggest its presence in the last common ancestor of the triploblasts and diploblasts. This key ocular gene mutated during a few tens of millions of years or less to lead to the *Pax6* gene that participates in the assemblage of an eye in the vertebrates and invertebrates.

Motility and bilaterality would require a head and sensory mechanisms. And the quests for nutrition, carnivory, predation, or reproduction would require directionality. Although vision was not the only mechanism of discovering directionality, it would have been one of the earliest, and perhaps the most likely.

Olfaction, the recognition of certain molecules in the water as toxic or attractive, was probably the first sense. Pax genes or closely related genes also organize it as well as vision. These two senses may have appeared at about the same time and would serve different purposes. Even worms would need vision, although their vision would have been rudimentary, blurry, and useful only at short distances, but quite directional.

These creatures would have had eyespots or early cups of pigment lined with photoreceptive cells for recognizing light and dark and encouraging directionality. At some point, lenses began to condense within tissue and create a more formed and focused image.

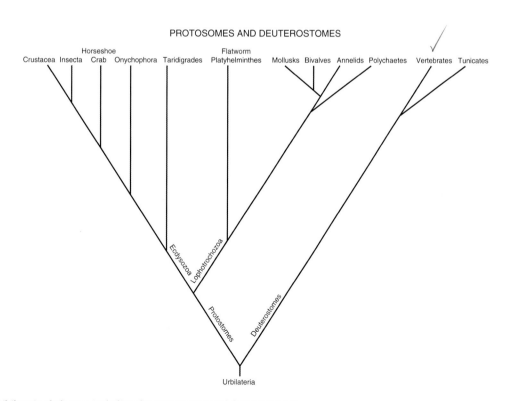

Figure 4.6 Uribilateria phylogeny including the protostomes and deuterostomes.

The first eye probably did not evolve with much of a lens. Nevertheless, although extant flatworms have had 500 million years or more to evolve and change, they illustrate how eyespots might have started in the Ediacaran.

Platyhelminthes (Polyclad Flatworms)

Pseudoceros dimidiatus. *Delicate, thin, and beautiful, polyclad flatworms, and flatworms are free-living (usually) marine organisms and simply defy imagination (Figures 4.7 and 4.8). Their ruffling edges cause them to move like the chiffon dress of a ballroom dancer. These animals are comprised of a single layer of skin cells with many beating cilia, a layer or two of muscle, and an internal layer of endoderm—triploblasts.*

These colorful flatworms are representative of the phylum Platyhelminthes and help us to understand the progress that eyes had made by this time. They also are basal and the closest extant representatives to the first worms of the Ediacaran, which brought eyes to the Cambrian explosion millions of years later.

These creatures have a head and a tail. The head contains the sensory organs and brain; the mouth is in the midventral section. The presence of a head and tail is an important evolutionary concept and is true of bilaterally symmetrical animals. Even most sea stars are bilaterally symmetrical in larval or embryonic form with a head and a tail. Although they become radially symmetrical in adult form, they are considered bilaterally symmetrical.

The process of organizing a head and a tail, called cephalization, permits the organism to have sensory organs and a brain placed together in the body to direct movement. The many flatworms that have eyes provide clues about the eyes of early arthropods, annelids, and mollusks. Although Platyhelminthes are not members of those phyla, they may be closer to the last common ancestor of all of them. And despite evolutionary changes in polyclad flatworms (they probably began in the Ediacaran or the Cambrian more than 500 mya), their eyes are so simple that they probably have not evolved substantially.

A flatworm's eye is a small, nearly round cup lined with photoreceptive pigment (Figure 4.8) that isolates the rhabdomeric cells within it. The cup's shape permits light to enter to provide a stimulus from only one direction, although most of these creatures have many individual eye cups. But polyclad flatworms' middle layer has given rise to muscles that move these primitive cups so they can "look" in at least a few directions. Early flatworms probably had the first extraocular muscles.

Polyclad flatworm eyes are scattered on each side of the "head," often at the base of what appear to be tentacles. The cerebral eyespot, if there is one, is in the midline on the dorsal surface, somewhat farther back than the pseudotentacular eyes. Even though it is unlikely that these worms radiated into the arthropods, it is easy to imagine that the last common ancestor had a similar arrangement in which the many eyespots consolidated into a compound eye in an early primitive arthropod. And the cerebral eyespot betokens the parietal (or third) eye that would appear in fish and find its real expression in reptiles.

There is another living creature that probably closely resembles the worms alive at that time in the Ediacaran, and this worm, *Platynereis*, provides us with insight into

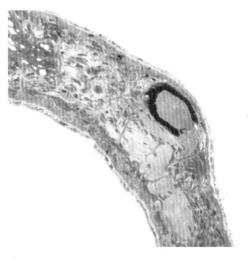

Figure 4.8 Eye of freshwater planaria. Although this is not a polyclad flatworm, it is a free-living flatworm with a simple eye cup, or ocellus, and a basic ocular anatomy similar to that found in a polyclad flatworm. The dark brown layer is comprised of pigment cells, and this layer limits the visualization of light from all but a few selected directions. This eye is capable of determining gross spatial details. Histology and image by Thomas Blankenship, PhD.

Figure 4.7 *Pseudoceros dimidiatus,* polyclad flatworm. *Image © Foster Bam.*

this early world of photoreception (Figure 4.9). The questions surrounding the various cells used in the construction of an eye may be partially answered with the understanding of this annelid.

The annelids, or more likely their predecessors, led to mollusks and arthropods and probably other phyla as well. Although there are other phyla of "worms" besides annelids, living annelids are perhaps as close as we will get to the primitive "worms" of the Ediacaran.

Annelids

Platynereis dumerilii

Worms seem taciturn and pedestrian, and yet, these creatures, particularly one rather well-studied polychaete (many bristles or legs) annelid (Figure 4.9), may hold the secrets to the differences between vertebrates' and invertebrates' eyes. Additionally, this ragworm may shed light on the origin of eyes—did eyes begin once (monophyletic) or many times (polyphyletic)?

Because bilateral symmetry is not an evolutionary "given," the first bilaterian helps us understand the development of two symmetrical eyes among other features. Molecular clock data, although controversial, hint that the first bilaterian diverged between 1.5 billion and 650 mya—quite a range. But the last common ancestor of invertebrates and vertebrates (already a bilaterian) probably lived between 600 and 543 mya in the Ediacaran period. Platynereis is probably a modern, but evolved, descendent of the animal that immediately predated the split into vertebrates and invertebrates. Consequently, Platynereis can teach us much about visual development even if it has evolved from the original bilaterian.

Figure 4.9 *Platynereis bicaniculata.* Note two individual sets of paired eyes. These four eyes all contain microvillus photoreceptive cells.

The uncomplicated eyes of Platynereis are best described as pigment cups behind a lens that is little more than a condensation of tissue. These pigment cups are burgundy or maroon principally because of orange pigment overlying black pigment in the supporting cells lining them. The rather profound secrets of these eyes are found with the visual pigments of the head, also called the prostomia.

Because Platynereis is an annelid, it is an invertebrate with the traditional rhabdomeric photoreceptors and rhabdomeric opsins (r-opsin) in all four of its eyes (Figures 4.9 and 4.10). Ciliary cells and ciliary opsins (c-opsin) exist within the brain somewhat closer to the head or front of the animal and between the two pairs of eyes. Remember that vertebrates have ciliary instead of rhabdomeric cells to house the photoreceptive visual pigment, c-opsin. It is doubtful that ciliary cells, found in the brain of Platynereis, and their photoreceptive pigment have any true visual function in this animal; they probably relate to setting the circadian rhythm. So, this worm has both the ciliary and rhabdomeric forms of visual pigment that are seen in vertebrates and invertebrates. But there is a further twist to this story.

Vertebrates may not be so far away from these annelids after all. To see some of the wormy secrets of our beginnings we have only to look to our inner retina. Investigators suggest that the invertebrate visual pigment, r-opsin, is present in vertebrate eyes, albeit in a few of the ganglion cells of the retina. That is, the vertebrate eye has retained the rhabdomeric photoreceptive cell with the visual pigment present in a small portion of retinal ganglion cells (the neuronal cells that transmit the signal from our photoreceptors to the brain)! A remnant of photoreception is present in these ganglion cells in the form of melanopsin, another photoreceptive compound. This visual pigment probably contributes to circadian rhythm control but has no visible light function that we consciously recognize. Melanopsin is also found in the pineal body and in the parietal eye of animals that have either of these tissues.

Vertebrates keep both forms of visual pigments (rhabdomeric, which is classically invertebrate, and ciliary, which is classically vertebrate) and have evolved an eye that unites them from two different anatomical sources.

Ediacaran worms and their progeny help us understand this process of ocular development in other ways. One possible scenario for how the photoreceptor in vertebrates and the photoreceptor equivalents in the invertebrates arose, first proposed by Purschke et al., fits the data well and has the potential to explain many extant eyes.

In the scenario, initially there were skin (or epithelial) cells on the exposed surface of metazoan organisms—collections of relatively uncomplicated but somewhat specialized cellular collections such as sponges or coral. Some of the cells would have contained photoreceptive

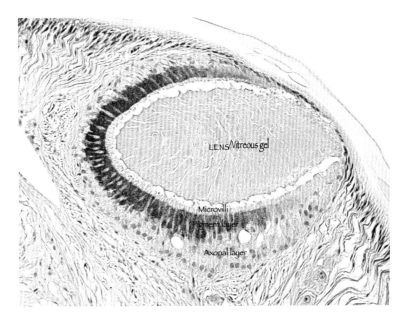

Figure 4.10 *Nereis virens* or king sandworm eye cup. *N. virens* is closely related to *Platynereis* and has a similar set of eyes. The lens, microvilli layer, and axonal layer are labeled. The microvilli layer is the layer of the photoreceptive portion of the cells. The axonal layer consists of the cell that receives the signal from the photoreceptive cell and sends the image to the brain. *Image by Richard Dubielzig, DVM.*

pigment—a rhodopsin or rhodopsin-like compound. A pigmented cell would have been immediately adjacent to the cell containing the photoreceptive compound. This combination may already have been engaging in a visual processing system—providing directional information by blocking light striking the photoreceptive compound from one or more directions.

These two cells invaginate or form a pit together, with the pigment backing the cell that contains the photoreceptive pigment, to create an ocellus. Note that either of the two principal cells of photoreception—ciliary or rhabdomeric—could form this pit. Note also that this simple method of invagination could cause this combination of cells to form an everse or inverse combination—the presentation of the photoreceptors to the incoming light. In other words, this explanation would be the source for the inverted retina in vertebrates, the everted retina in the invertebrates, and the various combinations in between. These cells and molecular compounds are so similar across all multicellular animals that they almost certainly had to have a common origin.

The two mechanisms could have been active in the worms because they contain both forms of photoreceptors, and this simple explanation encompasses all known forms of eyes. In the annelids and perhaps in other worms, the ciliary cells became more closely associated with the brain, and the microvillus cells became the photoreceptors for the eyes. In annelids, these ciliary cells became more active in other forms of visual response—such as circadian rhythm.

Furthermore, the eyes of that early worm as represented by this annelid may well have preceded that first bilaterally symmetrical animal—urbilateria. It is likely that these relatively complex eyes preceded the formation of the third body layer or that these eyes came along in a simultaneous geologic moment.

The annelids would radiate into other worms with different designs, and not all would use the ciliary and rhabdomeric cells in the same manner. The sabellid worms prove the point.

The sabellid worms are important because they illustrate an unusual compound eye unlike most other eyes, and unlike the eyes of *Platynereis*. Most compound eyes appear in the arthropods. The sabellid worms developed a compound eye that can help us understand development of the visual witness by illustrating that the annelids' separation did not inhibit the development of the compound eye—it just went in a different direction.

Sabellids: Megalomma *Species*

Worms are the earth's custodians. Some detoxify biotic pelagic debris by ingesting it. Sabellid worms are many of these stewards of the oceans (Figures 4.11–4.13).

Sabellids are polychaete worms found in all the oceans. Coral reefs are rife with them. Almost all sabellids are filter feeders, which means they remove organic refuse or plankton that float near the bottom or swirl on the currents. The head of sabellids has been transformed into a feathery "crown" of many radioles

Figure 4.11 *Megalomma* species. This feather duster worm can be seen from above. These are marine annelid worms and are generally sessile. *Image © Greg Rouse.*

Figure 4.13 *Megalomma* species. These radioles are tipped with compound eyes. *Image © Greg Rouse.*

that project from a protective tube into the water. This crown filters out the necessary food, but filtering takes time, which gives potential predators long exposure to sabellids, so the worm needs visual protection. This species has two compound eyes at the ends of the two radioles that project the farthest away, giving it a wide view (Figures 4.12 and 4.13).

Each sabellid "eye" consists of forty to sixty ommatidia (individual units or little eyes), but these "little eyes" are anatomically different from those in insects or spiders. The ommatidium is a tapered pigmented tube resembling an ice cream cone formed from a single cell, with a crystalline core as the "ice cream" (Figure 4.14). The apex of the cone contains a single photoreceptor that connects to an axon (Figure 4.14). The cone-shaped pigment cell isolates the receptor portion from stray photons and provides directionality.

Other sabellids, often equipped with at least twenty eyes on the crown, may have more eyes on each segment and even at their posterior tip.

Compound eyes that so typify invertebrate vision are almost exclusively restricted to the Arthropoda, but polychaete worms

Figure 4.12 *Megalomma* species. Lateral view of feather duster worm. They live in tubes they produce as shelter. The worms are filter feeders and can retract the crown of "feathers," called branchiae, and the two "eyestalks," called radioles, into the tube. *Image © Greg Rouse.*

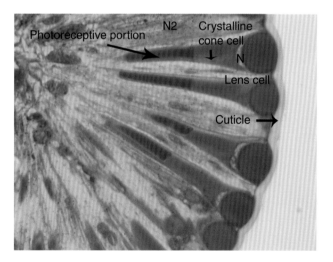

Figure 4.14 *Megalomma* species. Histology of compound eyes. Note the "ice cream cone"-like shape with the apex containing a single photoreceptor element—a ciliary cell. The most distal portion is the cuticle. Beneath that is the first of two cells. The "ice cream" ball is a lens within a cell, and the "N" is the nucleus. The cell more proximal is a crystalline lens, and the "N2" is the nucleus for that cell. That is a ciliary cell, and the photoreceptive element is labeled. Magnification 63×. Bar is 20 μm. *Image © Greg Rouse.*

and some bivalves (mollusks) also have them. The eyes of the sabellids, then, represent evolution's separate and disparate attempt at compound eyes. It also may be a fairly recent attempt because not all sabellids have them, and none of their close wormy relatives does. This suggests that eyes—compound, simple, and camera style—can evolve more than once. Nevertheless, the eye's evolutionary toolkit had to be present even from this early radiation into annelids, or the eyes would have been randomly different.

Arthropods, including insects, shrimp, and crabs, have exploited compound eyes since the Cambrian explosion. But polychaete worms are outliers, as their compound eyes have different structural and neurologic strategies, indicating that the evolutionary process actively sought other ocular designs well after the Cambrian. These different mechanisms offer clues into the early development and evolution of photoreception and visual processing.

The sabellids are exceptions to the rule of rhabdomeric photopigments in invertebrates because sabellids are invertebrates that have modified cilia to contain their visual pigment. (Remember that most invertebrates have rhabdomeric [microvillus] cells with photopigments as their photoreceptors, and most vertebrates have ciliary cells housing their photopigments.) Sabellid worms also have key neurologic differences that separate them from other invertebrates. These key neurologic differences mean that the photoreceptive cells act much like vertebrates' photoreceptive cells.

The last common ancestor of invertebrates and vertebrates clearly had ciliary and rhabdomeric cells. Different lineages used these two types of cells in different ways and for different reasons. Some lineages used both, and others used only one.

The early urbilaterian radiated into two distinct directions—the protostomes and the deuterostomes (Figure 4.6) as mentioned above.

These two different radiations are important to the visual witness in that deuterostomes would eventually lead

to the vertebrates and the protostomes, to the invertebrates. But some few invertebrates are deuterostomes, and these are important to the visual witness because they likely appeared in the Ediacaran. Although these creatures are invertebrates, they are the beginning of the lineage that would radiate into the early chordates and eventually the vertebrates. Most of these animals have only eyespots or ocelli, but one representative has interesting eyes. We will focus on this echinoderm, the brittle star, a contemporary creature that has a surprising visual witness. Brittle stars illustrate that eyes could be established in different ways and still be effective for that particular animal. There are eyes in those stars.

Brittle stars: Ophiocoma wendtii

Brittle stars are closely related to sea stars or starfish (which are not fish, so sea star is the better term) but are far more agile and mobile (Figure 4.15A and 4.15B). Sea star fossils date back to the early Cambrian, and they may have been present as early as the Ediacaran. Brittle stars diverged from sea stars in the Ordovician, and their eyes must have appeared very early— perhaps permitting divergence.

The visual witness runs through these animals, as they seem to have a sense of vision and are negatively phototaxic. Brittle stars are mainly nocturnal and will hide from light. Until recently we did not know how this was accomplished because brittle stars do not appear to have eyes. But in reality they do.

Each arm is covered with calcite lenses that focus light on nerves within the arms directly below the lenses (Figures 4.16 and 4.17). But these are no ordinary lenses. They are Huygens-Descartes correction lenses—a doublet-lens system, a pair of lenses that are exactly curved to eliminate spherical aberration (Figure 4.17).

One can appreciate this form of lenticular distortion (spherical aberration) by noting the distortion at the periphery of a clear glass marble. The doublet-lens system eliminates the distortion (see Chapter 5).

Furthermore, pigmented cells surround each eye, providing sunglasses for the creature. During daylight hours, these pigmented cells, called chromophores, migrate over the lens surface, causing the brittle star to darken and change color. This effective diaphragm decreases or eliminates the sensory input from the dorsal surface. Because all arms are covered in lenses, these brittle stars are, in effect, one large compound eye. The lenses are directed in slightly different directions, making the entire dorsal surface of the brittle star an "eye." The Huygens-Descartes correction lens will turn up again in trilobites, where it appears to have been independently achieved. This represents optical convergent evolution.

BOX 4.1 HOW PHOTOPIGMENT CREATES A SIGNAL

When light strikes the photopigment of the sabellid worms, it causes a change in the structure of the retinal-opsin combination. This configurational change initiates hyperpolarization of the membrane from a depolarized state, much as in the photoreceptors in vertebrates. But in most invertebrates, photopigment configurational change initiates depolarization from a hyperpolarized state. This represents a fundamental difference in ocular function and may be a clue to early photoreception.

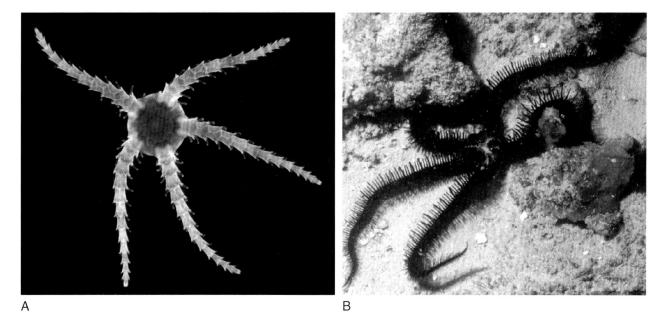

A

B

Figure 4.15 *Ophiuroidea* species. (A) Icelandic brittle star. Note the nummular central disc and long thin arms. *Image © Greg Rouse.*
(B) *Ophiocoma wendtii*. This brittle star species is all eyes, but not all species are. *Image © Gordon Hendler, PhD.*

We know that one mechanism to make an eye includes an evagination (bulging) of a surface with the photoreceptive elements directed individually. These evaginations show up as head shields in the few preserved impressions of the Ediacaran's soft-bodied creatures, suggesting that eyes were indeed in progress in the pre-Cambrian (Figure 4.18). Of course, that would be no surprise, as many of the optical designs of extant eyes probably appeared in the Cambrian or earlier.

We do not know much about the ocular design of these eyes, or even if these evaginations were eyes. One of the impressions of an animal of the Ediacaran—*Spriggina floundersi*—appears to be the predecessor of the trilobites (Figure 4.19). The three-lobed body closely resembles the body of trilobites. The impression has slight protrusions as in a shield that could be eyes, albeit simple ones. Are these the first eyes? Given the stage of development of eyes of

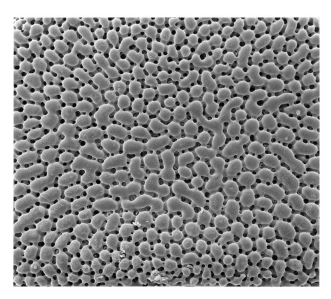

Figure 4.16 *Ophiocoma wendtii*. Scanning electron microscopic image of surface of an arm. Each aggregation is an eye. *SEM images by Pat Kysar.*

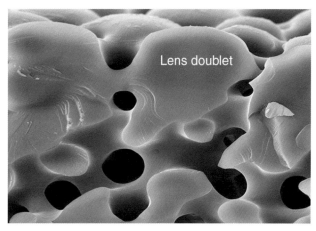

Lens doublet

Figure 4.17 *Ophiocoma wendtii*. A lens doublet system in profile. The lens system is welded together instead of being two separate lenses but optically functions as a lens doublet. *SEM images by Pat Kysar.*

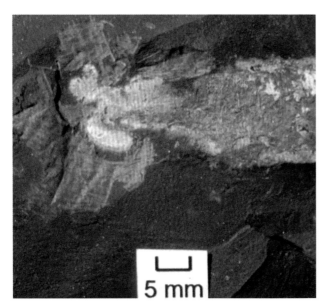

Figure 4.18 *Myoscolex*. Little is known about this Ediacaran animal, but it may have been among the first animals with eyes. Although a trilobite from the Cambrian is currently considered the first animal with eyes, *Myoscolex* preceded the trilobites. Investigators do not agree whether this animal had eyes. *Image © Jim Gehling, PhD.*

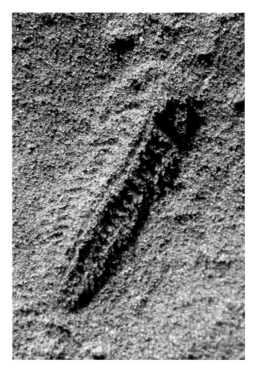

Figure 4.19 *Spriggina floundersi*. This bilaterally symmetrical animal was approximately 50 mm long and had a head shield seen in the image. It is tempting to consider the head shield as having eyes, but this cannot be determined. This creature is considered to be an ancestor to the trilobites or at least related to the trilobites. *Image © Jim Gehling, PhD.*

animals in the Cambrian, it is quite likely that true eyes formed in the Ediacaran, and these may be an example.

At this time, approximately 550 million years ago, the right ingredients were in place. Oxygen levels were perhaps two-thirds of what they are now but substantially greater than had been present as little as 200 to 300 million years previously. Some of the ocean floor was covered with microbial mats. The shallow seas were warm and full of life, both autotrophic (an organism that uses the sun to help produce its own food) and heterotrophic (an organism that must rely on the external environment for its food), with metazoan predators and prey beginning to hone their survival skills. The biological switches, genetic toolkit, and gene duplications had already created a great variety in multicellular animals.

The necessary steps and proteins were in place, and multicellular life had begun, but much had to change. A cataclysmic extinction took almost all of the Ediacaran fauna, leaving perhaps only a few diploblasts and robust triploblasts such as the worms to soldier on. When life emerged on the other side of the gate between the Ediacaran and the Cambrian, the atmospheric conditions were right. New niches were available. Higher oxygen levels would have been enough for the metazoans to begin to flourish. Just where would the lineage of worms lead Metazoa?

The Ediacaran worms probably also radiated into the mollusks, whose beginning we can follow from at least one creature found in that early fauna.

Mollusks Take the Stage

Kimberella quadrata and Acanthochiton communis

The mollusks are a highly varied phylum containing approximately ten classes, including two that are extinct. Mollusks are important to the visual witness because most of the extant classes possess eyes, so the early mollusks probably hold a key to the eye's development.

The hypothetical wormlike three-layered creature that was the animal now called Urbilateria began in the Ediacaran and radiated into the mollusks and the arthropods. Urbilateria gave the gift of ciliary (eventual vertebrate photoreceptors) and rhabdomeric (eventual invertebrate photoreceptors) cells to its various descendants. That was the gift of sight.

The Ediacaran had whisperings of extant phyla (although these may be so faint as to prohibit us from knowing exactly where to classify the organisms). The first known mollusk probably was *Kimberella quadrata*. This 10-cm (diameter) organism was originally thought to be a box jellyfish. But whether this animal was a jellyfish, a mollusk, or somewhere in between, this animal is important to the visual witness.

Kimberella is believed to have had a shell, although it may not have been mineralized as shells we know today. These early precursors of the mollusks almost certainly had some early visual abilities. Although this is speculation, many extant members of the phylum Mollusca have the opsins and photoreceptive cells similar to the annelids and arthropods, suggesting that the last common ancestor of these lineages and all of the radiated mollusks had the toolkit for eyes, if not actual eyes (Figure 4.6).

Kimberella (or *Wiwaxia* in the Cambrian) may have been the Precambrian ancestor of the chitons, as these mollusks appeared early and span from the Cambrian to today.

Chitons: Tonicella lineata

Tonicella lineata, *or the lined chiton, has multiple sensory structures called aesthetes lining the dorsal surface of the shell (Figure 4.20). This algal grazer is usually found pressed hard against a rocky surface in the tidal zone, along the coral reef, or even at some depth. The aesthetes are often numerous and bilaterally symmetrical, and they appear to have light-sensing abilities and to be a primitive eye. There is a clear layer on the shell, much like a cornea, with a lens beneath it. Each of several rhabdomeric photoreceptive cells in a cuplike distribution has a nerve leading away from the cell. These "eyes" are simple ocelli, but they are in the same pattern as the eyes of the octopus, a mollusk that would come millions of years later.*

The mollusks were here early, perhaps as early as the Precambrian, and may have been responsible for those early shells deposited at the cusp of the Cambrian—the small shelly fauna.

Small Shelly Fauna

Other Precambrian organisms appear to be early mollusks. *Halkieriidae*, now known as the small shelly fauna, a likely important early mollusk, was present at the cusp of the Cambrian, presaging the eruption of animals that would follow. Sabellid worms or other organisms, such as the

Figure 4.20 *Tonicella lineata.* A strikingly beautiful chiton, *T. lineata* is found on the west coast of North America. *Image © James Brandt, MD.*

mollusks, may have produced the shelly fauna, which is seen just before the start of the Cambrian. Speculation would suggest that these eyes resembled those of marine gastropods (the conchs to be discussed in Chapter 8).

Strong evidence suggests that the mollusks began before the Cambrian and were early offspring of Urbilateria. The first real mollusk probably had simple eyes and would further prepare the phylum for a continuing visual witness.

Odontogriphus and *Wiwaxia*, both mollusks from the Cambrian, may have been blind, but as the genetic toolkit was present, it is more likely that they had delicate, primitive eyes that did not leave any preservation or imprint.

Certainly, though, by the early Cambrian period, the mollusks—at least the bivalves, the gastropods, and the cephalopods—were established. Each of these lineages shows eyes with common features, suggesting that the last common ancestor had eyes, and perhaps not-so-primitive ones at that. The long, slow fuse of the Ediacaran was lit and smoldering for the once-in-a-planet's-lifetime biological explosion, which we see in the next phase of this story. Life was ready for its ascendancy.

VISION'S BIG BANG
BLAZES THE TRAIL

EARLY PALEOZOIC ERA
CAMBRIAN EXPLOSION
543–490 MILLION YEARS AGO

Eras do not start suddenly (although cataclysms often end them abruptly). Surely the creatures around at the beginning of the Cambrian Period (Paleozoic Era) would have noted no difference. No sonic boom of activity foretold it. Rocks dated to this time reveal the first recognizable creatures, which allowed us to define a beginning—543 million years ago. We define its beginning by recognizing fossilized hard body parts and even some soft-bodied animals that were preserved under special circumstances.

Within a geological flash of approximately 15 million years, there was an explosion of phyla. Evolution experimented with many body plans and laid the groundwork for the body morphology found in today's living taxa. By the Cambrian, morphologic divisions between the invertebrates and the vertebrates allow us to separate these two major groupings. The visual witness accompanied both major groups, and we will follow it in them, in separate chapters.

The Burgess Shale

Few fossil collections are as important to our understanding of early metazoan life or retained their secrets for as long as the fossils of the Burgess Shale (northwest Canada in the Canadian Rocky Mountains). Given enough time, though, such enigmas are usually unraveled.

Initially discovered in the Burgess Shale in 1909 (and more recently discovered in China's Chengjiang fossil beds), the fossils of the Cambrian provide us a taste of more complex animals, and what a smorgasbord. Within a few million years—a geologic flutter, really—evolution spawned many new phyla in a pique of fecundity never to be repeated. Probably these few million years were the absolute sweet spot with all the necessary elements: open niches, the genetic framework for Animalia, a few very plastic models and designs from the Ediacaran, oxygen in the atmosphere, and vegetable matter for nutrition.

Like a curious child with mounds of colorful clay, evolution assembled an astonishing variety of morphologies,

but many were crude and clumsy experiments. Just like an adolescent, however, evolution would abandon many of these novelties to focus on those with the power to survive.

By the early Cambrian, the visual witness was consolidated, had organized perception, and had moved forward with a direction probably driven by the sensory imperative of predation and competition. The wars between predator and prey began here or had been smoldering and exploded with the Cambrian's fierce carnivory.

To describe the visual witness in the Cambrian requires us to first consider the invertebrates, but this presents a problem. The vertebrate model in the hemichorodates, chordates, and probably even true vertebrates already existed, and we will visit that lineage subsequently. But, a few invertebrates are really in the vertebrate lineage, and these were important to the developing eye, too, and were discussed in the previous chapter.

For the remainder of this text, we will divide Animalia into the invertebrates and vertebrates. Much was happening in the Cambrian, but the vertebrates were not the only show, or for that matter, the most important one. So, we begin with the invertebrates.

The invertebrates were, and still are, the phyla that displayed the most creativity and variety. Although vertebrates would continue their maturation, the Cambrian was to be the age of arthropods.

Invertebrates in Ascendancy

The Cambrian, like the Ediacaran before it, benefited from several climate changes. Oxygen levels continued to rise to sufficient levels to support, even encourage, metazoan life. The tipping point was approximately 11–12 percent of the atmosphere (today's oxygen levels are approximately 21 percent). By this time, metazoan life had discovered adenosine triphosphate (ATP), an energy factory that represents metabolic coevolution at its finest. This metazoan energy factory requires oxygen and fuels the muscles of larger and more active animals—triploblasts.

Figure 5.1 *Parvancornia*, an animal of the Ediacaran. It is considered
a possible ancestor to the trilobites, as can be seen from its design.
This imprint of the soft-bodied animal shows a bilaterally symmetrical
pattern that may have been the precursor for the trilobites where the
first known eye has been found. This impression of *Parvancornia* does
not reveal eyes, although the frontal shield may have had the
precursor eyespots that would eventually evolve into eyes on the
trilobites.

Other factors besides oxygen helped to set the stage for
the metazoan explosion. The seas were shallower than they
are today, and life was mostly confined to the continental
shelf. The climate had warmed, and as a result the glaciers
were melting. Life was generally more abundant. The stage
was set.

Many phyla appeared—Mollusca, Arthropoda, Onychophora, and Echinodermata among many others. To
support them, sensory mechanisms were evolving. Vision
had to be high on the sensory ladder because many
organisms would have been responding to light. With
the visual mechanisms present and available, the visual
witness (photoreception) was ready to explore many other
morphologies.

The First Eye

The first eye probably had appeared well before the
Cambrian, but we cannot be certain of that because of an
inadequate fossil record. Nevertheless, several good
Ediacaran candidates for the first eye include *Parvancorina*
(Figure 5.1), *Spriggina floundersi* (Figure 4.19), and *Myoscolex*
(Figure 4.18). The figures illustrate what may have been
eyes. But not enough firm evidence defines any of these as
the first eye.

The trilobites, though, boast the first known and well-
accepted eye. The Paleozoic Era could well be considered the
age of the trilobite because these creatures burst onto the
scene approximately 540 million years ago to flourish in the
Cambrian and Ordovician seas. The number of trilobite
families peaked in the late Cambrian (variety in species
morphology probably peaked later, in the Ordovician).

In this lineage, the first known eye appears at the
beginning of the Cambrian—in the trilobite, Olenellus. The
trilobites illustrate at least one way it developed in the
Cambrian arthropods (Figure 5.2). Its peculiar optics is
relatively well understood. The lenses and the morphology
of the eye are preserved because these creatures had hard

parts that fossilized. The story of another early trilobite,
Phacops, will provide an additional glimpse into this first
eye and its remarkable optics.

Trilobites

Olenellus, Phacops rana milleri, and Phacops rana crassituberculata

In the seventeenth century two mathematicians independently solved the problem of spherical aberration in a lens.
Their doublet-lens system now bears their names: Christian
Huygens (1629–1695) and Rene Descartes (1596–1650).
Unbeknownst to them, however, evolution had already
solved this problem hundreds of millions of years earlier in
trilobites. This subtle, recently decoded, paleontologic tale
reveals the microevolutionary principles that are
continuously at work and capable of providing ingenious
solutions to the physical problems of vision.

Trilobites arose from the Cambrian explosion and lasted
for perhaps 300 million years (until the mother of all
dyings—the Permian extinction, 235 million years ago).
Much is known of these animals from the fossil record
because they had hard body parts, including calcite lenses,
which permitted preservation in stone and understanding of the visual capabilities and experiences of these

oldest eyes. Calcite, or calcium carbonate, can slowly form into crystal-clear structures known as rhombs. These crystals can be assembled intracellularly, as *Ophiocoma wendtii*, a brittle star (Chapter 4), has proven. The trilobites did just that.

The fossil record shows at least two morphologic types of these compound eyes: holochroal and schizochroal. A third, intermediary type known as abathochroal is less certain but should be considered a separate morphological form. The holochroal model was the simplest and first (Figure 5.2). The schizochroal eye of *Phacops* (Figures 5.3 and 5.4A) was optically sophisticated with a doublet-lens system. This remarkable design probably produced a sharp image on retinular cells containing only one visual pigment. Phacops may have been color blind and nocturnal, or at least benthic (ecological zone of the bottom of a body of water).

The arrangement of the lens doublet in these schizochroal eyes was the most advanced trilobite eye (Figure 5.4A and 5.4B). This lens system would have increased the contrast sensitivity by as much as fivefold, lowering the f-number (focal ratio) of the eye. The f-number is a photographic term that describes how successful a lens system or optical device is at gathering photons of light. The lower the f-number, the more sensitive the optical system is, and the better it can operate in low light levels.

Thanks to Riccardo Levi-Setti, the evolution of trilobites' optics of these creatures is becoming crystal clear. This investigator suggests that the first trilobite probably had

Figure 5.2 *Olenellus*, an early trilobite that possessed the earliest known eye because these eyes fossilized. It is doubtful that such an eye formed de novo and that predecessors from the Precambrian had precursors that led to this eye. The artist's rendition illustrates the conceptual image of the Cambrian seas and these early trilobites. *Image © The Natural History Museum, London. Artist Graham Cripps.*

The holochroal eye, the earliest and most common model, consists of small, closely packed ommatidia (individual eye facets) that are round or hexagonal (Figure 5.2). These facets are covered by a blanket layer of calcite that is, in essence, a "cornea." This layer could have been shed with ecdysis (molting), allowing for growth of the organism and the eye. The eye added identical new ommatidia below the existing ones in a hexagonal pattern, much as in a honeycomb. These eyes were probably assembled from individual simple eyes; the relative size of the lens varied from species to species.

The schizochroal eye, a unique variation in compound eyes, was found only in the suborder *Phacopina* (first seen during the Ordovician period and surviving until the end of the Devonian). These ommatidia were arranged in a hexagonal and rectilinear packing pattern, as in a honeycomb (Figure 5.4). They were generally larger and arranged differently than those in the holochroal eye. These ommatidia varied in size, becoming larger toward the ventral surface of the animal.

Lenses varied in shape. The first lens in each ommatidium was spherical; it was followed by an odd bow-shaped lens with a wave-shaped surface facing the first lens. These two lenses back to back are defined in the optical world as a "doublet" (Figure 5.5). This second lens, the Descartes-Huygens correction lens, provides for the correction of spherical aberration. This lens design is seen in only one or perhaps two other animals.

The abathochroal eye may have been a variation of the schizochroal eye, with a separate "cornea" of calcite covering the front surface of the lens. This can be seen in other species with compound eyes. Between the lenses was a layer of calcite "sclera" (Figure 5.6). A few trilobite species from the Cambrian had these unusual abathochroal lenses with a central nipple of lens material creating an aspherical inner surface that would help with spherical aberration and create lenses with bifocal capabilities. Most likely, however, this system would have created problems with internal reflection and limited light-collecting efficiency. But the shape could have facilitated a bifocal-like visual system.

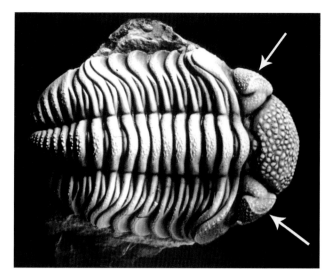

Figure 5.3 *Phacops rana milleri.* This dorsal view of *Phacops* illustrates the three-lobed nature of the trilobites and the eyes on the lateral aspect of the head (white arrows). *Image © Riccardo Levi-Setti.*

holochroal eyes with only one calcite lens in the cone of the ommatidium (individual facet of the eye) (Figure 5.2). This structure would create optical difficulties, including diffraction limitations.

Next, as the eye evolved, would come an abathochroal eye with a larger round lens, which would prove advantageous, especially as these animals extended into a dimmer, mid-ocean (mesopelagic) environment. As the lens enlarged, the optical system would be limited less by diffraction, but now by spherical aberration, producing a blurred focus.

The third and truly revolutionary step to be taken was the formation of a lens doublet with a perfected Huygensian profile (Figures 5.5 and 5.6). This step would allow for

A   B

Figure 5.4 (A) *Phacops rana crassituberculata.* Lateral view of a trilobite closely related to *Phacops rana millei.* Note the individual elements of the eye. These are individual ommatidia, and each one would have had a slightly different visual field. The trilobite brain would have "assembled" the image. *Image © Riccardo Levi-Setti.* **(B)** *Metacanthina issoumourensis,* a trilobite with schizochroal eyes. This animal's eye has the same ommatidial pattern as *Phacops* and is in the same suborder, *Phacopina.*

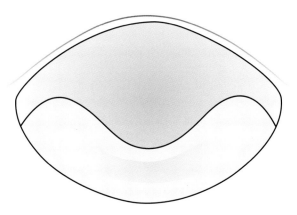

Figure 5.5 Diagram of doublet lens. Note the doublet nature of the lens, which would have decreased or eliminated spherical aberration, bringing the light wave to focus at a single point if the light waves were of a single wavelength. *Art by Tim Hengst.*

dramatically increased light-gathering capabilities and eliminate spherical aberration, as Descartes and Huygens would learn.

Chromatic aberration would probably still exist in such a doublet. It occurs when colors are bent or refracted more or less because of the different wavelengths of that particular color. But in a monochromatic world of mesopelagic or epipelagic (upper layer of the ocean water 0–200 meters) light, this would not matter much. And if the eye has only one photopigment, there would be no perceived color aberration.

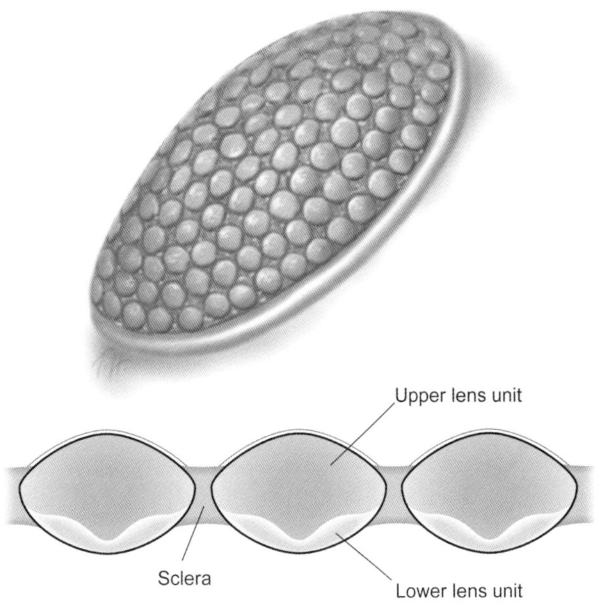

During their reign on earth, trilobites appeared that were remarkable in size—as big as a medium-sized dog—and ocular morphology. One trilobite, *Erbenochile erbeni*, developed peculiar eyes.

ERBENOCHILE ERBENI The three-lobed shape of trilobites shouts "Paleontology" to anyone with a passing interest in evolution, as these fossils are almost as familiar as dinosaurs. And these creatures were to their period what the dinosaurs were to theirs—diverse, widespread, common, and emblematic. The diversity led to various sizes and shapes of animals.

One of the weirdest trilobites was *Erbenochile* (Figure 5.7), 3.75 cm, with eyes shaped like grain silos that had been cut in half vertically (Figure 5.8). Each column that rose from the creature's head had 560 lenses whose visual field probably joined or overlapped anteriorly and possibly posteriorly. This would have given the animal the ability to see in all directions at once without turning its head.

Some 400 million years ago, this wondrous animal probably spent its days embedded in the mud of the sea floor off what is now the Moroccan coast. Its half-centimeter periscope-like eyes could have projected above its camouflage to protect the animal against predation and could have facilitated ambush hunting for morsels. This animal even had structures resembling sun visors, probably as a protection against the sun, although there may be other explanations for them.

But these twin towers were probably as much a hindrance as a help. The eyes would have been cumbersome and would have made ambulation more difficult and slow, especially on the seabed. But although the reign of such a creature was likely short, its eyes illustrate the extent of the creativity of evolution.

Upper lens unit

Sclera

Lower lens unit

Figure 5.6 Diagram of abathochroal lenses. Note the nipple-like morphology to the posterior or proximal curvature of the first or upper lens as another method to reduce or eliminate spherical aberration. *Art by Tim Hengst.*

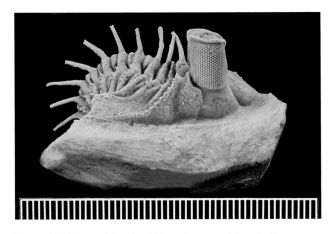

Figure 5.7 *Erbenochile erbeni.* Note the unusual "stacked" compound eye with the ability to protrude above the level of the sand. Ruler is in millimeters. *Image © The Natural History Museum, London.*

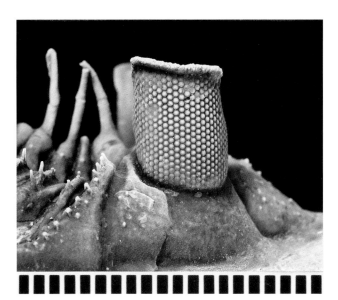

Figure 5.8 *Erbenochile erbeni*. A closer view of the silo-like eye with a sunshade. This may have been a hindrance to the animal at least in some circumstances, as it would have made swimming more difficult. Note the individual ommatidia. Bar is in millimeters. *Image © The Natural History Museum, London.*

More Invertebrates and Their Eyes Appear

The Cambrian produced a vast array of body plans in a biological blast. Some of these were never to be seen again after the Cambrian, but several phyla continue to the present, illustrating successful design—or simple happenstance and contingency. The rapid appearance of these animals during the Cambrian represents pubescent metazoan life, and as in humans, this time was one of great experimentation. The innovations were probably driven, at least in part, by the predator-prey duel that emphasized sensory modalities, including vision. As a result, it is likely that many of these new creatures were armed with eyes—and rather sophisticated ones at that. Much of the variation within the eyes of invertebrates probably began in the Cambrian, and what we know helps us understand where the development of the eye is headed.

Predation is a terrible and swift sword for prey species, and there is good evidence that to some extent, predation influences evolution. Sensory modalities drive predation—the predator must find prey, and prey must defend itself. But, vision did not drive the Cambrian explosion exclusively, although it likely played a role in species selection. Other factors, including climate, geography, water currents, oxygen levels, competition, and even chance must have played significant roles in the evolution of the Cambrian explosion.

Nevertheless, these early animals and their sensory mechanisms did influence selection. And although they have left us a relatively rich fossil record, much of the information, especially about the sensory mechanisms, must be gleaned by conjecture and circumstantial evidence.

Specific Examples

OPABINIA *Opabinia* (Figure 5.9), considered a close cousin of the arthropods, must have been a predator and an odd one at that. A 1975 monograph by Harry Whittington (a Cambridge paleontologist) described it as one of the most remarkable creatures in the history of science. He found five eyes with two stalked pairs and one smaller central eye in a creature approximately 50–70 mm in length. The eyes were almost spider-like in distribution, probably providing stereopsis (visual ability to have depth perception using both eyes).

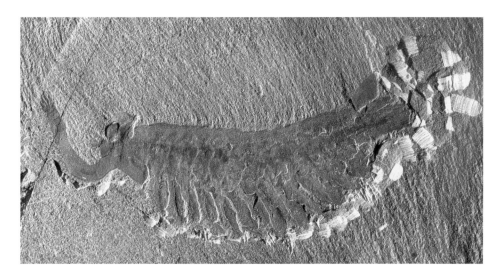

Figure 5.9 *Opabinia*, an unusual five-eyed animal of the Cambrian—perhaps ancestral to the crustaceans. *Courtesy of and © the Smithsonian Institution.*

Opabinia had a peculiar, flexible "elephant-trunk" proboscis with what was most likely grasping spiny lips at its end. This trunk probably delivered food to the mouth on the ventral surface. It is likely that *Opabinia's* large visual field, streamlined segmented body, and presumed aquatic speed made for an agile and successful predator.

ANOMALOCARIS If *Opabinia* was the resourceful, barracuda-like predator of the Cambrian seas, then *Anomalocaris* (Figure 5.10) was the seas' looming lord and master. It was the largest predator of the Cambrian seas, with some species known to be up to two meters in length. This animal was much more common and certainly less subtle than *Opabinia*. With large, laterally placed eyes, *Anomalocaris* must have had a wide field of view. Frontal views of the creature suggest the possible presence of stereopsis, which almost certainly would exist in the most efficient predators. Unlike the trilobite, no calcite lens system was found for the *Anomalocaris*, so all we know about the eyes is that they were not found to have facets, as compound eyes would.

Figure 5.10 *Anomalocaris,* a major predator of the Cambrian seas. This flattened raylike creature had a round mouth on the ventral surface and eyes on the front shield (arrows). With Permission of the Royal Ontario Museum and Parks *Canada © ROM. Photo Credit: J. B. Caron.*

Anomalocaris probably did not have *Opabinia's* agility. It is believed that *Anomalocaris*, a protoarthropod, more resembled a manta ray in its swimming. With its anterior grasping forelimbs designed for feeding and its circular mouth with massive teeth, it was accustomed to preying on less-well-defended or slower organisms, perhaps resembling sea cucumbers or their Cambrian equivalent. It would not have needed stereopsis.

WAPTIA FIELDENSIS Some Cambrian organisms represented extant phyla. *Waptia fieldensis* (Figure 5.11) was an arthropod, probably with compound eyes as in today's arthropods. This free-swimming but probably benthic species may have fed on organic debris. Its eyes were stalked and probably more sensitive to movement than possessing good visual acuity.

ODARAIA ALATA More evolutionarily pedestrian animals represent direct predecessors of contemporary phyla. *Odaraia alata* (Figure 5.12), an uncommon species in the Cambrian, is thought to have preceded crustaceans and is part of the lineage. Although the eyes are obvious and likely were compound eyes, we simply do not know whether they were or not. *O. alata* was likely a prey species, although little is known of its ecology.

These are but some of the unusual creatures spawned in the Cambrian, and animals alive today can help us bridge the gap from their imprints to the entire phyla. Close relatives of these large phyla have likely changed little since the Cambrian (Figure 5.13).

All of the aquatic and terrestrial species of arthropods, including trilobites, crustaceans, spiders, insects, and millipedes, are monophyletic (they share an ancestor). Although we do not know which animal was the first arthropod or when it appeared, we do know some candidates for early and perhaps basal arthropods.

The convex eyes of *Anomalocaris* (Figure 5.10) or *Odaraia* (Figure 5.12) certainly appear to have been compound eyes, although no imprints remain to prove it. *Anomalocaris* or other Cambrian denizens were probably quite basal and close to the last common ancestor of all arthropods.

Arthropods' likely wormy ancestor radiated into two groups dubbed Ecdysozoa (arthropods and other molting animals) and Lophotrochoza (invertebrates that do not molt).

Because the known Cambrian species are very different from each other, and some are thought to be different phyla, the arthropod stem organism probably preceded the Cambrian explosion. That stem organism had simple ocelli and perhaps early compound eyes. It had to have appeared

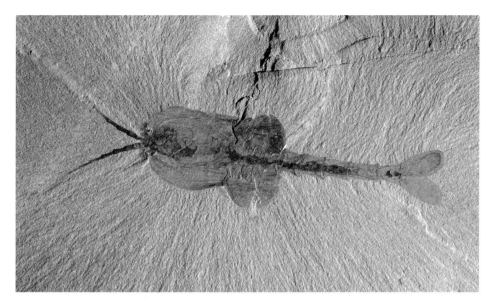

Figure 5.11 *Waptia fieldensis.* This 7- to 8-cm crustacean predecessor would swim in the water column but was probably a benthic feeder as well. The eyes are unmistakable and were probably stalked from beneath the bivalved carapace. There is evidence to suggest that this animal had a median eye as well. *Courtesy and © of the Smithsonian Institution.*

BOX 5.4 ECDYSOZOA AND LOPHOTROCHOZOA

Deuterostomes radiated into the echinoderms, hemichordates, cephalochordates, urochordates, and craniates, including the vertebrates. Protostomes diverged into the Ecdysozoa and the Lophotrochozoa (Figure 4.6). The Ecdysozoa are a lineage that periodically molt. These include the phyla Tardigrada, Nematoda, Onychophora, Chelicerata, Myriapoda, and Arthropoda, among a few others. The Lophotrochozoa include Mollusca, Annelida, Brachiopoda, and Nemertea, among others. These two principal divisions among the protostomes are important to the distribution of the visual witness.

at least by the early Cambrian, and perhaps in the Ediacaran. An example of what it may have looked like comes from the tardigrades—an extant phylum that is closely related to the arthropods and probably little changed since the Cambrian.

TARDIGRADES: ECHINISCUS TESTUDO Water bears are also known as tardigrades. These colorful and translucent creatures reveal a basal moment in the eye's evolution and the likely state of photoreception in the early Cambrian— at least for this phylum (Figure 5.14).

Tardigrades are quite small, with most measuring less than 1 mm. They are cosmopolitan in distribution, found almost anywhere there is water, including near the oceanic vents. Tardigrades have six legs, bilateral symmetry, and some astonishing abilities, including entering a latent non-aging state called cryptobiosis during unfavorable conditions such as desiccation or freezing. Neither desiccation nor freezing necessarily kills these creatures because they enter a state of inactivity called encystment. They can stay alive in a state of encystment for a year or perhaps longer. Although tardigrades do not have or need a respiratory system, they do have a brain and eyes that seem to "sprout" from it.

Tardigrades are in their own phylum (Tardigrada) but are believed to be a sister group to Arthropoda and Onychophora. Together these three compose a super group called Panarthropoda. The last common ancestor probably lived in the early Cambrian or in the mid- to late Ediacaran, with radiation into the tardigrades, onychophorans (velvet worms, which we will meet shortly), and arthropods occurring very early in the Cambrian (Figure 5.15). So tardigrade eyes reveal much about Panarthropoda and that last common ancestor, which fossil evidence dates to at least the early Cambrian.

Most tardigrades have eyes, and although they vary, they follow a similar pattern. These are simple eyes, consisting of a single cuplike pigment cell with one or two rhabdomeric cells (the photoreceptive cells found in most invertebrates). Additionally, each eye has a ciliary cell, although it is unknown if this cell responds to light.

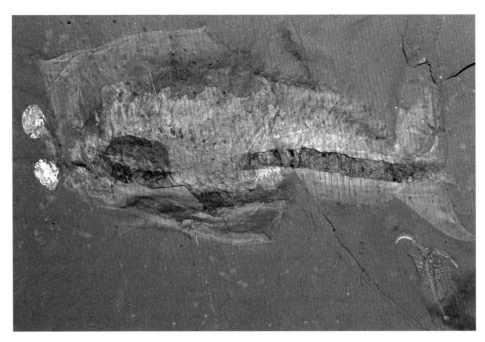

Figure 5.12 *Odaraia,* a crustacean predecessor, had frontally placed and large eyes. Little is known of its lifestyle, but it clearly used vision in its ecology. Note the impression of a *Marella splendens* to the lower right in the image. With permission of the Royal Ontario Museum and Parks Canada © ROM. *Photo Credit: Brian Boyle.*

Figure 5.13 Cambrian scene. *Anomalocaris* can be seen just left of the center rising up from the sea floor. The two spiked animals to the lower left of the image are *Hallucigenia.* An *Ottoia* can be seen in the lower right-hand corner in a burrow with its proboscis extending just above ground level. *Marella* can be seen above and to the left of the *Anomalocaris.* Another *Marella* can be seen just to the right of the *Anomalocaris.* *Pikaia* can be seen swimming in the upper left-hand portion. *Image © The Natural History Museum, London. Artist: John Sibbick.*

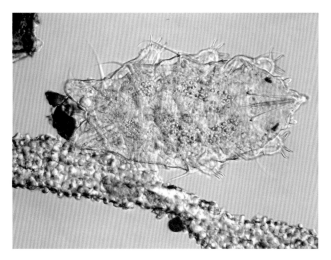

Figure 5.14 *Echiniscus* species from the Canary Islands. Body length ~0.3 mm The water bears (tardigrades) are sister groups to the arthropods and the velvet worms (*Onychophora*) and suggest that the most recent common ancestor to those three lineages (Panarthropoda) had simple eyes with a single or multiple pigment cups with or without lenses because at least two of those lineages still possess similar eyes. *Image © Martin Mach.*

Figure 5.15 *Epiperipatus biolleyi,* one of the velvet worms. Text describes ecology. *Image © Georg Mayer.*

That early ancestor probably possessed both rhabdomeric and ciliary cells, just like the early worms.

That last common ancestor of Panarthropoda probably had a primitive cuplike eye (with or without a lens) and rhabdomeric cells. The ciliary cells were nearby in the brain or in the eye itself. We know that the radiation into vertebrates would require these ciliary cells and that they were present in the first bilaterian animal (urbilateria), so it is no surprise that they would persist into this lineage.

At about the same time another sister group to the arthropods arose, and it provides us with hints about the status of the eye near the directional change into Arthropoda. This animal is so unusual that it remains a lonely and singular relict of its own phylum, although many believe it is a basal arthropod. Either way, it is so unusual that it merits discussion. Meet the velvet worm.

ONYCHOPHORA—THE VELVET WORM: EPIPERIPATUS BIOLLEYI Onychophorans (Figures 5.15–5.17), the closest relatives of arthropods, have changed little since the early Cambrian, as evidenced by their resemblance to certain fossils of the time such as *Aysheaia* (Figure 5.18). So in studies of arthropod evolution, velvet worms, are an important outgroup (not true arthropods but radiated from a closely related common ancestor as the creature that radiated into arthropods).

Although velvet worms rely less on vision than on other senses and usually avoid daylight, they do have a pair of simple lateral eyes (Figures 5.16 and 5.17). These lateral eyes

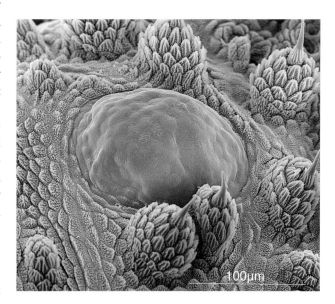

Figure 5.16 Onychophoran external eye. This image is of an eye of *Metaperipatus blainvillee,* a velvet worm and closely related to *Epiperipatus.* This image is a scanning electron microscopic. *Image © Georg Mayer, PhD.* Reprinted from Arthropod Structure & Development, 35, 231-245, Georg Mayer, G: Structure and development of onychophoran eyes: What is the ancestral visual organ in arthropods? © 2006, with permission of Elsevier.

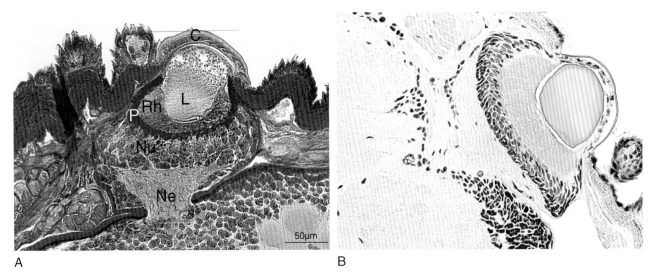

A   B

Figure 5.17 (A) *Epiperipatus biolleyi* eye. Histology of the eye shows a cornea, lens, and everted retina. The lens has been pulled apart in the preparation. C, cornea; L, lens; Rh, rhabdoms; P, pigmented layer; Nu, nuclear layer of retinular cells; Ne, nerve fibers leading away from the eye to the brain. *Image © Georg Mayer, PhD.* Reprinted from Arthropod Structure & Development, 35, 231-245, Georg Mayer, G: Structure and development of onychophoran eyes: What is the ancestral visual organ in arthropods? © 2006, with permission of Elsevier. **(B)** *Euperipatoides rowelli* (Onychophora). Similar to A, but the lens is whole, and the retina shows more detail. *Image by Richard Dubielzig, DVM.*

are in the approximate position of arthropod eyes. Recent study suggests, however, that these eyes are homologous to the arthropod median eyes (nauplius [larval stage] eyes of crustaceans and median ocelli of insects) because they are similarly innervated from the central part of the brain.

Compound eyes probably developed after the divergence of Onychophora but developed separately in other phyla.

So velvet worms are an evolutionary novelty of arthropods or a separate phylum that were a sister group to the arthropods, and we will see such novelties appear in other phyla.

So onychophoran eyes represent a starting point for the evolution of the arthropodian eyes and, in particular, the median eyes.

Figure 5.18 *Aysheaia pedunculata.* Remarkably similar to extant Onychophora, *Aysheaia* was a wormlike creature that grew to be up to 6 cm in length and is believed to be a grazer on coral. Rarely found as a fossil in formations of Cambrian, this creature may well have been a direct ancestor to these velvet worms. *Courtesy and © of the Smithsonian Institution.*

BOX 5.5 ONYCHOPHORAN EYE

An outer connective tissue layer analogous to our sclera encapsulates the eye. This layer extends to and connects with the cornea or cuticle, which is called a perikaryal layer. Inside that layer is a layer of pigmented cells analogous to the pigmented cells in our eyes. (As you may recall, by blocking or absorbing light, pigment increases discrimination.) Pigment is almost universally associated with eyes. Inside the pigmented cells are the rhabdoms from the retinular cells. The lens is beneath the cornea and focuses light onto the rhabdoms.

The anatomy of the onychophoran eye shows surprising similarity to that of a "simple" eye, an ocellus, but certain subtle features shine through (Figure 5.17). These eyes are more complicated than the ocellus. These perhaps counterintuitive features plant the seeds of differentiation in descendants.

Both the water bears and the velvet worms retain a simple eye from those early beginnings, although the velvet worm has a more sophisticated eye. It is not likely that this model advanced appreciably in these two groups, but it did in the arthropods. In addition to simple median eyes, arthropods developed compound eyes—a major step in the evolution of the eye.

THE AGE OF ARTHROPODS

A MAJOR PHYLUM BEGINS

PALEOZOIC ERA

CAMBRIAN PERIOD

543–490 MILLION YEARS AGO

6

The Cambrian had the stamp of the arthropods, although you might never recognize the animals of the Cambrian as arthropods. Whether *Anomalocaris* or *Waptia* or, more likely, a common ancestor to both started the monophyletic phylum, Arthropoda has been a major lineage since the Cambrian and certainly the most resourceful and creative of phyla. After all, the diversity of the aquatic, terrestrial, and aerial arthropods is stupefying if you consider that they all came from a single common ancestor.

Cambrian animals developed many of the prominent features found in modern arthropods. For example, stalked eyes were probably used much as they are by certain modern arthropods, such as crabs and lobsters, to enlarge the visual field without moving the head and to follow other animals as they move. Such stalks probably began so that moving animals could stabilize the visual world around themselves.

Fossils show us simple eyespots; the melanin that represents the backing of the photoreceptors is frequently preserved, showing dark black spots (see Figure 7.2). Additional detail shines though the ages in some of these fossils. The presence of a lens, which fossilizes poorly but distinctly from surrounding tissue, illustrates that simple lensed eyes appeared even in the early Cambrian. Some of those Cambrian phyla are alive today, and although the eyes have probably changed, these animals brought them through millions of years.

The sister groups of Onychorphora and Tardigrada (see Chapter 5) had simple eyes—ocelli. The development of some form of compound eye, which most arthropods have, was a major step in the evolution of vision. Compound eyes are also found in some annelids, a few mollusks, and, with distinct differences, in a few echinoderms (sea stars, brittle stars, and urchins and their kin).

The formation of compound eyes in other phyla represents convergent evolution, as seen in some basic differences (see Chapter 4), and suggests that ocular evolution probably occurred several times and in several directions.

Compound Eye Development

Evolution has provided more morphologic variations to eye design than most imaginations could generate. Ocelli (simple eyes with a skin covering over a lens with a shallow cup of cells beneath it) may evolve into camera-style eyes, but they often do little more than perceive the circadian rhythm. But in all of their forms, compound eyes evolved for vision, and in some animals it is surprisingly good vision.

Individual visual elements—ommatidia—have combined into a single visual and neurologic unit. A principal difference between the simple eye and a compound eye is the number of lenses in each visual unit.

The compound eye, an early aquatic development, set the stage for later changes in the invertebrate terrestrial eye. The groundwork appeared in the Cambrian or earlier, and probably so did the eyes.

Of several possible scenarios for the appearance of the compound eye, we will explore only one.

Eyespots or ocelli appear in many single- and multiple-celled creatures—for instance, *Volvox*. Consider that these same eyespots gradually invaginate to create a cup. With this invagination, the putative eye is capable of crude spatial vision, so it could be considered to be an eye. This eye goes on to evolve condensations of tissue to form a primitive lens. In this animal the primitive eyespots with primitive lenses gradually coalesce into a single eye with multiple units, perhaps though simple gene duplication.

When the compound eye had the elements of the individual ommatidia, the first and basic compound eye could have formed, perhaps by the early Cambrian. As the animal evolved, the compound eye would gradually evolve to improved sensitivity and resolution.

TABLE 6.1 *Types of compound eyes with examples*

Apposition
    Focal apposition—oldest and most common
        Horseshoe crab
        Mantis shrimp
        Many flies
        Bees, wasps, hornets
        Dragonflies
        Many beetles
        Other insects—grasshoppers, etc.
    Afocal apposition
        Lepidoptera—butterflies

    Neural superposition (unfortunate name because this is dissimilar to superposition eyes)
        Dipteran flies
Superposition
    Refraction superposition
        Some Beetles—especially nocturnal
        Moths
    Reflecting superposition
        Long-bodied decapods—shrimp, lobster, crayfish
    Parabolic superposition
        Some hermit crabs,
        Some amouran crabs
        Some free-swimming crabs

Compound eyes are found in two basic designs: apposition and superposition (Table 6.1). Within each basic design are several variations; how one form of eye radiates into another is controversial.

Apposition Eye

The apposition eyes can be classified into three groups—true or focal apposition, afocal apposition, and neural superposition (an unfortunate name because it is a subclass of apposition eyes, not of superposition) (Table 6.1). The superposition eyes are classified into three categories as well—refracting superposition, reflecting superposition, and parabolic superposition. None of these eyes fossilizes, but evidence suggests that true apposition eyes came first—probably in the Cambrian explosion or a bit earlier. The apposition eye requires substantial light, so it is restricted to shallower depths and, when in terrestrial animals, to diurnal lifestyles (mostly).

Apposition compound eyes show at least five subtly different varieties. Evolution has been tinkering with it over the last 500 million years or so and is likely to continue to do so. Although the apposition eye (Figures 6.1 and 6.2) is likely the oldest compound eye, the original morphology is probably lost forever. Perhaps the oldest known living

animal with a compound eye is closest to the original eye. We can review that eye as a model for it, the simplest compound eye.

The horseshoe crab is one of the oldest extant arthropods and very basal in the radiation of the arthropods. The basic compound eye, the apposition eye, had its start close to the birth of the horseshoe crabs, so their eye may be as close as we can ever get to the original apposition compound eye of the Cambrian animals.

Horseshoe Crab: Limulus polyphemus

The horseshoe crab teaches us much about evolution—and patience. With its helmet-shaped shell and clunky morphology, the animal's tenacity belies its clumsy, prehistoric appearance. This design has survived, little changed, for more than 400 million years. The original species probably arose early in the Cambrian. It is a generalist, capable of adjusting its lifestyle when tough times approach. It feeds on benthic organisms including animal remains, worms, and crustaceans and almost anything else that it catches.

Despite their name, horseshoe crabs are not crabs at all. They are more closely related to spiders. Along with sea spiders (not true spiders, but in another group closely related to arachnids),

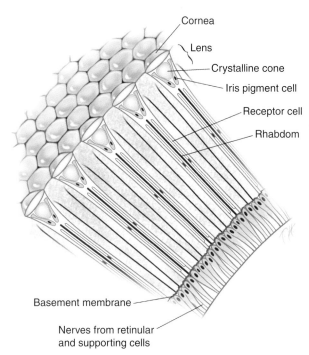

Figure 6.1 Apposition eye Anatomy of apposition eye. The basic compound eye with a corneal lens (cornea), a crystalline cone lens (Crystalline cone) to follow the corneal lens (the lens system is marked as "Lens"). These two lenses in combination focus the light rays onto the rhabdom. The iris pigment cells (labeled) will prevent stray light from interfering with the rays striking any individual ommatidium. The receptor cells surround the rhabdom as discussed in the text. The rhabdom is represented by the black line between the two retinular cells seen in each ommatidium. The basement membrane is produced by supporting cells. The neurons from the retinular cells run though the layer that includes the supporting cells. *Art by Tim Hengst.*

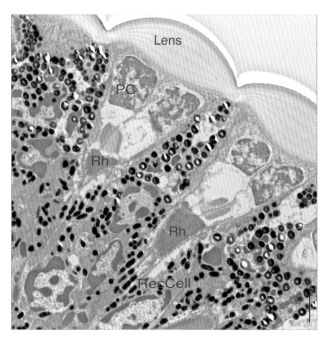

Figure 6.2 Example of apposition eye. Lens, the corneal lens; PC, the pigment cell that surrounds the crystalline cone; Rh, the rhabdom in two adjacent ommatidia; RecCell, the receptor cell that provides the microvilli creating the rhabdom (fairy wasp, see Chapter 21). *Scanning electron micrograph by Pat Kysar.*

each of which contributes a point of light from the image. So, at best, an image projected to the brain is composed of at least one thousand individual points of unfocused light.

The horseshoe crab has a unique slight variation on the apposition eye. This suggests that the last common ancestor may have had a relatively primitive compound apposition eye and radiated into chelicerates (arachnids, horseshoe crabs, and

horseshoe crabs are believed to be the most basal of the lineage that also leads to arachnids. Because both animals have changed little since the Cambrian, arachnids, like horseshoe crabs, also teach us much about the visual witness in the basal creature that led to arachnids, then, early arthropods.

Sea spiders have only simple eyes if they have eyes at all. These eyes are more like those of the Onychophorans (Chapter 5), but horseshoe crabs have two lateral eyes and five median eyes (Figures 6.3 and 6.4). The latter is a misleading number, though, because it includes two obvious median eyes, plus a median eye under *the carapace and a pair of median eyes* on the ventral surface *of the animal. To add to this confusion, this creature also has rudimentary eyes immediately behind the lateral eyes. The animal is studded with eyes, although vision is not particularly good in any of them.*

But the eyes confirm that the visual witness was well in progress when the horseshoe crab first appeared. The principal eyes, the two lateral ones, are composed of individual units— ommatidia. Each lateral eye has about one thousand of them,

Figure 6.3 *Limulus polyphemus*, horseshoe crab. This ancient animal is a survivor and has changed little over hundreds of millions of years.

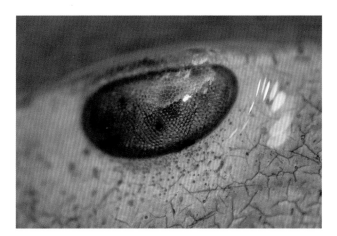

Figure 6.4 *Limulus polyphemus*. Compound eye to show multiple hexagonal ommatidia. This animal possessed one the earliest compound eyes.

sea spiders), crustaceans, and eventually insects (Figures 6.1, 6.2, and 6.5).

The compound eyes of horseshoe crabs are composed of the individual eye units that create the mosaic of the eye (Figure 6.4). Although an ommatidium is the smallest external unit for such an eye, each one may house four to twenty individual cells, with each providing a portion of the ommatidium's photopigment.

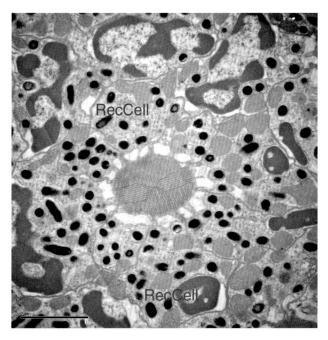

Figure 6.5 *Trichogramma* species, fairy wasp. Example of rhabdom. The rhabdom can be seen in the center of the image. RecCell, two of the receptor cells that surround the rhabdom. Each of these cells contributes microvilli to the rhabdom. *Scanning electron micrograph by Pat Kysar.*

Most arthropods with compound eyes have ommatidia that contain eight retinular cells arranged in a circle. The cells that compose the compound eye of insects are better organized than those of horseshoe crabs, suggesting that horseshoe crabs' eye is more basal.

The median eyes of the horseshoe crab are involved with the regulation of the circadian rhythm. These median eyes, as well as all other horseshoe crab eyes, besides the two compound eyes, are ocelli. Only the single pair of lateral eyes appears to be organized to permit image formation.

Perhaps the most interesting pair of eyes is that on the ventral surface. These eyes are probably involved with circadian rhythm. These may have developed to help the animal align right side up.

These numerous and widely scattered "eyes" suggest that the visual witness was beginning to organize as the horseshoe crab descended from its ancestor. That most recent common ancestor of the horseshoe crabs and other arthropods probably had a compound eye as well as median eyes that would continue through the other phyla yet to follow.

BOX 6.1 FOCAL APPOSITION COMPOUND EYES

The focal apposition compound eye (Figures 6.1, 6.2, and 6.6) consists of an array of ommatidia or "little eyes." Each ommatidium consists of an overlying "cornea" that focuses light onto a crystalline cone (lens) surrounded by a collar of pigment cells. These pigmented collar cells prevent light and the cellular signals from spilling across to adjacent cells. The receptor cells found deeper within each ommatidium surround a central core photoreceptive element called the rhabdom (Figure 6.5).

Each receptor cell in the within the unit is called a retinular cell. Along the central edge of each one of these retinular cells are micovilli that combine with those of the other surrounding retinular cells (Figures 6.5 and 6.6). The photopigment is housed in the membrane of the microvilli. The cells' photopigments combine in that central ommatidial unit—the rhabdom (Figures 6.5 and 6.6).

The retinular cells compose the "photoreceptor" element (rhabdom) in a complicated manner. The rhabdom contains the visual pigment that reacts to the photons and initiates the signal sent to the brain (Figures 6.5 and 6.6). The number of retinular cells that surround and contribute to the rhabdom usually falls between eight and thirteen depending on the animal. These receptor cells give off the receptor axons that send signals to the brain.

This fused element, the rhabdom, performs like a fiber optic cable; in optical parlance, it is a light guide. But the individual cells contributing to the rhabdom cannot discriminate individually. So the interommatidial distance determines the visual acuity because each unit presents a single spot (or pixel in computer terms).

It might seem, then, that the cells that comprise the rhabdom do not have a purpose, but the truth is otherwise. The visual pigment of each cell's microvilli is sensitive to a different wavelength. So this apposition eye has a mosaic pattern formed in the predominant color received by each rhabdom.

A focal apposition eye does not resolve the image beyond what each facet perceives directly. So each ommatidium receives a portion of the image directed toward it. The cornea and the crystalline cone focus the light at the distal tip of the rhabdom, which essentially represents a unit of color but no details of the image. The brain will assemble these units like a jigsaw puzzle or a mosaic to create an image.

The array of ommatidia in an appositional compound eye produces good sensitivity to movement. But acuity can be limited by diffraction and photon capture, so this eye performs best—or perhaps *only*—in a daylight environment. In this form of compound eye, the cornea (the first surface that light meets as it enters the ommatidium) acts as a lens that helps to focus light on the rhabdom. In essence each ommatidium corresponds to a single rod or cone in the vertebrate retina.

The five or more forms of the basic focal apposition eye are all variations on the above theme. The variation consists of the manner and style of corneal lens and crystalline lens behind it as described previously. As one would expect, the apposition compound eye in aquatic animals differs from that in terrestrial animals. In aquatic animals the corneal curvature cannot be used as a refractive surface, so the surface is often flat. The internal lens must be of a higher refractive index or constructed in a different geometric shape. Although I classify all of these eyes as basic apposition eyes, each different form could be considered a different style of eye.

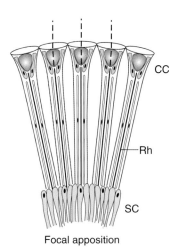

Focal apposition

Figure 6.6 Focal apposition eye anatomy. Note the corneal lens (not marked) and crystalline cone (CC) and the retinular cells around a rhabdom (Rh) or photoreceptive pigment. The rhabdom is represented by the central black rod. Supporting cells (SC) surround the neurons traveling from the receptor cells toward the brain. *Art by Tim Hengst.*

they may have been useful initially, but subsequently lost them.)

Compound eyes would have had a start with an ancestor to the horseshoe crab and probably diversified quickly. The aquatic apposition eye has evolved, matured, and become more organized over the eons since the Cambrian. At least one animal epitomizes how the apposition eye might have matured during those 500–550 million years.

The mantis shrimp's close ancestors must have been present during the Cambrian. Even though this animal's eye is the simplest compound-eye morphology, in some ways, it is the most complex eye in an invertebrate, and probably in the animal kingdom. Fossils closely resembling this animal have been found to be 400 million years old, and it is likely this animal has changed minimally over that range of time. Although living mantis shrimp species have radiated and matured since that time, and its eye was over 500 million years in the making, the eyes of the mantis shrimp are likely to closely resemble the eyes of its ancient predecessor.

Mantis Shrimp—Stomatopod: Odontodactylus scyallurus

Superman is alive and well in the warm, shallow seas of the coral reef. He is faster than a speeding bullet, more powerful than a locomotive, and capable of seeing into a realm we cannot begin to understand. The mantis shrimp, a misnamed stomatopod (Figures 6.7 and 6.8), has an array of talents and stealth so impressive as to make him a decathlete of unprecedented prowess. Eyes in these species should be considered the

The annelids, arthropods, and the chordates all developed median photoreceptive structures to assess, monitor, and direct circadian rhythm. The mollusks do not have those structures. (Perhaps mollusks had the structures, as

Figure 6.7 *Odontodactylus scyallurus* Note the "claws" or "arms" that hang on either side of its head. These appendages are capable of breaking the glass in aquaria or smashing a bivalve shell. *Image © James Brandt.*

arthropod equivalent of the avian eye in vertebrates—top of the line in the phylum.

The compound eyes of the mantis shrimp have a horizontal, beltlike region that seems to divide the eye in half (Figures 6.7 and 6.8). This equatorial belt is called a midband and has six specialized rows of ommatidia.

Figure 6.8 *Odontodactylus scyallurus.* Note the cells in the midband region that includes ommatidia that can see polarized light and ultraviolet light. *Image © James Brandt.*

Some of the cells in this midband region analyze polarized light and others ultraviolet light. This becomes complicated, as the light can be polarized in a linear or a circular fashion, and the mantis shrimp has to be able to recognize both. This skill of seeing two different forms of polarized light helps in identifying prey species.

Another set of cells within this beltlike midband contains ultraviolet-sensing visual pigment, a short-wavelength visual pigment (essentially blue) followed by a longer-wavelength visual pigment (such as green or red). So as light strikes the ommatidium, the ultraviolet light is analyzed first and removed, then the blue light is analyzed and removed, and then the longer-wavelength color light is analyzed and removed.

Another set of cells in that midband has colored droplets within the individual cells, and these change or limit the light that stimulates the rhabdom (the fused core photoreceptive element). The colored filters are much like wearing yellow sunglasses to filter out much of the blue light that comes to your eye.

As a result of this and other physiologic changes, the mantis shrimp's eye contains as many as sixteen different visual pigments across the entire eye, including the ability to perceive polarized light. This compares to three visual pigments for color vision in your eye, or four if you count the rhodopsin for nocturnal vision—essentially black and white.

The halves delimited by the midband are also highly organized. Some of the ommatidia in each half and in the midband are directed toward the same point in space that provides monocular range-finding abilities or even uniocular "stereopsis." All of that would be plenty, but there is more.

Because of the specializations of the ommatidia and heavy skewing of the ommatidial line of sight in the upper and lower eye regions, the visual field of the mantis shrimp is largely limited to a narrow strip in space. As a result, to achieve a complete visual image, it must scan its environment. Stalked eyes allow the peculiar ability to look in almost all directions and to rotate in all six possible axes, to the point of having one midband vertical in one eye and perpendicular to the midband in the other eye.

These abilities suggest a curious, even intelligent and gentle animal. But do not be fooled. This is a solitary rapacious predator with a thunderous attack. It can lie in wait patiently or prowl the reef like a street thug chasing down quarry.

To subdue its prey, the mantis shrimp uses its hammer-like arm with the force of a small caliber bullet, a force that can break the glass of an aquarium. This arm has perhaps the fastest acceleration in the animal kingdom. The mantis shrimp uses this ability to break open clams or snails, or to stun fish. Possessing one of the most complicated eyes on the planet, it relies on its sight for seeking its prey and its mates. It is a real-life superman.

Apposition eyes continue to be successful for the animals that use them in niches best suited to their design. But apposition eyes also radiated in other directions, including related forms in the terrestrial environment. The afocal apposition eye (in butterflies) and the neural superposition eye (an apposition eye form in certain flies as discussed above) are both terrestrial eyes. We will meet them in Chapter 21.

Superposition Eye

In the aquatic environment, evolution continued to work on the apposition eye model, producing an entirely different form of compound eye. The apposition eye radiated into the superposition eye, probably when animals invaded deeper—and dimmer—marine environments.

The apposition eye probably radiated into the parabolic, refracting, and reflecting superposition eyes in the late Cambrian or early Ordovician—quickly, as a dim, benthic environment would have required it.

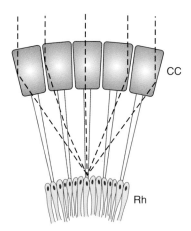

Reflecting superposition

Figure 6.9 Reflecting superposition eye anatomy. Compound eye found in long-bodied decapods. It is a superposition eye with a gap between the reflecting lenses and the rhabdoms. Incoming light is reflected from at least one and often more than one side of the ommatidium to focus on the rhabdoms. *Art by Tim Hengst.*

BOX 6.2 EVOLUTION OF THE APPOSITION EYE TO SUPERPOSITION EYE

Each form of compound eyes, both apposition and superposition, has unique features and fits a special niche. Of the three forms of superposition eye, the parabolic has more problems and limitations than the refracting and reflecting. This would suggest that the parabolic might have been the transition between apposition eye and the other two superposition forms. One model of development additionally suggests that the refracting superposition eye radiated into the reflecting superposition eye (Figure 6.9), but it could have gone the other way. The two types have similar sensitivity and acuity, but one may have some unknown selective advantage in certain environments.

Alternatively, it is also entirely possible that the superposition eye appeared from a similar mechanism as the apposition eye without the intermediate stage of an apposition eye. Various species of the decapods (such as shrimp and lobsters) possess all three forms of superposition eye. This suggests that evolution has led to other forms rather easily and that the last common ancestor of the decapods had the genetic potential for all three forms, even if unexpressed.

Whatever the sequence of evolution, the more sensitive, photon-concentrating superposition eyes evolved to use every scrap of light.

BOX 6.3 REFRACTING SUPERPOSITION COMPOUND EYES

Because of their anatomy and optics, the superposition eyes have much better light-gathering capabilities than apposition eyes (Figure 6.10). Animals that have colonized dimmer environments have evolved superposition eyes in response to lower light levels, and the refracting compound eye is the most common form.

Externally, the apposition eyes and the refracting superposition eyes are similar, both with hexagonal external facets. But internally, they differ greatly. Refracting superposition eyes have a corneal lens and a contiguous crystalline cone that is offset from the retinular cells (analogous to the photoreceptors of the vertebrate retina). Together the two lenses essentially make a telescope.

A clear zone between the refracting elements and the receptive elements could be considered analogous to the vitreous cavity of our eye. This space in refracting superposition eyes allows the lenses in each ommatidium to receive parallel rays and direct a focused image to the same point on the retinular cells. In refracting superposition compound eyes (as well as the other forms of superposition eyes), the image is upright, completely formed, and transmitted to the brain in that manner. The image is spread across many ommatidia and comes to focus on the rhabdoms, much as the image in a camera-style eye, although the retina is convex toward the lenses in the superposition eyes. The rhabdoms are again produced by a combination of interdigitating microvilli from surrounding retinular cells creating a sheet of rhabdoms in all forms of superposition eyes.

The optics of a classic refracting superposition eye reveals the clever, if complicated, design evolution has discovered. After parallel light rays enter the corneal lens, they are refracted to have a focal point within the second crystalline cone lens. Then these parallel rays are refracted within the second lens before exiting it parallel but bent toward a single point. This mechanism allows the lenses of a large number of ommatidia to focus light onto each rhabdom of the retina. According to the classic theory of Sigmund Exner (1891), this focusing requires the retinular cells to be at a set distance behind the lenses (thus resulting in a spherical eye) and the number of rhabdoms and facet lenses to be equal.

In superposition compound eyes, many more rhabdoms than facets are concentrated within the "retina," with the rhabdoms packed into the "retina" much more densely than the overlying facets would suggest. Denser rhabdoms in certain areas create a more detailed image wherever this occurs—in effect creating a fovea.

These areas of the eye also have the largest numbers of facets supplying light to each rhabdom. In a superposition eye, these regions have evolved to create a visual image in low light and, modified somewhat, a bright, well-defined image during daylight.

As various species of arthropods scurried deeper toward the abyss, evolution had to respond with better methods of light collection. Some of these adaptations are so clever that they astonish us. For example, long-bodied decapods found a peculiar approach, and the principle has become so important that astronomers rely upon it for x-ray telescopes.

Long-Bodied Decapod Crustaceans (Includes Lobsters and Shrimp but Not Crabs) Spot Prawn (Pandalus platyceros)

The order Decapoda includes crustaceans whose eyes generally are on stalks and regularly shed and regrow their corneas—a process

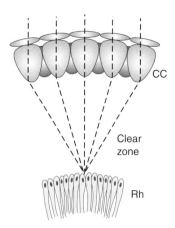

Refracting superposition

Figure 6.10 Refracting superposition eye anatomy. The lenses refract the light rays to be directed to the rhabdom. There is a space between the lenses and the rhabdoms. Retinular cells (Rh) that combine to create the rhabdom are also combined with supporting cells in the same layer. *Art by Tim Hengst.*

called ecdysis (or molting) (Figures 6.11–6.15). But the most interesting aspect about their eyes is the absence of lenses: instead, they have mirror boxes that focus the rays on the photoreceptor cells (see Box 6.4).

Many marine arthropods, such as shrimp and lobsters, developed a hard outer shell (carapace) instead of an internal skeleton. Such a shell protects soft parts from potential predators, but it could also inhibit growth. To solve this

Figure 6.11 *Pandalus platyceros*, spot prawn. Note the stalked eyes. *Image © James Brandt.*

Figure 6.12 *Pandalus platyceros*, spot prawn. Note the square ommatidia of the eye. The individual ommatidia are not this large, but the overall square pattern is seen as a composite. *Image © James Brandt.*

problem, these animals evolved the ability to shed their shells. The shell covers the entire animal, including the eyes, in a tight fit; so does the animal shed its eye?

Tasmanian Rock Lobster (Astacopsis gouldi)

While shedding hard shells, which lobsters (Figures 6.17 and 6.18) and shrimp, among others, must do to grow, the animals shed and replace the discrete outer layer of their eyes with

Figure 6.13 *Pandalus platyceros*, spot prawn. Surface of eye. Note the square ommatidia. *Image by Richard Dubielzig, DVM.*

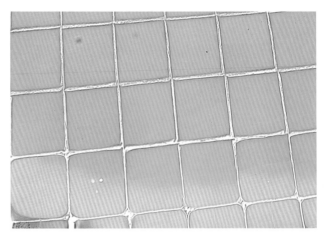

Figure 6.14 *Pandalus platyceros*, spot prawn. Horizontal section through the "mirror box" of the compound eye. Note that there is no lens, but the boxes are filled with an amorphous material that does not refract light. Note the square nature of the ommatidia. *Image by Richard Dubielzig, DVM.*

a new one. (Figure 6.18 illustrates just how complete the new one must be.)

The process of shedding (ecdysis) is controlled by a gland in the animal's eyestalk. Epidermal cells secrete the beginnings of the new shell as the old shell starts to separate from the epidermis (apolysis). When ready, the lobster pumps in water to force the old shell away from the newly formed, still incomplete shell. The lobster jettisons or eats the old shell (Figure 6.18). Then it expands further by ingesting seawater, allowing the new shell

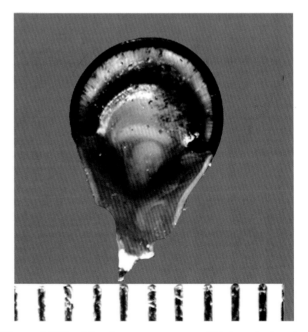

Figure 6.15 *Pandalus platyceros*, spot prawn This is sagittal section through the compound eye. Note the clear space between the reflecting elements and the photoreceptors. *Image by Richard Dubielzig, DVM.*

BOX 6.4 REFLECTING SUPERPOSITION EYES

Eyes of the long-bodied decapods are distinctly different from refracting ones because their ommatidia's external facets are square instead of hexagonal (Figures 6.9 and 6.11–6.16), and they contain no lens. Each ommatidium contains a clear gelatinous substance with no refracting power (Figures 6.14 and 6.16).

How it works: These square cylinders are lined with mirror boxes made of purines as a reflective compound. The cornea somewhat refracts a point of light before it strikes the mirrors. Almost all rays strike two parallel sides of the mirror box and are focused to a single point on the retinular surface.

The orthogonal shape (square and at right angles) of the mirrors does the trick. They act much as a clothing store's adjacent perpendicular mirrors, which reflect you almost any way you move. Parallel incoming rays strike ommatidia's mirror pairs to focus the image at a single point on the retinular surface. Similarly to the refracting superposition eye, after the rays enter and are reflected within these "mirror boxes," they exit parallel to each other to focus on the rhabdoms as an upright and "assembled" image. Ingeniously simple (Figure 6.9).

Because x-rays can be reflected but not refracted, the mirror-box concept provides a method to design a telescope that can examine x-ray emission from distant stars. Indeed, a "lobster-eye" telescope is currently circling the globe for just that purpose. A clever simple way to form an eye, it is one that could well have radiated from an early apposition eye.

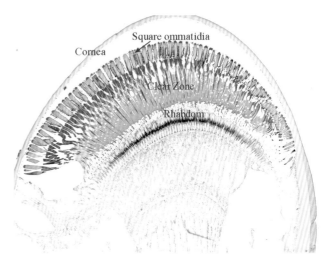

Figure 6.16 *Homarus americanus*, Maine lobster. Sagittal section through the eye stalk of this lobster. The cornea is part of the carapace and will be shed with each molt. The cornea is artifactually separated in this specimen. *Image by Richard Dubielzig, DVM.*

Figure 6.17 *Astacopsis gouldi*, Tasmanian rock lobster. Note the smooth surface of both eyes in this live specimen.

to complete itself with extra space to allow for some growth. During its molt, a lobster will grow about 15 percent in length and 50 percent in volume.

Ecdysis provides a gleaming new cornea, although the eye underneath remains the same. While waiting for its new shell to harden and new corneas to form, the lobster must hide because it is soft, vulnerable, and nearly blind (Figure 6.18).

Since the Cambrian, evolving crustaceans have modified and combined elements of the simple and complex compound eyes in remarkable ways. These combinations would lead to specialized and diverse uses of these different forms. *Dioptromysis paucispinosa*, which had not yet developed by the Cambrian, illustrates the diversification that crustacean eyes have followed.

Figure 6.18 *Astacopsis gouldi*, Tasmanian rock lobster. This is the shed carapace of the Tasmanian rock lobster. Note that the "cornea" has been shed as well.

Mysid Shrimp (*Dioptromysis paucispinosa*)

Diffraction depends on size and limits the performance of an eye. As creatures become smaller, visual mechanisms become more challenging if sight is to be one of the primary sensory modalities (especially if the intended prey is smaller still). A small predator risks sensory disability simply by being small. But resourceful solutions have been found and nearly perfected.

Mysid shrimp (Figures 6.19 and 6.20) are a family of small decapods ranging from 5 to 15 mm. Unlike the larger, long-bodied decapods with their reflecting superposition eyes, D. paucispinosa, one of the smallest of the mysid shrimp, has refracting superposition eyes similar to those of moths. This particular species has a second eye within the primary eye (as the name Dioptromysis would suggest). Moreover, this second eye is unique.

Many mysid shrimp live in shallow water on the floor of warm coral reefs. They prey on detritus and smaller living creatures, and they attract larger predators. As predators, mysid shrimp need to see well enough to spot their prey. As prey, they need a wide visual field and high sensitivity. These are demanding requirements for a creature that has a 0.4-mm eye.

For most diurnal insects, such as houseflies, apposition eyes are perfectly adequate because flies are active in bright light. In these eyes, each facet lens represents a single point of light to the brain. More points mean better acuity. To create them, the ommatidia must be smaller and more compact. This arrangement is limited by diffraction, and hence, the size of the ommatidia can be only so small.

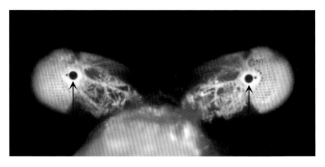

Figure 6.20 *Dioptromysis paucispinosa* binocular eye. *Image © Dan-E Nilsson.*

Superposition compound eyes, though, are optically different, as more space between the refracting elements and the rhabdoms creates an upright image from each ommatidium. Acuity improves not by narrowing the ommatidia in a superposition eye but by narrowing the rhabdoms and enlarging the optical elements, such as the lens. D. paucispinosa has done just that to the optical and physical limits of its size.

The frontal view of these shrimp seems to show a compound eye with multiple ommatidia, similar to that of other arthropods. But closer inspection reveals an unusual nipple-like projection from the lateral aspect of the eye (white arrows, Figure 6.19). The positively eerie rear view seems to show a backward-looking eye (black arrows, Figure 6.20). This looks like an eye because of the black spot in the center, appropriately called the "pseudopupil," a feature that absorbs all incoming light and reflects none back. The frontal view is an actual refracting superposition compound eye, and the rear view is a lens, combined with the necessary rhabdoms deeper in the eye to create a simple eye. As we have seen, a simple eye generally is a lens with photoreceptive elements located behind it, although these eyes can vary greatly in the quality of spatial vision. Our eye is a "simple" eye that has excellent acuity but an ocellus is a simple eye with poor spatial vision.

But this simple eye is different. It has evolved from the superposition eye to which it is attached and has a resolution comparable to insects ten times its size. The more conventional compound eye has better sensitivity but poorer acuity. This simple eye has evolved to let the creature maximize its acuity for such a small eye because its rhabdoms are light guides and diffraction-limited. It would appear that this animal's eye has reached the physical limits to achieve the absolute best vision in such a small package.

Why, then, would this evolutionary wonder point such an eye backward? Actually, it does not, despite its appearance. When it needs direct viewing, D. paucispinosa will rotate its eyestalks 130° to bring this acute zone simple eye into the frontal position. It can swing these "acute zone" eyes into proper position in approximately 0.16 of a second or less. It is as if it has added a low-vision device—like a pair of binoculars—to its

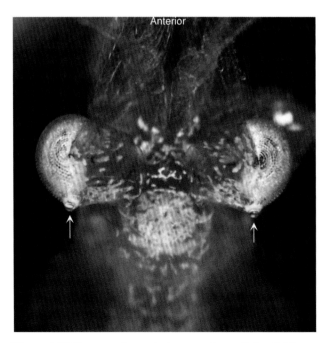

Anterior

Figure 6.19 *Dioptromysis paucispinosa* eye. *Image © Dan-E Nilsson.*

primary visual system. Furthermore, there is binocular overlap of approximately 15° with the contralateral eye creating binocularity and probably stereopsis.

In essence, Dioptromysis shrimp have a simple eye within a compound eye, the simple eye functioning much as binoculars would function as a low-vision device.

The Crabs

The apposition eye that established itself in the Cambrian radiated into the reflecting or the refracting superposition eye, both rather peculiar and unusual forms that are singularly different from the apposition compound eye. But at least one additional form of compound eye is intermediate. Although many of the crabs have apposition or refracting superposition compound eyes, some have a more unusual form of compound eye.

Hermit Crab (Strigopagurus strigimanus)

Within the diverse order Decapoda, the crabs have radiated into Brachyura, or true crabs, and Anomura, the hermit crabs and others. Traced to the Jurassic (206–144 mya), hermit crabs (Figure 6.21) probably evolved from a lobster-like ancestor.

The infraorder Anomura has a considerable variety of compound eyes within the group. Terrestrial hermit crabs typically have apposition eyes, large species of marine hermits have refracting superposition eyes, smaller species have parabolic superposition, and the squat lobsters have reflecting superposition. It is as if Anomura is an evolutionary workshop for compound eyes.

Some hermit crabs possess a compound eye that has the most complicated physiological optics on the planet. The eyes appear to have hexagonal ommatidia, much like apposition and refracting superposition eyes. Internally, though, this is a parabolic

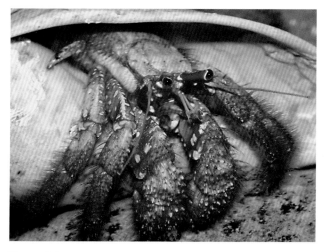

Figure 6.21 *Strigopagurus strigimanus.*

superposition eye. This eye combines two lenses and mirrors to produce an image projected on the photoreceptors (Figure 6.22).

The anomurans, and especially the hermit crabs, illustrate the evolutionary potential of the compound eye and its plasticity. The earliest compound eyes first arose in the early Cambrian and were of the inefficient apposition type. But apposition and superposition eyes are optically so different, it is hard to imagine how one can gradually evolve into the other. Parabolic superposition eyes provide the link (see Box 6.5).

The parabolic type represents less of an improvement in the eye than the step to the refracting and reflecting superposition eyes. It also appears that this eye has evolved several times in different taxa of unrelated crustaceans and at least once in insects.

In hermit crabs, this transition did not just happen in ancient evolutionary history. The present mixture of apposition and superposition types in these crustaceans suggests that it is happening right before our eyes.

BOX 6.5 PARABOLIC SUPERPOSITION EYE

This eye is unusual because it uses two lenses and mirrors to create a single image (Figures 6.21, 6.22, 11.2, and 11.3). Each ommatidium has a corneal lens, which is shed with each molt. This corneal lens refracts incoming light rays, bringing them into focus close to the tip of the crystalline cone, the second optical element within the ommatidium. The cone manages incoming rays differently depending on the incident angle.

Axial rays traverse the cone with little refraction and enter a light guide that crosses the clear space (analogous to vitreous cavity) within the human eye. Then they travel directly to the *rhabdom*, or retinal equivalent (Figures 6.22, 11.2, and 11.3). Oblique incident rays first strike the corneal lens and are refracted. Then they strike the side of the crystalline cone, which has a parabolic profile and a reflective coating.

After striking the mirrored surface, these rays are recollimated before emerging as a parallel bundle. They will cross the clear zone between the tip of the crystalline cones of many ommatidia. That means that these rays also will cross the light guide of these adjacent ommatidia without being deflected. The index of refraction of the light guides permits just that and brings the rays to focus on the retina. Like other forms of superposition eyes, these optics produce a real, erect image on the retina.

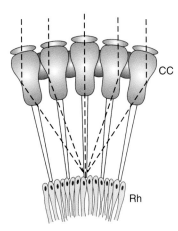

CC

Rh

Parabolic superposition

Figure 6.22 Parabolic superposition eye anatomy. Note the space between the lenses and the rhabdoms. The layer marked Rh represents the rhabdoms but also includes the supporting cells. The rhabdoms are composed of microvilli from several cells. The parabolic shaped lenses use both the refracting components of the lens and a reflective coating to direct the incoming light rays. (More on this eye in Chapter 11.) *Art by Tim Hengst*.

One of the more recent compound eyes, the refracting superposition eye, may have developed twice or more because it appears on land, not only in its watery birthplace. Possibly, this eye developed only once in response to the dark mesopelagic waters as the arthropods radiated into these depths. The similar aquatic and terrestrial versions could have come from the same common ancestor. Or they could have come from the same genetic blueprint that permitted such radiation—in other words, evolved again.

Superposition eyes, the most sensitive of the compound eyes, can be found in nocturnal terrestrial animals such as moths, fireflies, and certain nocturnal beetles, not only in the crustaceans. We will meet this eye again.

Another major phylum that would have profound effects on the earth and the development of the eye quietly appeared in the Cambrian. *Chordata* would eventually lead to vertebrates.

VERTEBRATES GAIN A FOOTHOLD

PALEOZOIC ERA

CAMBRIAN PERIOD

543–490 MILLION YEARS AGO

7

Vertebrates appeared in the Cambrian, but their predecessors had been evolving toward that moment for millions of years. The echinoderms laid the groundwork for the vertebrates via such animals as the salps and other tunicates. With this step, the visual witness would follow a different direction.

Phylum Chordata

Evidence increasingly suggests that vertebrate predecessors, such as the echinoderms and protochordates, expanded by the beginning of the Cambrian (although it was nothing like the party held by the invertebrates). Certainly, these early predecessors had eyes, albeit primitive ones. Some were jawless creatures that looked similar to lancelets (Figure 7.1), a cephalochordate. Lancelets have only a single median eye inside its brain, however, meaning that the cephalochordates are probably not more than a sister group to the line that led to true vertebrates. Gene duplication would be a likely suspect in the development into a median eye and then into two eyes. The protochordate *Pikaia gracilens* has long has been considered to represent the lineage that would lead to vertebrates. But there are other, better candidates known only in the fossil record, including *Haikouella*, *Yunnanozoon*, *Haikouichthys ercaicunensis*, or *Myllokunmingia fengjiaoa*.

These species had eyes, which may have been rather rudimentary (Figure 7.2) and camera-style (similar to those of the ancestors of hagfish and lampreys). They likely were among the first to have the neural crest—embryologic tissue that leads to the vertebrate head—which proves to be important in the growth of the vertebrate eye. It is likely that these early protovertebrates were contemporaries or predecessors of two other candidates for first vertebrate.

Not thought to be true vertebrates, hagfish are the oldest of living craniate species, and lampreys are believed to be the oldest living vertebrates. The two species probably radiated from a common ancestor such as *Myllokunmingia*, but hagfish alone are a sister group to the entire vertebrate clade. That is, hagfish are more like uncles than fathers to the vertebrate clade. Lampreys are considered a sister group

to the other vertebrate lineages rather than a true ancestor to the vertebrate lineage. This is a controversial area whose resolution may make a significant difference in the understanding of the development of the eye. Recent work suggests that hagfish are true vertebrates, but degenerate.

We will work from the assumption that hagfish, a sister group, had an early form of an eye that would become the vertebrate eye radiating from some common unknown ancestor related to *Myllokunmingia*.

Hagfish (Eptatratus stoutii)

Despite the absence of bone, tooth enamel, and other hard parts, hagfish and lampreys are still surprisingly well, but not plentifully, preserved in the fossil record (Figures 7.3 and 7.4). Because of the paucity of fossils, the phylogeny of these groups is incomplete and controversial.

Hagfish have been successful and have been present since at least the mid-Ordovician, about 470 mya. They or their predecessor were probably closely related to the common ancestor that would later lead to the vertebrate clade, and their eyes were similar to those of that first vertebrate. Hagfish and their eyes provide hints into the primitive nature of the last common ancestor of lampreys, sharks and rays (Chondrichthyes), ray-finned fish (Actinopterygii includes most living fish), lobe-finned fish (Sarcopterygii), and all tetrapods (some fish and all terrestrial vertebrates). Evolutionary evidence suggests that hagfish preceded the families of lamprey, with all three extant lamprey families coming from a common ancestor. And fossil evidence tells us that extant hagfish have changed little over the past 300 million years or much longer.

Hagfish would not seem to be good candidates to be a key species, let alone a controversial one, in the vertebrate lineage. But they are both.

Hagfish are classified with lampreys as cyclostomes, which refers to their round, jawless, toothless mouths. The two species are related, but probably not directly. Hagfish diverged from the more robust vertebrate line approximately 530 mya. Hagfish and lampreys may be related to a more basal protochordate—it is a controversial area, and the eye and its photoreceptors may help determine this phylogeny.

Figure 7.1 *Amphioxus* species, *Branchiostoma longirostrum.* Although there are paired pigmented spots at the anterior end, these are not true eyespots. The single median eye is located inside the "head" and has no pigment surrounding it. The pigmentation at the head end may be a clue to the development of the eye, as the single eye within the head contains ciliary cells. In the process of becoming two lateral eyes, the external pigment may have a role to play in the orientation of the (inverse) photoreceptors of the vertebrate eye.

Hagfish are primarily benthic or ocean floor–dwelling fish. They spend their lives tying themselves into knots in order to feed on decomposing marine creatures, such as whales. These cartilaginous scavengers are perhaps best known for their ability to generate slime to frustrate predators. The copious slime, secreted from specialized glands along their flanks, consists of

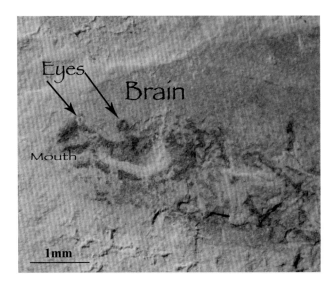

Figure 7.2 *Haikouella.* Note the two eyes and faint outline of the dorsal portion of the head. The eye is approximately 100 μm in diameter. *Image © Jun-Yuan Chen, PhD.* From Chen, J.-Y. (2003). Fossil sister group of craniates: Predicted and found. *Journal of Morphology 258,* 1–31.

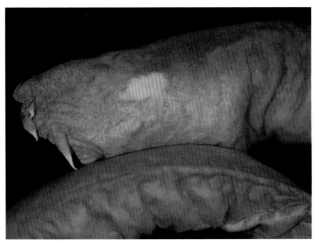

Figure 7.3 *Eptatratus stoutii,* Hagfish. Note the tan-colored skin covering the eye.

mucus that "explodes" on contact with seawater, and clogs the gills of predators.

Hagfish eyes are small and hidden beneath patches of translucent skin on the lateral surface of the head (Figure 7.3). They have no lens, iris, cornea, or ocular muscles (Figure 7.4). Like a vertebrate retina, the hagfish's retina has ciliated photoreceptors, but it (and the lamprey larva's retina) is less mature or less derived than those of true vertebrates.

Additionally, the ganglion cells of the hagfish retina that project to the brain project to the hypothalamus. The hypothalamus is an ancient brain area that controls circadian rhythm (as well as hunger, thirst, and temperature). In vertebrates, only a recently discovered small subset of ganglion cells travel to the hypothalamus. This subset contains melanopsin, a light-sensing compound that helps to regulate circadian rhythm.

Hagfish do not have a parietal (third) eye or a pineal gland, suggesting that, embryologically, this bilateral structure (the parietal eye complex) has evolved into the lateral eyes of this animal instead. By comparison, primates possess only the pineal gland as a remnant of their lost parietal eye. These and other differences hint that the lateral eyes of hagfish (Figure 7.3) are really parietal eyes that have positioned themselves laterally.

Hagfish have at least one other key basal anatomic difference that becomes important in the development of the visual witness. They have but one semicircular canal, which is vertically oriented when the animal is horizontally parallel to the ocean floor. Semicircular canals are crucial to balance, and primates have three of them. (The other two will become important as the bony fish develop.)

If hagfish are indeed a sister group to the vertebrate clade, the first vertebrate probably had a primitive eye resembling that of the hagfish. That first vertebrate radiated into lampreys

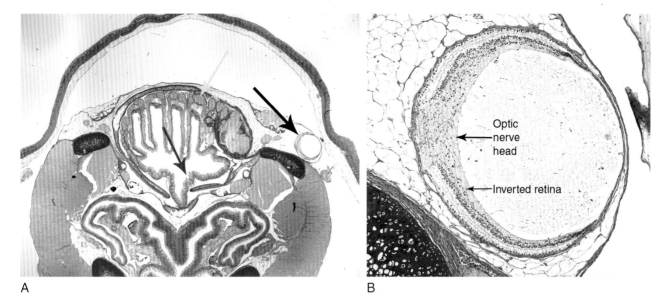

A   B

Figure 7.4 *Eptatratus stoutii.* **(A)** Head of hagfish. Note the eye (black arrow), the nasal mucosa (blue arrow), and the nerves surrounding the nasal mucosa (yellow arrow). Hagfish do not rely on sight for locating prey but rather rely on olfaction as represented by the nasal mucosa. This image shows the sensory tools. **(B)** Hagfish eye. Note the inverted retina, which appears less populated than the retina of vertebrates, but still the retina is inverted. The retina consists of only two layers of cells—the photoreceptors and the ganglion cells that lead to the brain. This section also goes through the optic nerve head with the axons of the ganglion cells collected to lead out of the eye. *Histology by Richard Dubielzig, DVM.*

and perhaps other eel-like ancestors, with one of those giving rise to the other vertebrates.

Reinforcing that possibility, lamprey larvae also have a primitive eye similar to the hagfish eye that develops into a much more conventional vertebrate camera-style eye. Their two retinae are also similar, but over several years, extra layers of cells appear in the larval eye of the lamprey between the photoreceptors and the ganglion cells—as in other vertebrate eyes. The adult lamprey eye has a lens, extraocular muscles, and other structures more familiar to vertebrate ocular morphology—all of which appear during metamorphosis. This well could be the source of the vertebrate camera-style eye.

But there is an alternate theory. The last common ancestor of the cyclostomes may have had a midline parietal-eye complex (descended from a basal protochordate), rather than "eyes" as we think of them. Early in embryological development, the parietal-eye complex consists of two identical halves, arising from two buds in the diencephalon (early brain). As hagfish evolved, early representatives of the hagfish clade retained both halves, which migrated to each side of the head instead of remaining midline.

In the lineage that gave rise to lampreys and living vertebrates, the parietal-eye complex remained midline. It followed a more typical course of development into a single parietal eye from the left portion and a pineal from the right portion, as we see later in some fish and reptiles. Through gene duplication or simple formation from the protochordate ancestor, two additional outpouchings or bulgings of the early neural tube

provided more conventional camera-style eyes that evolved into the vertebrate eyes present today.

First True Vertebrate

Haikouella and *Yaunozooan* are not classified as the first vertebrates but more or less as vertebrates-in-waiting. The first primitive vertebrate, which may have appeared in the early or mid-Cambrian, may hold a key to the understanding of the development of the vertebrate eye. Although some people would not classify it as a true vertebrate, we do. This animal was the long-enigmatic conodont (Figure 7.5).

Conodonts

Conodonts (Figure 7.5) must have been common in the middle Cambrian because their teeth fossils are found in large numbers. But the fossils were not understood until the 1980s when a fully formed fossil was discovered.

Several thousand species of conodonts over nearly 300 million years followed that first species. The first species appeared approximately 520 mya. These eel-like creatures also resemble the protochorodates mentioned above (although the phylogeny of these two creatures has not been determined). These animals clearly had rather large and well-organized eyes.

The musculature along the thorax of this animal resembles that of lancelets, suggesting that a conodont was an

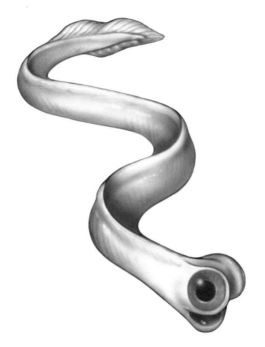

Figure 7.5 Artist's depiction of conodont. These are early eel-like vertebrates or protovertebrates. Although we know little about these eyes, we do know that the eyes were prominent, indicating early ocular development. *Art by Tim Hengst.*

Building a Vertebrate Eye

To build an eye, the first protochordate needed a condensation of collagen that would hold the eye cup. As chordates developed and the first vertebrates appeared, an outer coat developed around the photoreceptors, resulting in an integumentary sheath. Hagfish and lampreys have that first scleral outer shell, which must have developed in an earlier stem animal that is now extinct. The sclera is an outer tough coating that surrounds the eye in all vertebrates, including humans. The scleral shell likely developed as reinforcement around an invaginated cup that included photoreceptors and pigmented cells.

But the outer coat in hagfish and lampreys is thin and has the same (radius of) curvature around the entire sphere—much like a spherical marble. There would be no difference in curvature for the cornea. At this point, the sclera and the cornea were not well developed, and there would have been no established difference in the radius of curvature of the cornea that would distinguish it from the curve of the sclera.

The vertebrate eye is a simple enclosed cyst. At least in the lamprey, the eye moves freely beneath the clear surface ectoderm, although no sulcus surrounds the cornea or the eye to delineate the cornea from the rest of the eye (Figure 7.6). The cornea in hagfish and lampreys is very thin, with minimal epithelium. It is the first and most basal cornea known (Figure 7.4). The skin that covers this eye, called a primary spectacle, is translucent in hagfish and

early craniate (like the hagfish), if not the first true vertebrate. The best evidence suggests that this clade was more derived than hagfish, is a more closely related group to vertebrates than the hagfish, and likely is a vertebrate.

Conodonts' eyes, then, are possible early models of what would follow and illustrations of how quickly eyes can emerge. Good evidence suggest that eyes can appear in a geologic blink, perhaps as little as half a million years. But although these eyes could not have formed this well virtually overnight, they corroborate previous suggestions that the visual witness was organizing before the Cambrian explosion.

The few discovered whole conodont fossils reveal that the eyes were placed laterally on the head, much as in fish, lined with pigment, just as in fish, and were surprisingly large compared to the body. Animals whose eyes are large in proportion to the head or body are generally better sighted, faster, nocturnal, or live in murky or deeper water. Conodonts in particular had a relatively well-formed eye (Figure 7.5).

Hagfish-like animals or more likely a common ancestor radiated into conodonts. This relatively close relationship suggests that the conodonts probably had similar eyes and processing, including color vision. Although this was a sidelight to the real show of invertebrate development that was going on in the Cambrian, the vertebrate eye had its start.

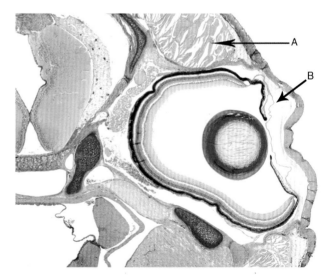

Figure 7.6 *Petromyzon marinus.* Lamprey eye. Note the arrow marked A. This points to the cornealis muscle, which can change the curvature of the cornea. The entire eye can move underneath the skin layer and is not attached to the skin. The cornea is marked with B and is considered the primary cornea. *Histology by Richard Dubielzig, DVM.*

clear in lampreys. This skin is completely separate from the underlying eye. Animals that have a primary cornea may have been common in this period, but the only extant animals with such a basal cornea and primary spectacles are hagfish, lampreys, and some aquatic amphibians.

The lamprey, then, appears to exhibit the early stages of the evolution of the vertebrate cornea (Figure 7.6) (Appendix D). The last common ancestor probably had a similar eye. As the ray-finned fish appeared, probably later in the Ordovician (490–445 mya) or the Silurian (445–415mya), the cornea likely developed a peripheral ridge, or sulcus, and a curvature change at the juncture with the sclera (Figures 7.7 and 7.8).

So although lampreys have basal eyes, these eyes illustrate the steps and ease with which evolution found a vessel for the precious neurologic machinery of vision.

The sclera and basal cornea appeared in lamprey, but what was happening with the retina at this time? Because lampreys are a sister group to other vertebrate lineages, they illustrate what may have been the status of that last common ancestor of the vertebrate clade. If the outgroups look like the basal vertebrates, the original vertebrate probably shared a similar arrangement.

Surprisingly, rather sophisticated visual processing and color vision was probably present. The story of the Southern Lamprey illustrates the direction of evolution and perhaps its sophistication, even by the early or mid-Cambrian.

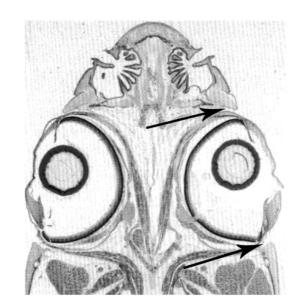

Figure 7.8 *Oncorhynchus mykiss,* rainbow trout. This section through the head illustrates the two eyes' placement. Note the two black arrows pointing to the sulcus described in Figure 7.7. This is not present in the lamprey. *Histology by Richard Dubielzig, DVM.*

Southern Lamprey (Geotria australis) *(Figure 7.9)*

Although lampreys have no bones or jaws, they are early cartilaginous fish with sucker-like mouths, and the lineage is 550–450 million years old. These are the only living jawless vertebrates and are reminiscent of Haikouella *(Figure 7.2).*

The southern lamprey, Geotria australis, *spends the first four to five years of its life with poorly developed eyes in the freshwater streams of Australia, New Zealand, South Africa, and South America. Although usually thought of as parasites,*

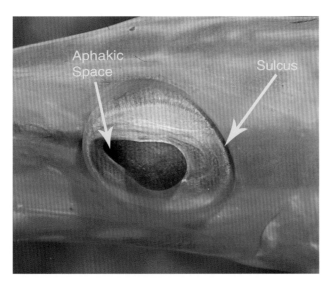

Figure 7.7 *Aulostomus maculates,* trumpetfish. Note the yellow arrow pointing to the sulcus, which permits the eye to move. If the eye were fixed in the head, the eye could not move in any direction, but the sulcus permits redundant conjunctival tissue to fold and allows movement of the slightly recessed globe. The yellow arrow with the label points to the aphakic space discussed in Chapter 10. *Image © James Brandt.*

Figure 7.9 *Geotria australis,* southern lamprey. *Image © Stephen Moore, Taranaki Regional Council, New Zealand.*

all lampreys actually begin life as filter feeders, feasting mainly on detritus and unicellular algae. During the lamprey's six-month metamorphosis into its pelagic phase, the eyes enlarge and develop the visual pigments necessary for it.

At the same time, the adult lamprey (but not lamprey larvae or hagfish) develops the cells within the retina that are interposed between the photoreceptors and the ganglion cells. These cells are important for the interpretation and modulation of the visual signal coming from the photoreceptors.

In the second phase of its life, the southern lamprey descends the fresh water rivers into the Southern Ocean, where it uses its sucker mouth and rasplike tongue to attach to fish, tear into their body tissues, and feed on muscle. During this phase, vision is critical, and the southern lamprey is prepared.

Recent molecular research suggests that the last common ancestor of jawless vertebrates in general possessed four of the five major classes of the visual pigments found in the radiation of jawed vertebrates. The southern lamprey has a pure cone retina with five visual pigments, three of which are orthologous (genes evolved from the same ancestor) to opsins found in jawed vertebrates. These three common cone opsins have peak sensitivities in the long wavelengths (red), short wavelengths (blue), and shorter wavelengths (ultraviolet). A fourth cone opsin with a peak sensitivity in the medium wavelengths (green) underwent an independent gene duplication within the jawless and jawed fish, leading to two different cone opsins in the southern lamprey and a different cone opsin and a rod opsin in the jawed fishes.

This research has several profound implications. First, rods and their rhodopsin are relative latecomers to the visual party, as they probably did not appear until at least the mid- or late Cambrian, approximately 500 mya. This makes sense, because an opsin with a peak in the blue wavelengths, or more likely in the ultraviolet wavelengths, was probably the first visual pigment (bacteriorhodopsin) and would have appeared with very early prokaryotes. Ultraviolet light would have been plentiful and would have delivered a high degree of energy when this visual pigment was first selected.

Second, this animal has an unusual extraocular muscle, called the cornealis muscle, which may change the cornea's curvature to change the focus of the image. Some birds and at least one fish, the sandlance, have independently evolved similar muscles. It is as if the cornealis muscle has appeared, disappeared, and then appeared again in a different form when needed.

The southern lamprey is a survivor, indeed, with five visual pigments in its retina, a bizarre lifestyle, and an unusual eye. And the lampreys illustrate the beginning of at least one other important addition to the visual witness, the tapetum (the eye layer that causes eyeshine, discussed below). These eyes illustrate the mechanisms that were present at the time because at least three of these five visual pigments and several anatomical features are also found in sharks and rays, a successful group that branches off the vertebrate lineage that would later lead to ray-finned fishes and lobe-finned fishes.

Short-Headed Lamprey (Mordacia mordax)

The only other lineage of lamprey in the southern hemisphere is Mordacia mordax, which is a nocturnal animal that has lost the variation in visual pigment and photoreceptor type. Its retina is composed of a single variety of large photoreceptor that is much more rodlike but retains some cone characteristics. This suggests that this family has traded diurnal vision for nocturnal vision. The M. mordax species have a tapetum as well to help maximize photon capture, helping further with their nocturnal lifestyle. The tapetum is a major step and echoes throughout the remainder of the vertebrate clade. This may be the oldest species, phylogenetically, with such a structure.

Tapetum

The tapetum lucidum (L. shining carpet) is a reflective structure found in different layers of different eyes of many diverse creatures and represents convergent ocular evolution solely for maximizing photon capture (also see Chapter 13). The tapetum is seen in the jawed fish including the elasmobranchs but probably arose separately, but not directly, from the lampreys. We do not know exactly when the tapetum first appeared, but it may have been as early as the late Cambrian. Whenever it appeared, the tapetum was an advance so important, and so useful, that evolution would return to this mechanism in a variety of strategies again and again. The techniques for the production of these reflective mechanisms are variable and, much like the crystalline lens, seem to draw from whatever materials the evolutionary process found at hand.

For example, guanine crystals provide biological reflection and camouflage by making a fish glisten and gleam, presenting confusing reflections to a predator. This coating probably appeared very early in the development of fish as a protective mechanism, perhaps as early as the Silurian.

Many fish, then, must have co-opted guanine as a biologic reflective coating for their scales as well as their tapetum. It remains a common and effective mechanism among fish.

Tapeta are composed of different materials and appear in different layers. Mordaciidae species have a retinal tapetum with the reflective coating immediately beneath and within the outer retina, not within the choroid, although the reflective compound is not known. More recently derived species have choroidal tapeta, and these will be discussed later. Although retinal tapeta appear in some bony fish, these do not appear to have descended from the lampreys, suggesting that the tapetum in Mordaciidae arose independently.

At least some animals during the Cambrian had full color vision, but why should this be so? After all, it may not have been necessary, or even worse; it would be "expensive," in that the process requires genetic "direction" and energy that could be devoted to other processes. The evolutionary development of structures and molecules requires energy, and wasted energy is a disadvantage to an animal.

Why Color Vision?

Monochromatic vision, which detects movement over distance, is not sufficient to distinguish approaching predators in the flickering reflected light of waves in a large enough body of water. If the water is shallow and reasonably clear, these waves become complex lenses and focus light as the type of fluctuating bright lines you see on the bottom of a swimming pool on a bright sunny day. To help with these challenges, early vertebrates had to develop color vision.

Color vision requires cones, particularly those with different opsins for peak absorption of different wavelengths. But color vision also requires the neurologic machinery to compare signals from cells with different "color" opsins. So the neurologic machinery must have come first or at the same time as color vision. Lampreys develop (bipolar) cells between the photoreceptors and the ganglion cells. These interneurons can inhibit or amplify a signal from one selected photoreceptor. Other brain cells appeared to help modify or modulate the signal from the photoreceptor and, as it turns out, to manage the flickering input that these waves would have created.

This is a complicated function of the retina—called color opponency—and it helps animals perceive color differences by "subtracting" the input from one cone from that of another. These additional "horizontal" cells laid the groundwork for color vision.

Lampreys developed five visual pigments that included the four visual pigments found in most fish. The last common ancestor that radiated into lampreys also would radiate into elasmobranchs and eventually lobe-finned fish or dipnoans. And this lineage would lead to the terrestrial tetrapods, which documents the path of color vision with four visual pigments to the amphibian and reptilian lineages.

The Ediacaran and Cambrian, then, were times of rich evolutionary experimentation. The vertebrate eye was established to include the scleral shell and a primitive primary cornea. The middle choroidal coat of blood vessels that would include the vascular supply for the retina (Appendix D) was established or matured. The retina's basic photoreceptor and neurologic mechanisms also were established. These basic ocular structures were surprisingly sophisticated and defined by this time. This period provided all the current phyla that have eyes with many of the model eyes.

Invertebrate eyes were similarly supplied with the basic form, although both vertebrate and invertebrate eyes would have to modify because of the adaptation to terrestrial life and to different niches.

The Cambrian ended abruptly, and the earth staggered into the Ordovician (490–445 mya) with cooling, glaciation, oxygen depletion, and species loss. The conodonts and trilobites, among others, would suffer considerable losses. But winning species would expand and fill the niches. By the end of the Cambrian, the hagfish and lampreys would have new relatives: ostracoderms and placoderms. These fish were descendants of the lamprey/conodonts or more likely a sister group to the lampreys. But before we can consider the fish, we must consider Mollusca—another important invertebrate group that had become established in the Cambrian but would thrive in the Ordovician.

SHELLY FAUNA RULE THE SEAS

PALEOZOIC ERA

ORDOVICIAN PERIOD

490–445 MILLION YEARS AGO

At the beginning of the Ordovician, the climate began to warm, and the atmosphere became wetter. By its middle, the ice packs had melted into warm, shallow seas. The conditions again favored life, and the creatures that had survived the Cambrian extinction flourished. Like the Cambrian, the Ordovician had an explosion of faunal genera. At least by late in the period, elementary plants began invading Gondwana, one of the two supercontinents that existed 600–180 mya. At the time it broke up, Gondwana was a landmass consisting of South America, Antarctica, Australia, Africa, Madagascar, India, and Arabia.

Although many of the strange and interesting creatures spawned by the Cambrian were lost, the early elements of the vertebrate eye and the basic compound eye survived the tempest of extinction. The trilobites, with one of the first eyes, continued and flourished in the Ordovician. Although they began to decline with multiple-species extinction at the end of the Ordovician, some of invertebrate phyla would succeed in the Ordovician and persist until the end of the Permian.

Like *Arthropoda*, its near contemporary—*Mollusca*—is a varied invertebrate phylum that began in the Cambrian, perhaps from that same basal urbilaterian. Although all major lineages of mollusks developed during the Cambrian, the Ordovician was to see their proliferation. Mollusks developed a shell as the arthropods developed a carapace; both developments arose as shelter from the Cambrian's predatory storm.

Mollusca

Bivalvia

The lineage of *Mollusca* may extend nearly to the beginning of the bilaterians in the Ediacaran. They competed directly with the arthropods during the Cambrian and split into the three major groups—Bivalvia (oysters, clams, scallops, and their kin), Gastropoda (snails, conchs, whelks, and their kin), and Cephalopoda (ammonites, nautiloids, octopus, squid, cuttlefish, and their kin).

The eye's toolkit had been present for at least tens of millions of years, and the metazoan eye has enough similarities to suggest that the last common ancestor of these three groups had surprisingly mature elements.

Of the mollusks, the bivalves were probably the first to establish themselves, but the true bivalves, which are more closely related to modern bivalves, did not become established until the Ordovician (Figures 8.1 and 8.2). Because of the known morphology and physiology of extant bivalve animals, these teach us about the visual witness of the day. Perhaps the best animal to set the scene for this diverse group is the scallop.

Scallops, Aequipecten irradians

The 1672 invention of the Newtonian telescope was a most dramatic step forward in the observation of the stars. In the Schmidt catadioptric telescope, a 1930 variation on the Newtonian telescope, incoming light rays first strike and pass through a corrector lens and then strike a concave mirror to be reflected. The reflected light is focused on a film plate within the column of the original light beam. One of the main purposes of the initial lens in the Schmidt telescope is to eliminate spherical aberration. But a simple bivalve mollusk had this elegant design as much as millions of years ago and still uses this system today (Figures 8.3 and 8.4).

The colorful eye of Aequipecten irradians *(and other scallops of the genus* Pecten*), as seen among the tentacles (Figure 8.3), is essentially the only Schmidt catadioptric-style telescope in the animal kingdom. The common scallop may have forty to sixty pallial (produced by the mantle of a mollusk) eyes around the edge of the mantle, each with a lens (acting as a corrector plate as in the Schmidt design), two separate tiered retinae, and a biological mirror lining a pigment cup (Figures 8.4 and 8.5). Two optic nerves leave each eye in opposite directions and join behind it. Much as in the jellyfish, there is no central processing or brain to interpret the image, so the animal does not "see" in the true sense of the word as we think of it.*

The lens of this biological Schmidt telescopic eye is aspheric with what would appear to be a curved bow on either side of the lens center (Figure 8.5). This is the lens design feature that

Figure 8.1 *Spondylus* species, spiny oyster. Not a true oyster, this exquisitely beautiful bivalve has eyes that rim the mantle and can be seen as small blue dots just beneath the shell. This bivalve genus is closely related to the scallops, and the internal anatomy is much more closely aligned with the scallops. One of the eyes is marked by the blue arrow. *Image © Foster Bam.*

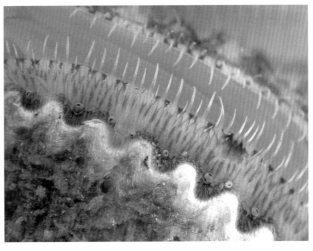

Figure 8.3 *Aequipecten irradians*. Scallops have multiple eyes just beneath the edge of the shell, which resemble blueberries attached to the mantle. *Image © Bill Capman, PhD, Augsburg College.*

reduces spherical aberration. But, with only 5000 photoreceptors in each retina, and only cerebral ganglia to process the image, one wonders why spherical aberration would be a problem for such a creature.

Here is how the scallop's visual system works. In the scallop's elegant model, light strikes the aspheric lens and is refracted toward the convex (cuplike) mirrored surface behind the lens. The photons pass through a layer of photoreceptor cells immediately behind the lens and a second more proximal retinal layer on the inner surface of the cup. The rays then strike the posterior surface, which is lined with guanine, a shiny biological mirror.

The guanine layer reflects the partially focused light (Figure 8.5), and these reflected rays come to focus on the retina

immediately behind the lens. Each photoreceptor of both retinas sends a fiber to the optic nerve with no synapses or interconnections. Each retina sends its fibers in a different direction with a separate optic nerve around the mirror and through the pigment cup to join into a single nerve behind to the eye. This optic nerve then travels to the visceroparietal ganglion.

The proximal retina lining the mirrored cup will never receive a formed image because the aspheric lens does not focus on the

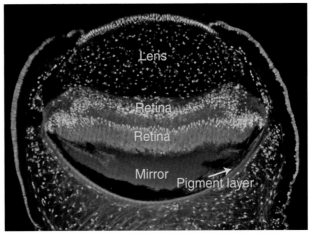

Figure 8.4 *Argopecten irradians* eye. With different stains and dyes used to color the nuclei and other intracellular structures of the cells, the eye of *Argopecten irradians* can be illustrated clearly. The red pigment layer is present to block any light that is not reflected by the mirror layer. The mirror layer reflects light to focus on the first retina immediately behind or below the lens in this image. *Image © Daniel Speiser. Related article by Daniel Speiser and Sönke Johnsen in bibliography.*

Figure 8.2 *Spondylus* species, spiny oyster. A closer view of the eyes of Figure 8.1. The blueberry-like masses are eyes. One is indicated by the yellow arrow. *Image © Foster Bam.*

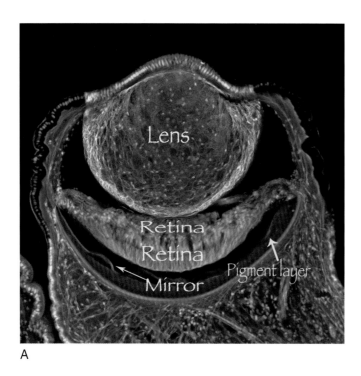

A

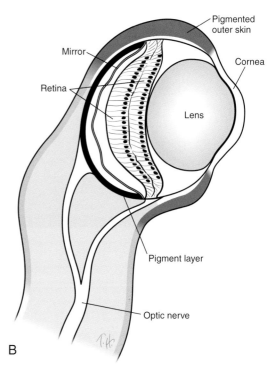

B

Figure 8.5 (A) *Placopecten magellanicus* eye. A second example of the eye of a different scallop species. Note the curved bow of the blue epithelium on the surface of the lens. That makes the lens aspheric, and this design helps eliminate spherical aberration. *Image © Daniel Speiser. Related article by Daniel Speiser and Sönke Johnsen in bibliography.* **(B)** Artist's conception of eye of scallop. *Art by Tim Hengst.*

mirror. Furthermore, this retina, which is farther from the external surface of the eye in the anterior-posterior direction, responds to a cessation of illumination. These photoreceptors fire when light goes off! The other retina (immediately behind the lens) fires, more typically, when stimulated by light.

It has been suggested that scallops use the information from the proximal retina to determine orientation toward or away from light. The information from the distal retina, immediately behind the lens, is sensitive to directional stimuli and responds to moving objects, which would require a more focused image. Predator avoidance would be the likely use because these species can "swim" by jet propulsion. Scallops accomplish this by using their adductor muscle to rapidly close their shells, thus producing a jet stream of water.

From an evolutionary and phylogenetic approach, these animals become even more interesting. The scallop's distal retinal photoreceptor cells are ciliary (usually true of vertebrates), whereas those of its proximal retina lining the cup are rhabdomeric (usually found in invertebrates). The potential of having both cells is probably carried forward through all lineages, including annelids and mollusks. Certainly, the potential of having both carried forward through the protochordates—Pikaia and later amphioxus. They were probably close to our ancestors and, like scallops, obtained their ciliary and rhabdomeric cells from a predecessor.

Although the bivalve lineage has undergone significant evolutionary change, some bivalves still have rather simple eyes, suggesting how the eyes began or at least how evolution has modified these eyes. The giant clam shows that several lineages of bivalves must have had eyes. These eyes suggest that such simple camera-style eyes were the order of the day as the mollusks arose.

Giant Clam, Tridacna gigas

Well-established classes within dominant phyla often produce outsized examples of that class, particularly if there are few predators and evolution has time to work. Immense and sometimes highly specialized forms of a particular class appear after prolonged stable evolution.

Among the bivalves, the giant clam (Tridacna gigas) is one such example of an enlarged animal within a specific class (Figures 8.6 and 8.7). This clam can weigh more than 500 lb (225 kg) and be as large as 4 feet across (122 cm). Native to the South Pacific and especially along the Great Barrier Reef of Australia, it has a gorgeous mantle, which may be purple, golden brown, yellow, or green (Figure 8.7). The mantle faces straight up into the shallow seas. When the clam is open, the siphon and the surrounding mantle often extend beyond the shell's lips. The clam prefers these shallow seas, as if seeking to expose its mantle to the sunlight.

Figure 8.6 *Tridacna gigas*. The giant clam is native to the South Pacific. Although the eyes are not easily visible at this magnification, they can be seen as fine black dots around the edge of the mantle just inside the rim of the shell. *Image © Foster Bam.*

T. gigas has an endosymbiotic dinoflagellate (Symbiodinium species) living deep within its tissues. These zooxanthellae (dinoflagellates) are also endosymbionts, serving the same nutritive purpose, in coral. These algae flourish because of the light and nutrients available in a shallow reef community. This symbiotic relationship is important to both the mollusk and the algae. T. gigas farms the algae as a food source because, as a filter feeder, it sieves the plankton in the seawater it takes in through its siphon.

Conversely, the algae use the clam as a safe home where it obtains the sunlight necessary for photosynthesis, consequently producing a carbohydrate food source for the clam.

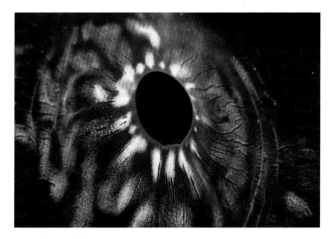

Figure 8.7 *Tridacna gigas*. The mantle is often a gorgeous purple but can be any of several colors described in the text. *Image © Foster Bam.*

Iridophores, the blue spots seen within the mantle, have lenses that focus light within the tissues to assist the algae. To some extent, the clam is able to regulate the amount of light striking the iridophores and the algae by moving the mantle toward or away from the surface of the water and by controlling the amount of screening pigment in the mantle itself.

Like its relative, the scallop, the clam has many simple eyes, called pinhole eyes and looking like small blue dots, scattered along the mantle's peripheral edge close to the lips of the shell (Figures 8.6 and 8.7). Although pinhole eyes do not produce superb images, they are sufficient for the clam. They can detect the approach of oncoming predators such as fish and measure light levels on the reef. Measuring the light levels might seem trivial, but the endosymbionts, being relatively delicate, respond best within a certain temperature and photic environment. Light and temperature beyond their range can harm or even kill the algae, so it behooves the clam to pay attention.

These pinhole eyes in extant animals give us clues to understanding the eye when mollusks diverged, around 600–550 mya. The photoreceptors contain visual pigments sensitive to blue-green (490 nm), blue (450 nm), and ultraviolet (369 nm)—the wavelengths of the world the clam occupies. These visual pigments illustrate that multiple visual pigments or the genetic tools for them were probably present early in the evolutionary process, perhaps as early as urbilateria (Chapter 4). It also illustrates the visual pigments likely to have been selected first, as discussed previously.

The photoreceptive cells of these eyes act like those of the vertebrates (ciliary cells) in sending signals to other nerve cells. Because other mollusks have cells that are rhabdomeric, both cells were probably present in the basal mollusks.

Perhaps the most durable mollusks are the nautiloids with their familiar coiled shells (Figures 8.8 and 8.9). Nautiloids appeared and flourished in the Mesozoic Era (251–65 mya). None of the original nautiloid cephalopods remains, but the earth still harbors one that is quite similar to those Cambrian pioneers—the *Nautilus*.

Cephalopoda, Nautilus

An evolutionarily basal species of mollusk, the Nautilus, is more than 110 million years old although close but extinct relatives originated nearly 500 million years ago (Figures 8.8–8.10). This invertebrate has changed little over that time. For example, it retains a basal eye illustrating its eye was sufficient for successful visual adaptation to its environment. The eye is located behind the tentacles, and the pinhole opening is its only external sign (Figure 8.9).

The Nautilus' basal eye probably has not changed much during the time their species and close relatives have been on

Figure 8.8 *Nautilus pompilus.* There are five individual specimens, each mirrored in the surface above them.

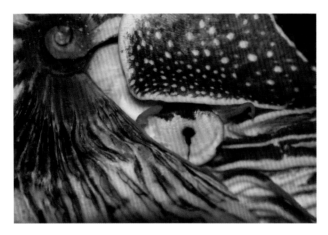

Figure 8.9 *Nautilus* with eye. A magnified image of the pinhole eyes of the *Nautilus.* Note that there is no cornea, and no lids.

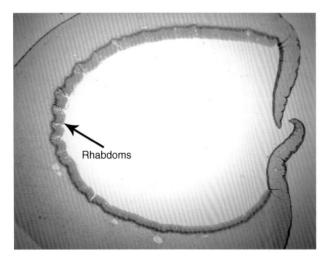

Figure 8.10 *Nautilus* eye. This histologic image shows the lack of a cornea, the pinhole pupil, and the absence of a lens. The central cavity is lined by retinular cells up to the point of flexion of the sclera. The rhabdoms are labeled. *Histology by Krisztina Matis.*

earth. *The eye is basal but not degenerate, which helps us understand how visual systems have evolved.*

The vertebrates that inhabit the same environment have an eye that is at least superficially similar. Both are camera-style eyes (in contrast to the compound eyes of many other invertebrates, such as insects and crustaceans).

This camera-style eye is also seen in other cephalopods, such as octopus and squid, which are related to the Nautilus. But most of those cephalopods have a lens and a cornea, as the eye has evolved to improve focus.

These eyes also show convergent evolution. Other lineages, such as vertebrates, have camera-style eyes, but the last common ancestor of mollusks and vertebrates, for example, did not. Mollusks have achieved similar results of ocular morphology but by very different mechanisms.

The Nautilus' eye has no lens, cornea, or any dioptric element whatsoever. Furthermore, the absence of a barrier to seawater allows for a continual exchange with the external pelagic environment. In life, a mucus or gelatinous substance fills the cavity. This probably resembles a vitreous body, which must contain seawater because of the open fistula.

The eye cup is almost spherical. The pupil weakly contracts to the diaphragm with a diameter range of at least 0.7 to 2.25 mm (Figure 8.10). The pinhole leads into a cavity 1 cm in diameter lined almost completely with photoreceptive cells called retinulae (singular retinula) that include microvilli facing the pinhole opening.

Adjacent cells' microvilli combine to create a rhabdom, or photoreceptor element. Supporting cells contain pigment granules that can migrate within the cell and are probably used to control light stimulus to the retinula. The axons lead from the proximal base of the retinula directly to the optic nerve without synapse.

Morphologically, the retinular cells extend their rhabdomal portion of the cell toward the inner cavity, allowing direct seawater contact with the photoreceptive element. In an evolutionary sense this photoreceptive eyespot has invaginated and developed supporting elements.

With no lens or cornea, such an eye is not subject to chromatic or spherical aberration. But it is subject to the diffraction limitations of a small opening. As the aperture size approaches the wavelength of light, diffraction causes the beam to be spread out and interferes with the waveform.

A small opening also brings other limitations. Because adequate sensitivity and resolution cannot occur simultaneously, as the pupil becomes smaller, resolution improves to the point of diffraction, but sensitivity decreases dramatically. Experimentally, the Nautilus favors adequate sensitivity over resolution. Looking at the mollusk's lifestyle may tell us why.

The Nautilus lives in the Pacific Ocean at depths of 150 to 600 meters along steep underwater cliffs where bioluminescence is a prime source of light. Maximum light transmission at that depth peaks at about 475 nm (essentially blue), correspond-ing to the absorption spectrum of its visual pigment. Spectral sensitivity has been determined to be maximal at about 470 nm (still blue).

Although Nautilus shows diurnal activity, it is most active as a crepuscular or nocturnal creature with circadian vertical migrations.

At these depths and in dim light, bioluminescence is also important—for locating decaying matter, the principal diet of the Nautilus. Animals with bioluminescent traits, such as some shrimp species, help to attract the Nautili to their common food source.

Nautili can see only a few meters. Although they feed slowly, they do not need distance vision for protection because they are encased in a hard outer shell.

Gastropoda

The gastropods are the third large and successful clade of shelled mollusks. They include snails (marine, freshwater, and later, terrestrial), slugs, and other mollusks that appeared in the Cambrian seas. Their approach to the vaga-ries of predation was a (nautiloid-like) often-coiled shell of hard calcium. Today, these animals fill many niches and rival the insects in their ability to adapt successfully to dif-ferent habitats. But some of them retain more basal charac-teristics, as conchs illustrates. Although the visual systems of all three lineages of the gastropods are related, it is they that have developed, or perhaps retained, a most interest-ing twist.

Conchs, Strombus galeatus

The adult S. galeatus is a robust and hearty vegetarian that can still be found browsing its way though the sea grass of the Costa Rican continental shelf of the eastern Pacific and else-where (Figures 8.11—8.14). The eyes of this heavily armored gastropod are at the tips of long stalks protruding from its shell's flared edge and will continue to grow with the animal through-out its life. Conchs can move these eyes independently.

These camera-style eyes, measuring 2.5 mm in a young adult, resemble those of not only the box jellyfish but also cephalopods as well as vertebrates. But, conchs' eyes differ from those found in vertebrates in embryology, physiology, and function.

Like many other gastropods these conchs have a thin bilay-ered cornea and a large, spherical piscine-style (fishlike) lens with an iris and a pupil (Figure 8.14). The outer cornea consists of clear epidermal cells overlying the ocular structures in an "eye-bulb." The inner cornea itself consists of two layers: an outer basement membrane that encircles the eye and an inner layer that transitions to, and becomes, retinular cells beyond the pupil.

Figure 8.11 *Strombus galeatus*. This is beautiful large mollusk with one stalked eye appearing at the right hand edge of the shell as it tapers; the other stalked eye appears less prominently to the left of the first eye around the curvature of the shell.

Immediately behind the lens is a cavity filled with fluid, analogous to vitreous. Cupping the "vitreous" is a concave pos-terior surface lined with rhabdomeric retinular cells with the rhabdoms facing the lens. (Figure 8.14).

Each of these retinular cells has pigment granules or vesicles within it that prevent light penetration. They also are retino-chrome packets that serve much like a tapetum (the anatomic structure that causes eyeshine—Chapters 7 and 13) and store retinal (the vitamin A derivative). This source of retinal replen-ishes microvillus visual pigment when needed, a system that is unique to the mollusks but not to the gastropods, as squid and other cephalopods have a similar system.

Figure 8.12 *Strombus galeatus* eye. Note the yellow-rimmed pupil, bluish eye surrounded by the red rim of the stalk. The greenish leaf-like attachment is the sensory tentacle used for olfaction. It is essentially the nose of the animal.

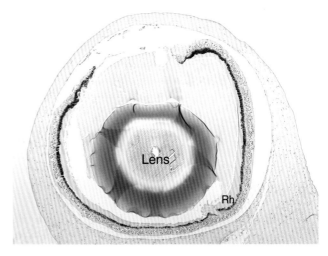

Figure 8.14 *Strombus galeatus* retina. An enlarged image of Figure 8.13. The lens is labeled. This cavity is analogous to our vitreous cavity. Note that the photoreceptive elements are supported by a pigment layer to block light from penetrating beyond that layer. The rhabdom (Rh), or photoreceptive element, is directed toward the lens with no intervening structure. *Image by Richard Dubielzig, DVM.*

Figure 8.13 Histology of strombid eye. These eyes can be of surprising size, ranging up to 17.5 mm (normal human eye is about 24 mm). The eye consists of a rather flat cornea whose purpose is protection of the remainder of the eye. Inside the eye is a lens, which in this preparation has dropped into the eye, although this is an artifact, as its position in life would be immediately behind the "iris." The retina lines the cuplike structure with a pigment layer behind the photoreceptive layer. The olfactory tentacle is visible to the left of the stalk containing the eye. Just as in many animals, the sensory elements are close to one another. The animal does not hear. *Image by Richard Dubielzig, DVM.*

Just what does an armored vegetarian—with an easily found, unlimited food supply—need with this elegant visual system? The shallow warm continental shelf seas are rich in the algae and sea grass meadows that form the mainstay of the animal's diet, so good vision is not necessary to find food. This environment, though, would be very bright with lots of confusing reflexes from the refraction of the waves on sunlight. Good spatial vision would assist the animal in its peregrinations, although this reason still does not seem to be sufficient. Prey avoidance seems likely to require good vision. The strombids are good diggers and, given some warning, can quickly bury themselves in the sand to frustrate predators, such as lobster, octopus, marine turtles, and sharks. These reasons may be sufficient for evolution, but they do not seem complete. Nevertheless, it seems clear that good vision is essential to the animal, for the most interesting aspect of this eye is its regeneration. Although some might consider the conch to possess a relatively basal and uncomplicated eye, this creature is one of only a few that can completely regenerate a very elegant slice of neurologic machinery. S. galeatus can retract its eyestalks most of the way beneath the shell, but eyestalks that remain even partially exposed can get injured or lost. So the animal regenerates them.

If the eyestalk is amputated below the eye, the optic nerve degenerates, but eight days later, new epidermal pigmented cells grow around the nerve toward what would have been the tip.

As the retinular cells (photoreceptor equivalents) are laid down, these cells send growth cones toward the degenerating optic nerve. Eventually, the growth cones unite to travel down the original optic nerve and reassemble correctly in a topographic fashion with the appropriate cerebral ganglion. At first the new eye is small, but in a few weeks, it grows to the approximate size of the original.

As the end of the Ordovician approached, Gondwana had drifted south toward the pole, and the earth had cooled. As the sea froze, sea levels fell.

The Ordovician ended with punctuation—perhaps a meteor strike, as these occurred every 60–90 million years, a gamma-ray burst or something else. This second-worst extinction to date for the earth caused the loss of perhaps 50–60 percent of all animal species, leaving the earth devoid of much life during the early Silurian (445–415 mya).

But the seeds of new species had been sown, and it would take the extinction's winnowing process to permit the expansion of the world of fish.

THE PISCINE EYE DEVELOPS

PALEOZOIC ERA

SILURIAN PERIOD

445–415 MILLION YEARS AGO

The Ordovician extinction decimated the marine life of the age, but its survivors would prove resourceful, and it would open new niches. Whatever life managed to get through the gate between the Ordovician and Silurian was fortunate, hardy, and decidedly piscinesque, perhaps well adapted even to cold water.

During the Silurian, conditions that support life's expansion would return. The climate warmed, and shallow continental seas returned, allowing expansion and diversity of marine life. The annelids and arthropods continued to evolve to species such as leeches and sea scorpions (eurypterids). New and varied corals, bivalves, brachiopods, gastropods, and shelled cephalopods appeared, increased, and/or flourished.

The first terrestrial life appeared as algae; liverworts, and mosses emerged onto or moved to the barren land. Shortly afterward, arthropods colonized land. Then came the first insects, scorpions, and spiders (all of which would prove important during the Devonian, in which these creatures and the vegetation around them would tempt the aquatic tetrapods to gain ground).

But it was life in the sea that saw the most expansion in the Silurian.

Vertebrate Expansion

The earliest fish, known as the Ostracoderms, would not look like fish to us. These long-extinct jawless fish arose at the same time as their extant relatives, agnathans, hagfish, and lampreys (Chapter 7).

Gnathostomes, the first fish with jaws, probably appeared around 410 mya as a now extinct lineage known as Acanthodii. Although they died out by the end of the Devonian and little is known of their eyes, these odd fish had impressive relatives whose eye we understand better—the placoderms. These animals help us understand the status of the eye at this stage.

Placoderms

Placoderms were a spectacular class of jawed, burly fish that arose in the Silurian. As jawed fish, these were unusual as perhaps the only such fish without real teeth. They were primarily bottom dwellers, or benthic fish, and were a sister group to the elasmobranchs (sharks and rays). Because the eyes of some Silurian placoderms were encased in bone (Figure 9.1), fossil evidence gives us much information about the maturity of the fish eye of the time.

Placoderms are important to the development and understanding of the visual witness. For example, placoderms had perhaps the earliest evidence of a parietal eye, although lampreys likely arose earlier, and they have a pineal gland, which is a counterpart to the parietal eye. This means that the midline outpouchings of the brain that led to the parietal eye and the pineal gland appeared early in the evolution of the vertebrate eye, perhaps even before the lateral eyes, as suggested by the hagfish eyes.

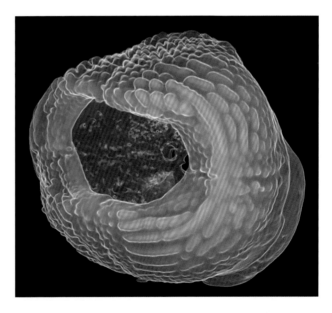

Figure 9.1 *Murrindalaspis wallacei* optic capsule. An early Placoderm from the Silurian, *Murrindalaspis wallacei* had an optic capsule, as did other placoderms. This bony capsule permits analysis of the extraocular muscles and the choroidal gland because of the molding of the bone. *Image © Gavin Young with associated work by Drs. John Long, Carole Burrow, and Allen Jones.*

The fossilized bony casing of the eye, though, is by far the most important and interesting information provided by this early fish, permitting understanding of the early intraocular and periocular anatomy. The bony capsule is homologous to the cartilaginous cup and probably the bony scleral ossicles that encompass much of the eye in many fish today (Figure 9.2). The cartilaginous cup of modern fish extends from the posterior aspect of the globe forward to encompass the posterior two-thirds of it (Figures 9.3 and 9.4). Scleral ossicles are bony plates ringing the eye, much like a napkin ring, found in some fish eyes in the sclera, anteriorly, near the corneal rim.

The homologous cartilaginous cup and scleral ossicles would continue, phylogenetically, through the fish to the tetrapods, reptiles, dinosaurs, and into the birds. Clearly, evolution found a design that works, as the cartilage and bony plates surrounding the eye are favored structures. But, the placoderms initiated the bony cup, illustrating that the eye was getting tougher, although the armored eye did not persist (Figures 9.1–9.4).

Fossils of these strange armored fish and the bony ocular cast provide evidence for the anatomy of the extraocular muscles of the Silurian vertebrate eye. Sophisticated eyes require sophisticated movement, which requires muscles, and these muscles would leave imprints in the bony casing.

Extraocular muscles (EOMs), which move the eyes, are fairly constant throughout the vertebrates, with a few valuable exceptions. Exemplifying the basal state of vertebrates at the time, placoderms had seven EOMs. Seven muscles is one more than some other vertebrates, including us, have. Most ray-finned fish, a major radiation of bony fish including almost all commercial fish, have only six EOMs. The seventh EOM missing from many vertebrates is the

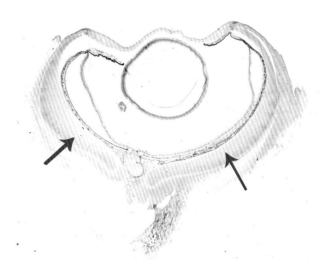

Figure 9.3 *Acipenser medirostris.* Green sturgeon eye. Note the cartilaginous cup marked by blue arrows. *Histology by Richard Dubielzig, DVM.*

retractor bulbi—a muscle directly behind the eye used to retract the globe toward the midline, often permitting lid closure. The retractor bulbi is important because it helps us understand how the eye evolved and its path to terrestrial animals as we will see.

The bony capsule also shows that placoderms were among the first, if not the first, fish to have an optic pedicel and a choroidal gland.

The optic pedicel is a cartilaginous rod extending from the skull to the back of the eye. This rod helps stabilize and

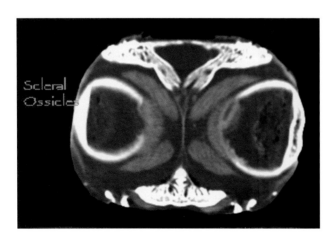

Figure 9.2 *Tetrapturus audax.* A CT scan of the head of a striped marlin. The scleral ossicles are bony supports around the eye and can be seen in this radiograph with bony density.

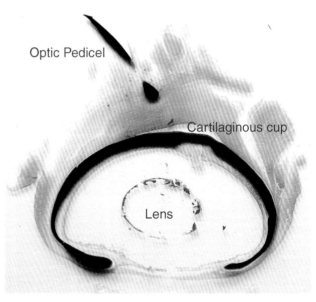

Figure 9.4 *Carcharodon carcharias.* Great white shark eye. Note the cartilaginous cup. The cartilage can be seen as a dense purplish capsule. The optic pedicel is also noted in the image. *Histology by Denise Imai, DVM.*

move the eye, and some fish, such as certain sharks and rays, even have a ball and socket joint at the articulation between the rod and the eye to permit stable movement (Figure 9.5). In Placoderms the optic pedicel ran through the bony case of the eye and attached to the eye. The optic pedicel survives in elasmobranchs (sharks, skates, and rays) and a few bathypelagic teleosts (deepwater bony fish). For species whose skulls are far from their eyes, such as the hammerhead shark, the pedicel must prove particularly helpful in stabilizing the eye.

Although optic pedicels are made of cartilage, the placoderms left fossil evidence of them. So did some of the extinct ray-finned (includes most living fish) and lobe-finned fish (includes the lungfish and the coelacanth) whose eyestalks seem to disappear in later radiations. But this adaptation does illustrate that the most recent common ancestor of the placoderms, elasmobranchs and the bony fish, probably had these eyestalks.

Enough detail was found in the bony case of placoderms to determine the likely method of retinal nutrition as provided by the choroidal gland in these fish of 400 mya.

The choroidal gland is not a true gland, although it was believed to be one when initially discovered. The organ is a plexus of blood vessels fed by a specialized gill, the pseudobranch, found on the inner side of the gill cover. This vascular plexus is present in many extant teleosts as a sole method to provide nutrition to the overlying retina (Figure 9.6).

So placoderms probably did not have direct retinal vascularization, suggesting a limited retinal thickness (Appendix C). Hence, the placoderms probably had relatively poor acuity without the need for a complicated retina. Still, placoderms likely had color vision and good motion detection. Some placoderm fossils have silver coloration on the ventral surface and red coloration on the dorsal surface, suggesting colored skin and hence color vision. Lampreys have five visual pigments and what must be vibrant color vision, so placoderms probably had color vision as well because they arose after the lampreys and some of the same visual pigments carry into more derived fish. Predators usually have good motion detection although not necessarily good vision for small objects—placoderms did not have to, after all. This creature ate other large animals, as *Dunkleosteus* illustrates.

Dunkleosteus

As a representative of this lineage, Dunkleosteus was the culmination of the class Placodermi with brute strength, massive size, and bony armor, as it ruled the seas in the late Silurian and Devonian (Figure 9.7).

At 6 meters and 3600 kg, this magnificent creature would have been the top predator in the seas, owing as much to its teeth-like bone plates and jaw as its size. These self-sharpening, razor-like bony plates could cut sharks in half, and these plates grew continuously.

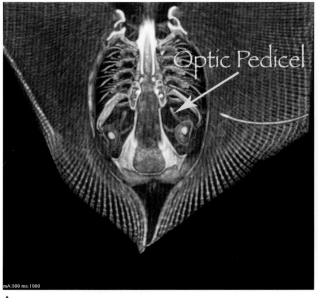

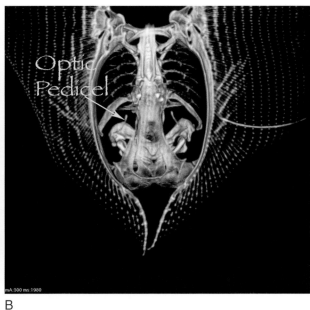

A   B

Figure 9.5 *Taeniura lymma.* Blue spotted ray computerized tomographic (CT) scans of ray. **(A)** Note the optic pedicel and the cartilaginous skeleton. The soft tissues have been false-colored as burgundy, and the cartilage as light yellow. **(B)** The soft tissues have been removed digitally, leaving only the cartilage. Again, note the optic pedicel. *Both images by J. Anthony Seibert.*

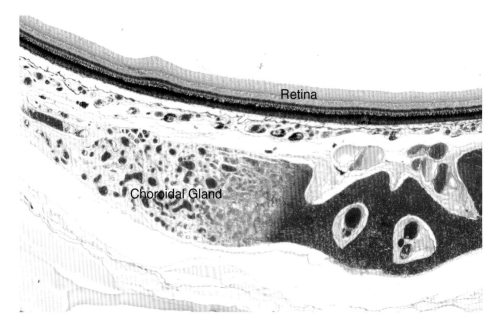

Retina

Choroidal Gland

The animal's lower jaw was hinged to permit a wider gape, and it was perhaps the strongest ever known, with a force of up to 5.6 million kg/m2 (8000 pounds per square inch). Its snaplike opening speed created suction to draw even large prey into its mouth. From scars on the bony-plated skulls, investigators believe that this fierce fighter probably even attacked and ate its confreres.

But for such a robust predator, this first real behemoth and other members of this lineage were relatively short-lived, as the

entire class Placodermi disappeared by the Devonian-Carboniferous border. These slower predators may have been outcompeted by less-armored, hence faster, more agile fish.

During the Silurian or the early Devonian, the initial fishlike creature (probably closely related to the placoderms or similar jawed fish) radiated into the extant jawed fishes including the elasmobranchs (cartilaginous sharks, rays, and ratfish), the ray-finned fish (including the teleosts,

Figure 9.7 *Dunkleosteus.* Fierce predator of the Silurian and Devonian seas. Armored with bony case for eye. See Figure 9.1. *Image © The Natural History Museum, London.*

which include almost all other fish), and then the lobe-finned fish, which eventually included the extant coelacanth and three species of lungfish.

Fish began to radiate into more benthic niches and murky depths, where they would need to develop the necessary neurologic machinery to salvage vision in these dimmer reaches. The first fish with rods as photoreceptors were likely the placoderms. But another very early line, the elasmobranchs, including the sharks, skates, and rays, are still with us, and these animals have rods. Although these animals have certainly evolved and are very different from the earliest representatives, the living elasmobranchs can help us understand the state of the eye at the time, the twists evolution follows, and the origins of rods in the visual witness. One such story illustrates the issue.

Elasmobranchs

Blue-Spotted Fantail Ray, Taeniura lymma

Evolution has discovered interesting shapes and colors to fill the potential niches of the earth, and rays are one of the most creative examples. Taeniura lymma, the blue-spotted fantail ray that lives in the shallow waters of tropical reefs, is a member of the Dasyatidae family of stingrays (Figure 9.8). Although it harbors a potential sting for the unwary, this shy ray will use it only in defense.

Like the other elasmobranchs, rays diverged from the vertebrate lineage before the bony fish were established.

Unlike teleosts, which have fixed pupils, most elasmobranchs, including rays, have highly mobile pupils. T. lymma displays a pupil that is conventionally round when conditions are scotopic (black-and-white or low-vision light levels) but a different shape when conditions are phototopic (color-vision light levels).

Figure 9.9 *Rhinobatos typus* eye. Shovelnose ray. The pupil is covered by this fringed operculum that creates multiple pinholes in a crescentic pattern when the pupil is constricted. This creates a mechanism for measuring depth of field with just one eye. The Scheiner's disc principle is responsible and is discussed in Chapter 15.

A flap, known as the operculum pupillare, *protrudes from the dorsal pupillary border that covers almost the entire pupillary aperture. As the operculum extends in response to strong illumination, it creates a crescent-shaped pupil, eventually leading to two small slitlike openings with one toward the front of the head and one toward the rear (Figure 9.8). As well as dramatically limiting the influx of light, multiple pupils may create a rangefinder effect, as the images from the two pupils will come to focus on the retina from only one point in the surround. This effect will help an animal determine distance to an object. Pupillary shape and morphology in this taxon are extremely varied, ranging from multiple pupillary apertures in some skates and rays to vertical or diagonal slits in some sharks (Figures 9.9 and 9.10).*

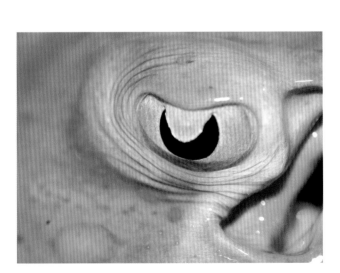

Figure 9.8 *Taeniura lymma,* blue spotted ray. The pupillary flap is called the operculum pupillare and can almost completely cover the aperture but with dilation is completed retracted.

Figure 9.10 *Heterodontus galeatus,* crested horn shark. The oblique pupil can close completely during sleep as the sharks do not have lids as such.

Accommodation (the process of changing focus between distant and near objects) is not accomplished in elasmobranchs by lens deformation (as it is in reptiles, birds, and mammals) but with a protractor lentis muscle. This muscle pulls the crystalline lens away from the retina and accommodates for near vision (see Figures 10.7 and 10.8). By contrast, the retractor lentis muscle in the eyes of ray-finned and lobe-finned fish accommodates for distance vision (see Figures 10.7 and 10.8 and Appendix E, Figure E2).

This is a conceptual difference, suggesting that the primitive creature that gave rise to the elasmobranchs and perhaps earlier to the placoderms had little or no accommodation at least initially. And this suggests that accommodation was a trait acquired when the elasmobranchs arose and would vary depending on niche requirements. (Appendix E provides more detail on accommodation in different species.)

Some rays, among other animals, possess a feature that allows them to focus on distant and near points at the same time. The feature is a multifocal retina, also known as a ramped retina, which is a conceptual and morphological oddity in which the posterior segment of the eye is curved in the shape of a vertical half of a pear, rather than an orange (Figure 9.11). It is almost as if the dorsal portion of the asymmetric eye were stretched upward or dorsally. This means that portions of the dorsal retina are further from the lens as compared to portions of the ventral retina, which has a more circular curvature. The dorsal retina will image the inferior field and is, in essence, myopic for the inferior field by being further away from the nodal point of the lens. To focus an image on this elongated

dorsal segment of the eye, the crystalline lens must also be asymmetrical and egg-shaped.

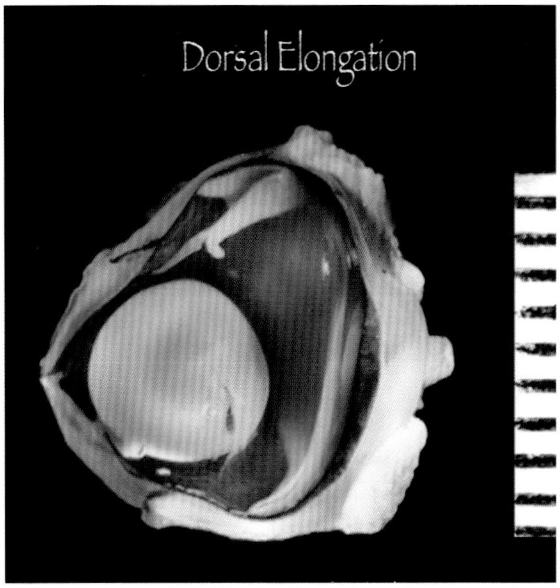

But there is one other remarkable feature of the elasmobranchs' visual system that is an important step in the assembly of the eye.

The elasmobranchs are probably the oldest extant taxon with rods as part of their visual system. One of the genes that had produced opsins that radiated into at least five distinct spectral classes in the jawless lampreys developed into the rod opsin found in most vertebrates. Because both bony fish and elasmobranchs have the opsin genes found in the remaining vertebrate lineages, their last common ancestor was probably the locus of this transformation. This transformation does occur in other lineages throughout evolutionary history, so it is not unique and may happen rather easily. Nevertheless, elasmobranchs are probably the oldest living lineage that has done so.

Many rays, particularly T. lymma and others inhabiting shallow reefs or coastal waters, have three spectrally different cones (trichromacy) as well as the rod visual pigment. So it is likely that the visual pigments for color and black-and-white vision were already present and evolving in the Ordovician and Silurian periods 500–430 mya. That would not necessarily provide good vision, however. Compared to humans, rays probably do not have good spatial acuity, although they may have a specialized area of better vision in their retina.

Early elasmobranchs not only had color vision and cone photoreceptors; some, if not many, probably had trichromacy. Furthermore, the eyes of elasmobranchs illustrate an important step in understanding the evolution of scotopic (nocturnal or lower light levels) vision in vertebrates.

Sharks are elasmobranchs as well, and they have perfected other sensory mechanisms. Although these mechanisms were developing by the time of the Silurian, they probably continue to evolve even now. Sharks manage to integrate their sensory input including olfaction, vision, temperature, proprioception, hearing, and even the sense of touch in ways we can only imagine.

Scalloped Hammerhead Shark, Sphyrna lewini

The class Chondrichthyes, which arose in the Silurian about 430 mya, includes the sharks, rays, and skates, all of which have a cartilaginous skeleton. The last common ancestor that radiated into the Chondrichthyes would later lead to the bony fish. The bony fish would follow and later divide into the ray-finned and the lobe-finned fish, and the latter would eventually led to tetrapods and terrestrial creatures. Extant cartilaginous and bony fish have similar eyes with many more similarities to, than differences from, each other, which provides a window on the sequence and state of ocular evolution at the time.

Figure 9.11 Dasyatis americana, southern stingray. In this sagittal cut, note the elongation of the vertical half of the eye creating a "pear" shape to the globe. Bar is in millimeters. Image by Richard Dubielzig, DVM.

Figure 9.12 *Sphyrna lewini*, scalloped hammerhead shark. Note the nostril to the left of the eye in this image. This postmortem specimen has a concave cornea instead of convex. *Image © Stephen Kajiura, PhD.*

Sharks always draw a crowd because of their commanding presence and predatory lifestyle, and the hammerhead shark evokes further interest because of the bizarre morphology of its head. What purpose would such a head serve and how does this creature manage its sensory input?

Sometime during the Oligocene (34 to 24 mya), the hammerhead family (Sphyrnidae, having eight species) arose within the elasmobranchs as an evolutionary anomaly (Figures 9.12 and 9.13). The narrowed, flattened dorsal-ventral aspect of its head creates what is called a "cephalofoil." This unusual head design probably evolved for improved sensory perception to include an exquisite sense of smell, the ability to sense and animal's surrounding electric field, and an improved visual field.

Some sharks have a thin cornea (approximately 160 μm thick) with epithelium (surface skin cells) making up one third of that thickness (Figure 9.14), and rudimentary endothelium (internal lining cells) on the undersurface. By comparison, the human cornea is 540 μm thick with a more derived endothelium. This suggests that the primary (primitive and initial) cornea was enough in some elasmobranchs and that the secondary layer we will meet later did not evolve until later fish appeared. The sharks, though, evolved a protective covering from an extension of the lower lid, much like a nictitans or third eyelid seen in terrestrial vertebrates, although this is not a true nictitans as compared to those of more derived vertebrates.

The retina of S. lewini contains rods and cones. Better-studied, related sharks (including the lemon, Negaprion brevirostris, and the silky, Carcharhinus falciformis) have a rod-to-cone ratio of between 5:1 and 12:1, which means these species have approximately the same ratio as humans (13:1). Investigators have shown that several species have eyes and visual fields well suited for twilight and daytime vision. This suggests that rods appeared early in the vertebrate lineage and probably represents evolution from one of the cone opsins.

The visual sensitivities of such sharks seem well adapted to match their visual environment and preferred niche. Recent work by McComb et al. suggests that hammerheads possess surprisingly large areas of overlapping visual field and binocular vision. These binocular visual fields are sometimes larger than those of more normal-looking sharks. For example, the wing-head shark (also in the hammerhead family) has an overlap of nearly 50°. And these sharks may move their heads from side to side as a behavioral compensation for an enlarged blind spot directly in front of them. Those skills imply that these hunters

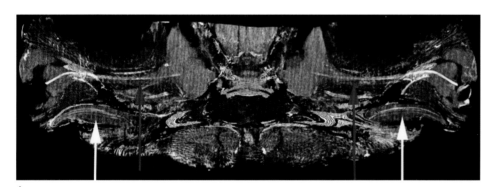

A

B

Figure 9.13 *Sphyrna lewini.* **(A)** Computerized tomographic (CT) scan of the head of *S. lewini* reveals large olfactory bulbs (white arrows) and long optic nerves (blue arrows). Soft tissue is false-colored to be a burgundy, and the cartilage has been colored a light yellow. *Image by J. Anthony Seibert, PhD.* **(B)** Computerized tomographic scan of the same head of *S. lewini.* All soft tissue has been removed digitally, leaving only the cartilaginous skeleton of the head. *Image by J. Anthony Seibert, PhD.*

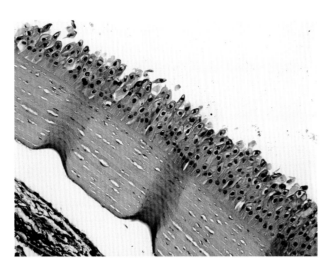

Figure 9.14 *Negaprion brevirostris*. Lemon shark cornea. Note the thin cornea and very thick epithelium or skin covering for the cornea. The endothelium is absent from this section but has probably been lost in preparation. Nevertheless, the endothelium is primitive and not completely responsible for maintaining corneal clarity. *Image by Richard Dubielzig, DVM.*

are equally well suited to the bright sun-drenched reef and the dim recesses of deeper waters. These skills have all developed as part of this species ecology.

The hammerhead's optic nerve may run a foot or more from the eye to the brain, on the dorsal aspect of the "hammer," with neither cartilaginous canal nor much protection (Figure 9.13). The chiasm is completely crossed, as it is in bony fish, suggesting that the last common ancestor in the shallow Ordovician seas had a similar anatomy.

The optic nerves lead to the optic tectum, where the signals combine with other sensory signals. The optic tectum in sharks (and other nonmammalian species) is homologous to a similar structure in mammals called the superior colliculus. In mammals the structure contributes information to the visual cortex, which does much of the visual processing. But in nonmammalian vertebrates, such as sharks, the optic tectum is the primary visual processing center.

The optic tectum also receives nervous input from auditory, mechanoreceptive, electroreceptive, somatosensory, and facial sensory nerves. The high degree of sensory input suggests that these creatures are very much tuned in to their surroundings with magnificent sensory perception and one very "cool" head.

The eye, as an optical instrument, must have become refined in later periods, as some superb piscine adaptations probably did not appear in the Silurian. Nevertheless, some of the astonishing optical refinements at this stage illustrate the range of possible adaptations, suggesting that the eye was surprisingly well adapted by this time.

The Silurian ended with a whimper when compared to the end of the Ordovician. This extinction process was probably caused by climate change, perhaps initiated by cometary impact. This blended transition would introduce the Devonian period, which saw the expansion of bony fish and equaled the Cambrian in the expansion of all manner of fish. The age of fish would be upon the earth.

THE PISCINE EYE MATURES

PALEOZOIC ERA

EARLY DEVONIAN PERIOD

415–398 MILLION YEARS AGO

10

The Silurian slid into the Devonian without the mass extinction seen at the end of the Ordovician, but the earth was changing nonetheless. Oxygen levels were rising, and the larger continents were moving south, leading to an eventual cooling of the earth toward the end of the Devonian. Continents were pushing apart as well, creating deeper pelagic habitats and new shallow freshwater aquatic habitats. These new niches would harbor new animals and encourage new modes of perambulation.

Bony Fish Appear

Unlike the rapacious placoderms, the bony fish had internal skeletons but no heavy, external armor, so they could be more agile and faster than the placoderms.

Some of these bony fish sought refuge in the shallows of the warming continental shelf seas, and especially in freshwater estuaries adjacent to the landmasses. There, with some potential to escape predation, these animals could survive and evolve.

We will review several living species that illustrate how evolution has solved the visual problems that would have faced fish of the Devonian or of modern times.

Described as the age of fish, the Devonian saw the rise of two major groups of bony fish we have met briefly before—the ray-finned fish (*Actinopterygii*) and the lobe-finned fish (*Sarcopterygii*). Here we get better acquainted. The ray-finned fish consist of two subgroups, *Teleostei* and *Chondrostei*. The first and larger subgroup includes the teleosts, which we consider most important. These fish, including tuna, salmon, cod, and the like, make up most of our commercial fisheries. The second and smaller subgroup includes the sturgeon. The eyes of both subgroups of ray-finned fish are similar, indicating that the most recent common ancestor of these two groups had such an eye.

Although the Devonian is described as the age of fish, our current age, with 28,000 known species, could better be so described because it probably greatly exceeds the number of fish found in the seas at the height of the Devonian.

During the Devonian, there was an explosion of the diversity of fish species, all with the basic piscine eye. A review of its basic anatomy will illustrate the development of the eye by this approximate time.

Anatomy of the Piscine Eye

Fish eyes are relatively and absolutely large. The camera-style eye is lined with a retina similar to ours but with different methods of nourishment to the inner and outer retina among other differences.

The Outer Coats of the Eye

SCLERA After the scleral coat evolved, as found in lampreys (and the craniate, the hagfish), the cartilaginous cup and the scleral ossicles (the overlapping bony plates embedded in the connective tissue of the sclera) mentioned earlier evolved to strengthen the eye wall, and the initial outer coating became established (Appendix D). The trial among the placoderms of a bony casing within the eye was over, although the bony scleral ossicles may be a remnant of that attempt. For example, large tuna have an eye that is nearly the size of an orange and contains demilunar ossicles, which provide additional support for the anterior portion of the globe. These are heavy periocular bones that form a "napkin ring" around the whole anterior segment of the globe (Figure 10.1).

CORNEA Most aquatic animals do not rely on the cornea for refraction because it has virtually the same index of refraction as water, and as a result, the cornea is flat. The primary cornea is retained from its original appearance in hagfish and lampreys primarily as protection. But this was not sufficient. Evolutionarily, as vision improved, the ocular "cyst" consisting of a sclera and primary cornea, an inner choroid, a retina, and a crystalline lens became more united with the overlying dermal layer. A second layer of clear corneal tissue appears and is now known as secondary spectacles. This extra corneal layer can be found in many species (Figure 10.2). In some fish these two layers are not well fused

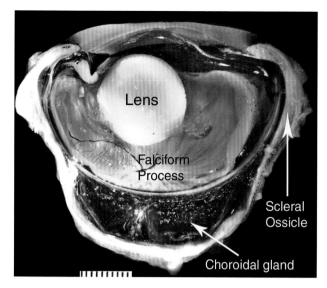

Figure 10.1 *Thunnus thynnus.* The demilunar ossicles are labeled, as is the falciform process. Note the large lens for such an eye. The scale is in millimeters. Also see Figure 10.2. *Image by Richard Dubielzig, DVM.*

and have a potential space between them. Optically, this double-layered cornea is not especially good, as it does not have to be (Figure 10.2). Some fish, though, make use of this space. For example, the pufferfish has taken advantage of the potential space between the two separate corneal layers (described briefly in Chapter 7 and Appendix D).

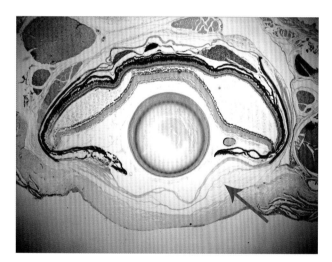

Figure 10.2 *Mastacembelus armatus*, zigzag eel. Some fish have two separate divisions of the cornea, and often these two layers do not unify. Much as in the lamprey there are a primary cornea and a secondary spectacle. Both layers are clear but have a potential space between them. This space can be seen clearly in this image (blue arrow). Some fish use this space to deposit pigment, creating sunshades in a bright reef environment (Appendix D). *Image by Richard Dubielzig, DVM.*

Spiny Puffer, Diodon holocanthus

Pufferfish, members of the family Diodontidae, comprise nearly one hundred species of mostly epipelagic fish, so named because they can expand their stomach to nearly a hundred times its original size, resulting in marked enlargement of the body (Figure 10.3).

The spiny puffer illustrates at least two peculiar twists in the development of the visual witness. It has orbital (bony) eye-shades that help block direct rays from the sun. It also has corneal pigmentation that reduces glare during diurnal foraging in its shallow-reef habitat. Unlike many other fish species that have corneal pigment, the spiny puffer has corneal iridescence, a rarer form of pigmentation. Some fish that have corneal coloration actually have pigment granules that migrate in fine channels between the two corneal layers with the stimulus of increased light.

Corneal iridescence has not been extensively studied in puffers, but theirs resembles that of the eyes of several other fish with coloration created by stacks of multiple layers of connective tissue or collagen fibrils. These layers or plates provide constructive interference and reflect a specific wavelength of light. They are particularly effective when the angle of incidence of light is oblique, in the way that bright downwelling sunlight would appear in a shallow reef. These reflected rays have a characteristic multicolored appearance because only a restricted band of wavelengths is reflected for any single viewing angle. Investigators have calculated that the iridescent layer in the fish they have studied provides a significant increase in underwater visual range without sacrifice of sensitivity.

Other fish have modified the cornea as the need arose, indicating that the cornea can be easily molded in an evolutionary sense.

Figure 10.3 *Diodon holocanthus,* puffer fish. Note the pigment flecks in the superior half of the cornea.

The Flyingfish, Hirundichthys albimaculatus

Flyingfish occupy an unusual niche (Figure 10.4). This winged animal feeds in the ocean's surface, which is rich with plankton but dangerous. Predators can use the sky as a backlight, patrolling beneath the surface of the water column looking for prey. So a surface fish must be primed to effect a hasty escape. To do this, the flyingfish has chosen to "fly," which means it must be prepared to navigate epipelagic waters as well as the aerial world. It does actually become airborne for significant distances, using pectoral fins as wings for gliding.

Exhibiting an "S" trajectory, the flyingfish genera have a spectrum of flight capability that can be quantified based on the ratio of wing area to body mass (Figure 10.4). This ratio probably correlates with visual capability out of water. In the Sargasso Sea, at least one species of flyingfish prefers consistent sites for launch and reentry, so some element of visual recognition is required both in water and air, especially for takeoff and reentry.

To accomplish this recognition, the flyingfish has evolved a pyramidal cornea with an apex at its center. This creates a three-windowed cornea, unique in the Phylum Chordata. One window of the cornea is directed posteriorly, one anteriorly, and one inferiorly. It is logical to consider that the flyingfish needs clear lines of sight posteriorly for possible predators, inferiorly for gauging distance to the water's surface and possible predators, and anteriorly to look for a safe place to land. The flyingfish has been documented to be emmetropic (neutral) in air and only mildly hyperopic (farsighted) in water.

In an examined specimen the cornea was large, with a diameter of 28 mm, 8 percent of the animal's total length of 36 cm. The axial diameter of the eye in that specimen measured was 19 mm. Essentially, the cornea extends almost to the equator of the globe, and much of the outer ocular coat of the eye is cornea.

Certainly, few animals would proceed in any medium without the sensory mechanism to interpret their surroundings, especially if they are moving with any speed. Because the species has not developed other sensory mechanisms to manage problems such as aerial obstacles, reentry sites, and predators, vision through these separate "windows" is a logical sensory mechanism. But, our understanding of the sensory modalities of the flyingfish is incomplete.

The Lens

CRYSTALLINE LENS The piscine crystalline lens and its position reveal much about ocular evolution, and the crystallins (the water-soluble proteins that fill the individual cells that comprise the lens) reflect the morphologic and molecular evolution of the eye.

Koran Angelfish, Pomacanthus semicirculatus

Native to the Indo-Pacific coral reefs, the koran angelfish captivates us with its colorful designs and can help us understand the crystalline lens and its role in accommodation (Figure 10.5).

All of the refraction in the piscine eye comes from the almost perfectly spherical lenses (Figure 10.5), which protrude through the pupil to nearly touch the cornea.

The koran angelfish (Figure 10.5) reveals the iris and lens adaptations for binocularity that most predatorial fish have adopted. Like many others, this fish has a pyriform-shaped pupil whose narrower portion is directed toward the mouth (Figures 10.6 and 10.7). A close look at the image reveals an edge

Figure 10.4 *Hirundichthys albimaculatus,* flyingfish. *Image by Robert Pitman, NOAA.*

Figure 10.5 *Pomacanthus semicirculatus,* koran angelfish. This alluring fish has rather "flat" eyes in an anterior-to-posterior direction but still has the cornea protruding beyond the profile of the body. Note the cornea and the lens within the eye. The anterior portion of the lens protrudes beyond the iris.

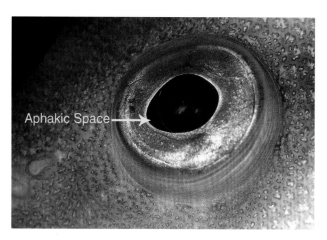

Figure 10.6 *Aphakic space*. Note the lens edge seen in the pupil. Hence, light travels through the eye without passing through the lens for focusing. *Image © James Brandt, MD.*

of the lens visible within that pupil. The space in front of the lens's edge—the "aphakic space"—has no focusing element. The lack of crystalline lens as a focusing element means that the image will be unfocused as it passes through this space.

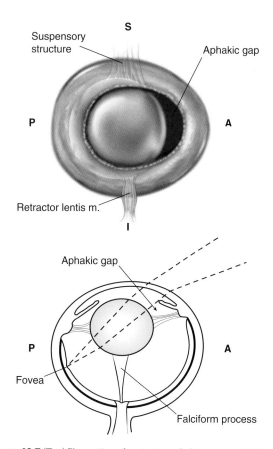

Figure 10.7 (Top) Illustration of a piscine aphakic space with the crystalline lens illustrated behind iris. (Bottom) Illustration of ray tracing through aphakic space to fovea near periphery of retina. *Art by Tim Hengst.*

A most interesting retinal adaptation explains this aphakic space. The space is created because of the notch in the pupil and allows an image to pass through and be focused by the lens near the anterior edge of the pupil (Figures 10.6 and 10.7). To achieve binocularity within the design limitations of an eye on either side of their heads, predatorial fish evolved a very temporal fovea near the ora (the peripheral edge of the retina found anteriorly near the lens of the eye). When the koran angelfish finds prey, it rotates the eyes into the most forward position possible. The image crosses the anterior notch in the pupil, through the lens, and strikes the temporal foveae in both eyes, allowing binocularity and perhaps stereopsis (Figures 10.6 and 10.7). Without the space, the nasal edge of a circular pupil would interfere with that transmission.

The fish eye grows throughout life, and even in the same species, a larger fish will have a larger eye. If the enlarging globe is not to outgrow the position of the retina and temporal fovea, the retina must also continue to grow, and it does. Stem cells at the periphery of the retina continually make new photoreceptors to match the growth of the eye. This retinal growth corresponds exactly to the eye growth so that the fovea remains in the direct line of the primary rays to maintain properly focused binocularity. What requirements have driven this adaptation?

Fish have evolved to be hydrodynamically thin with laterally placed eyes. To be a predator, though, requires that both foveae work together for binocular or stereoscopic vision. Angelfish and other predatorial fish have evolved a morphology that permits binocular foveal image capture despite their lateral eyes. Retinal stem cells and a pupillary notch in fish illustrate evolution's alternative solution to a preexisting design limitation.

ACCOMMODATION In bony fish, lenticular accommodation is accomplished by contraction of the *retractor lentis* (one or two muscles) (Figure 10.8 and Appendix E). In almost all fish this muscle originates at the distal end of the falciform process (a vascular structure extending from the optic nerve nearly to the lens) or at the ventral inner surface of the eye (the pars plana analogue) and inserts into the periphery of the lens capsule (Figures 10.1 and 10.8; accommodation cladogram, Appendix E, Figures E1 and E2).

In most bony fish, when the muscle contracts, the lens retracts toward the retina with both an axial and nasal component (Figures 10.6–10.8). In many fish the muscle extends from the falciform process to the inferior pole of the lens (Figure 10.8). On the opposite superior pole, the lens is suspended from the internal aspect of the eye (the pars plana), and this combination allows the lens to be retracted posteriorly for accommodation. This method of accommodation means that fish have a "neutral point" at near and

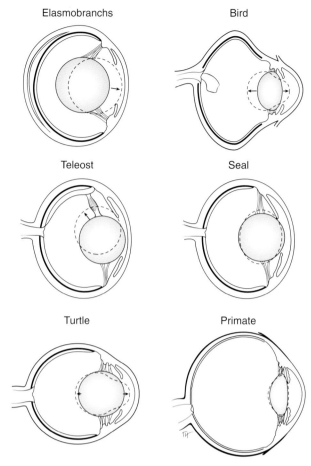

Figure 10.8 Accommodation panel. Accommodation is the ability for an animal to change focus from distance to near or vice versa. Elasmobranchs and teleosts as well as amphibians (not shown) accommodate by lens movement. Birds, seals, turtles, and mammals accommodate by lens deformation. Among those examples shown, only teleosts accommodate for distance. Further description in text and in Appendix E. *Art by Tim Hengst.*

change focus for distance vision. Mammals accommodate in a reverse fashion, and most change the lens shape rather than move it. Movement of the entire lens in fish is rather clumsy as compared to lenticular deformation. Hence, piscine accommodation is neither fast nor particularly accurate, but it is useful enough for fish. Fish do not require a fine degree of accommodation to focus on a distant object, as most of their world is at near. Distance vision is often limited in water anyway because of turbulence or debris or loss of light.

MATTHIESSEN'S RATIO Fish, with few exceptions, have an interesting optical relationship within the eye. Matthiessen's ratio describes the distance from the center of the lens to the retina as 2.5 times the radius of the lens. The cornea is optically neutral, so the crystalline lens must provide all the refractive power for the eye and will consistently obey this ratio. If the lens always has the same index of refraction, and it usually does with some variation, and if there are no other elements in the visual axis, the piscine eye can be described in terms of this ratio and will describe the size of the eye. Almost certainly this ratio was established very early in the evolution of the aquatic piscine eye because it equates to focusing of the image.

Matthiessen's ratio predicts and describes the size of most fish eyes, but a few fish have fulfilled their evolutionary destiny with lifestyles that are on the edge and require different lens styles. In some ways these lenses will not satisfy Matthiessen's ratio. The "four-eyed" fish are just one such group.

Cuatro Ojos, Anableps anableps

Most animals have evolved eyes that focus either in air or in water. Some have even evolved eyes that can do either at different times, but Anableps anableps *and two other closely related species are among the few to have evolved a design that can focus in both media simultaneously.*

Anableps anableps *seem to be so odd as to be unbelievable. Although they are often described and named as having four eyes, actually they do not. They have two eyes that on first glace appear to be four eyes (Figure 10.9A–C).*

A. anableps has evolved the necessary adaptations to prosper as intertidal specialists in fresh, brackish, or even pelagic waters.

The family of Anableps *species as represented by* A. anableps *have similar environmental needs for large, multiple eyes that see simultaneously in air and water (Figure 10.9). As "cuatro ojos," their Spanish nickname, suggests, they have four eyes with one pair for aerial and one pair for aquatic viewing, but in reality, they have but two eyes. Each of the two eyes divides into aerial and aquatic halves that at first glance appear to be two separate eyes. The division corresponds to a finger-like projection of the iris that divides the pupillary aperture into two. This overlapping projection of the iris corresponds to the pigmented ridge on the cornea (Figure 10.9A–C). The iris and corneal ridge create separate pupillary apertures for both the aerial and the aquatic images. When dilated, the two separate pupils form a single, dumbbell-shaped aperture.*

The lens is pyriform and asymmetric with the more rounded, circular portion of the lens in the ventral half of the eye corresponding to the aquatic portion and the flatter portion in the dorsal half corresponding to the aerial portion (Figure 10.9B,D). As a result, the lens diameter is smaller along the aerial visual axis than along the aquatic one. In equatorial cross section the lens is oval instead of round and has the accommodative mechanisms of anterior-posterior movement along the visual axis of the pupil (as do lenses in other fish species).

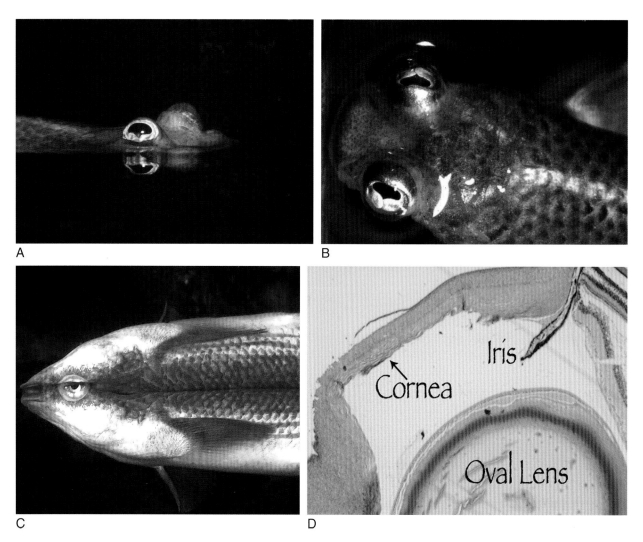

Figure 10.9 *Anableps anableps.* Four-eyed fish at the surface. These fish live at the surface with the top half of each eye protruding out of the water and the lower half looking into the water column. Note the unusual pupil seen in A, B, and C. Note that there is a pupil above and below the waterline. D shows the histology of an oval lens with a dilated pupil. Most piscine lenses are spherical. *Histology by Thomas Blankenship, PhD.*

The inner portion of the retina (nuclear layer) in the ventral half of the globe, subserving the aerial field, is thicker and contains more interconnecting (bipolar) cells than the dorsal half. And, the ventral half of the retina has perhaps 30–50 percent more cells in the innermost layer leading to the brain (ganglion cell layer) when compared to the dorsal half of the retina. The ventral and dorsal retinae are also distinct in that there is a separation between the two halves that subserve the aerial and aquatic fields. But input from these two retinae combine into a single optic nerve, which receives projections from each separate retina.

This must be confusing for the brain. How is this reconciled? Electrophysiological methods have been used to map the area that receives the signals of both the aerial and aquatic visual fields. That area is called the tectum (or optic tectum). The visual fields have been found to be similar to those of other freshwater fish except for the horizontal band above and parallel to the water line. That band has a greatly enlarged magnification in the tectum that correlates with the increased ganglion cell density in the ventral retina. In other words, the horizontal region above the water's surface is magnified, and the detail is enhanced, in the Anableps family of fish. This is somewhat analogous to our fovea. The threshold for movement is lower in the aerial visual field than in the aquatic field and seems correlated with a much higher cone density in the ventral as compared to the dorsal retina.

This unusual fish has a most unusual set of photoreceptors with nine cone visual pigments and one rod visual pigment for a total of ten different visual pigments in the photoreceptors. These visual pigments' peak sensitivities range from ultraviolet to blue, green, and red. Although this distribution is not completely understood, this fish may be adapted to two worlds with one world of aerial vision being bright and clear and an aquatic one that is murky and dimmer. The different visual pigments may make vision in these two different worlds easier. Some of these cones contain oil droplets, although their function is not completely understood either. Such oil droplets probably contribute to the ability of discrimination. Oil droplets may also expand the range of colors perceived by an individual cone or restrict glare, although any of these reasons is difficult to prove.

BOX 10.1 OIL DROPLETS

Many diverse species of vertebrates, including fish, frogs, certain lizards, turtles, birds, monotremes, and some marsupials, all have an unusual anatomical accessory called an oil droplet within the photoreceptive cell. These droplets, found in the body of the cones but not in the rods, can be brightly colored or colorless.

Colored oil droplets decrease overall photon capture, shift the peak sensitivity of the cones to wavelengths longer than the peak absorbance of the visual pigments they contain, and narrow their spectral bandwidth. By doing so, these brightly colored droplets are believed to reduce glare, minimize chromatic aberration, remove unwanted short-wavelength sensitivity, and participate in the range of color vision. The clear drops may be involved with the perception of polarized light, although this is not completely understood.

Composed of dietary carotenoids, oil droplets represent short-wavelength-cutoff filters. The presence of an oil droplet between the photopigments and the incoming light that strikes them restricts the sensitivity of that photoreceptor. Depending on the species, the droplets' distribution varies among retinal areas.

The current distribution of oil droplets across phyla suggests that these biological devices preceded the radiation of ray-finned and lobe-finned fish but are not found in any elasmobranchs. Many of the ray-finned fish do not possess oil droplets, although some more evolutionarily basal ray-finned fish, such as the sturgeons, do possess them. It is not completely clear when these structures were lost or never gained in most fish. The lobe-finned fish retained their oil droplets, at least enough of them to transmit these structures to the tetrapods, for they can be found in many of the terrestrial lineages (Figure 10.10).

Anableps eats insects above and below the water line using the more magnified and detailed area of vision corresponding to this portion of the visual field. They may also leap out of the water to attack aerial insects and use their aerial vision for this purpose.

Anableps feeds on the tidal flats, sifting through the mud with unusual teeth that act as a sieve. The aerial vision with the superior cornea and superior half of the lens provides for excellent acuity and is used to find prey and to alert the species to predators—such as birds. They must maintain a clear cornea to retain good vision and hence must keep their cornea moist. Because, like all fish, Anableps has no eyelids, they will frequently submerge the aerial eye to wet the corneal surface.

The crystalline lens is the principal refractive element of the piscine eye and must be as clear and as precise as possible. Evolution has found a remarkable adaptation to this problem.

Although the bluefin tuna did not arise in the Devonian, Carboniferous, or Permian, the *Scombroidei* (the suborder that includes tuna) did. Some of the predatory Scombrids developed a most unusual mechanism to cope with the effects of cold water and as well as perfected vision. This tuna helps us understand how.

Bluefin Tuna, Thunnus thynnus

The bluefin tuna, a wide-ranging pelagic predator with enviable visual mechanisms, is an endothermic marvel of the piscine world, having one of the best crystalline lens optics with perhaps the sharpest focus in the entire vertebrate world.

Most fish are poikilothermic, meaning the temperature of their bodies is the same as that of the sea around them. (By contrast, mammals, being endothermic, maintain a steady higher body temperature.) A lower body temperature may penalize such fish with muscular sluggishness, neurologic lethargy, and slower, poorer visual abilities.

To solve this problem, some piscine predators have evolved ingenious thermal mechanisms, perhaps none so curious as those of the bluefin tuna, which is nearly at the apex of the food chain.

Bluefin tuna are partially endothermic. The predatory tuna species sustains body core temperatures that average 23–26°C, which is sometimes as much as 21°C above ambient levels. Although these are not mammalian levels, they are sufficient to ensure that in colder water, the predators will be energetic, especially when compared to their prey. Several specialized features, including a countercurrent heat exchange, enable them to isolate the core of their bodies thermally and eliminate the loss of heat at their gills and external body surfaces.

Swordfish, certain species of mackerel, and some sharks have some features of endothermy, although the strategies of each

OIL DROPLETS

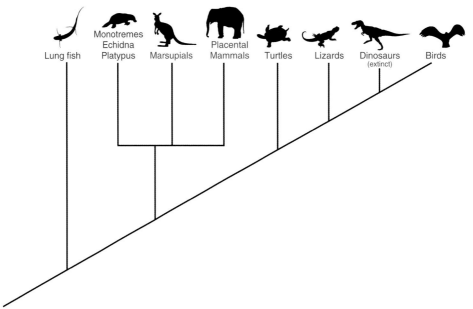

Figure 10.10 Oil droplet cladogram. Oil droplets are found in the photoreceptors of some primitive ray-finned fish and the lobe-finned fish. Only the radiation of the lobe-finned is shown in this cladogram, but some few ray-finned fish radiated from the same lineage and still possess oil droplets. Within the terrestrial tetrapods, some lineages have lost their oil droplets, and these can be seen in red. Lineages that retained the oil droplets in their photoreceptors are in black, including some dinosaurs.

are different. This would suggest that the ability to adopt endothermy has evolved at least twice in teleosts and at least once in elasmobranchs. Each time, this evolution coincided with the species moving into colder waters, providing a locomotory advantage over any prey species that could not warm its musculature, its brain, and its eye.

The bluefin tuna may range 25,000 miles or more during a year, and it dives as deep as 1000 meters, and both skills require superb visual mechanisms. The increased body temperature of the bluefin (and other species of tuna) improves muscular activity by increasing endurance and enhancing power output for both cruising and bursts of speed, and this, too, will require excellent vision.

Weighing up to 750 kg, the bluefin tuna needs a crystalline lens that can be more than 2.5 centimeters in diameter. A lens of this size risks spherical aberration because the periphery of the lens supports light transmission. And spherical aberration would be significant if the lens were of one uniform index of refraction. Aquatic vision requires that the crystalline lens be almost entirely responsible for the total dioptric power of the eye. Spherical aberration would represent a significant visual problem requiring very special lenses.

The tuna's solution is a lens that has a higher index of refraction in the nucleus and a lower one toward the periphery.

The spherical crystalline lens grows throughout life (as much as one thousandfold, in some species). As the lens grows and adds new lens fibers peripherally, the more central layers slowly

lose their water content, and the fibers become more compact. The index of refraction of the peripheral lens fibers changes from about 1.38 to 1.56 as they become more compact. This dynamic mechanism will virtually eliminate spherical aberration, even in a large crystalline lens such as the tuna's.

Although this lowering of the index of refraction from center to periphery is seen in most other fish and many nonaquatic animals, it is never so important as in the tuna, given its crystalline lens whose enormous size would make it particularly prone to spherical aberration.

The crystalline lens, however, was not the only ocular structure that would need to change to address the ecological challenges faced by the bony fish. The extraocular muscles were inherited from the early vertebrates with little or no change from the placoderms. These would play important roles in many fish. But why have exraocular muscles at all? What purpose do they serve?

EXTRAOCULAR MUSCLES Evolutionarily, vision had become the principal sensory mechanism, especially for distant prey and predators. But for both, a visual panorama or visual field, not just a single point of fixation, is essential for concealment or hunting. Muscles move the eyes to increase the total visual field or to permit visualization of a portion of it without moving the body or head.

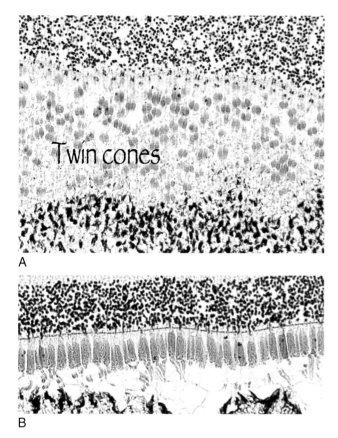

A

B

Figure 10.11 *Thunnus thynnus,* Atlanta bluefin tuna.
(A) Vertical and somewhat tangential cut of a tuna retina. This cut
transects the photoreceptor nuclear layer at the upper portion of the
image, the twin cones that are labeled, and the retinal pigment
epithelium at the bottom of the image. The twin cones are identical,
and this variety of photoreceptors is seen only in fish. See Appendix A
for further orientation. **(B)** This vertical cut of the retina from inner to
outer retina shows twin cones in the middle of the image with the
photoreceptor nuclei directly above the cones. These photoreceptor
nuclei are in the outer nuclear layer (Appendix A). At the bottom of
the image, the pigmented cells are retinal pigment epithelial cells.
Images by Richard Dubielzig, DVM.

Many prey species (fish, reptiles, birds, and mammals)
have eyes on the side of the head. Others (amphibians,
turtles, and some fish) have moved their eyes to the top of
their head. In all, the extraocular muscles further assist in
enlarging the visual fields, so the visual field imperative
may have increased the impetus for development of the
extraocular muscles. (Further discussion and explanation
of the evolution of the extraocular muscles and the semicircular canals can be found in Appendix B.)

In vertebrates, muscular movement has become a key
mechanism for bringing the area of best visual acuity (usually the fovea) to focus on a particular object. And the center
of gaze, which must be moved, is often done symmetrically.

Neurologically, though, the extraocular muscles have other
functions.

If an object continuously shines photons onto the same
set of photoreceptors, these cells will "fatigue" of that
image, and the image will essentially disappear. Because of
this potential, evolution has provided fine micromove-
ments of the eye called microsaccades, which are fine, fast,
almost continuous eye movements that prevent most
vertebrates from absolutely steady gaze on any object.
These movements are so fine that they are almost invisible
to us and could be considered like a very fine tremor.

Six (of the seven potentially present) extraocular
muscles are involved in creating these microsaccades. The
seventh muscle—the retractor bulbi, present in many
vertebrates—is used to pull the eye into the head to help
frogs swallow and lizards to protect themselves, among
other uses.

Some fish have co-opted these or some of these seven
muscles for other purposes, as evolution often does. The
Northern Stargazer helps to illustrate the plasticity of
extraocular muscles and where evolution can take them.

Northern Stargazer, Astroscopus guttatus

*Astroscopus (Latin for "star seer") has evolved extraordinary
uses for its extraocular muscles by developing an electric organ
from them (Figure 10.12).*

The electric organ of A. guttatus *is composed of portions
of four extraocular muscles, including the analogue for the
superior, medial, and temporal recti as well as the superior
oblique. The organ begins to develop when the larva is*

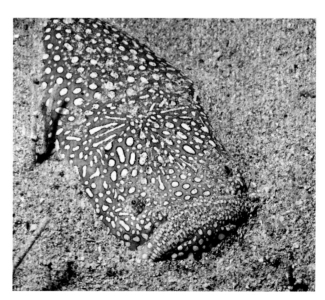

Figure 10.12 *Astroscopus. Image © David B. Snyder; marine
biologist, CSA International, Inc., Stuart, Florida.*

12–15 mm long. At this stage, the four muscles begin to grow to six times the size of the normal extraocular muscle cells. These muscles contribute the outer layers of each to form a syncytium (a multinucleated mass of cytoplasm contributed by many cells that have lost their cellular membranes). This syncytium becomes an electric organ that is histologically different from the ones in other electric fish and is distinct from the extraocular muscles.

Located behind the eyes and innervated only by the oculomotor nerve (cranial nerve III), this electric organ can discharge up to 50 volts. It consists of approximately 200 layers, whereas the more powerful electric organ of the electric eel (Electorphorus electricus), *Astroscopus's distant cousin, may have 30 times as many. Astroscopus's discharge does not appear to be strong enough or timed well enough to stun prey. But the electric organ is used to frighten off a potential predator and to attract and even capture prey. Talk about "a withering glance."*

Extraocular muscles and the nerves that innervate (stimulate) them may develop new functions relatively easily, just as *Astroscopus* has shown. Other fish, especially those that live on the edge, find other uses for their extraocular muscles.

One of the most illustrative is *Periophthalmus*, a south Pacific native.

Silverstripe Mudskipper, **Periophthalmus argentilineatus**

Although it is epipelagic and amphibious, Periophthalmus argentilineatus, *in the goby family, spends nearly all of its time on the exposed mudflats of African, Asian, and Polynesian tropical mangrove swamps. This requires some ocular adjustments that include the extraocular muscles.*

Much like a periscope, Periophthalmus' *eyes can be raised and lowered (Figure 10.13). To raise the eyes for spotting prey, two stiffened extraocular muscles (inferior oblique and inferior rectus) form a hammock-like structure. Conversely, when these retractile eyes are drawn closer to the body for protection, puckered folds of skin (pseudolids) cover them, as a method to wet the corneas.*

Ocular motility in Periophthalmus *is unusual as these fish have both stalked eyes and the ability to move the eyes in all directions as if they were on a turret. Because the eyes are stalked, the direction of gaze can even be directly inferior. Eye position can vary from both eyes in the frontal plane, permitting perhaps 10–15° stereopsis, to back to back in the sagittal plane (imaginary plane bisecting the body vertically from the dorsal surface to the ventral surface), creating two 180° fields above, below, and on each side of the fish. This helps to overcome the lack of head motility that necks provide most terrestrial vertebrates. The third nerve (oculomotor) innervates the inferior*

A

B

Figure 10.13 *Periophthalmus argentilineatus,* mudskipper. **(A)** Head-on view of the fish showing the elevated position of the eyes. **(B)** Lateral view with head and one eye visible.

oblique and inferior rectus to elevate the globe, although this is a substantially different innervation pattern from other vertebrates. (See Appendix B for more information regarding the extraocular muscles.)

Other changes have occurred in this fish because of its semiterrestrial niche. P. argentilineatus has powerful and rapid accommodation (for a fish), because of the iris's functional sphincter. In most fish, the iris has little or no musculature, and the pupil is usually fixed. The sphincter of the pupil of P. argentilineatus allows constriction and probably plays a role in accommodation in this fish (Figure 10.13). The lens is spherical and provides emmetropia for the fish during aquatic forays. When the eye is not accommodated—when it is used for near vision—the lens protrudes substantially through the pupil as it does in most fish.

The cornea has changed in this fish as well, as it is steeply curved and much more like that of a terrestrial vertebrate than a fish. This suggests that the steepening curvature we will see in frogs and reptiles is relatively easily accomplished on an evolutionary basis.

The retina is structured to help its owners feed during the day. Cones populate the retina, almost exclusively, at least in the inferior half of the retina. In the superior half, the retina gradually transitions to more numerous rods, but the cones still represent 80 percent of the photoreceptors of the entire retina. This retina will allow diurnal feeding, as the excess light of the mud flat will not dazzle the fish. With such a retina, though, the fish is limited at night and will tend to seek its shallow burrow in sand at night. The retinal pigment epithelium (the layer underlying the retina) is thick and pigment-rich; it presumably protects its photoreceptors from the constant light exposure since the fish has no true lids. At its peak in the fovea, the retina contains 225,000 photoreceptors and 90,000 ganglion cells per square millimeter, a much greater number than in many predatorial fish. This concentration of photoreceptors rivals that of the human fovea.

So mudskippers greet an ebb tide with elevated eyes and by breathing air through their skin as they prowl the sunny mudflats. But with the flood tide, they lower their eyes and settle into their burrows to breathe water with their gills, like any other fish.

Other species have used their extraocular muscles for different purposes to fill their particular visual niche. The billed fish would not evolve on earth for another 300 million years or so after the end of the Devonian, but they illustrate the plasticity of the extraocular muscles and how evolution may co-opt structures in surprising ways.

Striped Marlin and White Marlin, Tetrapturus albidus *and* audax

"Brain-freeze headache" describes the cephalic sensation that comes from drinking a supercooled beverage too quickly. This odd sensation represents vascular and neurologic compromise from the sudden temperature reduction. Any predatory fish that would move from warmer epipelagic depths to near-freezing ocean depths would face a formidable challenge. If the fish is poikilothermic, such changes would impede muscular activity as well as neurologic and visual mechanisms, including photoreception, as mentioned for the tuna. The visually directed predatory billed fish, including the four species of marlins (Figure 10.14), the swordfish, and the sailfish, all solve these problems with unique evolutionary adaptations (Figure 10.15).

As all visual predators do, T. audax *relies on stealth and speed in varying degrees and on visual mechanisms that are usually superior to those of its prey. But the billed fish also need these processes to function in all prey habitats, including cold and dark waters.*

Figure 10.14 *Tetrapturus albidus*, white marlin. Growing to over 4 meters long, this spectacular fish may weigh more than 900 kg. These marlin are principally pelagic fish and feed on prey in epipelagic waters but also dive to feed in much deeper waters. *Image © Masa Ushioda/SeaPics.com.*

T. audax *and especially swordfish spend as much as twelve to nineteen hours of each day at depths of 300 to 700 meters. At the deepest of these depths, the water temperature is a "brain-freezing" 2–8°C. Such temperatures impede neurologic, visual, and muscular physiology, so predators that live or feed permanently at these mesopelagic depths must accept these limitations or have adaptive physiology to improve their performance.*

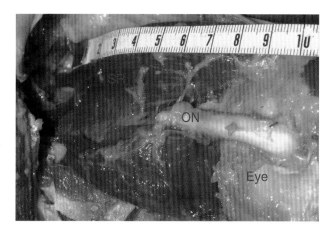

Figure 10.15 *Tetrapturus audax*, striped marlin. Orbital dissection of the right orbit of T. audax. Note the huge extraocular muscles. The eye is the size of a tangerine, and the muscles are very large. SR, superior rectus; ON, optic nerve. The scale is in centimeters.

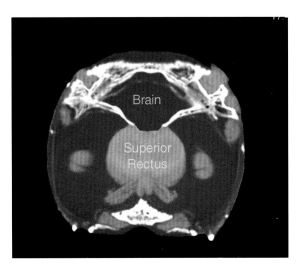

Figure 10.16 *Tetrapturus audax*, striped marlin. Computerized tomographic (CT) scan of marlin head. Note the combined superior rectus from each eye. These two muscles are very large and adjacent to the brain and just anterior to the eyes.

T. audax *and the other billed fish have evolved the most functionally distinct and enlarged extraocular muscles in the animal kingdom (Figures 10.15 and 10.16). These muscles possess a mass of special fibers directly beneath the brain. These fibers have lost the myosin and actin fibrils (the proteins that permit muscle contraction) of neighboring extraocular muscle cells. But, in a dramatic illustration of the plasticity of muscle, the fibers have elaborated more enzymatic and mitochondrial engines for energy production per cell than any other skeletal fiber known. The mitochondrial volume as a percentage of cellular volume is two to three times that of most active aerobic muscle cells.*

Because the energy produced cannot activate muscle activity, heat is created and dissipated to the vascular plexus that permeates these powerful energetic knots. These cells have evolved thermogenic tissue at the expense of contractile ability. The circulation comes directly from the carotid artery through the gills where it is oxygenated and directed to these "heaters." There, the blood is warmed considerably with a countercurrent system and passed directly through the choroid of the eye and then through the vascular system of the brain.

Even the belly of the superior rectus muscle coincident with these "heaters" has direct access to the brain because the muscle belly at this point has a very thin overlying bone between it and the meninges (linings of the brain) (Figure 10.16). This allows for improved potential energy capture and for the brain to maintain temperatures 20–30°C higher than the surrounding water temperature, facilitating smooth neurologic and visual operation. Possessing eyes the size of tangerines, this visual predator requires good vision and a consistent ability to see movement at its hunting depths. Maintaining the elevated temperature of the retina and brain for faster and more accurate

processing of the visual signals preserves these skills. So the brain remains alert, and the eyes' ability to see movement remains intact—important skills for a predator at this depth.

RETINA The retina, the inner layer in the eye, has evolved since the ray-finned and lobe-finned fish arose, but the similarities and differences help us understand the composition of the retina of the last common ancestor. The original and novel adaptations and innovations reveal the evolutionary forces at work in the piscine eye. Four of the five cone visual pigments from the lamprey and the rods from the elasmobranchs remain in many fish. The neurologic organization and arrangements remain similar, indicating that these were well developed when the lineages of bony fish arose. Nevertheless, new designs appeared in response to the problems presented by the new niches.

PHOTORECEPTOR DESIGN Fundamental photoreceptor design evolved in at least a subset of ray-finned fish, as many of these fish have paired cones in two different arrangements: (1) double cones with one cone being distinctly larger then the other, and (2) twin cones with both elements being identical.

RETINAL VASCULARIZATION The inner layer of the eye—the retina—probably evolved rather quickly in the early to mid-Cambrian in tandem with the choroid

---

BOX 10.2 DOUBLE CONES
Double cones, found in most fish, reptiles, birds, and the monotremes, are two unequal cones supplied by a single axon leading away from the cell. The longer and slimmer cone, which frequently has an oil droplet, is described as the chief cone. Independent of that oil droplet, the accessory cone of the two cones may have an enlarged paraboloid, an oval droplet in the inner portion of the photoreceptor cell, which contains glycogen. The paraboloid is involved with cellular metabolism and serves as a nutritional reserve. Double cones often appear in a standard pattern, like a square checkerboard. The function of such paired cells is unknown, although there are hints that these cells assist the fish in assessing the speed of a viewed object or perhaps improve discrimination or even improve acuity. But this is not known.

Twin cones, found only in teleost fish, are two similar cones that are joined for much of their length and supplied by a single axon. Twin cones do not usually have oil droplets (in either cone). In most fish these cones contain the same opsin and respond to the same stimulus, but in some fish they may contain different opsins and respond to different stimuli. In some species, twin cones outnumber single cones. Speculation on function centers on motion detection or improved discrimination, but the true purpose is unknown.

surrounded by the sclera as previously discussed. During embryology, the sclera grows to enclose the eye dorsal to ventral—top to bottom. The line of closure is "zippered" closed ventrally or inferiorly. This "fetal fissure" never quite closes completely in some fish, leaving a small gap. The falciform process (a plexus of blood vessels) grows through it, as if squeezed out, from the underlying layer of blood vessels—the choroid. The falciform process is essential for retinal nutrition and as an origination site for the muscle of accommodation in some fish. Most teleosts have a falciform process, but it varies greatly in size, shape, and distribution. Some fish have an almost vestigial falciform process, and a few, such as batfishes and puffers, have none at all, only a network of "hyaloid" or "vitreal" vessels (Figure 10.17) distributed across the surface of the retina. (See Appendix C for further discussion of the vascular supply to the inner retina.)

RETINOMOTOR PIGMENT MOVEMENT Many species of bony fish evolved, perhaps very early, another interesting physiological mechanism that is found sporadically

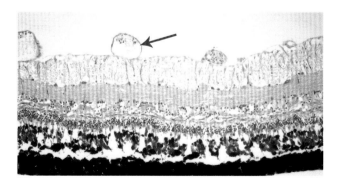

Figure 10.17 *Pygocentrus natterri.* Piranha retina. Note the blood vessels on the surface of the retina. The blue arrow points to one of the vessels. *Image by Richard Dubielzig, DVM.*

throughout Animalia. They (passively) sheath their retinal rods as daylight approaches and expose them at night. The retinal pigment epithelium (RPE) is a layer of cells deep to the retina and closer to the sclera (Appendix A). The photopigment portion of the retinal cells sits in the RPE cells whose long "arms" extend around the photoreceptors. At night, these arms gradually empty of pigment granules. As daylight approaches, the pigment granules rise and surround the photoreceptive portion of the rod photoreceptor. This protects the more light-sensitive portion of the rod during daylight and preserves it for use at night. This pigment exchange averages twenty to thirty minutes.

ARGENTA The argenta, a layer of a reflective compound on the *outside* of the choroid (or, less commonly, outside of the sclera), probably developed early with the evolution of fish and could have been present in the Placoderms. Found in many living, less-derived fish, the argenta is composed of guanine, the same reflective material found in most fish scales. It is most useful for fry, or baby fish, many of which are transparent or translucent for camouflage. If not for the argenta, their pigmented layer of cells that are found behind the photoreceptors in the eye would create starkly black dots moving though the water, thus creating a target for predators. The argenta reflects the light so that the fry look like light reflecting off waves. The argenta even extends forward onto the iris surface and is responsible for the shimmering effect of the iris.

Neurology and Optics

When fish radiated into the shallow freshwater niches, they discovered vegetation and eventually prey at the water's edge, sometimes tantalizingly close but not in the water column. As Arthropoda came ashore and radiated into aerial insects, they became fair game for the fish that remained in the water. But the difference in habitats created a problem that piscine predators had to negotiate because when they look out of water, refraction occurs at the air-water interface. Some fish solved this problem through simple trial and error. Each generation relearns the solution. Other fish solve this by redesigning their eye.

Because water is 300 times denser than air, light travels more slowly in water and is refracted by the encounter. Because of scattering, an infinite number of rays strike the water surface from all directions, even when the sun is directly overhead. When light rays are nearly parallel to the surface of the water, light is reflected instead of refracted, and rays do not penetrate from that angle or any greater one. The refraction of light creates a cone of light penetrating the water at a well-recognized angle known as Brewster's angle, about 97° (Figure 10.18). The same phenomenon,

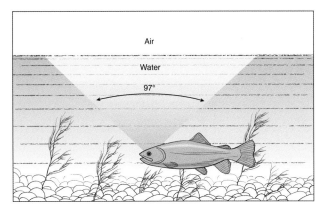

Figure 10.18 Snell's window. Any animal looking up from underwater has a limitation to what can be seen above the surface. That limitation is defined by Snell's window. This cone of visualization, approximately 97°, occurs because of the refraction of light in water. Further discussion can be found in the text. *Art by Tim Hengst.*

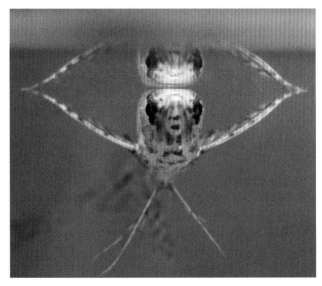

Figure 10.19 *Pantodon buchholzi*, African butterfly fish. This fish has been described as a freshwater flyingfish because it leaps out of the water as an escape behavior. It has a bizarre appearance for a fish, with distinctive dorsal and anal fins that act almost like vertical stabilizers. Its lateral, or pectoral, fins are especially wide, like butterfly wings. Although it is a strong leaper and may jump to a height well beyond its body length, it does not "fly" in the same manner as the pelagic species of flyingfish. *Image © William Saidel.*

seen below the water surface when looking up, is known as Snell's window, described by Willebrand von Snell in 1621. (This window, well known to SCUBA divers, can be seen as the bright circle of light directly above them as opposed to the darker waters outside of the 97° circle of light.)

At least one fish has managed the problem of simultaneous aerial and aquatic vision with the refraction of light with a unique evolutionary solution. The African butterfly fish has an asymmetric eye with two different retinae in the same eye. But these fish still have to learn to overcome refraction by trial and error.

The African Butterfly Fish, Pantodon buchholzi

Found only in West Africa, the African butterfly fish occupies a highly specialized ecological visual niche just below the water's surface. It is an obligate surface feeder and will prey on insects and other creatures that it sees on or above the water line. It must simultaneously see into the air and in water, yet it lives just beneath the surface (Figure 10.19).

To succeed in its niche, this tropical novelty has evolved a most peculiar eye with a set of intriguing adaptations that are reflected in its visual pathways.

Pantodon's eyes are most unusual and have adapted for simultaneous vision in air and in water. A falciform process composed of, and covered with, heavily pigmented epithelial cells extends from the optic nerve to the posterior capsule of the crystalline lens (Figure 10.20). This pigmented shelf divides the eye horizontally and delimits the dorsal and ventral retinae. The dorsal retina supports aquatic vision. The ventral retina supports aerial vision and has its own nucleus in the brain to which it sends fibers. Functionally, the falciform process prevents intraocular light scatter, particularly from the intense aerial light that could otherwise dazzle the ventral retina *(Figure 10.21).*

Study of the geometric organization of the retina reveals that the photoreceptors in both dorsal and ventral retinae are directed simultaneously along independent optic axes. Cones in both retinae are directed toward the pupillary aperture to exploit special properties of light. The anatomy and direction of

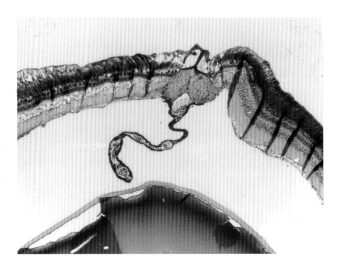

Figure 10.20 *Pantodon buchholzi*, African butterfly fish. Falciform process extending from optic nerve toward piscine crystalline lens (yellow arrow). *Image © William Saidel.*

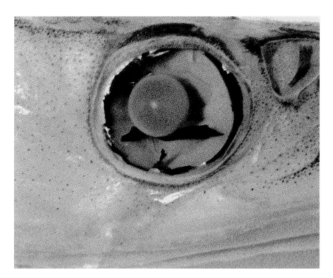

Figure 10.21 *Pantodon buchholzi* eye. The tip of the mouth in this image is to the right, and the anterior one third of the eye has been removed to visualize the position of the falciform process. In most fish this structure would be vertical and perpendicular to the linear axis of the fish. *Image © William Saidel.*

each cone capture the light in a tube effect called light guiding, which provides the brightest possible image—like fiber optics. This prevents the light from escaping as it travels along the length of the cone.

Cones in the ventral retina are narrower than those in the dorsal portion of the eye, suggesting improved aerial acuity as compared to aquatic acuity. This characteristic difference in cone width reflects an evolutionary adaptation within one portion of the retina to the different physical constraints of image formation in air and water.

But the brain connections of the ventral retina are the most interesting. Saidel and Butler have traced the fibers from the ventral retina to a distinctive (diencephalic) brain nucleus called the nucleus rostrolateralis (NRL). As an aerial or surface target is presented to one eye, the information is conveyed to the optic tectum (which processes visual information into images) and NRL.

These two brain areas and their complex neurologic connections make up part of a "feeding circuit" that includes identifying and acquiring prey. This circuit is recognized in only a few other teleost species among all vertebrates. A. anableps, discussed above, is one such species and must also be able to interpret simultaneous vision above and below the water line (Figure 10.9).

Pantodon's eye and visual system have evolved to the niche it fills and are closely tied to its behavior. The ventral retina and visual pathway are primarily designed for feeding, and the dorsal retina and its visual pathway are primarily for predator recognition and avoidance. Curiously, the ventral retina has different neurologic reflexes, but the dorsal does not, almost as if the upper and lower halves were different eyes. There is more

to the story. Most teleost species, including Pantodon, *have a lateral line in the skin surface along the sides of the animal. These are sensitive hair cells, similar to those in the mammalian cochlea in the ear. These cells are supreme mechanoreceptors, sensing vibrations and bioelectric currents. The wave action induced by the movements of an insect on the surface is sensed by the lateral line system of* Pantodon. *Much as pit vipers integrate infrared and visual stimuli (Chapter 19), the input of the lateral line is integrated with the visual input at the level of the optic tectum. So this marvelous fish actually "sees" with its lateral line, as this input probably goes to the equivalent of a visual cortex in these fish.*

The neurologic connections and mechanisms to control the extraocular muscles have been driven by vision, and in most fish vision has become the most important sensory modality. So bony fish that rely on vision have been tuned to maintain good acuity at almost any cost—witness the flatfish.

Sanddab, Citharychthys *Species*

Bilateral body symmetry in animals came very early, as shown by the metazoan imprints from the Ediacaran period (600–543 mya). Even now, symmetry represents a sign of fitness because it suggests biological success at a structural level.

But about 500 species of flatfish (of the order Pleuronectiformes) discarded such a predominant characteristic primarily with the eye and orbit. Instead, it has an asymmetric body that has been directed by its eyes. Its eyes-up, bottom-dwelling orientation allows it to spot and capture prey that frequent the near bottom and to avoid predators, which come from above.

Flatfish hatch from eggs into bilaterally symmetrical, transparent fry that live near the ocean surface more or less as plankton and have the same fusiform shape as most other fish (Figure 10.22). These young flatfish feed on smaller planktonic creatures high in the water column until their metamorphosis, which is vastly different from that of other species. During this process, flatfish become flattened, compressed, and asymmetric: One of the flatfish eyes, with corresponding bony structures, appears to migrate to the opposite side of the body to join the other eye. This fish, having transformed into a bottom-feeder, then lies on its eyeless side on the ocean floor.

This odd transformation also leaves only one anal and dorsal fin, modifies the gastrointestinal tract, lightens the color of the fish's downward-facing side, and changes the bony structures to accommodate the migrated eye. Although the transformation causes both eyes to migrate (at least to some degree), only the eye that changes sides gets a complete bony orbit, as the eye that did not migrate from the opposite side has an incomplete orbit (Figure 10.22B).

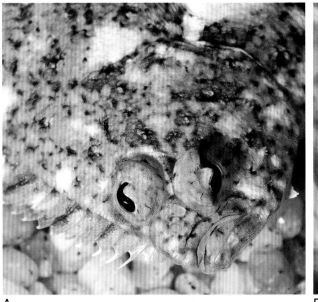
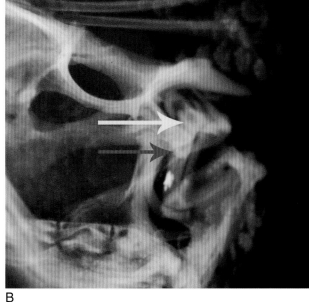

A   B

Figure 10.22 *Citharychthys* species. sanddab. **(A)** Note that both eyes are on the same side of the body. Note the projection of the superior iris—called an operculum. *Image © James Brandt, MD.* **(B)** Achirus lineatus, lined sole. X-ray of head of related flatfish. Note the incomplete orbit of the eye that did not change sides, marked with a blue arrow. The complete orbit of the eye that changed sides is marked with a yellow arrow. *X-ray performed by John McCosker, PhD, and thanks to the California Academy of Sciences.*

When compared to the extraocular muscles of other fish, the adult flatfish's are similar in their neuronal input and compensatory eye movements. Their eyes have the same spatial orientation to each other as before metamorphosis, although some changes in extraocular muscle sizes occur.

But little change occurs in the semicircular canals and otoliths of the vestibular system. This means that metamorphosis must also include substantial changes to the brain pathways in order to maintain a normal vestibulo-ocular reflex (brain to eye reflex circuit that keeps the eyes stabilized with movement—see Appendix B). Symmetry around the visual axes and visual fields is maintained. It is almost as if the eyes retained their position and the body of the fish rotated beneath them. To satisfy this extreme makeover, the brain rearranges itself in response, suggesting that it is visual symmetry that is ultimately important.

The transformation also brings change in the photoreceptors and visual pigments of the eyes. The fry have only one class of photoreceptors in the form of single cones. In at least some flatfish, the opsin in those photoreceptors of the fry has peak sensitivity at approximately 520 nm, which is in the "green" range. With no other visual pigment, these fish live their young colorblind lives in a monochromatic world, but because their environments are full of light, they have little need for rods. Like the cone distribution of the adults of many other fish, theirs concentrates these cells in the temporal aspect to permit the fish to visualize and discriminate objects, such as prey, directly in front of them.

Driven by the thyroid hormone, metamorphosis brings ecological change that the retina must accommodate. In a dramatic transformation of the retina, flatfish develop rods and three different varieties of cones, including double cones discussed above.

In the lower light levels of the ocean floor, rods would be required to see forms and movement. Additionally, in bright light, adult flatfish can change colors, at least to some extent, and will do so to match their background. Cones with visual pigments sufficient to permit color vision are required to interpret the color of the surround. Flatfish illustrate that the visual system commands the transformation of metamorphosis—at least part of it.

Habitat Expansion

As the number of fish species expanded with the basic eye as a principal sensory mechanism, they would find, explore, and exploit deeper, darker habitats. Different conditions required specialized eyes; we can review what extant lineages have done to meet these challenges.

As fish radiated into midwater (mesopelagic: 180–600 meters), deep water (bathypelagic: 600–3000 meters), and very deep water (abyssal: 3000–6000 meters), their eyes needed to adapt in several ways. The appearance of rods would be essential in eyes of fish that would swim toward the abyss.

Formation and refinement of the eye with rods permitted further diversification. These habitats have few if any occupants, so they would require stranger methods of viewing the world and different strategies for obtaining food and mates. For example, animals living on the ocean floor need not look down, and only rarely forward, so they may direct their eyes upward. Barreleye illustrates the dilemma and the solution.

Barreleye, Macropinna microstoma

The barreleye (Macropinna microstoma) is a member of the Opisthoproctidae *family, a bizarre collection of approximately eleven oceanic species that are mesopelagic to bathypelagic in habit. They evolved many unusual ocular adaptations to master this biologically forbidding environment.*

The barreleye is an odd, translucent (even to its visible brain) small fish with a maximum length of about 14 cm (Figure 10.23). Living at depths of 900 meters or more below sea level where darkness, extreme pressure, and numbing cold will winnow the weak, this translucent bottom dweller reveals that evolution can squeeze specialized vision from the most unusual habitats. Relatively enormous eyes and several adaptations allow them to maximize photon capture, prey recognition, and perhaps protection against predators.

To begin with, barreleyes can rotate their tubular eyes from looking upward to forward, which permits them to be benthic at rest but forward looking when hunting. The fish have large spherical pigmented crystalline lenses that are bright green (Figure 10.23). This pigmentation is an adaptation believed to improve acuity by reducing chromatic aberration of the shorter wavelengths. Some mammals, such as ground squirrels have yellow lens pigmentation for much the same reason.

Lens pigmentation, at least in humans, may also occur as an age-related degenerative phenomenon, resulting from ultraviolet exposure. But not all age-related pigmentation changes of the crystalline lens are degenerative. For example, starting in middle age, some mesopelagic fish (such as Argyropelecus affinis *or hachetfishes in the Sternoptychidae family) have an abrupt pigmentation of the lens fibers. These fish accumulate pigment only in the more peripheral lens cells and not the nucleus, which remains clear throughout life. And these carotenoid-like pigments are associated with the crystallins of the lens fibers and are not diffusible between cells.*

The unique lens pigmentation of M. microstoma *probably represents a peculiar form of convergent evolution because it seems that lens pigmentation has evolved several times and for different reasons. Furthermore, lens pigmentation is uncommon in mesopelagic or bathypelagic species.*

Although the pigment is freely diffusible throughout the crystalline lens of M. microstoma, *it has different spectral qualities from the pigment found in studied fish or terrestrial vertebrates. And any pigmentation will, at least to some extent, limit photon capture and be a hindrance at this depth, but it must have evolved for a reason.*

The green/yellow lens pigmentation has several functions. These lenses would filter out at least a portion of the downwelling light, especially on the shorter end of the wavelength spectrum (shorter than 450 nm). This would increase acuity by decreasing chromatic aberration—limiting short-wavelength light scatter within the eye—and by increasing contrast sensitivity by eliminating the "blue haze." But wavelengths longer than 450 nm never reach the eye at these depths, having been filtered out by the water column. So, such pigmentation would

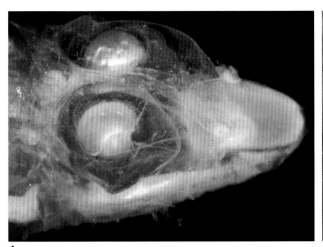

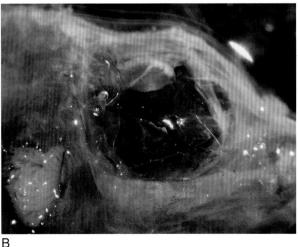

A   B

Figure 10.23 *Macropinna microstoma,* barreleye. **(A)** This specimen is looking up, but in life this fish can rotate its eyes to look forward. The yellow lenses are green in life, probably to break camouflage of its prey. **(B)** Lateral view with the eye looking upward. *Specimen courtesy of Scripps Oceanographic Institute.*

probably not offer enough evolutionary advantage to sacrifice bandwidth and photon capture.

But in a well-considered manuscript, Muntz offers evidence for a clever alternative. Many mesopelagic prey (and sometimes predator) species have photophores (bioluminescent organs) on their ventral surface. These organs generally produce light at approximately 475 nm, the peak wavelength of the downwelling light from the surface. So any prey species with photophores producing this wavelength would have almost perfect camouflage by not providing a silhouette that would betray its presence and path.

But Muntz predicts that this camouflage will not be effective against a predator that has a high-pass lens filter in the form of a yellow lens because this filter will make the photophores (or any other form of bioluminescence) appear brighter. And that is because bioluminescent light often has a broader spectral composition than the background (downwelling) light. Furthermore, if the photoreceptor visual pigment is tuned to a wavelength such as 480 nm or longer (and generally fish at this depth do have such visual pigments), this bioluminescent target will be even brighter and more visible over a greater distance.

Muntz also suggests that such predators must be looking upward to maximize such benefit. If this species has two visual pigments and, hence, does not live in a monochromatic world, these pigmented lenses would be of increased service by allowing the fish to distinguish the wavelength and intensity of a bioluminescent signal. M. microstoma (Figure 10.23) is looking straight up with very large eyes and green lenses. Bioluminescence is common among the smaller invertebrates that are part of M. microstoma's diet. Perhaps breaking the camouflage of these invertebrates is important to limit the time and space of the search for food and maximize capture in such dim and hostile environment.

The crystalline lens in this species and their close relatives is useful for other reasons because they approach what is called the evolutionary "quit" point for vision. There is no downwelling light below about 1000 meters, even in the clearest of ocean waters. So at these depths, some fish (and invertebrates) have begun using bioluminescence to communicate and even to illuminate, much like a torch.

Nevertheless, evolutionary pressures must respond to the lower light levels by making every photon count. So in mesopelagic and bathypelagic fish, the pupil is fixed and dilated, not constricted. The large lens is almost perfectly spherical, and the pupillary aperture matches the equatorial diameter, thus using the entire lens for photon capture.

Because the eyes look upward, at least some of the time, and the lens protrudes through the pupil even to the point of affecting the contour of the dorsal surface, the visual field could be predicted to be almost hemispherical. This tubular eye does have photoreceptors rising on the sides of the "tube." This ensures the sensitivity of that hemispherical field, including detecting the movement of any bioluminescent object, even if the far peripheral portions of the retina would be too close to the lens to be in proper focus.

Similarly, when the eye is rotated horizontally, a similar broad field would come into view. This fish could sight a prey, swim toward it by directing its eyes horizontally, and feed on it with a small fluted mouth that would suck the prey into its mouth. Extraordinary adaptations for an extraordinary environment.

Other fish also have dramatic adaptations that maximize photon capture at great depths. Perhaps, if an animal were in midwater, it would want to be able to look up and down. Light levels at mesopelagic depths (200–1000 meters) are limited, creating unique problems for predators and their prey living at such depths. Evolution has responded with profound adaptations. *Bathylychnops* illustrates the potential.

Javelin Spookfish, Bathylychnops exilis

The slender javelin spookfish (Bathylychnops exilis, Figures 10.24 and 10.25) is a translucent mesopelagic fish with evolutionary adaptations so unusual as to seem unimaginable. Specimens have been taken at depths as wide-ranging as 30 meters to 1000 meters. Evidence suggests that this fish is capable of nocturnal upward migration, especially among the juveniles, and that the adults may live at depths of 600 meters.

For this species, endemic to the northern Pacific Ocean, prey probably consists of small crustaceans or coelenterates. The javelin spookfish probably remains motionless until prey

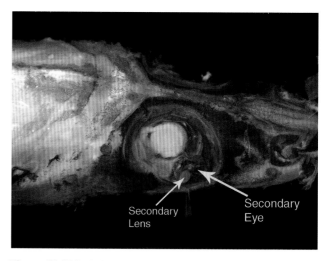

Figure 10.24 *Bathylychnops exilis.* Labeling illustrates the secondary eye and extra lenses in the sclera. The secondary lens is only a lens but not a complete eye. It is embedded within the sclera, as are two other similar masses along the lower edge of the eye. *Specimen courtesy of Scripps Oceanographic Institute.*

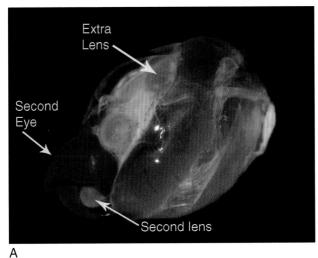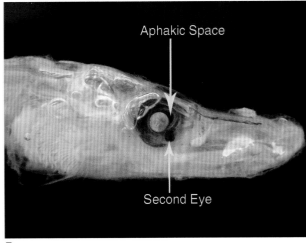

A

B

Figure 10.25 *Bathylychnops exilis* pathology. **(A)** Note the secondary eye with a lens and the extra lens within the sclera. There is another auxiliary lens immediately adjacent to the second eye. **(B)** Juvenile of the species with forming secondary eye but no extra lenses at this point in development. Note that animal is transparent with visible gills and brain. Note the aphakic space pointing at an upward oblique angle. *Specimen courtesy of Scripps Oceanographic Institute.*

comes within range. Then, with short bursts of speed, this predator catches and crushes the prey within its mouth.

The most startling anatomic aspect of this species is the smaller ventrally located eye that buds directly off the limbus of the primary eye (Figures 10.24 and 10.25). Both principal eyes are very large in proportion to the head size, an expected feature, as the animal is trying to maximize the capture of photons in a permanently nocturnal world. As is seen in many avian species, the septum between the primary globes is thin, and the orbital structures are rudimentary, suggesting that evolution has tried to maximize the relative and absolute globe size (Figures 10.24 and 10.25).

The head of this fish illustrates the translucent tissues that surround the eye (Figure 10.25). The eye shows the large primary and smaller second completely formed eye on the lower left portion of the globe (Figures 10.24 and 10.25). In adult specimens, the secondary globe may be as large as one half the diameter of the primary globe. This secondary globe is a retinal diverticulum with a completely formed lens and an independent cartilaginous coat resembling other fish eyes. The ganglion cells project their axons to the primary globe and the single optic nerve exiting it. The transparent retina allows light to pass through the secondary lens to stimulate cells in the retinal diverticulum, which lines the secondary globe and connects to the primary eye through a small slit in the primary globe. Covering the slit is an operculum (flap or cover) that blocks excess stray light that might scatter through the smaller secondary eye and enter the primary ocular cavity, although it is hard to imagine stray light at this depth.

The primary eyes face anteriorly and dorsally with a 35° angle with the horizontal. This provides a large binocular field directed somewhat vertically above the horizontal.

Two dense intrascleral masses along the lower orbital rim of the primary globe line up distal to the secondary eye. These structures resemble the secondary lens and may allow for photoreception directly parallel to the horizontal (which otherwise would be a blind spot for the more dorsally focused primary eyes). This would permit a complete panorama of the entire inferior 180° of visual field and probably most of the frontal field, although discrimination would be limited in much of that field.

The retina is believed to be an all-rod retina. Presumably this single photopigment is tuned to about 475 nm, the frequency of bioluminescence, as it is in most known fish predators at this depth. Other fish at this depth have only rarely evolved more than one pigment because the wavelengths of light that penetrate to this depth are so limited.

In fact, the quantity of light is so limited that the eye has to seek other methods of capturing photons to register a signal. One method is to enlarge the number of photoreceptors that contribute a signal to the brain. This technique is called summation and consists of adding many photoreceptive cells that contribute their signals to a single ganglion cell (this cell brings a signal to the brain for interpretation). This helps to create a positive signal but decreases the ability to discriminate. This technique will blur the vision substantially. Deepwater fish such as B. exilis exhibit a high degree of summation, and this will limit the quality of the image that goes to the brain.

The binocularity of the fish suggests that it is a deepwater predator. The optic lobes and the olfactory lobes are described as large and bulbous, and the acoustic system is also hypertrophic. (Because the dwindling light at that depth limits vision, other predators, especially at this depth, have evolved excellent olfactory abilities presumably to supplement that sense.)

Little functional significance can be ascribed to the secondary eye or additional scleral masses. Presumably, the secondary globe is used to locate food or to avoid predation, and the scleral bodies may increase the visual field by focusing light from a benthic perspective, if they are lenses at all. Because the primary globes are believed to be used for prey detection, the secondary eyes may be used to avoid predation from below.

But the evolutionary implications for this fish and its second pair of eyes are profound. Other deepwater fish possess retinal diverticula (side pouches) without a secondary lens. They often also have tubular eyes, described as "telescopic," to maximize light collection at mesopelagic or bathypelagic depths (Figure 10.26). Fish within the same taxonomic group and related to B. exilis *have retinal diverticula, but they do not have anything resembling this secondary eye, suggesting that evolution can and does produce eyes rather quickly—in no more than 1 million years. Perhaps* B. exilis's *second eye budding off its first represents its effort to extend the evolutionary quit point or to expand its scavenging abilities. This is powerful and profound evidence for those who would doubt natural selection.*

Extremes of morphology distinguish themselves in extreme environments. The mid- and deepwater habitats are particularly challenging for any animal but especially visual ones. Here is another example of an extreme modification of the eye to suit its domain.

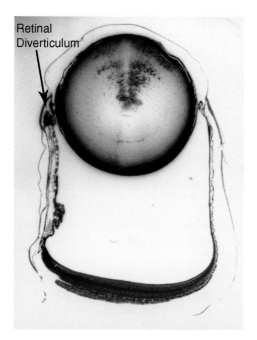

Figure 10.26 *Opisthoproctus soleatus.* A fish closely related to *M. microstoma, O. soleatus* is another Opisthoproctid and has a similar eye. Note the tubular design and a "pocket" of extra retina nasally, known as a retinal diverticulum. Its optics and visual capabilities are not completely understood. *Image © Shaun Collin.*

Mesopelagic fish (200–1000 meters) encounter challenges most other fish do not. Such fish must find and catch prey that are usually above them but can be anywhere. To complicate matters, they must also avoid predation that comes from all directions but usually below. All this happens in a very poorly lit environment where overhead shadows and winks mean everything.

Possessing a long, thin seemingly compressed body, Stylephorus chordatus *possesses a bizarre mouth-oropharynx combination (Figure 10.27). This animal can enlarge its inner mouth cavity up to thirty times or more with a hinged mechanism to protrude its lower jaw. It sucks the seawater in through a straw-like mouth creating substantial force and pulling small organisms in with the aspirate. The animal then closes its mouth-pouch to force the water out through its gills and trap the small copepods and bioluminescent shrimp.*

Like those of M. microstoma, *the lenses of Stylephorus are colorful, although probably yellow. Probably also like* Macropinna, *the lenses evolved to permit this fish to break the camouflage of bioluminescence. Like two extraordinary telescopes, the eyes look up to find the small bioluminescent shrimp in the dark blue above them (Figure 10.27). The tubular eyes have become elongated, but the sides of the sphere have been cut off, leaving only the center tube of what would have been a larger globe. The eyes have a large pupil and, like other fish, a large spherical lens. The yellow lenses break the camouflage because the downwelling light is different from the bioluminescence produced by detritus and small shrimp. That difference makes the shrimp stand out distinctly.*

The eyes have at least one other surprise—a retinal diverticulum or pouch adjacent to and attached to the tube as an accessory retina, very much like Bathylychnops's *eye, but this retinal outpouching does not contain a lens. The diverticulum*

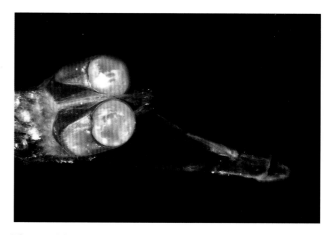

Figure 10.27 *Stylephorus chordatus.* Note the tubular-shaped eyes and the potential for elevating and depressing the globe. *Image © Danté Fenolio/www.anotheca.com.*

probably enlarges the field of vision narrowed by the tube-shaped eyes.

As we have seen, predation and protection from it are principal reasons for sensory improvement, but they are not the only ones. Mate selection enters into the evolution of sensory mechanisms, and because even finding a potential mate at extreme depth is challenging, novel solutions become necessities for survival.

Dragonfish, Malacosteus niger

Because light does not penetrate much below watery depths of 1000 meters, and then at that depth, only in the clearest of tropical seas, darkness continuously shrouds nearly two-thirds of the earth. And the sun's rays illuminate any earthly surface only during half the day. So evolution has selected many animals that make their own light.

As we have seen in the text, at depths of a few hundred meters or more, sunlight is restricted to a narrow spectral range of between 460 and 490 nm. Bioluminescence emitted by almost all animals at least at this depth has a similar spectral range with a peak at approximately 475 nm. Evolutionarily, the exception proves most interesting.

M. niger, the deep-sea dragonfish, has two different forms of bioluminescence (Figure 10.28). The retro-orbital light organ produces the usual bluish range of emission, but in addition to this organ, an infraorbital organ beneath the eye actually emits red bioluminescence (Figure 10.28) This is almost like "red-vision goggles." This unique appendage allows M. niger *not only to view prey that do not see these wavelengths but also to communicate with others of its species. It also helps* M. niger *live and compete in a sunlight-less realm, dominated by animals that produce and see bioluminescence in the "blue" range, 460–490 nm.*

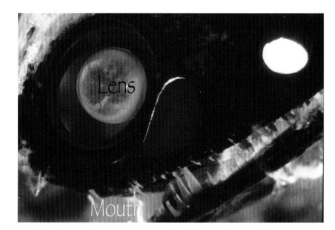

Figure 10.28 *Malacosteus niger*. Note aphakic space with a small spherical lens noted just above the red light organ. The mouth is labeled for orientation of the image. *Image © Justin Marshall, PhD, ARC Professorial Research Fellow.*

Such an unusual lifestyle requires several adaptations. M. niger, *and two other similar mesopelagic fish, are members of the three deep-sea dragon fish genera that have two photophores, one producing bioluminescence in the shorter wavelengths between 460 and 490 nm. This is a common spectral range for bioluminescence in deepwater fish. Nearly 80 percent of such fish produce light in this range almost exactly coincident with the spectral range of downwelling light from the sky. In addition, these three genera have bioluminescent organs that produce light with much longer wavelengths. For* M. niger, *that spectral emission peaks at wavelengths beyond 700 nm. (The longest wavelength humans can see is approximately at 700 nm, a color we would call deep red.)*

Emitting light with two restricted spectra requires different visual pigments and other adaptations to interpret those signals, and evolution has responded.

M. niger *has a heavily pigmented lens that appears yellow to our eye, at least when brought up from its home in the deep. The absorption profile of this unusual lens has two maxima—one at 429 nm and one at 460 nm, although the absorption appears to be lens size dependent and, hence, age dependent. As the fish and eye grow larger, the lens has a higher optical density at 429 nm. These unknown lenticular pigments restrict the shorter wavelengths, perhaps to enhance the perception of bioluminescence or to break the camouflage of it as discussed above.*

M. niger *has an astaxanthin-based (a carotenoid) or vitamin A–based red pigment retinal tapetum (eyeshine, Chapters 7 and 13). This deep-red tapetum (Figure 10.28) presumably will further enhance the sensitivity to the longer wavelengths that are emitted from the infraorbital photophore by giving the photoreceptors two chances to respond to incoming photons.*

But most fish at such depths, including M. niger's *predators, do not have the requisite visual pigments and cannot see the longer wavelengths at all, so the tapetal reflex does not risk exposing the fish to predator or prey.*

M. niger *has developed two visual pigments that respond to longer wavelengths but have only a maximum absorbance of approximately 517 nm and 542 nm. That would not seem to be high enough to maximize the reception of the red stimulus. Further testing reveals that the outer segments of the retina of* M. niger *do have a pigment that absorbs light in the longer wavelengths we would call red, although it does not send a neurologic signal on that basis.*

Douglas and others have shown that this third retinal photopigment is a photosensitizer, much as certain medications or foods act in humans. The photosensizing pigment is derived from chlorophyll. Investigators have convincingly shown that this chlorophyll-derived photosensitizing agent will respond to the longer wavelengths and result in excitation of shorter wavelength visual pigments, making it a special form of blue light. This photosensitizing pigment is coupled with the two true visual pigments to secondarily stimulate them to respond to the longer (red) wavelengths. This allows M. niger to use these longer wavelengths emitted from its own photophore to find prey or to communicate with its closest relatives

Although no vertebrate is known to produce chlorophyll, certain copepods contain it, and M. niger feeds primarily on copepods. Yet M. niger's chlorophyll is uncommon in Animalia, including in its prey. It is found only in certain green sulfur bacteria that live only in the subtidal marine environment, so it is not clear where M. niger gets it.

Evolution of the piscine eye has continued throughout the millions of years that followed the Devonian, and many unusual visual mechanisms must have come and gone. Most of these we will never know. One novel fish has appeared in the Cenozoic with an unusual predatorial style and unusual eye to match.

Sandlance, Limnichthyes fasciatus

Ambush predators possess the advantage of surprise, which, when coupled with speed, is devastatingly successful. Speed and accuracy often require excellent vision, which for vertebrates in particular usually means large eyes. But evolution has found creative, often unique, solutions to help smaller vertebrates (with smaller eyes) compete. The sandlance, Limnichthyes fasciatus, has just such a solution.

No longer than 20–40 mm, this benthic predator hides in the loose sand of the Indo-Pacific's shallow reefs. It will explode from its camouflaged hideout amongst the coral rubble in a brief and rapid attack on small planktonic prey.

Unlike most fish that have all their eyes' refractive power concentrated into a spherical crystalline lens with little or no refractive effect from the cornea, the sandlance has a very thick cornea representing about one-seventh of the diameter of the approximately 1-mm-long eye (Figure 10.29). Unlike any other fish, its cornea contains a lenticle (little lens) that functions as a lens. Using a striated retractor cornealis muscle (Figure 10.30), it can unleash approximately 180 diopters of accommodative potential in a fraction of a second. Although seen in many birds, corneal accommodation is unique to this fish in the piscine world. (The lamprey's cornea muscle is not used in the same manner or derived in the same way.)

The sandlance's crystalline lens, which is flattened, is also peculiar for a fish. The combination of a corneal lenticle with a flattened but hyperopic lens creates a magnified image on the fovea. Although the complicated optics is beyond the scope of this text, this arrangement permits monocular parallax and a form of depth perception. Most vertebrates, like us, must move their heads or even their bodies to gain monocular distance cues through parallax, and a robin or pigeon does so by bobbing its head, but the sandlance merely moves its eye to achieve the same degree of parallax.

Figure 10.29 Limnichthyes fasciatus, sandlance. Image © Shaun Collin, PhD.

Because of this unique system, the sandlance can function perfectly well as an ambush predator in a three-dimensional world with only one eye sighting prey and can dissociate its optokinetic (forced visual response to movement) response from one eye to the other.

The retina consists of a peculiar matrix of a single cone surrounded by four double cones. The ganglion cells are tightly packed with a foveal density of 150,000 cells per mm², among the highest densities of fish studied, rivaling birds' highly specialized retinae. The fovea is convexiclivate (steep-walled

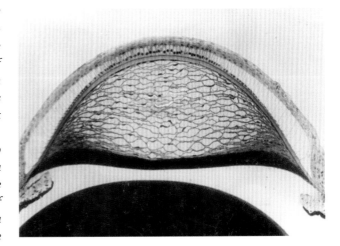

Figure 10.30 Limnichthyes fasciatus. Sandlance cornea. Note the intracorneal lens as well as a crystalline lens making a two-lens system. This, combined with a convexiclivate fovea, produces a telescope. Image © Shaun Collin, PhD.

with a deep central depression). This anatomical arrangement probably provides image magnification because of the difference in the index of refraction between vitreous and retina combined with the steep curvature of the sides of the foveal pit.

The animal burrows slightly below the sand with only its eyes exposed and waits, scanning with both eyes—independently. The eyes move independently and alternately (to avoid the simultaneous acquisition of different prey by each eye). When a planktonic prey species, such as a copepod, comes into striking distance, the sandlance wastes little time in its attack, which, at 4 meters/sec and ~200 body lengths/sec, rivals the strike of a chameleon's tongue. Coupled with an oral "cage" that allows the fish to engulf prey if it is not precisely on target or if the prey should move erratically, the sandlance rarely misses.

The sandlance's unique optics, which also resembles those of a chameleon, allows it to target prey monocularly.

Many visual adaptations favor the predatorial species enhancing vision for the hunt or capture. But prey species have contributed to the evolution of the eye by making recognition more difficult. Fooling the visual system is an old technique.

Piscine Camouflage

Camouflage continues to be a tool for prey species, as it probably has been at least since the Cambrian, but some of the known subtleties evolved in the Cenozoic. It is an important counterpoint and constant companion to our visual witness. Evolution has provided that mysterious sleight-of-hand ability to many successful herbivorous species. Yet the camouflage cannot be so good that it stills communication is within the species, limiting social interactions such as courting and mating. The parrotfish illustrates one of evolution's cleverest yet almost incidental solutions.

Bullethead Parrotfish, Chlorurus sordidus

The bullethead parrotfish uses the technique of pointillism to hide in plain view at a predatorial distance while keeping its harem in order at a closer distance (Figure 10.31).

Pointillism is a painting technique that uses small dots of "pure" colors to create an overall effect of "blended" colors when viewed from the proper distance. Theoretically, the technique increases luminosity to achieve an optical blending at the retinal level. Although Georges Seurat was not the first to use this artistic idea, he demonstrated it in his 1884 painting, A Sunday Afternoon on the Island of La Grand Jatte.

The technique relies on the concept of receptive fields (the neo-impressionist painters were unaware of this aspect of visual pathway organization). At a certain distance, a "pure" colored dot stimulates one or more cones, contributing to

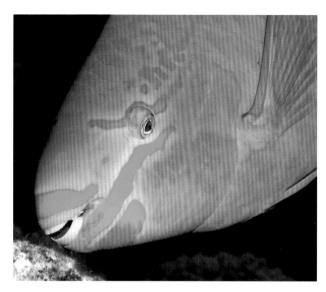

Figure 10.31 *Scarus vetula*, queen parrotfish. Note the complementary colors. This fish blends well in a colorful reef despite its bright colors. Look closely and you can see the lens protruding through the pupil. *Image © James Brandt, MD.*

a center-surround receptive field. The brightly colored dot is seen in its original color unless the painting is viewed from a more distant point.

With increasing distance, the adjacent dots stimulate the same receptive field with two or more colors. If those colors are complementary, the resulting neural signal will be an achromatic dull intermediate color—the reason many artists eventually abandoned pointillism. But becoming duller at distance is exactly what parrotfish need for survival.

Most male parrotfish and their kin perform a disappearing act by having two adjacent complementary colors on their bodies. When viewed at close range, the males appear bright and colorful, which is important for communication and mating. But when viewed from the distance of a predator cruising the reef, the combined complementary colors create the illusion that the fish is gray-blue, in essence disappearing into the backlight of the reef.

The parrotfish's disappearing act would work for any color vision system, but it is critically dependent on the resolution of the eye. Most reef fish, even predators, have relatively poor acuity, perhaps because it is not possible to see more than a few meters even in the clearest waters. From distances of as little as 1 to 5 meters, the colorful pink, blue and green parrotfish would simply appear to blend into the reef background as a form of biological pointillism.

The parrotfish family (Scaridae) is evolutionarily quite young, originating approximately 12 mya in what would have been a geologic precursor of the Indo-Pacific tropical waters. The transformation into the highly colorful and numerically dominant coral reef fish has been rapid with a profound selection pressure directed toward male coloration.

In reef fish, rods are initially seen in juveniles but are not found in the shallow-water larval forms because the larval forms do not need them. They increase in the adult forms as the fish settle into deeper waters. As the eye matures, additional photoreceptors are added, and the cones in particular become more organized to improve acuity. The adult fish has a retinal mosaic with single and double cones conferring a visual potential of approximately 10 arc minutes (human 20/20 vision equals a visual angle of one arc minute when viewing a spatial pattern), which is a considerable improvement over the larval and juvenile forms. Although these fish have a predominantly cone retina, they also have some rods for lower light. And at least juveniles may recognize ultraviolet and polarized light.

Finally, the last disappearing act to mention is a sexual one. If a harem loses its male, the dominant female of the harem will transform into a male, including the development of the bright coloration. To further mystify, this male can return to being a drab female if conditions require it.

As the fish were gaining an aquatic foothold in the seas, the arthropods were coming ashore. The vegetation was surprisingly plentiful and attracted the vegetarians including the clade that would be the insects. The arthropod predators including dragonflies and eventually spiders would follow as well. Animalia was enriching terrestrial life.

INSECTS ARISE TO FLY

PALEOZOIC ERA

EARLY DEVONIAN PERIOD

415–398 MILLION YEARS AGO

11

Animalia's expansion onto land during the Devonian began with the arthropods. They came following the vegetation. These vegan invertebrates helped to draw the predators ashore. Animals at the water's edge would have represented a food source, and their assumption of land was an important step.

Arthropods Come Ashore

The early Devonian (415–398 mya) was a dramatic time for both vertebrates and invertebrates. Although arthropods in the form of primitive mites and myriapods (millipedes) may have come ashore before the Devonian, all but one of the centipedes and all of the millipedes and related creatures (Myriapoda) have simple eyes and represent a separate lineage from the insects. Only the house centipede (Figure 11.1) achieved a compound eye, and it is quite rudimentary.

The arthropod lineage that led to the myriapods split from the lineage that would eventually lead to the insects

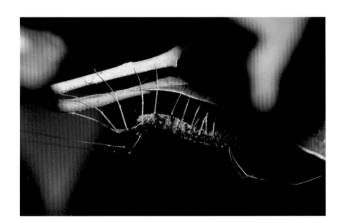

Figure 11.1 *Scutigera coleoptrata*, house centipede. The house centipede is a myriapod with compound eyes and is the only myriapod to have any eye beyond a simple eye. Even this eye contains a cluster of ocelli on each side of its head. These compound ocelli resemble a true compound eye, although they are less well integrated. The eye is a pseudofaceted eye, and such an animal may have been the link from ocelli to compound eyes.

perhaps as long as 500 mya, and this suggests that the arthropods came ashore more than once. Myriapods are not believed to be as closely related to the insects as are the crustaceans. Most investigators believe insects arose from a crustacean lineage, with these crustacean ancestors bringing their eyes along with them.

The first insects were probably vegetarians feeding on the alga mat spreading across exposed land during the late Silurian and early Devonian. These first terrestrial vegetarians were close ancestors of extant bristletails and silverfish, with fossils dating to at least 400 mya in the early Devonian. These pioneers had early and very basal apposition compound eyes as the first terrestrial insect eye (Chapter 6), which they retain today, although surely a somewhat refined version of the original. This morphology would require very little adaptation to see on land. But those insects that would follow the bristletail relatives jumped quickly from an aquatic environment to an aerial one.

Powered, winged flight lifted off in the early Devonian. Pterosaurs would not discover the skies for at least another 100 million years; birds would not fly for another 250 million years. Insects were the first aviators.

The aerial frontier forced further refinement. Very directly and early in the Devonian, arthropods arose from the thick medium of water to fly as insects in a much thinner medium. This would have required good vision.

Doubtless, several arthropodian species came ashore in a relatively short time, although little fossil evidence remains of these transitions. Superposition eyes were probably brought by one of the more crepuscular or nocturnal arthropod species.

The Superposition Eye

Early predecessors of the mayflies (and probably some other lineages of terrestrial insects) were the first to fly in the Devonian. Because these delicate creatures are basically crepuscular, it is likely that their ancestors brought the parabolic superposition and the refracting superposition eyes into the mid- to late Devonian's terrestrial environment.

In the early to mid-Devonian, insects began first tentative flight, although for the visual witness *how* flight developed matters little. Flight in insects is almost certainly homologous across all insects, starting from a single (extinct and unknown) common ancestor. Although mayflies, damselflies, and dragonflies begin in water and still spend most of their lives there, they illustrate the variety in early flight. Prey species, such as mayflies, first took to flight as an escape mechanism, and their predators, such as the dragonflies, soon followed.

In addition to a means of finding nutrition, flight also improves an insect's ability to find mates and to travel. But flight is challenging, requiring better vision and faster image processing. Given visual development in other animals, these early flyers probably quickly developed the visual abilities that flying demanded.

Mayflies, Baetidae *Species*

Mayflies, the most basal winged insect and somewhat less derived than dragonflies, were among the first, if not the first, winged insects to face the aerial environment on earth. Trace fossils from the Devonian do not provide evidence of the compound eye that came along from a crustacean background, but these animals have changed little over the ensuing 400 million years. Likely the parabolic superposition eye came along with them for the ride. We have reviewed this compound eye in some detail previously (Chapter 6, Figure 6.22), but mayflies exhibit an interesting ocular twist. As illustrated (Figures 11.2 and 11.3), this specimen has both parabolic superposition compound eyes and apposition eyes. And it is only the males that have the superposition compound eyes. Presumably these eyes have evolved to assist in finding the females. Leading ephemeral lives for a few days or even a few hours in some cases, mayflies must have tightly scripted lives to locate the opposite sex, mate, and lay eggs. The mass and synchronous emergence is a survival strategy in that predators cannot anticipate or consume all of the adults, as there may tens of millions of them.

The young, or naiads, will molt up to thirty times over months to years. These naiads are mostly aquatic vegetarians, but there are predatory species. Most of their lives, then, are aquatic, not aerial, and these species are closer to water than air or land. These ancient aquatic insects helped bridge the gap between water and land (Figures 11.2 and 11.3). The parabolic superposition eye found in the adult male mayfly illustrates how easily the transition occurs from apposition to superposition compound eye or, alternatively, illustrates that the superposition eye came with the crustacean ancestor to these insects.

The Apposition Eye

The crustacean transition from an aquatic life to an aerial one eventually produced the apposition eye with the largest

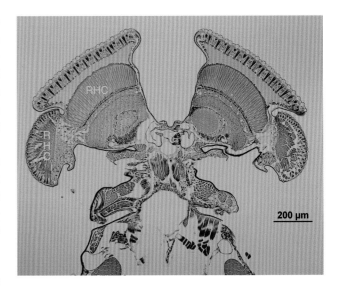

Figure 11.2 *Baetidae* species, mayfly. Note primary parabolic superposition eye found only in the male. The secondary eye lateral to each primary eye is an apposition eye. This illustrates how the different varieties of eyes may coexist in the same animal although the female only has apposition compound eyes. RHC, rhabdomeric cells in both sets of eyes. The parabolic superposition eye is described in Chapter 6 in more detail. *Image by Richard Dubielzig, DVM.*

number of ommatidia and the best acuity. Soon after the mayflies, other early adaptors, dragonflies and damselflies, would appear, and the dragonflies would perfect the apposition eye (Figures 11.4–11.6).

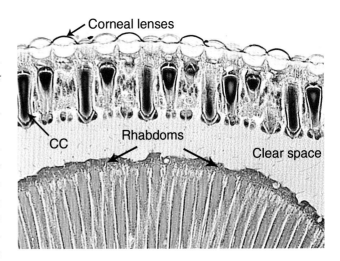

Figure 11.3 *Baetidae* species. Mayfly parabolic superposition eye with higher magnification. Corneal lenses, clear space, and rhabdoms are labeled. CC, one of the crystalline cones responsible for the unique nature of this eye. The parabolic superposition eye is described in Chapter 6 in more detail. *Image by Richard Dubielzig, DVM.*

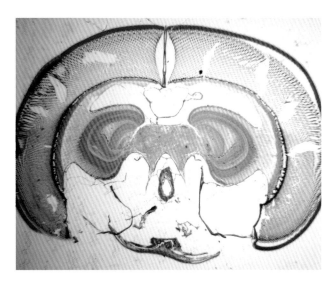

Figure 11.4 Dragonfly eye. Note the apposition compound eye with thousands of ommatidia. *Image by Richard Dubielzig, DVM.*

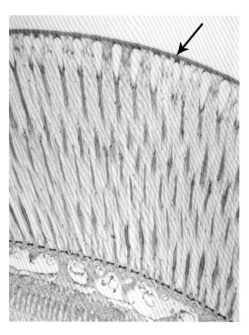

Figure 11.5 Dragonfly eye. Note the corneal lens marked by yellow arrow. The eye is cut across many ommatidia, giving the pattern to the image.

Dragonflies: *The Green Darner,* Anax junius; *The Blue Dasher,* Pachydiplax longipennis; *and The Flame Skimmer,* Libellula saturata

Dragonflies are beautiful, voracious predators that have taken the apposition compound eye to its fullest extent, at least with current oxygen levels (Figures 11.4–11.6). Each eye has approximately 30,000 individual ommatidia, with the two eyes fused in the dorsal midline (Figures 11.7–11.9). This represents the greatest concentration of ommatidia of any insect species and as a consequence, the best vision of any insect with a compound eye.

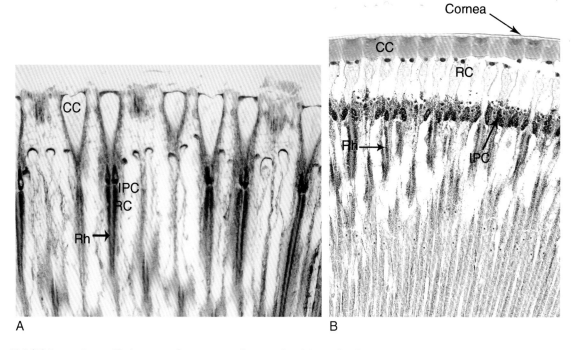

A B

Figure 11.6 (A) Dragonfly eye. Higher-magnification view of ommatidia of dragonfly. The corneal lens has been lost. The remainder of the eye is intact showing the crystalline cone (CC), the iris pigment cell (IPC), the receptor cell (RC), and the rhabdom (Rh). *Image by Richard Dubielzig, DVM.* **(B)** Damselfly eye. Note slight differences in anatomy of this eye as compared to the dragonfly eye, although the pattern is very similar. The labeling is the same, but in this specimen the thinner cornea is present. *Image by Richard Dubielzig, DVM.*

Figure 11.7 *Anax junius,* the green darner. *Image © Giff Beaton.*

Although predecessors of dragonflies arose in the Devonian, the zenith for these predators was in the Carboniferous. During that period, these early carnivores became the size of a large woodpecker with a wingspan of 70 centimeters or more, and probably had comparable or better vision than modern dragonflies. These insects were the griffenflies, a group containing the largest insect ever. Why would these insects have become so large at this time?

Figure 11.8 *Pachydiplax longipennis.* Dragonfly compound eyes. Note that the eyes meet in the midline, and note the change in density of the ommatidia. The smaller and more densely packed ommatidia are frontally faced and represent better acuity for these hunters. These areas would be the equivalent of our macula or region of best vision (Appendix A). The black area on the right compound eye is a pseudopupil. Those ommatidia are receiving light directly placed in the ommatidia but are not reflecting any light back to the camera. This indicates that these are the ommatidia that "see" the camera. The other ommatidia across the eye are reflecting light toward the camera and are viewing other portions of the horizon. *Image © Thomas Shahan. www.ThomasShahan.com.*

Figure 11.9 *Libellula saturata,* flame skimmer. Note the light reflexes. These are hexagonal, although these reflexes represent many ommatidia, not individual ones. Note that the eyes meet in the dorsal midline.

The Carboniferous's higher oxygen levels permitted insects—especially highly visual insects—to flourish because of the eye's need for oxygenation.

Most insects with compound eyes deliver oxygen to their eyes by drawing oxygen into their bodies though a series of spiracles (openings along the thorax). It is distributed through a series of ever-smaller tracheal tubes. These final, smallest branches, called tracheoles, are so narrow that their diameter is in the range of the wavelength of light. They send oxygen throughout the body but especially beneath the eye. This is like delivering water from a broad flowing river to a small inland plot by smaller and smaller branches of irrigation channels.

The positions of these tracheoles permit the reflection of light back through the eye, maximizing photon capture for the insect and creating eyeshine (the same seen in moths and similar nocturnal insects at night if illuminated directly by a torch). The reflection is maximized and can be colored because of the separation distance between the tracheoles (see Figure 21.6).

Insects such as dragonflies would have profited greatly from the higher levels of oxygen in the atmosphere because oxygen permits the growth of larger eyes. Some dragonflies had eyes the size of golf balls, with probably the best visual acuity on Earth at the time.

An insect presence with a frighteningly large wingspread and the ability to hover and fly backward would have meant that they would have ruled the skies in the Carboniferous Period. The top speed of modern dragonflies is up to 35 miles an hour, and the griffenflies would have had similar or better abilities.

Today's dragonflies are not as big as those of the Carboniferous but remain supreme insect predators. Being hawklike hunters, dragonflies have good spatial vision and good processing to see prey while both the hunter and hunted are on the move.

Within the family of dragonflies, A. junius have several hunting strategies, including perching, hovering, and hawking. Different arrangements of their ommatidia assist each hunting strategy.

A. junius, like most other dragonflies, can see in all directions at once, as their eyes wrap around their head. To permit this vast field of vision, they cannot move their eyes, only their head. Like the other fast hunters, A. junius has sacrificed some detail in their wide field for better stereopsis and acuity in the frontal field as well as larger frontal binocular fields than most dragonflies.

Thinner ommatidia in the dorsal frontal area of their eyes and those of other hawklike species have their optical axes directed more closely together with more overlap than other areas of their eyes (Figure 11.8). This translates into better detail of the area of binocular overlap directly in front of the creature. Because dragonflies cannot move their eyes, these multiple ommatidia have an almost identical direction. Focused on the same point, they add detail and sharpness to the image—in stereo. The position and concentration of the ommatidia are key to its hunting skills. These grouped ommatidia are the equivalent of our fovea—the critical area of best vision.

The larva of A. junius also has good vision. Confined to water for one to three years, the larval form undergoes several molts. At some point, the larva climbs out of water, begins breathing air, and molts for the last time into a dragonfly.

All dragonfly larvae are predaceous, eating anything they can catch and subdue, including tadpoles, frogs, and even fish. They are also cannibalistic and will eat their siblings. Although it is not known how the lineage that includes dragonflies arose, it probably began as an aquatic insect much like the larvae and eventually took flight when the potential prey of terrestrial insects appeared.

Flight is more than wings and lift, however, because the visual witness requires neurologic processing to form and analyze an image. In all but the most derived animals, eyes alone cannot form an image without the brain making sense of the signals it receives from the eyes. Motion adds to the brain's challenges because the image must be stabilized on the retina or the signals become a blur.

Primates and many other vertebrate species, to varying degrees, achieve such stabilization using a sophisticated system of extraocular muscles and input from the vestibular, or balance, system of the body. This vestibulo-ocular reflex system is accompanied by an optokinetic reflex system in the human brain. Both are involuntary. The optokinetic reflex causes the eye to follow movement and to stabilize an image on the retina (Appendix B).

This same reflex directs the eye to switch from one object to another, such as how your eye manages telephone poles seen from a moving car. Additionally, this reflex permits the eye to rotate somewhat in a circular fashion. This rotational movement is called torsion. Together the vestibulo-ocular reflex and the optokinetic reflex permit vertebrates to rapidly move their gaze from object to object in our local environment with eye movements known as saccades (fast and slow) and pursuit (slow).

Well developed in primates, these movements extend back to fish and probably the first vertebrate. These functions depend on the extraocular muscles, their innervation, and the vestibular system of the vertebrate brain.

Dragonflies, hoverflies, stalk-eyed flies (among other flies), and the rock crab, a rather dissimilar arthropod, have been found to have similar methods of image stabilization. Because their eyes are fixed, these flies and crabs move their head and body to substitute for the extraocular muscles. (Indeed, the way a dragonfly uses head movements to follow its prey resembles our pursuit movements, and its head movements probably mimic saccades.)

Like all animals that attack while flying (and often attack other flying animals), dragonflies must be able to detect motion against a nonmoving background. They even seem to be able to detect motion while they themselves are moving. This task requires that the dragonfly be capable of recognizing the three-dimensionality of its world and measuring distance effectively even monocularly.

Remarkably, dragonflies orient their bodies, and hence their eyes, to maintain the prey in its own "crosshairs" formed by the visual midline and this concentrated area of ommatidia in both eyes. These two areas of concentrated ommatidia, one area in each eye, essentially blend into each other across the midline; they must integrate together in stereopsis. The insect will catch its prey by anticipating where its prey will be at intercept.

This insect, then, uses its head movements, and possibly body movements, to compensate for the lack of extraocular muscles. Sophisticated neurologic processing is required, and dragonflies were up to that task hundreds of millions of years ago.

Other successful species of predators appeared in the late Silurian or early Devonian, but they used a different set of strategies—silk and stealth. The arachnids would be on the move.

STEALTH, SPEED, AND PREDATION

PALEOZOIC ERA

DEVONIAN PERIOD

415–362 MILLION YEARS AGO

12

Arachnids were among the first colonists to arrive early in the Devonian and may have colonized the land more than once. These ancestral organisms radiated into the other arachnids and the order *Araneae*, the true spiders, which would exploit the vegan arthropods that had preceded them on dry ground.

Arachnida

Arachnids include the spiders, scorpions, ticks, mites, and harvestmen. They are part of a larger group (*Chelicerata*) that includes the horseshoe crab (Chapter 6), sea spiders (which are not spiders at all), and extinct sea scorpions. Trilobites (Chapter 5) are closely related to *Chelicerata* as a sister group.

Spiders began as hunters and are strictly carnivorous, with perhaps a rare exception. Early spiders did not spin webs, so these spiders relied on speed or stealth. Animals that rely on such skills almost always depend on good vision. So does the wolf spider.

Wolf Spider, Lycosid Species

Distributed worldwide, wolf spiders (Figures 12.1 and 12.2) descended from the first terrestrial predators from the Devonian period. These itinerant hunters are in one of only a few spider

Figure 12.1 *Hogna* species, lycosid spider or wolf spider. Note four pairs of eyes. *Image © Thomas Shahan, www.ThomasShahan.com.*

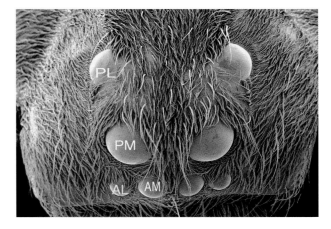

Figure 12.2 *Hogna* species, lycosid spider. Scanning electron microscopy image of wolf spider. Note the four pairs of eyes. PL, posterior lateral pair; PM, posterior median pair; AL, anterior lateral pair; AM, anterior median pair. *Scanning electron microscopy by Pat Kysar.*

families that do not spin webs (although some in the family use silk to line their burrows or to wrap their eggs).

Without the stationary death traps of their kin, these species must rely on mechanical and visual senses for all aspects of their behavior from courtship to prey capture and predator avoidance. In 400 million years, the wolf spiders have honed razor-sharp sensory abilities that allow them to "see" their prey with numerous organs although lycosids are not that old.

Being "simple," all spiders' eyes are different morphologically from most insect compound eyes. The first surface is the principal refracting surface (Figure 12.3). It is called the cornea but is, in essence, a single lens. This cornea/lens is a biconvex structure bulging into the center of the eyecup in direct contact with the multiple individual vitreous cells (Figure 12.3).

Wolf spiders have four pairs of eyes (Figures 12.1 and 12.2) with the eyes being divided as follows: the anterior row include two pairs that can be described as anteromedian or AM that point forward (principal eyes that point forward) and anterolateral or AL that point frontally and laterally. The larger single pair positioned just above the anterior row are called the posteromedian or PM, and the pair seen highest on the image (Figure 12.2) are the posterolateral, or PL, eyes.

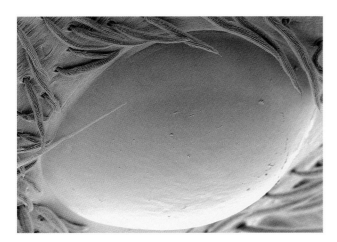

Figure 12.3 *Hogna* species, lycosid spider. The surface of this posterior median eye shows that there are no individual ommatidia. This surface is the "cornea" but is part of a cornea-lens unit, as there is no separation of the cornea and the lens. This is a large simple eye. *Scanning electron microscopy by Pat Kysar.*

Each pair probably has different functions, not just different directions. The AM eyes provide orientation and homing by interpreting the linear polarization of light. As the principal diurnal eyes, they lack a tapetum (eyeshine), unlike the other three pairs. Each of the two PM eyes is larger than the other eyes, measuring 400 µm in diameter. These larger eyes would suggest that they are probably specialized for nocturnal vision, although these spiders may also be crepuscular or diurnal. The principal eyes, the PM pair, are different embryologically and anatomically from the secondary eyes.

The principal eyes have microvillus cells that provide rhabdoms. These microvilli contain the photopigment (and are analogous to our photoreceptors). The visual pigments are distal to the cell body and immediately behind the individual vitreous cells. By preventing the light transiting the lens from being scattered by the proximal nucleus and axon, this arrangement provides a sharper image.

In the secondary eyes the nucleus is distal (instead of proximal as in the PM eyes) to the photoreceptor pigment. This allows for closer approximation of the tapetum to the photosensitive portion of the rhabdom. The secondary eyes have a complicated tapetum, which, in effect, doubles the effective length of the receptors and markedly increases photon capture. As a result, the secondary eyes of wolf spiders are approximately one hundred times more light-sensitive than those of the jumping spiders, which probably have the best visual acuity among the arachnids. The retinal arrangement of the wolf spiders' secondary eyes suggests that they may have sacrificed acuity for sensitivity. The wolf spider's photopigment spectral sensitivity ranges from 360 nm (ultraviolet) to 510 nm (yellow-green), but different eyes have different ranges, and the wavelengths in between are covered, too.

All eight wolf spider eyes enjoy a stunningly low f-number, which is essential for nocturnal predators. Ctentid spiders, which are closely related to the wolf spiders and have similar eyes, have been recorded as having an f-number of 0.74 for the principal eyes and as low as 0.59 for the secondary eyes. These would have the light-gathering potential to provide astonishingly bright images even at night.

The eyes of wolf spiders are necessarily quite small and have a very short focal length. Although they have no accommodative abilities, they do have relatively good vision and depth of field within their short range. The interreceptor angle is approximately 90 arc minutes (1.5°) in the PM eyes, even at night. The interreceptor angle of the photoreceptor array in the human fovea in full sunlight and optimal conditions is 1–2 arc minutes for comparison. But human vision drops off dramatically in dim or nocturnal light, and, depending on light levels, the wolf spider will have better vision at certain times over short distances.

But good vision at night is not all that is needed to hunt. Helping these spiders "hunt" their prey is the ability to run very quickly over short distances, at speeds of up to 2 feet per second. Nevertheless, they usually remain stationary for long periods, waiting for unsuspecting insects to happen along.

Mechanoreception plays a role in the wolf spider's lifestyle (as it does in certain vertebrates, such as crocodiles). The legs and bristles along the body detect vibrations. The vibratory recognition allows for agonistic display, mate detection, and ritualized fighting. These bristles are astonishingly sensitive, especially to the wing movement of nearby insects, maybe even to the wave front of air pushed forward by a moving insect. The hairs may interact in some way with the visual system neurologically although this is not understood.

The spiders that brought simple eyes to land evolved and radiated over the next 200 million years or so into other spiders, and visual processing evolved as well. *Portia fimbriata* is one such example.

Jumping Spiders, Portia fimbriata

In Shakespeare's Merchant of Venice, *Portia* is the clever, beautiful heroine who fulfills her role as a master of disguise. Her namesake spider has similar qualities and is the undoubtedly the queen of the Salticidae—the family of jumping spiders that includes 5000 known species.

Portia fimbriata is a spider's spider. All of the Salticidae are aggressive carnivores, even of other spiders. Portia, though, goes a step beyond, as it prefers to hunt other jumping spiders, for which it has acquired special adaptations.

Jumping spiders are 3 to 17 mm long. They have four pairs of eyes, with the large anterior median (AM) pair the largest and most obvious. These fixed circular eyes provide an "attentive child" appearance because they are large relative to body size *(but tiny on an absolute scale). These placid eyes, which will watch an approaching person closely, belie the organized complexity and evolutionary genius that lies beneath the carapace (Figures 12.4–12.8).*

Figure 12.4 *Portia fimbriata*, jumping spider. Jumping spiders are charming and common creatures that are spectacular predators. Note the large anterior median eyes. *Image © Robert Jackson, PhD.*

The AM eyes are Galilean telescopes with a corneal lens fixed to the carapace and a second "lens" at the end of a small tube immediately in front of the retina. The corneal, or first, lens has a wide field of vision at about 45°. The second lens consists of a steeply sloped fovea-like pit (Figures 12.7 and 12.8) with an index of refraction gradient between it and the vitreous-like

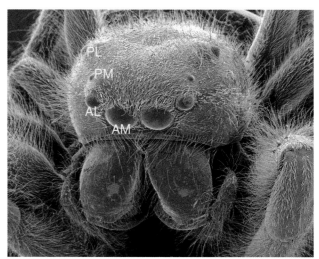

Figure 12.6 *Portia fimbriata*. Scanning electron microscopic image of the face of the spider. Note the four pairs of eyes. PL, posterior lateral pair; PM, posterior median pair; AL, anterior lateral pair; AM, anterior median pair. The AM eyes are directed forward and give the spider the image of curiosity. *Scanning electron microscopy by Pat Kysar. Specimen by Robert Jackson, PhD, Duane Harland, and Aynsley MacNab.*

amorphous fluid in the tube. This tube contains the retina and as a result limits the field of view to approximately 2°. This pit and its optics create a minus lens and a Galilean telescope much like the convexiclivate fovea of many raptors (Figures 12.7 and 12.8).

A creature this small cannot have much retinal area, so to maximize the number of photoreceptors struck by photons, the retina is tiered, with four stacked layers of retinal cells for light

Figure 12.5 *Portia fimbriata*, jumping spider. Remarkably, this spider hunts, and hunts other spiders as this image reveals. *Image © Robert Jackson, PhD.*

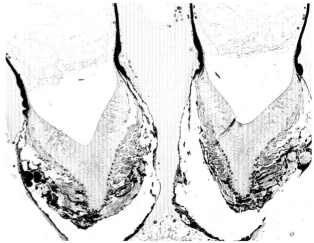

Figure 12.7 *Portia fimbriata*. Note convexiclivate fovea. This fovea creates the second element of the Galilean telescope. *Image by Richard Dubielzig, DVM.*

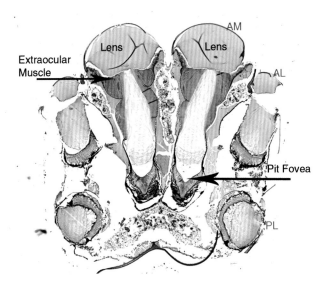

Figure 12.8 *Portia fimbriata*. Note the "tube" for the anterior median eyes. The pit fovea is labeled at the end of the "tube." Note the extraocular muscles along the "tube" used to move the eye back and forth in the scanning movements required to produce an image. One such muscle is marked. The crystalline lenses are labeled. AM refers to the anterior median eyes, which are the principal eyes; PL refers to the simpler posterior lateral eyes; AL refers to the anterior lateral eyes. The other pair of eyes, the posterior median eyes, are not in the plane of section. *Image by Richard Dubielzig, DVM.*

to traverse. Each layer of retina extracts information from the light passing through the rhabdom into the next tier, until it strikes the fourth and final one. There are at least three visual pigments, including one that senses into the ultraviolet, in these different photoreceptors. This distribution of visual pigments in this spider's eye means that the ancient ancestor to the insects (Hexapoda) and the spiders (Chelicerata) had at least trichromatic vision. In these spiders, the span of separation perpendicular to incident light from the center of one rhabdom to another is approximately 1.7 μm, creating a mosaic slightly more than three times the wavelength of light. This affords excellent short-range visual acuity.

This unique visual system begins with the aforementioned corneal lens connected to an elongated tube within the cephalothorax (first segment of the spider's body). Six muscles move this tube (Figure 12.8), although they are not homologous to our six extraocular muscles. Because the external cornea/lens is fixed to the carapace, it creates an image that is fixed at one point within the tube.

Despite the excellent acuity achieved by this compact telephoto lens system combined with the tiered retina, its tiny field of vision represents a problem. So to increase it, this optical marvel moves the tube housing the retina and attached muscles with a variety of mostly conjugate eye movements, including pendular and rotary, in a form of scanning eye. Portia scans its

world, although the retina is boomerang shaped, not linear. The other, nonscanning, pairs are used mostly as motion detectors to find other animals for the AM eyes to decipher.

With the AM eyes, jumping spiders have the finest discrimination of all arthropods, and probably all invertebrates, as they are visual hunters. But because Portia *hunts other Salticids, sharp-eyed predators with vision nearly as good as its own, it must have more than just good vision. It also must have stealth, patience, crypsis, and cunning.*

Portia *uses several techniques. As it stalks its prey, it avoids alerting or alarming the intended victim who may turn to see the predator. If* Portia *is exposed and out of range, it freezes, and, because it is cryptically colored, it will look like a piece of detritus. It hides its palps, as these and its legs apparently identify it to other spiders.*

If Portia *is hunting an orb-weaver spider, it plucks the web as if it were a caught insect. This plucking lures the poor-sighted orb-weaver to its death. If necessary,* Portia *patiently uses wind gusts to ply the silk.*

Heuristic Portia *learns to detour if it cannot easily approach the prey, even when the prey will be obscured by the detour. To do so,* Portia *must remember exactly where the prey was located.*

After detouring, when Portia *gets as close as perhaps three to four body lengths, it leaps, bites, and injects venom that rapidly kills the prey.* Portia *even can kill larger and much more venomous spiders using its eyes and brain. This spider has taken ocular minification to a fine art.*

As we have already seen, spiders have discovered small and unexploited niches, but the peculiar ecology of these niches often calls for extraordinary solutions. Some spiders have evolved to find clever solutions that approach the limits of what an eye can do, as the net-casting spider can show us. This is a true visual specialist.

Net-Casting Spider, Deinopis subrufa

Deinopis subrufa *(Figure 12.9) is restricted to eastern coastal Australia and can be found as an ambush hunter in a very specialized niche—nearly total darkness. To do this, she spins a web of cribellate (a sticky form of silk) into a trap. The spider holds the web with four of its eight legs and hides under vegetation (Figure 12.10). She often dangles, head down, under a leaf just above ground level, with its long body and legs resembling a stick. She defecates onto the ground below, creating a "target." When an insect travels across the "target," she drops the sticky net and wraps the insect. She bites and paralyzes the insect to consume or store. Tonight's dinner becomes tomorrow's target.*

The net-casting spider (Figure 12.9) has probably maximized the limit of light collection in such small eyes and, as a result, has filled an unusual niche. And although these eyes are small,

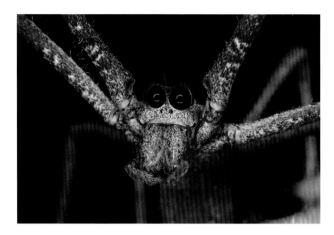

Figure 12.9 *Deinopis subrufa*, net-casting spider. Note the large eyes. This animal has the lowest f-number of any animal on earth. *Image © Mick Turner.*

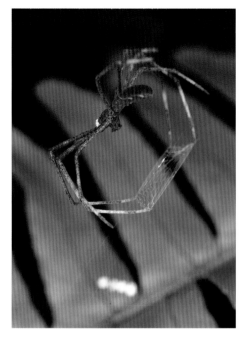

Figure 12.10 *Deinopis subrufa*. Note the position of the spider. She has positioned herself beneath a leaf hanging by a thread. She has defecated onto the leaf below to create a target, which is seen as a series of slightly out-of-focus white spots. She will position herself directly above the white spots of her feces. The odor will attract insects, and as these insects come to the feces she will "snap" down and loop the net around the insect for capture. The insect captured tonight will be part of the target for tomorrow night. *Image © Mick Turner.*

they are among the largest simple eyes in terrestrial invertebrates, attaining as much as 1.4 mm in diameter.

In each of the posterior median eyes, the net-casting spider has a steep cornea and a round lens that is very large (for a spider), with a graded index of refraction from the center to the periphery of the lens. The index of refraction of the lens as a whole is about 1.68 (much higher than our lens, at about 1.41). Because of the cornea and a small highly curved lens with a high index of refraction, the focal length is very short in this compact eye (similar to a fish lens but with a higher index of refraction). This spider may have the lowest f-number in the animal kingdom—about 0.57 (as compared with the lowest human f-number of about 2).

The decreasing index of refraction of this lens from the center to the periphery alleviates spherical aberration, much like a piscine lens. The large and long photoreceptive elements, with diameters of 16–20 μm, limit the spatial acuity or resolution. Investigators believe that this eye will absorb up to 2000 times more photons than the human eye will.

As we have seen, evolution provides clues to ocular development and the imperative of visual development to fit a niche. A broad clue is staring at us, like headlights, from Figure 12.9.

So the net-casting spider or, more likely, a closely related predecessor, found its niche deep in the warm nights of the early Mesozoic, following the Permian extinction (251 mya). This family was established before the breakup of Gondwana (~200 mya) and before the ancestor of the jumping spiders and the wolf spiders, suggesting that enlarged eyes and improved vision evolve as needed with either set of eyes.

While choosing to capture what few photons it could find under dense foliage, the net-casting spider created a triumph of ocular evolution—a visual instrument with superb light-gathering capabilities in a small package.

Vegetation was becoming more robust, and various arthropods were coming ashore one way or another. The vertebrates would see the opportunities of both foliage as well as animals as a new food source. The lobe-finned fish would start to come ashore.

THE AGE OF TETRAPODS AND TERRESTRIALS

VERTEBRATE ANIMALIA COMES ASHORE . . .

PALEOZOIC ERA

LATE DEVONIAN PERIOD
385–362 MILLION YEARS AGO

13

Generally warm with a high degree of tectonic activity, the late Devonian must have been a dramatic and exciting period. Bony fish and bizarre sharks held sway in the seas. Trilobites plied the seas. Various arthropods with different and unusual body types, in the form of insects, scorpions, and spiders, filled the oceans and occupied the land (Figure 13.1A).

But the most profound step, literally and figuratively, came from the early tetrapods. These odd amphibious creatures, some possessing seven or eight digits on each limb, filled estuaries, river deltas, and other freshwater habitats. Then, following the prey species and escaping the predators, they marched ashore, although, of course, not quite that dramatically or simply.

Their story illustrates the state of the aquatic eye as such species assumed land and exemplifies how the eye adjusted to a new terrestrial environment.

Vertebrate Quest for Land

The Devonian is described as the age of fish, but other portentous events were occurring, including the vegetative colonization of the land. The diffuse algal mats of previous periods that had covered exposed land were evolving. Instead of just coating the surface, structural innovations allowed—and competition for sun's rays demanded—plants to begin to grow upright into liverworts and grasses. Eventually, small shrubs and bushes appeared, and these would lead to medium-sized, seed-bearing plants. In time, small trees dominated the land and pushed toward the interior of the landmasses. By the end of the Devonian, considerable numbers and variety of plants had created new habitats, including shallow-water ones akin to current mangroves.

The late Devonian was mostly temperate with moist warm conditions across equatorial Gondwana, the large supercontinent filling the southern hemisphere of the globe. These conditions would create a substantial expanse of shallow, warm, freshwater habitat—numerous shallow floodplains with sluggish, and perhaps relatively anoxic, fresh or brackish water rich in aquatic and terrestrial vegetation. As plant life developed, arthropods coevolved to capitalize on a new food source and a new unexploited niche. Amphibious predators, evolved from the piscine world, would follow the herbivores, and the arms race toward sensory perfection would continue. The terrestrial visual witness would prosper.

Lobe-finned fish (*Sarcopterygii*) diverged into multidigit amphibious-like creatures that would eventually be classified as tetrapods. These were amphibious creatures but did not resemble living frogs or salamanders (Figure 13.1). Straddling the gap of the aquatic-land interface, these early tetrapods range from T*iktaalik* (probably one of the first animals with a neck), to *Ichthyostega, Acanthostega*, and *Tulerpeton* (as well as various others that live on only as partial fossils).

It is impossible to identify the initial animal that led to terrestrial vertebrates because tetrapods may have come ashore in the form of several related, or even unrelated, lineages. After all, even today fish are still trying to come ashore, as seen in living fish lineages (such as *Periophthalmos*) that are partially terrestrial.

During the Devonian, lobe-finned fish, considered direct and immediate ancestors of the tetrapods, could survive periods of drought by burrowing or encasing themselves in mud. Lobe-finned fish of the Devonian, like the lungfish of today, probably could travel from one shallow pond to another as the water sources dried up. If fish were to colonize these new niches, they would have to be prepared for shallow pools interconnected by marsh, or even dry land because some of the shallow pools would dry up completely.

A

B

Figure 13.1 (A) Devonian scene. Rich in diverse life, the Silurian and Devonian seas were likely as creative as the Cambrian's, but less recognized. A placoderm, described further in Chapter 9, is seen in the lower left-hand corner of the image. An unusual animal, pterapsis, is to the right of the placoderm along the sea floor. This jawless fish has a pointed, hornlike rostrum with bony armor over its head. The large tetrapod in the upper right-hand corner of the image with six digits resembles what *Tulerpeton* must have looked like. *Ichthysostega* had seven digits but would have many similarities with this animal. Sharks, rays, and numerous different fish can been seen present in this Devonian scene. Near the tetrapod is an ancient scorpion-like creature, and to the right of the scorpion is a myriapod species—millipedes and centipedes. The green fish in the middle of the image at the water's edge trying to come ashore is the artist's rendition of *Eusthenopteron*. *Image © The Natural History Museum, London. Artist John Sibbick.* **(B)** *Tetrapods came ashore in the Devonian.* These artistic images represent Ichthysostega, which were fishlike creatures that spent much of their time roaming the marshlands. These were probably more fishlike than terrestrial animals but represent the step between water and land. *Image © The Natural History Museum, London.*

Fish, such as the lungfish, or its relatives, that could cope with such conditions, would flourish. Lobe-finned fish shared their skills with the tetrapods, which permitted them to survive not only drought but also the extinction process at the end of the late Devonian. Eventually, the early tetrapods that arose from the lobe-finned fish learned to have an uneasy, if incomplete, relationship with terra firma. These were the first amphibians, although the term is misleading for they did not resemble modern ones (Figure 13.1B).

Early Tetrapodian Eyes

The ocular characteristics in the lungfish, including photoreceptor oil droplets, four visual pigments, scleral ossicles, and a falciform process for inner retinal nutrition, give us a splendid window into the developing, or possibly the developed, eye of the late Devonian animals. It is likely that the eye of the successful early tetrapods, at least the predators, was actually better than that of the lungfish described below. Because these tetrapods were primarily feeding in shallow fresh water, competing with glare, reflections, and the air/water interface, they needed better visual function. Certainly evolution would have changed these eyes, this is as close as we can get to tetrapodian eye of the time. Nevertheless, the lungfish eye illustrates much of the "state of the art" as the tetrapods were scrambling ashore.

Lungfish, Neoceratodus forsteri

When vertebrates lumbered ashore, perhaps 370 million years ago, they used modified swim bladders as lungs. The creatures that crossed this biological Rubicon were probably part of the clade that includes the Dipnoan order, and some of these are still with us.

Lungfish have three surviving genera (on three Gondwanan remnants—South America, Africa, and Australia). The most basal of these survivors is the Australian species Neoceratodus forsteri (Figures 13.2, 13.3, 13.4A and B), which has changed little over the last 100 million years.

Although N. forsteri *has been observed walking across land, much like a seal, it is unable to caulk itself into a mucus-lined mud case during drought as its African and South American fellow genera can. The lack of this ability suggests that* N. forsteri *is more basal and hence closer to the most recent common ancestor that led to Ichthyostega, one of the very early terrestrial tetrapods (Figure 13.1).*

Resembling eels, although more stocky, these lungfish are stout and aggressive omnivores. Vision is not as important as electroreception and olfaction to their hunting, but they do use vision. Reflecting its small relative importance to N. forsteri, *its eye is small in relation to body size (Figure 13.2). These eyes are of some use when the fish feed in slow-moving pools of clear*

Figure 13.2 *Neoceratodus forsteri,* Australian lungfish. *Image © Shaun Collin.*

fresh water or come to a river surface, which allows better visibility.

N. fosteri's eyes are well adapted for the lower light levels of a benthic lifestyle. To maximize photon capture, the eyes possess the largest photoreceptors described to date, measuring as much as 21 μm in diameter (Figure 13.3). For comparison, some birds' photoreceptors are about 1 μm in diameter, and those of humans are only slightly larger than those of birds. Because of their huge size, these lungfish photoreceptors do not offer much discrimination, but vision is not a high priority in fish that will

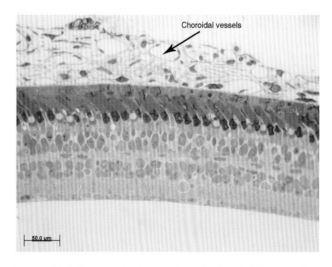

Figure 13.3 *Neoceratodus forsteri,* Australian lungfish. Image is the retina of *N. forsteri* showing the large photoreceptors that are perhaps the largest in the animal kingdom. Some of these photoreceptors are approximately 20–24 μm in diameter. The choroidal vessels are labeled—see Appendix C. *Image © Shaun Collin.*

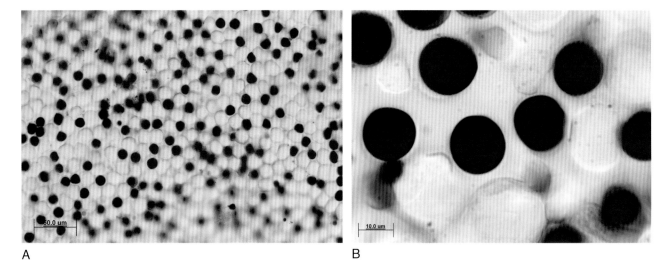

Figure 13.4 *Neoceratodus forsteri,* Australian lungfish. **(A)** Image of flat mount of lungfish retina showing oil droplets. These oil droplets will accompany the lobe-finned fish phylogenetically to the reptiles and eventually birds. It is likely that the tetrapods also had oil droplets because both major divisions of the terrestrial reptiles have oil droplets. **(B)** Higher magnification of oil droplets. Note the size of the photoreceptors. Some lungfish photorecepetors are nearly 20 μm. For comparison, the smallest human photoreceptors approach 1 μm, and some photoreceptors in birds are somewhat smaller yet. *Images © Shaun Collin.*

eat anything and that do not depend on vision as a principal mechanism of prey detection. The fish and its eye grow together throughout life, as do almost all piscine eyes. Although the overall cone density decreases as the eye ages and grows, the piscine eye actually adds photoreceptors to the peripheral retina as the photoreceptor diameter increases

The lungfish retina has at least four cone types: three with an oil droplet and one with a yellow pigment (called a myoidal pigment) in the inner portion of the photoreceptor. It limits the shorter-wavelength photons that enter those photoreceptors, although its function is not completely understood.

Phylogenetically, N. forsteri *is the oldest fish known to have colored oil droplets and one of the few fish that has them at all. One droplet type is brightly colored (red), but the two other types are colorless (Figure 13.4).*

To maximize photon collection, this lungfish evolved a tapetum to reflect light after it passed though the photoreceptor. The tapetum creates a pinkish reflex, unusual for an aquatic animal. Tapeta are useful and important for those tetrapods who inhabit a dim murky environment or who are nocturnal.

But unlike other members of the dipnoan clade, N. forsteri do not have an oil droplet in their rods, which suggests that,

---

BOX 13.1 TAPETUM

The tapetum appears in fish after first appearing in lampreys and then elasmobranchs, as discussed earlier (Chapter 7). The tetrapod that radiated into the two major terrestrial vertebrate lineages on earth probably had a tapetum, although the lampreys that have a tapetum appear to have evolved it independently. Furthermore, neither frogs nor salamanders have a tapetum, and even some nocturnal lizards such as the tuatara do not have a tapetum. So, at least one tetrapodian lineage did not bring a tapetum to land. Perhaps tetrapods came ashore in different waves, and some did not have a tapetum. But for many tetrapodian species, the tapetum is important and probably came from the lobe-finned fish.

Mechanisms for light reflection are similar but varied. Crocodiles have a retinal tapetum with the reflective coating immediately beneath and within the outer retina, *not* within the choroid.

Mammals must have arisen without tapeta because the least-derived extant groups, such as the monotremes—the echidnas and the platypus—do not have a tapetum, although many of the crepuscular marsupials do. This suggests that the tapetum evolved convergently in mammals after being lost in the lineage that would lead to them.

Later, mammals, especially the hoofed mammals such as goats, antelopes, and wildebeest and their predators, the major carnivores such as the felids (cats of all varieties), developed tapeta (Table 13.1). These tapeta, however, are

TABLE 13.1 *Tapeta*

ANIMALS	LOCATION	SOURCE
Lampreys	Retinal	Unknown
Elasmobranchs	Choroidal	Guanine
Bony Fish	Retinal	Guanine
	Choroidal	Guanine
		Fibrosum
Amphibians	No ?some toads	
Reptiles		
Tuatara	No	
Crocodilians	Retinal	Guanine
Turtles	No	
Lizards	No	
Snakes	No	
Birds		
Goatsuckers(?)	Retinal	Unknown
Monotremes	No	
Marsupials	Choroidal	Fibrosum
Opossum	Retinal	Riboflavin
Placental mammals		
Fruit bats	Retinal (?)	Unknown
Some rodents	Choroidal	Fibrosum
Carnivores	Choroidal	Cellulosum
(Cats and dogs)		
Pinnipeds	Choroidal	Celluosum
Ungulates	Choroidal	Fibrosum
Cetaceans	Choroidal	Fibrosum
Primates		
Owl	Choroidal	Fibrosum
monkey		
Others	No	

A

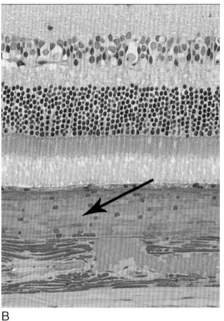

B

Figure 13.5 (A) *Leopardus pardalis*, ocelot southern pantanal. Note tapetal reflex. **(B)** *Felis catus*, domestic cat (choroid stained with toluidine blue (×20). Note regular tilelike cellular structure beneath retinal pigment epithelium (RPE) and photoreceptor outer segments. These cells in tapetum cellulosum are filled with many smaller platelike structures or rodlets seen with ultrastructural investigation.

choroidal and can be divided into pigmented and nonpigmented. Pigmented tapeta use light-scattering pigments, such as lipids, astaxanthin, and melanoid compounds, to create a mosaic colored with brilliant reds and blues.

The nonpigmented choroidal tapeta can be further divided into cellular (the tapetum cellulosum is composed of stacked reflecting cells; Figure 13.5) and fibrous (the tapetum fibrosum is acellular and composed of stacks of densely packed collagen fibrils; Figure 13.6 and 13.7). The types, seemingly evolved in tandem, create constructive interference and reflect light of a specific wavelength.

Although the tapetum reflects much of the light that crosses the photoreceptive element directly above it, the animal pays for the privilege. The reflection can scatter to adjacent photoreceptive elements and, depending on shielding of adjacent photopigment, may inadvertently cause that receptor to fire. This would blur an image and degrade acuity. So, most predators, needing better acuity, have protected adjacent photoreceptors with pigment sheaths. Invertebrates have put pigment sheaths around the ommatidia in a similar protective fashion.

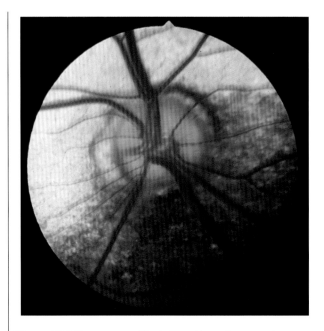

Figure 13.6 Goat tapetum. The brilliant blue color occurs because of the taeptum fibrosum. *Image by Nedim Buyukmihci, DVM.*

Figure 13.7 *Nilgiritragus hylocrius,* nilgiri tahr. Note the same brilliant blue coloration as seen in Figure 13.6 in the camera reflex of this rare goat of India.

These mirrors, then, are biological oddities cobbled together as needed, reflecting the creativity of random evolutionary mechanisms given enough time.

when they split from the other dipnoans, N. forsteri *lost the rod droplet.*

From investigation into other types of animals that have colored oil droplets, we know that they help color vision by increasing the amounts of visible colors. This suggests that color vision and photon sensitivity are important to lungfish.

Further ocular evidence that the dipnoans eventually led to the tetrapods that first walked the earth: photoreceptor morphology in the dipnoans is closer to that of urodele amphibians (such as newts, salamanders, and others that keep their tails into adulthood) than it is to most fish.

The first tetrapod, perhaps Eusthenopteron, Tiktaalik, *or* Ichthyostega, *using its ancillary lungs, may have inherited one very large rod and at least four large cones and brought them ashore in the Devonian. The stage was set for terrestrial development and further evolution.*

The Challenge of a Terrestrial Environment

The lobe-finned fish that were the ancestors of the early semiterrestrial tetrapods are long since extinct. But once the tetrapods had gained ground, they had to develop a terrestrial eye. The anatomy of the eye needed to change to surmount the new challenges of an aerial interface. Current amphibians can help us understand the changes that were necessary.

Frogs, Phyllomedusa bicolor

Members of the Class Amphibia (Greek for "double life") represent a transition between an aquatic to a terrestrial life, and the frogs and salamanders are a living model for early tetrapods. Somewhere between the modern coelacanth's ancestors and an animal like Ichthyostega *(Figures 13.1 and 13.2), vertebrates first came ashore. These early amphibians breathed air but needed water for their eggs, like most members of all three major orders of the Class Amphibia (Anura, frogs; Urodel,: salamanders; and Caecilia or Apoda, nearly blind wormlike animals). Although these living members of Amphibia are only a sister group to terrestrial vertebrates, they share a common tetrapod ancestor with all other terrestrial vertebrates.*

Current evidence suggests that frogs (Figure 13.8) evolved from other more primitive amphibians, with a common ancestor preceding both the Anurans and the Urodelans. Although both orders have many piscine-like qualities, suggesting earlier steps in this process, their eyes are intermediate.

Figure 13.8 *Phyllomedusa bicolor*, monkey frog. Peruvian frog found in the Amazonian basin.

Figure 13.9 *Carcharias taurus*, gray nurse shark. Note the pseudonictitans that begins to close over the eye as the shark is about to bite.

The early tetrapods were not frogs, as we know them, although they were amphibious. But frogs are as close as we can get to examining the soft tissue of these early tetrapods, and frogs have to face prolonged periods away from water, as did the early tetrapods. To be sure, frogs have evolved since they arose, but there are at least some similarities to be considered.

Eyelids and the Lacrimal System

The terrestrial eye would need lids, which fish do not have. Some fish have only "vertical" lids, perpendicular to the direction of swimming, that never completely close—folds of skin composed of skin and fat positioned in front of the eye that help streamline the swimming profile. Some sharks have a pseudonictitans (third eyelid), and it is similar to what the frog has evolved but is not homologous to a true nictitans as found in a reptile or bird. Sharks' third eyelid, which extends from the lower lid, more resembles the so-called "adipose" lids of some of the faster fish mentioned above. In sharks, this rudimentary third eyelid is not of much use except perhaps to protect the eye from injuries that struggling prey might otherwise inflict (Figure 13.9).

Fish also do not have a lacrimal system either. There is little use for a tear film when the eye is immersed in water. Many fish produce only a thin patina of mucins, produced by the surface cells, to help the interface between the outer integument (skin) and the aquatic interface. But in the aerial environment, air would have dried the surface cells, including those of the cornea, in the eyes of the early amphibious tetrapods (Figure 13.1). Perhaps the first step for these newly forged terrestrial amphibians would be to develop a lacrimal system that would produce some form of lubrication, preceded or followed by lids to spread these new "tears."

At least two of the lobe-fin fossils, *Eusthenopteron* and *Osteolepis*, had a lacrimal system in the form of a tear duct draining tears into the nasal cavity, but since lids do not fossilize, we do not know the status of the lids in these animals.

Not surprisingly, the lids of frogs and salamanders are more rudimentary than we see in lizards or mammals, but frogs and salamanders do have lids. These lids are essentially folds of skin with a terminal ridge that can be used to close the eye, although another method is generally used. Eyelid closure is accomplished by the third eyelid, or nictitans, moving from beneath the lower lid toward the upper lid (Figures 13.10 and 13.11). The upper and lower lids then close together to meet in the midline over the nictitans.

The third eyelid or nictitans of extant amphibians is not homologous with that of reptiles and mammals. The third eyelid is so important that it evolved convergently at least twice in two different ways.

Extraocular Muscles

Frogs, salamanders, and newts have a retractor bulbi muscle immediately behind the eye. Although phylogenetically this is not the first appearance of this muscle, it had been in the shadows since its beginning in the placoderms (Chapter 9).

Figure 13.10 *Litoria infrafrenata,* white-lipped tree frog. Australian frog found in the wet tropics. Note the nictitans partially raised from the lower lid.

The last common ancestor of the placoderms and the lineage that would lead to the ray-finned fish probably had such a retractor bulbi. And the coelacanth, closely related to the lungfish and the early tetrapods, has a homologous muscle, called the basicranial muscle, used in feeding and skull movement. In the coelacanth (and the placoderms), the muscle is innervated by the sixth cranial nerve, as is the retractor bulbi in frogs and other species, illustrating its homologous origins.

The muscle is used to pull the globe beneath the folds of skin that are phylogenetically developing to cover the eye. The retractor bulbi muscle connects to the ventral and medial aspect of the globe via muscle fibers and to the rim of the lower lid via a tensed cord of tissue that pulls the globe down and inward. When the globe is retracted enough, this activity closes the lower lid and effectively pulls the eye inward. When a frog swallows a fly, the globe seems to

Figure 13.11 *Nyctimystes dayi,* Australian lace-lid. Note the nictitans that covers the eye has a lacelike appearance. It is not understood why the nictitans has this pattern. *Image by Danté Fenolio, www.anotheca.com.*

disappear, so in essence, the eyes help the frog to push the food into the stomach. In addition to feeding, the retractor bulbi is used in globe protection and surface lubrication.

Tear Glands

Some of the salamanders have a movable lower lid that contains separate sets of secretory glands nasally and temporally. These probably are the predecessors of the Harderian gland (which secretes a mucous or sebaceous solution and can be found in most terrestrial vertebrates) nasally and the lacrimal gland (which secretes a watery tear solution) temporally. These secretory glands will prove important in the development of the adnexa of the terrestrial eye. The secretory adnexa differ in frogs and in salamanders. Frogs are perhaps closer to the Devonian tetrapods, as most frogs have a Harderian gland situated at the nasal aspect of the globe. This gland develops at metamorphosis, as tadpoles must become at least partially terrestrial as frogs.

Cornea and Lens

In addition to a tear film and lids, the early amphibians had to develop a much smoother corneal surface and a more transparent cornea. The potential for, and need for, better distance vision would require a clearer cornea.

Terrestrial life enabled and required better optics and focusing. Even in the clearest water, there are limits to distance vision. Any particulate matter in the water would degrade the image. Air is clearer, however, and this permits vision at greater distances and better clarity. Optics and accommodation would have to evolve.

Fish have a rather rough texture to the surface of their corneas because water contact neutralizes surface irregularities. The purely aquatic form of the frog, the tadpole, has a piscine-like (two-layered) cornea as described (Chapter 10 and Appendix D). The tadpole globe has a clear cornea with a sulcus where the cornea changes its radius of curvature to become opaque sclera. A layer of skin, called the secondary spectacle, overlies the cornea (Figures 10.2 and 13.12A) in both (many) fish (Figure 10.2) and tadpoles. During metamorphosis, the skin layer of the tadpole eye unites with the primary cornea to create a thicker, more robust cornea, which becomes a terrestrial eye (Figure 13.12B). Metamorphosis also changes the optics of the eye. The cornea becomes smoother and steeper––much steeper––to become the predominant mechanism of refraction in the frog eye. This collects more light and produces a sharper image. The cornea would have to be steeper, more compact, and more transparent for aerial vision. Additionally, the frog's corneas are steeper and smoother than those of fish

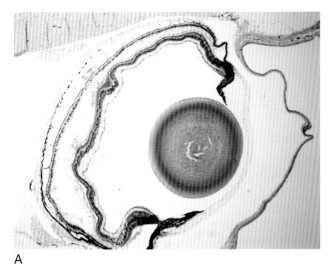

A

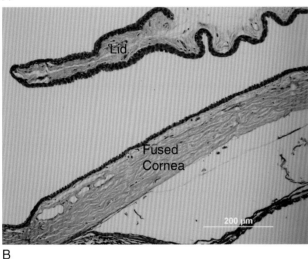

Lid

Fused
Cornea

200 μm

B

Figure 13.12 (A) *Rana catesbeiana,* American bullfrog tadpole. Note the space between the external skin layer and the thin cornea. This is very similar to the fish cornea we see in Figure 10.2. *Image by Richard Dubielzig, DVM.* **(B)** *Phyllomedusa bicolor,* giant wax frog. Note that the two layers of the cornea seen in the tadpole have united into a single layer during metamorphosis. The lid and the cornea are marked. *Image by Richard Dubielzig, DVM.*

so as not to degrade the image, although perhaps not as steep as the reptilian cornea. This suggests early and only partial adaptation to terrestrial life, but then, frogs are semiterrestrial. Of course a completely aquatic animal would have no impetus for a steeper cornea, which would bring the greater refractive ability needed for focusing in an air-eye interface. But terrestrial vertebrates would gain substantially if the cornea assumed much of the refractive "heavy lifting."

This is precisely what happened, first in the early amphibians and later to greater degree in the reptiles. Terrestrial species' increased corneal refractive power permits the

intraocular crystalline lens to decrease in size and assume different shapes besides spherical. That size change leads to a larger and probably more defined image, permitting better discrimination. But frogs retain the spherical lens of fish because they still need the refractive power of the spherical lens underwater.

Both frogs and salamanders retain and use accommodative mechanisms (Appendix E) resembling those of cartilaginous fish (sharks, rays, and their allies), which suggests that lobe-finned fish, hence tetrapods, had a last common ancestor with the cartilaginous fish. Or, more likely, accommodative mechanisms may be very plastic and able to change easily. The accommodative mechanisms in frogs are exactly opposite to those of the bony fish, which move their lenses forward, rather than backward like the frogs (Figure 10.8).

Retina and Vision

Retinal evolution is best discussed in the context of other amphibians because the differences among these groups best demonstrate how the tetrapodian eye may have changed as species diverged.

Salamanders (urodela) and frogs (anura) diverged in the Triassic, perhaps 250 mya, and illustrate the changes that have occurred in the opsins and neurologic changes over that period of time. Presumably, these changes represent adaptations to the different niches of each lineage, although other explanations are possible. For example, the original salamander lineage has gone through further and more extensive changes than the frogs.

Frogs lack the visual acuity of salamanders, although both are nocturnal. But frogs have gained a more acute sense of movement detection. Frogs have no fovea, and only a few species even have a concentrated area sophisticated enough to be considered an area centralis, an analogue of the macula. Many frog species seem to have lost the range of color vision that salamanders retain today, but both lineages have cones. Perhaps it is because frogs adopted a more or earlier nocturnal lifestyle and evolution adheres to the "use it or lose it" principle. Both salamanders and frogs have an unusual rod called a "green rod." These rods are morphologically different from other rods, and certain anatomical features suggest that they have transmuted from cones into rods during the trek into nocturnality. "Green rods" are misnamed, though, because the opsin is tuned closer to blue (short-wavelength-sensitive) than to green. In some salamanders these short-light-wavelength-sensitive rods have transmuted back to being cones. Subtle changes such as these indicate that the ecologic differences can easily affect color perception even in related species.

Pupillary differences evolved as well. Salamanders have round pupils, whereas many frogs have pupils that are vertical or horizontal, oval, or even heart-shaped. Presumably these peculiar pupils relate to improved diurnal vision for a predominately rod retina, or they may relate to prey capture.

Neurologic Changes

These newly terrestrial animals faced other visual challenges and opportunities. Efficient terrestrial predators would need stereopsis, and that would require a change in facial morphology and the neurologic connections. Stereopsis, the ability to see in three dimensions, requires integrated visual input from both eyes viewing the same object from slightly different perspectives.

Fish have a visual system in which the visual input from the right eye is sent to and interpreted by the left side of the brain, and vice versa (as far as we understand). In fish, the optic nerves cross completely in the brain at what is called an optic chiasm. True stereopsis is possible only when visual signals coming through two different neurons (one from photoreceptors in each eye) sampling the same portion of the visual field, but from slightly different angles, come together to compare and integrate their inputs. Hence, evolution had to find a way to integrate each eye to both sides of the brain in a more seamless manner and provide for stereoscopic vision, although this mechanism is not well understood in fish.

Facial morphology changed from the thin profile of a fish to a broader body and skull, and the eyes moved to a more forward and separated position in the skull as vertebrates became more accustomed to a terrestrial lifestyle. A streamlined profile was no longer as much of an asset, and a broader skull would enhance the amount of simultaneously viewed visual field and hence binocular vision. This would permit and encourage stereopsis, but changes in the brain would also be needed.

The amphibian predecessor that first came ashore brought many piscine visual and neurologic adaptations. For example, whereas most fish have a completely crossed chiasm, higher terrestrial vertebrates, especially predators, have a partially crossed chiasm with splitting of the fibers from each eye to each side of the brain. So, where did the change occur?

Frogs begin as eggs and hatch into tadpoles. A tadpole is an entirely aquatic phase, and, like fish, it has a completely crossed chiasm. When tadpoles develop, they send the neuronal fibers from the eye to the contralateral optic tectum, similarly to fish. The optic tectum, the principal site for visual processing in frogs, is the homologue to a mammalian structure called the superior colliculus. In higher mammals there is a more derived and complex system that goes beyond the colliculus called the visual cortex. In most other vertebrates, though, the optic tectum is the site of visual interpretation.

But the transition from tadpole to frog during metamorphosis requires that the tadpole send new neuronal fibers from the retina toward the chiasm, where something astonishing occurs. At the chiasm, genes (called netrins, ephrins, semaphorins, and slits) produce proteins that direct the migration, connectivity, and growth cessation of these new neurons as they pass through the chiasm. (At this stage, the new neurons are called neuronal growth cones.) The temporal fibers are directed to the ipsilateral brain, and the nasal fibers to the contralateral side. These genes must turn their proteins on and off at the correct time to stimulate crossing or to prevent recrossing and must do so only in the appropriately selected fibers. Other related genes must tell the fibers to stop at their appropriate rendezvous, where they must also direct the correct topographical distribution within the tectum to provide the correct representation of the scene on the retina.

Specifically, the ventral nasal fibers of the left eye, for example, must connect with the next neuronal layer in the tectum precisely adjacent to the connections of its retinal neighbors that the patterning of binocularity needs. This would be substantially different for the developing frog because the tadpole's completely crossed chiasm would have had very different signaling in its chiasmal and tectal development. But, during the process of metamorphosis, that is exactly what happens.

Other physical changes occur when a tadpole metamorphoses into a frog, of course, but these neurologic changes are the most profound for vision and signal the development of true stereopsis. It is a metaphorical step as well. During metamorphosis frogs change from water to land, and the skull, eye, and brain all assume the necessary changes to meet their terrestrial environment. This is the animal whose evolution illustrates the step from water to land.

Amphibians are versatile and have continued to evolve over hundreds of millions of years, although there is evidence that the eye has not changed substantially. Nevertheless, frogs have shown the ability to face the challenges the environment would deliver to them. The North American wood frog illustrates just what the eye can and will endure to persist.

North American Wood Frog, Rana sylvatica

Species that are ill-prepared for polar latitudes, among the earth's harshest climates, sometimes pay with their lives. Witness the humans who were bent on finding the Northwest

Passage in the early 1800s: explorers such as Parry and Franklin lost ships, supplies, men, and, in some cases, their own lives.

Some other vertebrates prepare for such extreme conditions in a peculiar manner. They allow up to 65 percent of their body water, including their eyes, to freeze during winter.

Rana sylvatica, the North American wood frog (Figure 13.13), is the only frog found north of the Arctic Circle. Its survival there is an extraordinary feat. Although this ectothermic anuran evolved very recently, these adaptations illustrate evolution's capabilities to adapt to almost any conditions.

The first signs of freezing north of the Arctic Circle, approximately 66° latitude, often appear as early as September. When these early freezes arrive, the wood frog quickly begins its descent into a cryobiological state. The frog will burrow to a set point below the permafrost. There, the first formation of ice crystals on the skin of the wood frog triggers genetic expression of a gene in the liver that participates in the production of the cryoprotectant glucose.

As glucose levels rise, the cells lose free water, even shrinking to some extent. The peripheral limbs freeze first, beginning with the fingers and toes. Freezing progresses centrally until it reaches the core. All organs freeze in an almost scripted fashion, ending with the three principal ones (liver, heart, and brain); they freeze simultaneously.

The eyes freeze as well—but not uniformly. The lenses freeze first but late in the overall freezing progression. As they freeze,

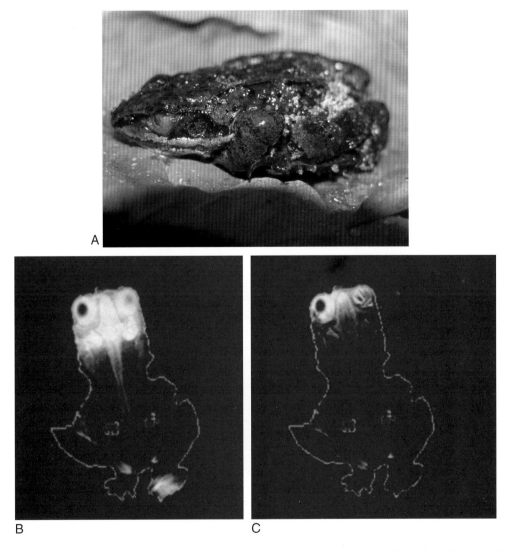

Figure 13.13 *Rana sylvatica,* American wood frog. **(A)** Note that the frog, including the eye, is frozen. *Image by J. M. Storey, Carleton University.* **(B)** Magnetic Resonance imaging (MRI) of *R. sylvatica.* The frog is being frozen and is nearly completely frozen. Note that the lenses are frozen but the eye and brain are not yet frozen. *Image by J. M. Storey, Carleton University.* **(C)** *R. sylvatica* MRI, Note that the frog is nearly completely frozen with one eye nearly completely frozen and one eye with the lens frozen but the retina and brain still active. *Image by J. M. Storey, Carleton University.*

they become cloudy and opaque. It would appear from MRI images that the retinae are among the last tissues to freeze. In that respect, the retina is very much like the brain (Figure 13.13B,C). The entire process usually takes no more than twenty-four hours.

Each of the individual eukaryotic cells in the frog's body must experience suspended frozen animation by itself. Any rupture in the cell membrane or alteration in cellular metabolism (be that cell an astrocyte brain cell, a corneal endothelial cell, or a hepatocyte) would result in that cell's death. A cryoprotectant is essential to assist cells to pass this biological trial. In essence, the frog becomes diabetic by converting glycogen stores to glucose and flooding the system with levels a hundredfold greater than normal. Cardioacceleration during the freezing process ensures penetration of the cryoprotectant to the cellular periphery.

Glucose makes an excellent protectant because it is so readily available and can be generated from glycogen within minutes. Each cell, then, crenates, losing water molecules that, as crystals of ice, would disrupt the cell membrane. The concentration of glucose rises even further in the core organs, which actually shrink. All cell functioning ceases—completely.

As astonishing as this process is, there are limits. If the temperature falls to −30°C, which is entirely possible in the Arctic, this degree of freezing is lethal. To protect against death, before beginning its winter freeze, the frog burrows a few inches below the surface leaf litter and snow where the temperatures are more constant and less severe, even if frozen.

But as difficult as it is to imagine an animal freezing as a form of hibernation, reentry to the world of the living is even more bizarre and precarious. The heart, brain, and liver thaw concurrently, and the entire thawing process proceeds as the reverse of the freezing process. If the periphery were to thaw first, it would undergo necrosis without the blood supply from the frozen cardiovascular system. The whole process is much quicker than freezing, and all tissues thaw nearly concurrently, perhaps in as little as one or two hours.

The core organs thaw at a relatively lower temperature than organs with a lower level of glucose. Heartbeat is the first physiologic function to be restored. The lenses do not become clear again until the thaw is nearly completed, after heartbeat and breathing are reestablished. The ocular melting process is not a mirror image of the freeze, as the lenses thaw later than would be expected.

Life can evolve to meet the worst of conditions.

___

So with amphibians, tetrapods' leap into a terrestrial and predatorial lifestyle forever changed the visual system. Other ocular adaptations to the lifestyle and harsh terrestrial conditions would follow.

Consolidation of the Assumption of Land

So by now, in the late Devonian, the optics of the cornea was smoother, and external eyelids (Figures 13.8 and 13.10) and the third eyelid (Figures 13.10 and 13.11) had appeared, as had a rudimentary lacrimal system. The cornea had become a single layer instead of two (Figure 13.12).

Tetrapods came ashore because of two unexploited primary food sources—vegetation for the vegans and a thriving community of potential prey animals for the carnivores.

The Devonian, the birth time of many of today's phyla, came to an end with its significant marine extinction. One of five major extinctions in the history of life, this was a stuttering affair with several events that would eventually lead to significant cooling and thus to enlarged ice caps and lowered sea levels. The events wreaked havoc on the agnathans or jawless fish in number and variety, for example, and severely depleted the trilobites. The placoderms went extinct completely, and the door would open for many new species to expand, and many stepped up to the task.

TERRESTRIAL LIFE FLOURISHES

PALEOZOIC ERA

One of life's greatest extinctions ended the Devonian, but the creatures that entered the Carboniferous would soon find warm shallow seas edged by vegetation on broad continental shelves. Conditions were once more ideal for life, and it flourished. Plant and animal life exploded on land, almost symbiotically. Oxygen levels that were nearly double today's levels accompanied the warm climate. These conditions permitted the evolution of giant arthropods and other invertebrates as well as amphibians now freed from limitations imposed by poor tissue oxygenation. Great swarms of large insects appeared, with the large griffenfly as the dominant aerial predator.

Vertebrates

The Carboniferous had an event that was only slightly less providential than the Devonian's tetrapod invasion. The cleidoic egg appeared.

The cleidoic egg has a leathery or calcareous shell that permits animals to roam far from water without the need to return to water to lay eggs. It probably evolved because of microbial or invertebrate predation on the jelly-like egg masses of the amphibians or the drying of land-based eggs. Such a cleidoic egg would open the door for all manner of terrestrial vertebrates. But this egg was more than just a protective shell—it includes an amnion, or membrane with a surrounding chorion and allantois (in other words, some form of placental-like organ that nourishes the embryo). Amniotes appeared probably in the early Carboniferous after the tetrapods had established themselves. This would eventually include any animal that lays eggs with an amnion, such as reptiles and birds, and placental mammals with their eggs inside their body.

Lineages that would lead to living reptiles, as well as all other living terrestrial vertebrates, appeared approximately 350–340 mya as amniotes in the early Carboniferous Period,

but they did not resemble modern lizards. They were all amniotes, however.

Early tetrapods laid eggs that required water—much as frogs do today. It was not until the hard-shell cleidoic egg that life's eggs could move from the encumbrance of water because they would not desiccate and could withstand the vagaries of terrestrial incubation—at least in some form. The common amniote and the cleidoic egg were major and key biological steps.

Reptiles began as cotylosaurs or stem reptiles and diverged into at least two different lineages—the synapsids and the sauropsids (or diapsids). The last common ancestor of the diapsids probably did not resemble any of its descendants, but this creature had two holes in its skull behind the orbits, distinguishing it from the synapsids, which had but one. This was a principal difference.

The synapsids diverged from the most recent common ancestor in the mid-Carboniferous and radiated into mammal-like reptiles, called pelycosaurs, and eventually the mammals (see Figure 15.1). The pelycosaurs and their descendants would go on to dominate in the Permian. At some point, at least a twig of this synapsid radiation would eventually become warm-blooded, sprout hair, and radically change their facial and ear bones. They were destined to radiate into the stem mammals and be a sister group to the other reptilian amniotes.

The diapsids radiated into the turtles, lizards (eventually snakes), crocodiles, and eventually birds (Figure 14.1).

Synapsids and Their Eyes

The intraocular structures suggest that the synapsids diverged from the same, or closely related, ancestor that radiated into frogs, and diverged early, perhaps soon after the lineage that would lead to frogs. Because frogs and all known synapsid lineages (including humans) have smooth muscle in their irides (singular iris) and ciliary

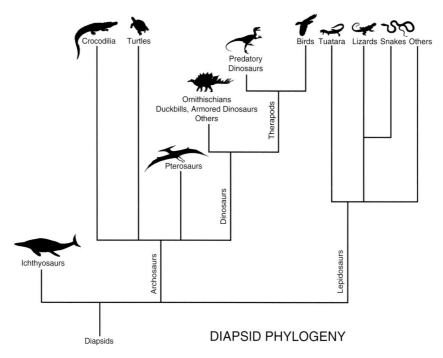

Crocodilia Turtles

Predatory
Dinosaurs

Birds Tuatara Lizards Snakes Others

Ornithischians
Duckbills, Armored Dinosaurs
Others

Therapods

Pterosaurs

Dinosaurs

Ichthyosaurs

Archosaurs

Lepidosaurs

Diapsids

DIAPSID PHYLOGENY

Figure 14.1 Cladogram of diapsids.

body, pupillary contraction and accommodation occurs more slowly than it does in reptiles and birds and is a key difference in synapsid eyes, including ours (Appendices A and E).

The synapsids would have inherited a very similar eye to the lobe-finned fish of the time from which they descended. These early synapsids probably had oil droplets in their photoreceptors, four different cones, each with a different visual pigment, along with rods, and similar intraocular muscular arrangements—all inherited from their tetrapodian ancestors. These early synapsids, such as *Dimetrodon*, almost certainly would have inherited this or similar ocular anatomy.

Dimetrodon is often linked with the dinosaurs but was not their contemporary or even of the same clade. *Dimetrodon*, however, was a major diurnal predator and probably a dominant species of the times. But this would change, and these particular synapsids would die out. Eventually, the surviving members of this lineage would become small, secretive, and nocturnal. Visual adaptation to nocturnality required more emphasis on rods than on cones, providing the ability to see in limited light rather than in color (color vision being irrelevant for most animals at night). The mammalian eye would never be the same. But, that story would come later.

Diapsids and Their Eyes
This last common ancestor of the diapsids continued with an eye that must have been similar to the eye of a lobe-finned

fish as well. This included double cones (Chapter 10), scleral ossicles, and an epiretinal vascular net that would radiate into the conus (reptilian retinal vascular system) to nourish the inner retina (Appendix C). Some of the characteristics that remained would be modified.

Most bony fish, including the lobe-finned, have little or no pupillary movement, and accommodation occurs by lenticular movement—specifically by lens retraction. But these mechanisms would not be sufficient for the reptiles (Figure 10.8).

The sauropsidian, or reptilian, eye changed to meet the different challenges of a diurnal, terrestrial lifestyle as it was freed from the requirements of vision underwater. The cornea had become steeper and smoother with better optics, as mentioned previously for the early amphibians. The steeper cornea would assume more of the eye's refractive power, and morphologically, this would permit the lens to move posteriorly in the globe and to become thinner.

The disappearance of these elements of piscine corneal and lens morphology from terrestrial eyes would hint at significantly improved design for several possible reasons. Optically, the configuration would create more magnification and the potential for more discrimination. It also would permit finer accommodation, a brighter image (more photons would be gathered with the steeper cornea), and a larger image projected onto the retina. These changes would provide a mechanism to provide a brighter and more detailed image. Whatever the reason or reasons, the terrestrial design is unquestionably better because the piscine lens and accommodation system do not exist in any living terrestrial creature (Appendix E).

For the first time, phylogenetically, in every reptilian lineage except snakes (which are a special circumstance discussed later), accommodation is achieved by lens deformation rather than the to-and-fro lens movement used by fish.

The diapsids developed a different intraocular muscular system than the synapsids, although the morphologic changes of the lens were probably similar. All living lineages of diapsids have striated muscles in their irides and ciliary body for accommodation, but none of the synapsid descendants do. This has made for faster and more accurate focusing in these lineages of diapsids. It is possible that both the early synapsids and diapsids developed striated musculature for accommodation, but the more parsimonious explanation is that only the diapsids developed such striated musculature.

The protoreptile that radiated into the diapsids and synapsids possessed the first fully terrestrial vertebrate eye in the Carboniferous. At that point they had adapted many changes begun in amphibians for this new life on land and made some of their own. We have extant representatives that probably resemble those early stem creatures, including tuataras, crocodiles, and turtles. Some of these have changed little since that time.

The Carboniferous more or less simply blended into the Permian. A considerable ice age did cause the extinction of some species, but it was mild by comparison to its bookend extinctions of the Devonian and Permian periods.

The Permian saw consolidation of almost all the land masses into a supercontinent, Pangaea. This continent straddled the equator, with lower sea levels, and this alone would squeeze some of the new tetrapods into extinction. Trees continued to evolve, and some of these from this age, such as the cycads, still survive. Although the Permian began with an ice age, the landmasses warmed during this period, terrestrial plants flourished, and species again exploded with dramatic changes.

The vertebrates continued to diverge, and other key lineages, especially among the invertebrates, appeared and flourished, including the beetles.

Invertebrates of the Permian

An important invertebrate lineage arose in the Permian as the terrestrial plants would have needed help with pollination and provided provender in exchange. Plants coevolved with insects and by this time would feel the insect burden. Insect predators would follow.

Beetles appeared by the early Permian, approximately 299 mya, with apposition and refracting superposition eyes. These were the early pollinators, recyclers, and plant pests.

Beetles are the largest order of insects, with more than 350,000 known species, probably only a small fraction of the actual total. In fact, those estimates may be off by an order or two of magnitude. This diverse and adaptive group represents a fifth of all living organisms and a fourth of all animals. In general, beetles have been successful because they are particularly well equipped to overcome extinctions, climate change, and heavy predation. Many rely on vision.

The ceiba beetle, the lantern click beetle, the tiger beetle, and the whirligig beetle represent the terrestrial insect radiation that began in the Devonian and led to the beetles of the Permian.

Giant Ceiba Beetle, Euchroma gigantea

Giant metallic ceiba beetles are forest jewels that begin their lives as eggs in rotting stumps or logs. They hatch into a larval form, remaining for a year or more in the rotting vegetation without exposure to light. There, they feed on wood, digesting cellulose with the help of bacteria in their gut and turning fallen trees into compost. The larvae pass through various stages, known as instars, to maturity, when they exit the stump capable of flight.

The structural color of this 5- to 6-cm-long beetle provides a glistening metallic appearance and may function in camouflage and communication (Figure 14.2). This dramatic color also attracts humans, who have used "jewel-like" beetles such as Euchroma gigantea *in jewelry and as decorations at least since the pharaohs. Amazonian tribes have used carapaces of E. gigantea as body ornaments, and eastern Ecuadorian tribes have cherished them as decorations for shrunken heads.*

The diurnal beetles such as E. gigantea *possess a focal apposition eye described in Chapter 6. These eyes require more light than the superposition eyes and thus are restricted to diurnal lifestyles.* E. gigantea *is the largest wood-boring beetle and as a*

Figure 14.2 *Euchroma gigantean,* giant ceiba beetle. Note the large eyes. These are focal apposition eyes.

result has large eyes with many ommatidia to improve the image (Figure 14.2). Apposition eyes were brought ashore by the crustacean lineage that gave rise to the dragonflies (Chapter 12).

Lantern Click Beetle, Pyrophorus phosphorescens

Nocturnal beetles, as represented by the click beetle (Figure 14.3), have an entirely different compound eye (the small black structures in front on the image).

These beetles have a refracting superposition compound eye that is like the diurnal beetles' apposition eye externally but completely different internally (Chapter 6). The crystalline cones, acting as lenses, are set above the rhabdom layer (like our retina) with a clear region that is similar to our vitreous. These superposition compound eyes function as simple inverting telescopes, creating a single erect image on the rhabdom layer. Superposition eyes are much more sensitive to light than apposition eyes and provide better acuity. Moreover, these eyes permit the beetle to see well in dim environments, especially at night (Chapter 6). The eyes of the click beetles have visual pigments that have spectral peaks in the range of "ultraviolet" and "green," especially in dim environments, although the peaks are broad. With these visual pigments, there is a "yellow" sensitivity among these beetles that is important for communication and based on bioluminescence—a key to this beetle's success.

Beyond their glistening appearance the nocturnal click beetles, much like North American fireflies, have evolved an important visual function for communication and, in this case,

subterfuge. The yellow spots on the carapace (Figure 14.3) are bioluminescent organs, and there is a similar organ on the abdomen. They are not eyes, although predators may assume so. These pseudoeyes will make the insect appear larger. These organs on the thorax of Pyrophorus produce light with a combination of luciferin and luciferinase, much as fireflies do. This beetle generally uses the light for nocturnal illumination and to signal a potential mate, with each species within the click beetles having its own pattern and style of presentation. When active at night, these beetles provide an ethereal, comfortable feel to the warm moist equatorial forests. They are jewels indeed.

Tiger Beetle Larvae, Tetracha carolina

The Tiger beetles are especially voracious (as the word Adephaga, a Greek word describing their suborder meaning gluttonous suggests). They eat live ants, flies, lice, and fleas, as well as offal. Their eyes—even in the larval stage—support their appetite (Figures 14.4–14.6). In many species of beetles (and other insects), the eyes become sophisticated only in the adult (imago) stage, not in any of the three preceding stages (embryo, larval, or pupa). Tiger beetles are exceptions. Although the beetle clade arose 299 mya, Adephaga suborder arose in the Triassic (251–208 mya) but is considered here.

The tiger beetle larva prefers sandy soil and frequents areas near streams or water. It burrows into the soil and builds its home as a long vertical tube (Figure 14.4). As the larva grows, it

Figure 14.3 *Pyrophorus phosphorescens,* click beetle. Arrows point to eyes. The yellow spots on the dorsal pronotum are phosphorescent, much as in fireflies. This bioluminescence, though, is for communication and intimidation of predators.

Figure 14.4 *Tetracha carolina.* This tiger beetle larva is waiting in ambush. Note the pair of eyes looking directly up, and an additional pair can be seen just below the edge of the carapace (*yellow arrows*). *Image © Giff Beaton.*

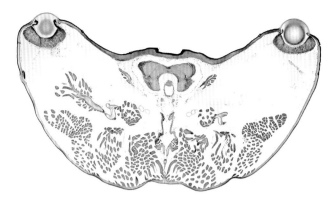

Figure 14.5 *Tetracha carolina*, tiger beetle larva. This section through the head of the larva transects the pair of eyes illustrated in Figure 14.4 by the yellow arrows. *Image by Richard Dubielzig, DVM.*

The large stemmata *have a large cornea-lens and make up a single refractive unit. The underlying retina is flat, and the peripheral portion of the retina is farther away from the lens, which creates some retinal disparity when light rays pass through the lens and focus on the retina. This disparity helps the larva to determine distance and location—key to its ambush skills (Figure 14.4). The tiger beetle remains lodged in its tube with its head just slightly below ground level. Prey must come within its feeding range of 10–15 mm. Movement, shadows, or other perceived threats beyond that range cause the larva to scurry deep within its tube. This is a surprisingly aggressive and gluttonous larva, illustrating the different eye styles that beetles have perfected.*

Some beetles even have both aerial and aquatic apposition eyes and illustrate the versatile nature of the apposition eye.

Whirligig Beetle, Gyrinidae

Commonly found on small, acidic ponds, whirligig beetles can see in two worlds at the same time. To suit their livelihood, their paired eyes have evolved in an unusual manner: they see above and below the water simultaneously (Figures 14.7–14.9).

The beetles usually seek living prey, such as mosquito larvae, but they also scavenge dead insects on the surface or in the water column. Although they are not wing feeders, they, like most beetles, can fly, but they usually dive to chase prey. They remain underwater for several minutes (breathing air they have stored beneath their "wings"), so they need to be able to see both in the water column and above it.

As the whirligig swims or floats, the water level evenly separates the aquatic and the aerial eyes on each side of the head. The separate eyes are apposition style with similar optics,

lengthens the tube, to as long as 30 cm. In case of attempted predation, the larva can even secure itself in the tube by using two posterior spines to dig into the soil of the wall.

The larva is fast and armed with powerful pincer-like jaws. It leaps from its tube, attacks and subdues other small insects, drags them into its tube, and consumes them.

Purely a visual ambush hunter, the larva possesses one of the largest simple eyes on earth. Its six pairs of simple eyes, called stemmata, *include two large and four small, almost dotlike, pairs. The large* stemmata *share a large binocular field and a slender uniocular field on each outer edge of the binocular field. The binocular visual field of the second pair overlaps slightly with the first pair, thus enlarging the first visual field for the animal.*

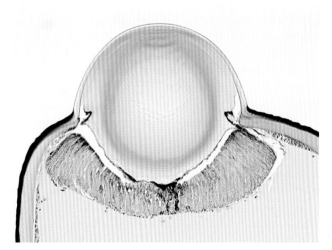

Figure 14.6 *Tetracha carolina*, tiger beetle larva. A higher-power view of the eye. Note the concentric layers of the lens with each layer have a slightly different and lower index of refraction to eliminate spherical aberration. *Image by Richard Dubielzig, DVM.*

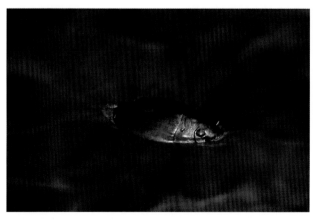

Figure 14.7 *Gyrinidae*, whirligig beetle. This common amphibious beetle swims erratically on water surfaces, usually in groups of a dozen or more. *Image © Giff Beaton.*

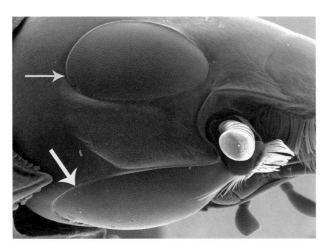

Figure 14.8 *Gyrinidae*, whirligig beetle. The animal has a pair of eyes above the water line and a pair that peers down into the water column. The right eye of each pair is illustrated with white arrows. *Scanning electron microscopy image by Pat Kysar.*

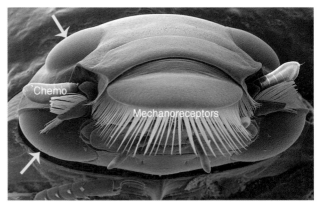

Figure 14.10 *Gyrinidae*, whirligig beetle. White arrows point to eyes on the right side of the animal. The upper arrow points to the aerial eye, and the lower eye points to the aquatic eye. There is a pair of chemoreceptors, with the right one marked with the word "Chemo." The whisker-like projections are mechanoreceptors and are so labeled. *Scanning electron microscopy image by Pat Kysar.*

requiring relatively high light levels; it is a versatile eye suitable for underwater and aerial vision (Figure 14.9).

These beetles can usually be found in groups of a dozen or so but sometimes as many as a hundred. Their name comes from their behavior: They swim in semicircular patterns at a frenetic pace. Although this behavior makes predators notice them, the beetles have a special mechanism for protection (in addition to their rather mundane color and shape). They are unpalatable. These invertebrates produce a toxic and repulsive compound from paired glands, called pygidial glands, at the rear of their body. Secreted with any attack, this compound is so toxic that fish that have been sensitized to the secretion will spurn whirligigs thereafter.

Other sensory mechanisms besides vision cause this creature to appear quite alien to our world. The "whiskers" on the labrum

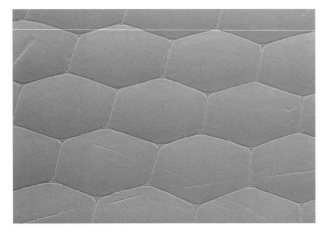

Figure 14.9 *Gyrinidae*, whirligig beetle. Hexagonal ommatidial units. *Scanning electron microscopy image by Pat Kysar.*

or "upper lip" and the palp (the odd fringed triangular tissue at the lateral edges of the labrum) are mechanoreceptors that sense what is happening at and below the water's surface (Figure 14.8). They enable beetles to know instantly if an insect is struggling on the surface as well as permit the beetle to know where his mates are and what they are doing.

The clublike antennae, extending laterally just above the aquatic eye on each side, are in contact with the water surface as chemoreceptors to "smell" their prey (Figure 14.10).

So these beetles have three highly developed senses—vision, mechanical, and chemical receptors—directed toward understanding the surface and water column. They are skillful indeed.

Whirligig fossils as recent as 150 mya have been found in the late Jurassic, but the order that contains them, Coleoptera, is probably twice that old. The whirligig beetles were by no means the first insect to gain land, but they illustrate the ease with which the apposition eye could have reached it, as this animal does on a daily basis. It is as if such an insect is a bridge between the two environments.

Permian Extinction

As the end of the Permian approached, both the marine and terrestrial life were rich and the variety stunning. But a sea change was coming. The conodonts (with the "first" vertebrate eye?), the trilobites (with the first known eye), and the ammonites (the early group of cephalopods related to modern octopi) were three key species in the story of the evolution of the eye. As the earth approached the end of the Permian, these species were all in decline but still extant. They would be obliterated in a massive great dying the likes of which the earth had not seen before or has since. This extinction process was most likely a prolonged affair with multiple causes or events over a few million years. It was a constellation of destructive events that produced profound numbers of deaths and probably included climate change, volcanism, methane release, and likely a meteor strike. Some combination or all of these processes resulted in a massive anoxic event that, while predominately marine, would leave only 5 percent of the original species living. Even the insects were severely affected.

Several reptilian lineages made it through the portal of the Permian extinction into the Triassic, including the turtles, the crocodiles, the tuatara (in its own phylum or as a basal lizard), and at least one other group of reptiles. Some of those reptiles would eventually lead to the dinosaurs. Other stem organisms for key radiations made it through that terrible process of the Permian extinction, including the protomammal, whatever that species looked like at the time.

The rich marine biota of sea scorpions, trilobites, most brachiopods and many others, though, were extinguished or severely affected. It would take the earth millions of years for Animalia to recover, but that recovery would lead to new species and a new order. The promise of the Mesozoic and its mighty reptiles would follow.

REPTILES PUSH THE OCULAR
ENVELOPE

15

THE AGE OF REPTILES

MESOZOIC ERA

TRIASSIC PERIOD

251–208 MILLION YEARS AGO

The Permian extinction was so profound as to silence 95 percent of species and 96 percent of animals of that period. Whether geologic or extraterrestrial, a single calamity or a combination of misfortunes, this pivotal event in life's history exterminated all animals larger than a wolverine and many much smaller.

By clearing the Paleozoic biological slate, The Permian-Triassic extinction (probably a combination of events over a few million years) made way for the Mesozoic Era. Life would expand again, and it would be largely reptilian.

Synapsids

Our mammalian predecessor, a synapsid, crawled though that curtain of death with a slightly more upright gait than that of the more basal cotylosaur. This synapsid, a mammal-like reptile, had separated from the vertebrate line that would lead to lizards, turtles, and crocodiles in the Carboniferous. This creature, though, looked nothing like the mammals that would evolve in the Cretaceous and Tertiary, more than 200 million years later. It followed its own trail with few tracks for us to follow during this period, so we know little about its eyes. This lineage had been almost obliterated by the Permian extinction, but some feature, or combination of features, such as size, habits, or the physiology of the animals in this clade permitted their survival. The lineage probably was a generalist in its niche and food requirements and diurnal or already nocturnal in its habits. This synapsid lineage, though, would have to wait for hundreds of millions of years to become a dominant group as mammals.

Once the Permian had ended, those few species that had survived would find many niches, old and new, open to them. This permitted and even encouraged further speciation.

Already honed by hundreds of millions of years of evolution, the vertebrate eye of the creatures that entered the Mesozoic era probably had surprisingly good vision. Ocular

development must have progressed quickly once land had been gained in the Devonian, a process driven by predation, carnivory, and natural selection. The giant amphibians, only vaguely resembling modern ones, that had dominated during the Carboniferous and helped confine the reptilian expansion during the Paleozoic, were gone. As cooler temperatures and lower oxygen levels of the late Permian or early Triassic restricted large amphibians, though, reptiles began to hold sway. Because of the cleidoic egg (Chapter 14), these first truly terrestrial reptilian clades were able to colonize the land.

Generally, the land was drying out and rising, literally and figuratively. This terrestrial freedom permitted radiation into a brighter and more defined environment. Much of the neurologic machinery necessary for terrestrial life and vision had been established in the reptilian or amphibian predecessor. By the Triassic, those stem reptiles had few competitors and many possible niches to fill. To fill them, the reptiles needed to improve their terrestrial sensory mechanisms, and vision would be of prime importance.

Sauropsids

The sauropsids diverged into the diapsids and anapsids, with the anapsids having been either a subgroup of the diapsids or a separate lineage (Figures 14.1 and 15.1). Diapsids and anapsids are so named because of the number of major holes or "fenestrae" in their skulls behind the eye sockets: anapsids have no holes; diapsids have two.

Anapsids

The anapsids are probably older and more basal than the diapsids. They consist of several now-extinct groups of early reptiles including the mesosaurs and placodonts. The anapsids' only possible extant order is the Testudines, or the tortoises, terrapins, and turtles. But more current

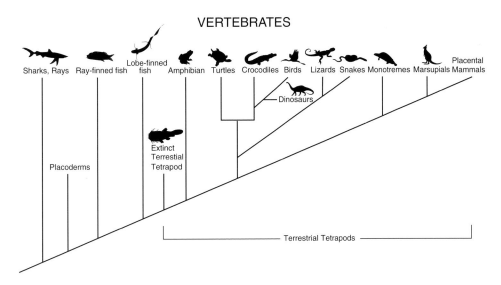

VERTEBRATES

Sharks, Rays   Ray-finned fish   Lobe-finned fish   Amphibian   Turtles   Crocodiles   Birds   Lizards   Snakes   Monotremes   Marsupials   Placental Mammals

Dinosaurs

Placoderms

Extinct Terrestial Tetrapod

Terrestrial Tetrapods

Figure 15.1 Cladogram of Vertebrate Phylogeny.

evidence suggests that turtles are diapsids that only *appear* to be anapsids because they convergently evolved by closing the fenestrae, so we will consider turtles to be diapsids. But whether anapsid or diapsid, turtles are ancient reptiles (Figure 15.1).

Diapsids

In the late Permian, the diapsids radiated into two broad groups, the Lepidosauromorpha (lepidosaurs) and the Archosauromorpha (archosaurs) (Figures 14.1 and 15.1), with both lineages entering the Triassic.

The Triassic diapsids displayed magnificent differentiation. They lived during a rich time of ocular evolution. The extant reptilian (sauropsidian) eye helps us to understand both the general differentiation and the state of ocular evolution during the Triassic. Although we cannot be certain of the morphology or physiology of the reptilian eye at the time, certain characteristics are seen in most diapsid (and probably synapsid) lineages. Snakes, an exception, are latecomers. They illustrate that neither animals nor their eyes are static, as we will see. Evolution continues, and the eye has continued to evolve as well (Chapter 19).

Many of the features of the reptilian eye are common to all lineages of diapsids, whether lepidosaurs or archosaurs, and were probably common to the synapsids of the period too.

The Adnexa of the Reptilian Eye

EYELIDS All reptilian lineages possess external eyelids, and most have moveable ones inherited from their amphibian ancestor. Chameleons have eyelids that are partially fused, leaving only a circular opening smaller than the corneal diameter—about the size of the pupil (Figure 15.2). Chameleons do not have a nictitans but rather regularly shed their skin including their corneal surface to keep it clear and clean.

Snakes, most geckoes, and some burrowing lizards have completely fused lids that have then become transparent. This creates a "spectacle," also called a brille. These fused lids are clear, creating a window of skin covering their eye and are considered a tertiary spectacle. These snakes and geckoes with a brille shed their skin and some of their spectacles. Geckoes that have movable lids (*Eublepharines*—only a few species) do not have spectacles.

So-called lidless geckoes (but really fused-lids geckoes not lidless ones) have no system to moisten the external

Figure 15.2 *Calumma parsonii,* Parson's chameleon.

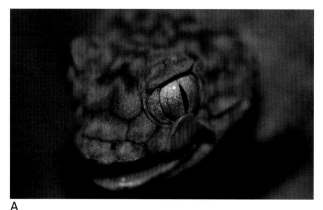

A

B

Figure 15.3 *Nephrurus asper*, rough knob-tailed gecko. **(A)** This gecko has no lids so must use its tongue to clean and clear its cornea. **(B)** *Nephrurus asper* targeting prey.

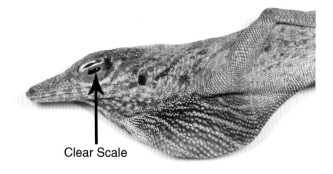

Clear Scale

A

B

Figure 15.4 *Anolis argenteolus*, Cuban anole. **(A)** Note the clear scale in the lower lid. The anole can bring the lower lid to the upper one and still see through the spectacle. **(B)** When the lid is open, the clear scale is an inconspicuous line beneath the lower lid margin. *Images* © *S. Blair Hedges.*

lids (spectacles), but when they need to moisten the ocular surface, they simply lick the eye (Figure 15.3).

Some lizards have a transparent window in their lower lid that may be useful for protection and yet is visually acceptable when the lower lid is closed. For desert-dwelling lizards, these windows also may protect against sandstorms or dry humidity (Figure 15.4). The same genetic mechanisms that permit clarity of the lower lid in some lizards must be at work permitting clarity of the fused lids in snakes and geckoes.

Within the reptiles, the majority of lizards that have movable lids also have a cartilaginous plate, called a tarsus, in the lower lid. Extant amphibians have no tarsus in either lid, suggesting that the lids were just forming as the amphibians came ashore and that there was little

stimulus to gain them. The tarsus would develop in more derived species for substantiation. For example, most mammals, including humans, have a tarsus in both the upper and lower lids, although the tarsus in the lower lid is smaller.

The lower eyelid is brought *up* to the upper eyelid in reptiles with movable lids, and the upper one moves very little if at all. Crocodiles do not have a tarsus in the lower eyelid but do have a bony plate in the upper lid shielding the eye to some extent, but this bone is part of the skull. Although they do not close their upper lids, these biological submarines can raise and seal their lower lids rather tightly to the upper lid, much like the hatch on a submarine, presumably for protection against flashing hooves of prey (Figures 15.5–15.7).

NICTITANS Even in the reptiles that have lids, the external lids do not blink often and are not used for frequent cleansing or wetting as our lids are. Reptiles that have lids, including a few geckoes, have a third eyelid, called the

Figure 15.5 *Alligator mississippiensis*, American alligator. Note the vertical slit pupil.

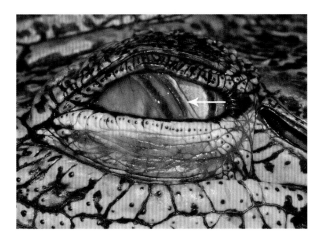

Figure 15.6 *Crocodylus porosus*, Australian saltwater crocodile juvenile. The lower lid can be raised to meet the upper lid, but there is no tarsus in the lower lid. The bony tarsus is in the upper lid, but it does not close. The white arrow illustrates the leading edge of the nictitans. *C. porosus images by Denise Brudenall, MA, VetMB.*

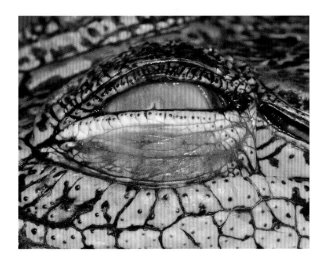

Figure 15.7 *Crocodylus porosus*, Austrailian saltwater crocodile. Note the closure of the eye with the nictitans beginning the process of the blink followed by the elevation of the lower lid to meet the upper lid. *C. porosus images by Denise Brudenall MA, VetMB.*

nictitans or nictitating membrane (Chapter 13). The nictitans originated in reptiles, although frogs do have an analogous structure that is also called a nictitans. In these frogs, though, this is not the same nictitans as found in reptiles but rather a fold of skin that is not homologous to the nictitans of reptiles. In amphibians, this "nictitans" is drawn from the inferior to superior aspect of the globe, although it serves much the same purpose—to wet and clean the eye—and appeared convergently. Snakes are an exception, as they have lids, albeit fused, but their nictitans disappeared during their fossorial period.

LUBRICATING GLANDS Movable lids usually contain lacrimal and/or Harderian glands to provide lubrication. The Harderian gland is specialized to provide an oily, sebaceous secretion along the leading edge of the nictitans, whereas the lacrimal gland produces salty tears. The Harderian gland, usually found in the ventral or ventral medial aspect of the orbit, is maintained in evolutionary descent even through birds. With only a few exceptions, the lacrimal and/or Harderian gland has appeared when lubrication is needed. *Eusthenopteron*, a lobefin fish (Chapter 13), the first known animal to have a lacrimal (or Harderian) system, appeared in the late Devonian, perhaps 100 million years before the Triassic. This early prototype of the nasolacrimal system may have begun to help lubricate food, a possible sign of the transition to a terrestrial lifestyle because completely aquatic animals would not need lubricated eyes or food. Some orders of reptiles, such as crocodilians and the tuatara, have only the Harderian gland, suggesting that the initial stem reptile (cotylosaur of Chapter 14) did not have a lacrimal gland either. But some lineages, such as the crocodiles, may have lost the lacrimal gland because of their proximity to water.

In reptiles with a lacrimal gland, it generally appears where ours does, in the upper temporal portion of the orbit. Snakes, an exception, have only a Harderian gland, whose oily secretions lubricate the space between the spectacle and the thin cornea (see Figure 19.1). These oily secretions then drain into the oropharynx, where they help to lubricate food, which may have been their initial primary function.

EXTRAOCULAR MUSCLES The retractor bulbi muscle (mentioned in Chapter 9 and Appendix B) persists in reptiles for protection (not to help with swallowing, as in frogs). To prevent some forms of trauma, the muscle can actively and forcibly retract the globe into the orbit.

In reptiles and birds, a muscle in the orbit called the pyramidalis muscle draws the nictitans across the eye.

This muscle and related muscles were likely derived from and are homologous to the retractor bulbi, first found in the Placoderms from the Silurian. That means that the retractor bulbi was retained in the lobefin fish (lungfish lineage) through the early tetrapods and into both radiations of the stem reptiles. Eventually, it would continue on to the birds and the synapsids that would lead to the mammals. But only some mammals retain the muscle. (See more information in Appendix B.)

Reptilian Eye

The reptilian eye evolved as a diurnal eye with a globe that is generally round, especially when compared to the variations found in the piscine and avian eye (see Chapters 10 and 20). In lizards and some other reptiles, the sclera is thin and must be supported by the cartilaginous cup (Figure 15.8). Much as in fish and birds, there is a cartilaginous cup encompassing the posterior two-thirds or more of the globe. Also as in fish and birds, most reptiles have bony plates, called scleral (or bony) ossicles, adjacent to the cornea. Scleral ossicles are bones that help support the eye and encircle the cornea like a skirt.

By the beginning of the Triassic age of reptiles, reptilian eye sockets, or orbits, were already proportionally large, especially compared with the brain. This indicates that much ocular development had already taken place in the Carboniferous and Permian in one or more of the stem organisms as they adapted to terrestrial life. These basal animals brought relatively well-formed eyes into the Mesozoic Era.

Figure 15.9 *Cyclura lewisi*, blue iguana. This beautiful reptile from the island of Grand Cayman has red sclera and a mostly cone retina. It is a diurnal reptile with few rods in its retina. Although incompletely studied, the animal probably has four visual pigments, and its visual range extends into the ultraviolet. Note the bright reflex from the cornea. *Image by John Binns, http://www.ircf.org/. Copyright International Reptile Conservation Fund.*

CORNEA The cornea is quite clear and steep, even when compared with that of amphibians. In the larger lizards (Figures 15.9 and 15.10), it is relatively thin, with distinct layers resembling our own cornea (Figure 15.11). Although the reptilian cornea differs from ours in some respects, this

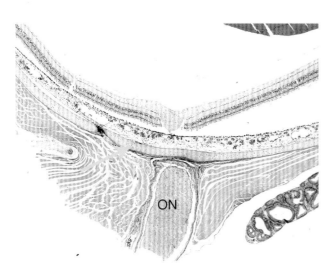

Figure 15.8 *Sphenodon punctatus*, tuatara. Note the cartilaginous cup (yellow arrow). The optic nerve is also marked (ON). *Image by Richard Dubielzig, DVM.*

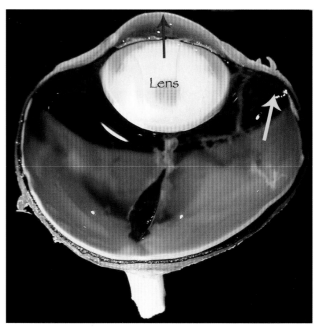

Figure 15.10 *Cyclura lewisi*, blue iguana, whole globe. Note the more tapered lens (labeled) and the thin cornea (blue arrow). The yellow arrow points to the ciliary body, which assists in accommodation. *Image by Richard Dubielzig, DVM.*

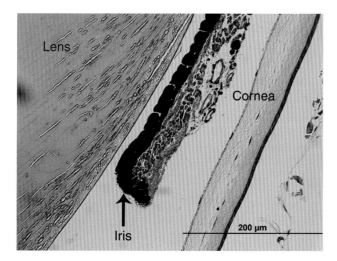

Figure 15.11 *Calumma parsonii*, Parson's chameleon. Note thin clear cornea. Iris and lens are labeled. Thanks to the San Diego Zoo for the specimen.

model is seen in almost all terrestrial vertebrates and had to have been present in a similar fashion in the original cotylosaur of the Devonian before that lineage divided into synapsids and sauropsids.

UVEA The eye's middle coat or vascular tunic, called the uvea, is thin, especially compared with that of birds. This morphology was probably present in the stem cotylosaur, as visual mechanisms for a terrestrial life were just beginning to evolve. This thinner vascular tunic would suggest rather thin retinae—thus fewer nutritional needs because the retina requires nutrition—in the earliest cotylosaurs and the continuing evolution of visual processing. As visual demands increased, the retina would thicken, and the uvea would have to thicken to feed it.

The evolution of retinal nutrition parallels the complexity of visual processing (Appendix C, Figure C1). A brief overview can be seen in the diagram (Figure C1).

The ciliary body (Appendix A; Figure A1), which is the front of the uvea just behind the iris, has at least two properties. It secretes the aqueous, the fluid that fills the anterior chamber, the space between the cornea and iris (see Appendix A), and is the principal organ of accommodation in most terrestrial animals. This muscular structure squeezes the lens directly or indirectly to help change its dimensions and shape, resulting in accommodation (Appendix E).

Phylogenetically, reptiles exhibit the first step in the process of using lens deformation for accommodation, one of the major changes in the visual system that the march to terrestriality required (Chapter 14 and Appendix E). The ciliary processes lengthened and fused to the lens capsule, permitting rapid lens deformation. As an accommodative

mechanism, deformation of the lens occurs only in reptiles, birds, and mammals, with the first two doing so by squeezing the lens directly.

CRYSTALLINE LENS Larger crystalline lenses increase photon capture, so eyes with them are generally more effective in dim light or at night. A nearly spherical lens, though, can require a disproportionately large globe and increase the anterior-posterior (AP) dimension. But the tapered lentil-seed-shaped lens of the newly crowned terrestrial creatures phylogenetically could enlarge more easily along the equatorial axis of the lens without filling the entire eye in the AP dimension (Figures 15.10, 15.12, and 15.13). Furthermore, the lenses became softer and more ductile as the lens proteins (crystallins) evolved to permit more rapid accommodation.

The lens of most reptiles is relatively flat on its anterior surface. Marine turtles and sea snakes are exceptions, with more fishlike lenses because these animals cannot rely on their cornea for refraction in a watery environment. But these turtles and snakes do not accommodate as fish do. This suggests that lens morphology can change relatively easily on an evolutionary basis but that the more complicated mechanisms of accommodation cannot.

Chameleons are ancient, but it is not clear when they diverged from the other squamates (the group that includes the lizards and snakes). Nevertheless, chameleons illustrate the common, shared features of the Reptilia and the divergent mechanisms, especially those of accommodation and lenticular anatomy.

Parson's Chameleon, Calumma parsonii

The Parson's Chameleon (Calumma parsonii) is one of the largest and most robust chameleons, with males growing to a maximum length of 65 cm (Figure 15.2). Madagascar is home to it and to more than two-thirds of the world's chameleon species.

These lethargic reptiles are stealth predators that can await their prey motionless and unnoticed for hours, moving only their turret-like eyes, which have special qualities. Their ocular adaptations for predation are most startling, yet appropriate.

Chameleons have a unique ocular device—a minus crystalline lens (the principal internal lens for focusing) (Figures 15.11 and 15.12)! All other animals have a plus internal lens. A plus lens causes light rays to converge and become focused; a minus lens causes them to diverge. The combination of a high-powered plus lens (the cornea) followed by a minus lens some distance away creates a Galilean telescope (Figure 15.13). The evolutionary trend toward increased reptilian corneal power combined with decreased crystalline lens power led animals toward better visual acuity and better discrimination. Chameleons have taken

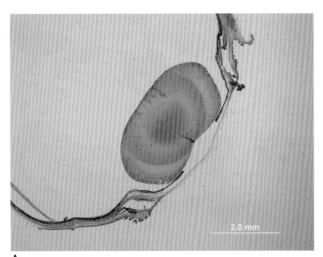

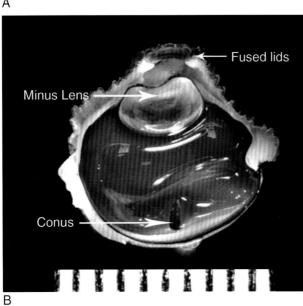

A

B

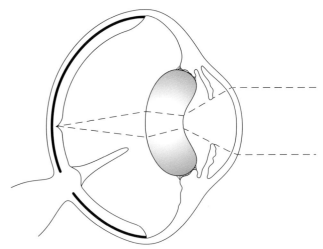

Figure 15.13 *Calumma parsonii*, Parson's chameleon. Artist rendition of optics of a chameleon eye. The cornea is the plus lens, and the crystalline lens is the minus lens. *Art by Tim Hengst.*

Figure 15.12 *Calumma parsonii*, Parson's chameleon. **(A)** The crystalline lens of chameleons is a minus lens and is unique in that regard. This creates a Galilean telescope when combined with the plus lens of the cornea and the minus crystalline lens. The convexiclivate fovea is a second minus lens, adding to the magnification. Specimen courtesy of the San Diego Zoo. **(B)** *Calumma parsonii*, Parson's chameleon. Note the conus and the fused lids discussed in the text. Note the position and shape of the minus crystalline lens. *Image by Richard Dubielzig, DVM.*

this to an evolutionary extreme, leading to a myopic lens. This optical configuration provides for a larger retinal image size on a comparative basis. In addition, chameleons have added a rather sophisticated eyepiece with a fovea that also magnifies the image.

These eyes appear bizarre, showing only the pupils, because the lids are fused open as well as to the eye itself, allowing for a comparatively small opening. Chameleons perform independent, uniocular scanning saccadic (rapid to-and-fro) and pursuit

(slower, more directed) eye movements. One eye may remain still and unmoving while the other eye disjunctively scans the environment until it finds potential prey, generally insects for smaller chameleons and small birds or mammals for large ones. Especially when the roving eye finds prey, it performs uniocular accommodation to assess distance during the scanning process.

These large and independent saccades are produced by the same set of extraocular muscles that other tetrapods have, which are very similar to our own. Human eyes obey a neurologic pairing of eye movements (called Hering's law of equal innervation). Because our eyes are yoked together neurologically, they move together, which means that the neurology of the extraocular muscles of the chameleons must be different. The neurologic management of these movements has evolved to permit these independent movements without sacrificing the ability to bring the eyes together for binocular vision, although there are no saccadic or pursuit movements in the same manner as with more conventional animals.

Once potential prey is located within the known capture distance, simultaneous binocular activity begins until both eyes are yoked neurologically. But stereopsis does not occur, at least as we understand it. Stereopsis, as we may recall from Chapter 13, is that precise neurologic overlapping of the image from each eye, especially from the fovea, that provides three-dimensionality to our vision. Shortly before the strike, coupled and identical accommodation in chameleons occurs to confirm the correct distance from the tip of the mandible. Accommodation ranges are large—up to 45 diopters—and very fast—up to 60 diopters per second—resembling those in birds or other reptiles. (By contrast, the human accommodative potential is about 20 diopters for an infant, declining to about 1–2 diopters for a sixty-year-old.)

Chameleons accomplish rapid accommodation using striated intraocular muscles found in reptiles (Chapter 14). Humans use

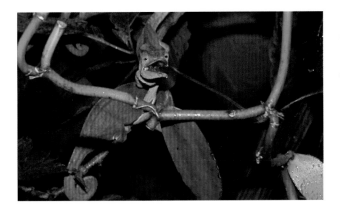

Figure 15.14 *Calumma parsonii*, Parson's chameleon. Note the capture; the bony element in the tongue is apparent in this image and in Figure 15.15.

smooth muscle for accommodation, which is much slower. The large and very rapid range of accommodation allows focus on near objects with great precision, coinciding with the projection length of the tongue. The retinal image plane is the most important cue used to measure distance to prey, and this projection onto the retina is achieved through precise accommodation.

The chameleon is a visual predator with a well-developed fovea that is convexiclivate (steep-walled) to help provide magnification of the image (Figures 15.14 and 15.15) and mediate increased visual acuity. But the chameleon may use binocular, not foveal, primary target fixation. It achieves binocular fixation by projecting the target onto the retina just temporal to the fovea with a slightly divergent gaze position.

The retinal ganglion cell maps of a chameleon reveal a horizontal visual streak and a second visual area temporal to the macula (area of best vision within the eye) where the image is probably located during this phase of prey capture. The ganglion cells are the last cells in the retina that receive input from the photoreceptors, summarize these signals, if necessary, and send them to the brain directly.

Chameleons use the cues from independent binocular accommodation rather than stereopsis to determine and confirm distance. When the chameleon prepares for the strike,

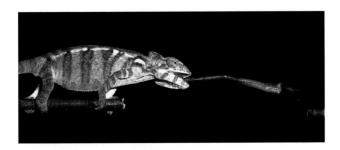

Figure 15.15 *Furcifer oustaleti*, Oustalet's chameleon. Note the capture of a cricket and how the tongue engulfs the insect. *Image by Kristiaan D'Août, Functional Morphology, University of Antwerp.*

it strikes with a most spectacular tongue (Figures 15.14 and 15.15).

The chameleon's tongue represents a physiologic masterpiece. For prey capture, it uses a most unusual muscular system to often exceed the chameleon's body length. In effect, the tongue is "shot" out of the chameleon's mouth as we might squeeze watermelon seed to project it.

IRIS The reptilian iris is often colorful, sometimes even with coloration that continues in stripes across the skin as if it were a single tapestry (Figure 16.10). The iris is composed of striated muscle. In the lizards, snakes, and turtles, the iris is thick. In some species, it has a complicated muscular arrangement for the dilation and constriction of the pupil, especially a slit or corrugated pupil. When the pupil is slit or complicated, the muscular system must be as well and permits rapid and more or less voluntary pupillary movement.

In some aquatic turtles, the iris is so stiff that it is used to retain the periphery of the lens during accommodation. The turtle lens is so soft and ductile (perhaps the most ductile in the vertebrate world) that the central portion of the lens squeezes through the pupil quite easily to create the anterior, or accommodative, lenticonus as discussed in Appendix E.

As the reptiles became more diverse, some would inevitably annex the night. Adaptations would precede that annexation. The pupil provides some understanding of those evolutionary changes.

The pupil is usually round in diurnal lizards and slitlike in nocturnal reptiles such as crocodiles and nocturnal lizards such as geckoes. Crocodiles can close their pupils completely, like a shade or horizontal drapes, which is useful in bright environments (Figure 15.16).

Many of the nocturnal lizards are exceptions to the round pupils seen in diurnal lizards, which tells us much about the evolution of pupillary functions. Just as other structures within the eye have evolved, so has the pupil, especially for nocturnal vision. Different niches have different needs and different solutions. With its peculiar pupil, the tokay gecko illustrates that principle.

Tokay Gecko, Gekko gekko

The pupil of the tokay gecko (Gekko gekko) creates an arresting image of this remarkable nocturnal reptile (Figure 15.17). At night, this noisy and aggressive gecko has a large pupil that is nearly round when fully dilated. But when this (occasionally diurnal) gecko must be active during the day, its pupils constrict to a vertical ellipse with four diamond-shaped pinholes decorating the vertical slit. These bizarre pupils with their multiple pinholes probably serve a variety of functions, but the most important function is employment as a Scheiner's disc.

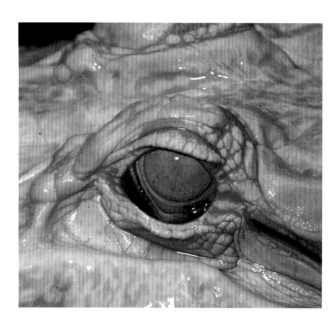

Figure 15.16 *Alligator mississippiensis*, American alligator. Note the pupillary closure in this albino alligator.

Figure 15.18 *Uroplatus phantasticus*, leaf-tailed gecko. Native to Madagascar, this beautiful gecko is occasionally diurnal and has a similar pupil as *Gecko gecko*, but this one has four or possibly six pinholes when the pupil constricts. Note the overlying operculum at the midpoint of the slit closure.

The Scheiner's disc optical principle states that an opaque disc containing two or more small holes will have a focus of light traveling through these holes at only one point on the retina. Murphy and Howland provided clear and convincing evidence that this principle allows the gecko to use these multiple pinholes as a focus indicator or range finder by forming four images of an object on the same point on the retina.

If the object lies precisely within the plane of focus, with a relatively shallow depth of field around that plane, the image will be sharply in focus. If the object is closer or more distant, the four images will not be in proper focus nor coincident. In that case the gecko will know that distance for capturing prey or avoiding predators, and each pinhole would provide relative image clarity if the brain could suppress the multiple images produced by the other apertures.

Figure 15.17 *Gecko gecko*. Note the four pinhole pupils that would result as the pupil constricts further

With the pupil constricted, the vertical polycoria (multiple pupils) combined with the slit would produce an image of considerable clarity without accommodative adjustment and would significantly limit the light flux to the retina. The vertical pupillary or stenopeic slit, an optical device common to nocturnal creatures, especially if they are occasionally diurnal (Figures 15.17 and 15.18), provides other potential advantages. A slit pupil tends to improve focus in the direction perpendicular to the slit, so a vertical pupil provides the best view of the horizontal meridian relative to the animal's head (and the horizon of a gecko, which is often on a wall or ceiling, may be different from yours). Many carnivorous predators, such as snakes and cats, also have vertical pupils, presumably to assist in the hunt on a horizontal plane.

By contrast, prey species (such as herbivorous mammals) have horizontal slit pupils, presumably to allow the animals to see an approaching predator in the horizontal meridian when their heads are lowered to graze.

The multiple apertures of the gecko eye and the slit provide a relatively larger visual field than a circular pupil of equal surface area. In darkness, even full pupillary dilation produces a hexagonal characteristic, leading some observers to believe the pupil may be under "voluntary" control.

But the pupil is only part of this animal's interesting ocular adaptations. Because the species is primarily nocturnal, some investigators consider that the retinal photoreceptors would be best described as rods, although these cells may be phylogenetically unique. Ultrastructurally, the outer segments of the single photoreceptors are large and resemble cones.

The retina contains two visual pigments that appear to be more like cone pigments with longer-wavelength absorption characteristics. Gecko photoreceptors have an unusual spectral sensitivity curve, which is similar to the human scotopic curve

(sensitivity during low light or nighttime levels) with one visual pigment having a maximum sensitivity at 521 nm and a shorter-wavelength maximum at 467 nm. The latter pigment is probably responsible for the ultraviolet capabilities. (For reference, rhodopsin in your rods has a maximum sensitivity at about 500 nm.) These two pigments may be the residua, or vestige, of complete color vision in a diurnal phylogenetic ancestor.

Otherwise, it is more difficult to speculate on the presence of two photopigments in these nocturnal predators. Geckoes are not the only species with two photopigments (whether described as rods or cones). The best evidence suggests that geckoes arose from a diurnal line of lizards well suited for bright light with such retinal cells as double cones. And that suggests the last common ancestor had an all-cone retina, and at least two of these cones evolved to resemble rods. These nocturnal geckoes would have had cone photoreceptors better suited for daylight, but although these cones were transmuted to be rods, they still resemble cones—shards of the ancient ancestry of the geckoes. Some of these geckoes, notably Phelsuma, have even become secondarily diurnal, and the trail of nocturnality can still be found in the morphology of the eye.

RETINA Perhaps the most important advance in the visual witness during the period of lizard development was in the lizard retina. Although we cannot know exactly when the lizard retina evolved to its current degree, it was likely to have occurred very early in the evolution of these lineages. We know this because the lizard retina is only somewhat less well organized than that of birds (Chapter 20) and very similar. The basic plan for the avian retina had to have been set long before the avian lineage appeared during the late Jurassic. The lineage that would ultimately lead to birds (archosauria) and the separate lineage that would lead to lizards (lepidosauria) diverged from a common ancestor in the late Permian. The retina had to have been quite complex at the time or was simultaneously evolved in an example of neurologic convergent evolution.

The lizard retina is thick, highly organized, and complex with well-defined layering, very much like that of the birds. Both the lizard and avian retinae permit their owners to receive and process images quickly with transmission of a more highly organized image to the brain. When compared to the thinner, simpler, and less-well-organized retina of the more basal chelonians (turtles), the lizard retina suggests much greater reliance on vision as the principal sensory modality (Figure 15.19). There are some exceptional turtles we will meet later, however (Chapter 16), and these illustrate the key role these animals play in understanding the evolution of the eye.

With some exceptions, the visual acuity in lizards is excellent, and most diurnal lizards as well as the nocturnal

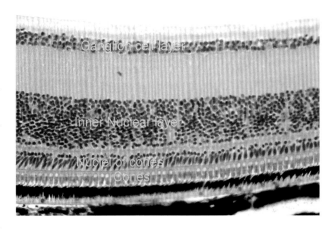

Figure 15.19 Gecko retina. Note cone-rich retina and well-organized, labeled cellular layers. Note the oil droplet within each of the cones. Consult Appendix A for further explanation on retinal layers. *Image by Richard Dubielzig, DVM.*

tuatara have a single fovea in each eye. Many lizards possess a deep and steep-walled fovea that further improves vision. A deep, steep-walled fovea (called a convexiclivate fovea) will cause light rays to diverge as these rays strike the sides of the walls so that the image is enlarged slightly (Figure 15.20). Many lizards, such as chameleons, have a larger and more densely packed fovea than humans. The density of the photoreceptors in the fovea of chameleons and most other lizards permits better discrimination and color detection when compared to most other vertebrates.

Most diurnal lizards and turtles have kept the single and double cones with oil droplets (Figures 13.3 and 13.4) as seen in certain fish, including the predecessors of the tetrapods (Chapters 10 and 13). Even one of the smallest, legless *burrowing* lizards, the California legless lizard, has oil droplets in its retinal cells. As seen in Chapters 10 and 13, the oil droplets probably assist in discrimination.

But not all reptiles necessarily have good vision. Only a few snakes have even a shallow, poorly defined fovea, and other snakes have none at all. Vision is not usually the principal sensory mechanism in this lineage, and none of the crocodilians has a fovea, as they do not rely on vision as a principal sense either.

In most diurnal lizards, the retina is almost always cone-rich. Many chameleons, such as *C. parsonii*, described above, have an all-cone retina with a high density of foveal cones. Except for those lizards that have become nocturnal, such as some of the geckoes and the tuatara, the lacertilian clades have kept four visual pigments. These characteristics are similar across all lineages of lizards but not necessarily the other reptiles (Figures 15.19 and 15.20).

Evolutionarily, as the retina thickened and became more complex, visual processing also improved. These

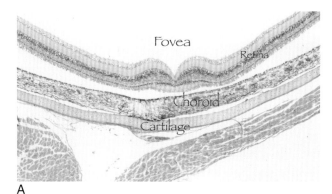

A

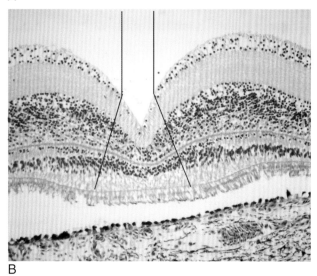

B

Figure 15.20 *Sphenodon punctatus, tuatara.* **(A)** Convexiclivate fovea in Tuatara. Note the steep walls of the fovea, which cause rays to diverge and hence enlarge the image by spreading it across more photoreceptors. **(B)** Convexiclivate fovea with overlying diagram of light path. The diverging rays illustrate how this principle helps enlarge the image at the level of the fovea. The image covers more photoreceptive cells, and finer detail is perceived. *Image by Richard Dubielzig, DVM.*

changes would demand improved retinal nutrition and encourage and empower lizards as a terrestrial vertebrate species.

Retinal nutrition mechanisms have matured in lizards, so they may provide a hint to the evolution of this tribe (Appendix C). With few exceptions, lizards have a nutritive organ, called the conus (Figure 15.21). Embryologically, this plexus of blood vessels is derived from neuroectoderm (from the neural crest), just like the tissue in their eventual descendants, the birds. Birds have a very similar but more sophisticated version of the conus, called the pecten (Chapter 20). The conus and the pecten are so similar as to suggest a more or less direct descent with the intermediate organisms (such as the therapods) almost certainly having had a similar structure.

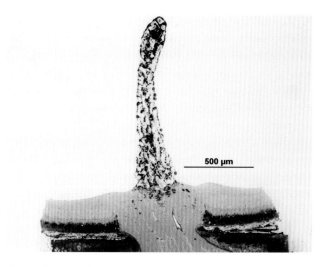

Figure 15.21 *Basiliscus galeritus,* basilisk conus. The blood vessels come directly through the optic nerve to form a plexus. *Image by Christopher M. Reilly, DVM, University of California, Davis.*

One lizard illustrates the differences in—and importance of—retinal organization and nutrition.

Basilisk, Basiliscus galeritus

Basiliscus galeritus (Figure 15.22) represents the lizards and the complexity of their lifestyle. This mostly carnivorous reptile must accomplish visually complex tasks as it is a diurnal predator of invertebrates and other vertebrates. It also illustrates the visual imperative placed on lizards as they became terrestrial. They are visual hunters, but they are also prey. These creatures must develop unusual strategies requiring sophisticated visual abilities. This lizard has a particularly unusual skill.

This animal can walk on water. The physics of this ability are understood but are complicated and beyond the scope of this text. Nevertheless, this animal relies on speed and agility to

Figure 15.22 Basiliscus galeritus, basilisk. Image courtesy of Joe Burgess, The International Reptile Conservation Foundation, http://www.IRCF.org.

locate and dispatch prey as well as escape predators. This requires considerable visual skill.

A robust and complicated ecological niche, such as the one filled by B. galeritus, would require changes to the visual system that would not have been present in the early semiaquatic tetrapods that preceded the reptiles. Although it is not clear when these changes occurred, and B. galeritus is a recent descendant, the demands of a completely terrestrial lifestyle suggest that the lacertilian lineage would have had to face these dramatic changes early in the transition from water to land. These changes would have been driven by the need to eat and the risk of being eaten. At the same time, other anatomical demands were being made to the evolving globe as the retina thickened and visual processing became more complex.

Such a terrestrial niche would require an improved image quality, a faster mechanism of accommodation, and improved inner retina nutrition. All of this was accomplished in the evolution of the reptilian eye.

Figure 15.23 *Sphenodon punctatus*, tuatara.

Lepidosaurs

During the Triassic or early Jurassic, Lepidosauria radiated into several extinct orders and two living orders—Sphenodontia and Squamata. The order Sphenodontidae, which arose at least by the early Triassic, now consists of only one remaining genus, the tuatara. Later, the lepidosaur lineage radiated into Squamata that includes the lizards and worm lizards. Several of Squamata have been discussed above in the description of the reptilian eye. Snakes are considered part of Squamata too but as arising much later from within the lizard lineage (Chapter 19).

The tuatara, then, is the oldest living member of the lepidosauria lineage and teaches much about the early reptilian eye.

Tuatara (Sphenodon punctatus)

Although the dinosaurs are gone, living lepidosaurs provide clues to the development of photoreception among the clades that preceded the dinosaurs. The tuatara (Figure 15.23) remains as the last member of its order and was at least contemporary to other less derived tetrapods of the time.

The ancient order Sphenodontia, which includes Sphenodon as its only family with two surviving closely related species, had its heyday between 200 and 100 mya. After the lepidosaurian line gave rise to Sphenodontia, the line continued to radiate into lizards, crocodiles, turtles, snakes, and eventually birds. The tuatara is an ancient lineage, and the eye and visual mechanisms may tell us about the ocular anatomy and physiology of the time, but we do not know the degree of ocular evolution since this animal arose.

Approximately 100 mya, as other members of this order succumbed to extinction, Gondwana broke apart, and New Zealand

and its islands broke free from Australia, thereby isolating this last member, presumably without competition. The tuatara survives as a relict.

A nocturnal lizard-like animal from remote islands of New Zealand, one of the tuatara's important features is the third or parietal eye (Figure 15.23–15.25). In the tuatara, this eye is

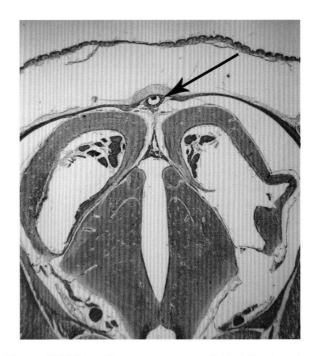

Figure 15.24 *Sphenodon punctatus*, tuatara. Vertical slice through the head of a tuatara. In the midline is the parietal eye (black arrow) with a subtle thickening of the tissue above the eye for cornea concentration. Image courtesy of Ung, C. Y-J. and Molteno, A. C. B. (2004). An enigmatic eye: The histology of the tuatara pineal complex. *Clinical and Experimental Ophthalmology, 32*, 614–618, © Wiley-Blackwell.

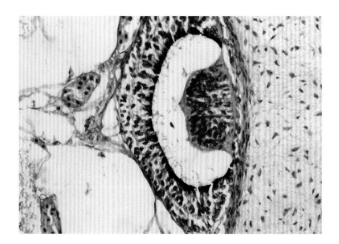

Figure 15.25 *Sphenodon punctatus*, tuatara. The parietal or third eye of a tuatara. Note the lens/cornea combination as the lens unites with the layers of skin dorsal to it. The photoreceptors are not inverted as they are in the lateral eyes of the reptile (or like the photoreceptors in our eyes), and there are no processing cells as there are in the retinae of the lateral eyes. The photoreceptor cells synapse (connect) to the next cell, the ganglion cell layer, directly, and these cells send axons out of the eye. Image courtesy of Ung, C. Y-J. and Molteno, A. C. B. (2004). An enigmatic eye: The histology of the tuatara pineal complex. *Clinical and Experimental Ophthalmology, 32*, 614–618, © Wiley-Blackwell.

Figure 15.26 *Tiliqua scincoides scincoides*, eastern blue-tongued lizard. Reptilian external third eye. The central dorsal scale illustrated with the arrow shows a small clear window that is, in effect, a cornea for the third eye.

robust, which suggests that the early lepidosaurs relied on it to determine circadian rhythm.

The third eye has been lost in the radiation into later orders, such as crocodiles and birds, although remnants can be found in most of these. Because mammals also have a remnant, even the synapsid radiation must have had these organs. This is not a surprise as we saw a parietal eye in the lampreys. The third eye represents evolution's earlier approach to photoreception and circadian rhythm.

In tuataras, as in the other reptiles that have the third eye, it is a dorsal midline structure just beneath a clear parietal plug. This small translucent or transparent window appears—and is—like a cornea (Figure 15.26). Immediately beneath the parietal plug is a lens that is surprisingly similar, at least on a histologic basis, to lenses of the lateral eyes. The vitreous cavity analogue is behind the lens within the eye and above (dorsal to) the retinal layers. Pigment cells are at least analogous to the retinal pigment epithelium, but the pigment in these pigment cells in the tuatara is distributed differently. The pigment cells surround the ciliated ends of the photoreceptors that include the visual pigment. These primitive (compared to other such cells in the lateral eyes) ciliated photoreceptors, which are everted (our photoreceptors are inverted with the photopigment on the end of the retinal cells furthest away from the lens), point directly toward the lens and have no cellular elements between them and the incoming light (Figure 15.25).

Ganglion cell layers are proximal to the photoreceptors (ventral or beneath), much more like the retina of an octopus rather than that of a vertebrate (Figure 15.19). These ciliated photoreceptors of the parietal eye are similar to those of the lateral eyes of Sphenodon *and are sensitive to blue or green light just like those photoreceptors found in vertebrates that existed hundreds of millions of years before the tuatara.*

Why should the photoreceptors be everted in the third eye but not in the lateral eyes? As explained by Eakin in The Third Eye, *the answer relates to embryology. In vertebrates, all eyes begin as evaginations (outpouchings) of the diencephalon (early brain) to create optic vesicles, and the lateral eyes proceed to invaginate, forming optic cups. But the third eye never invaginates. It is lined with the ciliated epithelium that becomes the photoreceptors; the ciliated portion extends inward toward the center of the vesicle or cyst. The most distal (and dorsal) portion of the third eye's evagination condenses to become a lens, leaving the cyst lined with everted cells. The portion of the developing third eye that condenses into a lens, then, is homologous to a portion of the retina in the lateral eye.*

Furthermore, the ciliated cells that line this cyst in the third eye differentiate into pigment epithelium and everted photoreceptors. It is as if some stimulus is missing, and a complete "normal" vertebrate eye never forms. This explanation is not entirely satisfying, but it is the best one we have.

The parietal eye in reptiles, including *Sphenodon*, is not laterally symmetrical, which in itself is unusual in the world of bilaterally symmetrical organisms such as vertebrates. The reptilian third eye is actually believed to have developed as a pair of diencephalic evaginations, with the more rostral (towards the anterior portion of the head) and left-sided portion becoming the parietal eye and the more caudal and right-sided portion becoming the pineal sac. Animals that have lost the parietal eye, including mammals, retain the pineal sac and condense it into the pineal gland. In reptiles and humans the pineal synthesizes melatonin (and many other neuroendocrine regulatory compounds). Melatonin influences vertebrate thermoregulation as well as the circadian rhythm by acting as a somnifacient.

Frogs and even some fish have a homologous photosensitive organ, suggesting that this central dorsal third eye is much older than reptiles and probably belongs to our watery beginnings as chordates. The fact that the lamprey, a basal fish, has the asymmetrical pineal complex, suggests that the structure probably leads as far back as the protochordates. But because the pineal complex seems to be in decline even among the reptiles and in subsequent radiations, it may be going the way of the appendix.

The tuatara is a nocturnal wisp of early Reptilia and probably does not represent the main trunk of reptilian evolution. Still, when compared to other living branches of lepidosauria, we can better understand the anatomy and physiology of the early reptiles.

Archosaurs

The Archosaurs radiated into several groups that are now extinct, including dinosaurs and pterosaurs, and other groups that survive, including the turtles, crocodilians, and, later, birds. Turtles appear to have radiated from the early Triassic stem reptiles, with the occlusion of the major fenestrae in the bony skulls as a form of convergent evolution. Thus, turtles resemble the anapsids but did not descend from them. Recent evidence suggests that turtles are closer to birds and crocodiles (archosaurs), and the turtle eye would fit with that designation.

Plesiosaurs, later in the Triassic, also came from that same stem reptilian ancestor. They and the ichthyosaurs represent a different order from the dinosaurs and pterosaurs.

Ichthyosaurs may have evolved in the late Permian, but the first such fossils are from the early Triassic. As marine reptiles, the ichthyosaurs must have been stunning and dramatic animals. These whales of the reptilian world evolved into fearsome and enormous reptilian predators. Although no ichthyosaur survives to show us its ocular anatomy, we know that this order claims several records in the visual world, and one of them is the largest eye. Archosaurs are worth a visit.

MARCH OF THE ARCHOSAURS

16

The Triassic and the Jurassic saw the rise and divergence of Reptilia—principally the anapsids and the diapsids. The diapsids diverged into the lepidosaurs (Chapter 15) and the archosaurs. The archosaurs blossomed into multiple storied lineages in the Triassic and Jurassic.

Archosaurs

Archosaurs radiated into turtles, crocodilians, pterosaurs, dinosaurs, and, eventually, birds. Most researchers see turtles as part of the archosaurian (diapsid) radiation (Figure 14.1), although some consider turtles to be a sister group, much like the ichthyosaurs. The turtles are interesting because of the variety of visual mechanisms illustrating the maturity and plasticity of the eye.

The Order Testudines (Turtles, Terrapins, and Tortoises)

Tortoises are believed to have differentiated early from the terrestrial vertebrate line. Recognizable members of the order Testudines are believed to have begun as terrestrial animals. The oldest fossil, which looks like current tortoises, is from the late Triassic (approximately 215 mya). Some of this order returned to water as terrapins (amphibious) and turtles (virtually entirely aquatic).

The tortoise is likely the most basal of the extant reptiles (Figures 14.1 and 15.1), so it follows that its eye is likely the most basal existing one in the reptilian lineage. A review of this eye reveals the commonality and the contrasts with the other reptilian orders. The law of parsimony helps us know the eye's state at the time of divergence. The term "turtle" will be used to describe all three lineages (tortoise, terrapin and turtle), as the eyes are similar.

TURTLE EYE In general, the eyes of turtles (also known as chelonians) are similar to one another and to those of lizards (Figure 16.1). In most cases, though, the chelonian eye is simpler and more basal. This less-developed eye hints

at what must have been present in that cotylosaur (early stem reptile), before modifications (ocular evolution did not stop with the first turtle). Although some researchers believe that several turtles lost some reptilian adaptations, especially as those turtles became aquatic, vision would have played a considerable, or even major, sensory role in most chelonian lifestyles. The chelonian eye has evolved during the ensuing 200 million years depending on the adopted niche and the relative importance of vision in that niche.

Like that of other reptiles, the chelonian eye is basically spherical, but unlike that of a lizard, it has little differentiation from the curve of the cornea to the curve of the sclera

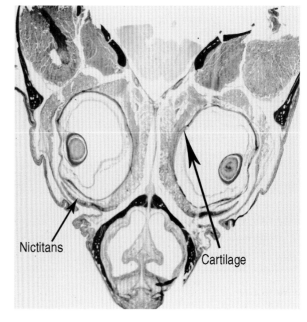

Figure 16.1 *Chrysemys picta*, western painted turtle. A section through the western painted turtle's head. Note cartilage and nictitans—both labeled. *Image by Richard Dubielzig, DVM.*

(Figures 16.1 and 16.2). The turtle eye is considered diurnal with a relatively small crystalline lens (Figures 16.2 and 16.3), although some marine turtles are active in oceanic waters so deep that it might as well be night. Like the eye of other reptiles and fish, the chelonian eye contains a cartilaginous cup (Figures 16.1 and 16.4). Scleral ossicles, or bony plates, surround the rim of the cornea and are located just behind the cornea (Figure 16.5). These plates overlap to support the outer aspect of the globe.

Some of the internal anatomy of the turtles' anterior segment is midway between those of amphibians and lizards. For example, like frogs (and fish), some turtles have an intraocular transversalis muscle that moves the lens. The two species' intraocular muscles are probably homologous (of the same origin). Like lizards, turtles use lens deformation to accommodate (change lens focus), probably showing that the most recent common ancestor of all reptilian clades could do the same (Figure 16.6).

In most reptiles and all turtles, the crystalline lens is very soft, ductile, and easily molded, permitting rapid accommodation (Appendix E). Striated muscles in all reptiles (and birds) make for rapid accommodation, too (Appendix E). But the protoreptile that diverged into the synapsids and diapsids was unlikely to have had that capability of rapid accommodation, and accommodation

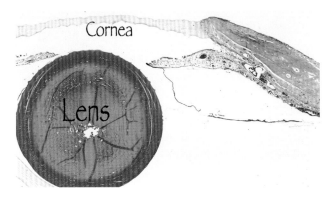

Figure 16.3 *Dermochelys coriacea*, leatherback turtle. Note the small lens and thin cornea. The iris is seen above the lens and is relatively muscular. *Image by Richard Dubielzig, DVM.*

proceeded in two directions at the divergence of the synapsids and diapsids.

The mammalian radiation from the synapsid line had smooth muscles for accommodation and a different physiology for accommodation but still arrived at lens deformation as a mechanism of accommodation. Although this represents convergent evolution within the eye, there are enough differences to cast doubt on a common ancestor with these mechanisms (Appendix E).

Most marine turtles accommodate by using ciliary body muscles that press on the iris while using zonular fibers to help deform the lens, although some muscles press directly on the lens (Figure 16.6) (Appendix E). Some turtles use the transversalis muscle to move the lens not to accommodate

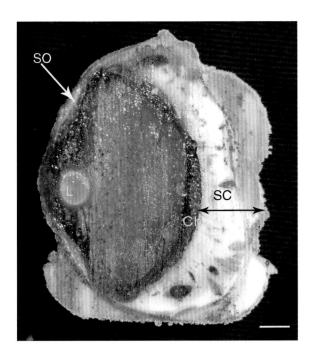

Figure 16.2 *Dermochelys coriacea*, leatherback turtle. Cross-section of deep-frozen right globe illustrating small, round lens in relation to globe size. L, lens; SC, scleral cartilage; thickness indicated by double-ended arrow, SO, scleral ossicle; Ch, choroid. Bar = 5 mm. *Image by Denise Brudenall, MA, VetMB.*

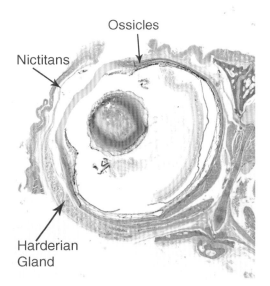

Figure 16.4 *Emydura subglobosa*, red-bellied short-necked turtle. Note the cartilaginous cup, the scleral ossicles and the nictitans. The Harderian gland can be seen within the base of the nictitans. *Image by Richard Dubielzig, DVM.*

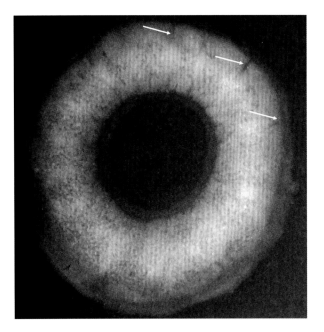

Figure 16.5 *Dermochelys coriacea*, leatherback turtle. Scleral ossicles in turtle eyes. Radiograph of scleral ossicles showing almost complete ossification. Radiolucent lines are suggestive of the previous outline of individual plates (arrows). The ossicle ring is oriented with the dorsal aspect uppermost. *Image by Denise Brudenall, MA, VetMB.*

but to permit binocular and stereoscopic vision. Turtle eyes are lateral, so for stereoscopic vision (Figure 16.7), the eye must turn forward, and the lens must move to achieve it.

Although the chelonians probably began as terrestrial animals, some later transformed into marine animals that would fill pelagic niches by adopting different lifestyles and visual mechanisms.

Marine turtles illustrate the challenge of different niches and evolution's response.

The Leatherback Turtle, Dermochelys coriacea

The superfamily of fully marine turtles, which arose by the late Jurassic, would produce some monster specimens. Although endangered, one of these behemoths remains alive today.

The leatherback marine turtle (Figures 16.2, 16.3, and 16.8) is the sole member of the family Dermochelyidae. *It also is closely related to the family* Protostegidae, *which contained Archelon, at 2500 kg, the largest of these monstrous creatures.*

The leatherback is the largest extant reptile. At its peak, it weighed almost 1000 kg, though human predation probably has eliminated animals that big.

In contrast to our image of turtles as slow, these marine turtles not only grow quickly—more so than any other reptile—they also swim quickly. Furthermore, they can outmaneuver a shark and dive to at least 1200 meters, and they rival the great whales for their panoceanic, pole-to-pole migrations. The leatherback owes these accomplishments to its relatively warm muscle temperature (like the tuna's, Chapter 10), due to fatty insulation and a countercurrent heat exchange system in its limbs. Its body temperature can be as much as 18°C higher than the surrounding seawater, providing the necessary metabolism for such odysseys.

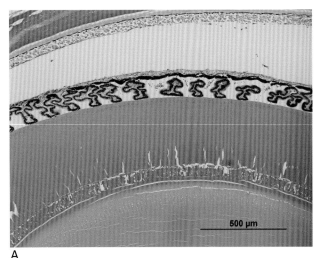

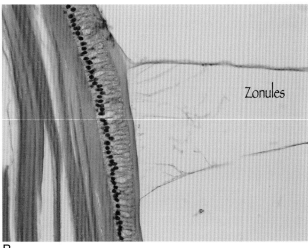

A

B

Figure 16.6 (A) *Geochelone elegans*, star tortoise. Note the muscular fibers with direct pressure on the lens. **(B)** *Dermochelys coriacea*, leatherback turtle. Note zonular fibers attaching to lens capsule of turtle lens. These two mechanisms permit the turtle to change focus by changing the shape of the lens. Phylogenetically, turtles are the first animals to use lens deformation for accommodation. Many tortoises and some turtles use direct muscular pressure via the ciliary processes, but marine turtles use these zonules much as do mammals. *Image by Richard Dubielzig, DVM. Accommodation panel Figure 10.8.*

Figure 16.7 *Chelodina longicollis,* eastern snake-necked turtle (Australia). Note that both eyes are turned forward to view binocularly.

Leatherbacks primarily feed in open ocean, unlike other marine turtles, which tend to feed in epipelagic or coastal waters. For leatherbacks, as for most animals, visual ecology generally determines visual specialization. The leatherback's retina has relatively wide visual streaks. These are horizontal concentrations of ganglion cells that indicate concentrated photoreceptors, providing improved visual acuity in that area. To help it feed on jellyfish (Figure 16.8), the leatherback also has an "area temporalis," another concentration of ganglion cells, in the retina's superior temporal portion. The turtle uses these concentrated visual cells to spot the jellyfish aggregation in the water column below it.

Figure 16.8 *Dermochelys coriacea,* leatherback turtle with lion's mane jellyfish in its mouth and trailing behind the animal. The leatherback's principal food source is jellyfish, and with this diet the animal consumes a great deal of salt. *Image © The Canadian Sea Turtle Network.*

The visual acuity of leatherback is not particularly sharp (the optics of the living animal has been analyzed only in hatchlings), nor does it need to be. After all, jellyfish are large and easily visible. But the leatherback and its vision are flexible in that they show both diurnal and nocturnal activity and feeding behavior. They contrast with other marine turtles, which are diurnal.

Leatherbacks feed almost exclusively on jellyfish (Figure 16.8), a high-salt diet indeed. Because jellyfish are diploblasts, they have virtually the same concentration of salt as the surrounding seawater. Salt also comes in the form of seawater, which the leatherback swallows with the jellyfish. But the turtle's throat is designed to eliminate some of that seawater. Backward-directed esophageal papillae, or enlarged spikes, retain the jellyfish while allowing the esophagus to expel much of the excess water with a strong muscular contraction. (Before the seawater leaves, however, virtually the entire gut will extract oxygen from it.)

But enough salt remains from the food and water to leave the turtle with a salt problem. The answer to this problem lies in the lacrimal glands, which are nearly twice the size of the leatherback's brain. They must be the largest lacrimal glands for body weight of any creature. These glands are used to concentrate and excrete salt so efficiently that a hatchling deprived of fluids will drink seawater to rehydrate. The lacrimal glands can excrete fluid that is six times the concentration of blood and twice that of seawater. Evolution has found an elegant and extraordinary solution to a difficult physiologic problem.

The fact that terrestrial turtles are more visually active than their pelagic cousins suggests that marine turtles may have lost some of their visual abilities. After all, marine species do not rely on visual acuity to the same degree as many other turtles, such as those that live in clearer shallower waters or semiaquatic turtles. Furthermore, even though terrestrial turtles are the most basal reptilian radiation, they may have subsequently evolved many of these abilities. By continuing to evolve and adapt, turtles have achieved a remarkable eye. Because they rely on visual mechanisms, freshwater turtles reveal the most about the visual potential and mechanisms of the turtles, as is illustrated by this one.

Red-eared Slider, Trachemys scripta elegans

As a class, birds have the best and most complicated visual perception and acuity, but a turtle may have the most complicated retina. The red-eared slider (Figures 16.9 and 16.10) has four different colored oil droplets and four different visual pigments (617 nm or "red," 515 nm or "green," 458 nm or "blue," and 372 nm or "ultraviolet") included in a retina that has rods and both single and double cones. Pigmented oil droplets act

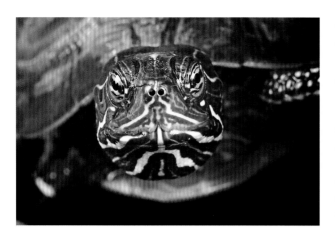

Figure 16.9 *Trachemys scripta elegans,* red-eared slider.
Image © Michael D. Kern, www.thegardensofeden.org.

as colored sunglasses placed in front of that photoreceptor.
*The compound's absorbance determines the principal color that
crosses the droplet. This absorbance effectively changes the distribution of light striking the visual pigment contained in the
other segment. The droplets also narrow the spectral sensitivity
function of the cone and reduce the overlap with adjacent
spectral types. The droplets are optically dense, so essentially no
light of a shorter wavelength gets through them, thus shifting
the spectral sensitivity.*

*The red-eared slider has seven cone populations and one rod.
The photoreceptors have four different cone visual pigments,
with four oil droplets creating a wide range of color stimulation
and discrimination.*

*Like lungfish, some amphibians, most lizards, all birds, and
even some mammals, turtles have oil droplets in their photoreceptors. The presence of oil droplets in these diverse groups*

Figure 16.10 *Trachemys scripta elegans,* red-eared slider. Note the
eyes are set high on the head so the animal can rest near the water's
surface but keep most of its body submerged. Note the black streak
diagonally through the pupil, probably for camouflage.

*suggests that the early tetrapods and the stem reptile (cotylo-
saur) had them as well. The lobe-finned fish like the lungfish
(Chapter 13) also bequeathed four visual pigments and a superficial retinal vascular system to the turtles.*

*Turtles share other retinal adaptations with birds. Both
turtles and birds have double cones that are neurologically and
optically coupled photoreceptors. These cone pairs probably help
both types of animals with motion detection and discrimination. The commonality of the oil droplets, visual pigments, and
double cones in the retina suggests descent from lungfish radiating to the stem reptile to the synapsids and diapsids.*

Other explanations are possible, but the lineage of turtles is old and suggests that the eye was surprisingly
sophisticated at least by the time the tetrapods radiated
into the turtles.

Crocodilians

The saurian triumph of the Mesozoic ended 65 mya with a
cosmic missile slamming into the Yucatan. In the dark ooze
that remained after that catastrophic explosion, opportunistic creatures survived by capturing live prey, feeding on
carrion, or even going long periods without feeding at all.
These grisly veterans—the crocodiles—are so old that they
can trace their lineage to perhaps the mid-Triassic as
another radiation of the archosaurs. One of these basal
reptile-like creatures led to the crocodilian lineage, although
it probably did not resemble modern crocodiles until the
Jurassic. From those tough and resistant predecessors
came our modern-day crocodilians with few, and mostly
conservative, morphologic changes in their skeletal
structure and visual mechanisms. Why should they change?
They were successful. These animals had become mostly
nocturnal with less emphasis on vision but with many of
the original characteristics. The eye followed suit, becoming
nocturnal with the emphasis on rods and a tapetum
(Appendix H and Chapter 10). A look at the order Crocodilia
will illustrate the evolution and radiation of reptiles into
the night.

American Alligator, **Alligator mississippiensis**

The American alligator (Alligator mississippiensis) *is one of
the twenty-three species of extant crocodilians (Figures 15.5–
15.7, 15.16, 16.11–16.13). Living in swamps and bayous in the
southeastern United States, this species radiated from the principal line rather recently, in the Tertiary. Their eyes are so
similar to those of crocodiles that the eyes must have been stable
for hundreds of millions of years.*

*The crocodilians, represented here by the American alligator,
have several ocular amphibious adaptations that shed light
on ocular evolution. The eyes are set above the principal dorsal*

Figure 16.11 *Alligator mississippiensis,* American alligator. The lens is photographed from the perspective of looking out toward the world. The vertical pupil can be seen through the translucent lens. Note the large spherical lens similar to a piscine lens. The billowy retina can be seen detached from the peripheral sclera as an artifactual consequence of cutting the globe at the equator.

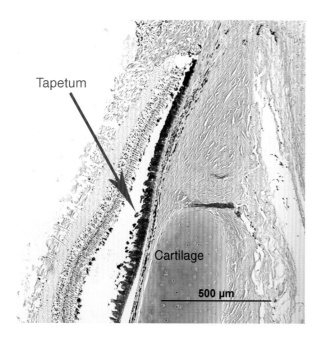

Figure 16.12 *Alligator mississippiensis,* alligator. Retina of alligator with retinal tapetum noted. *Image by Richard Dubielzig, DVM.*

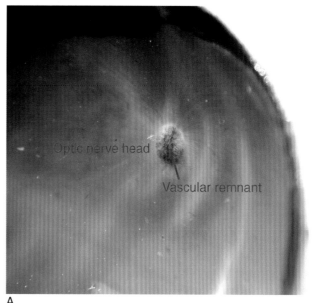

A

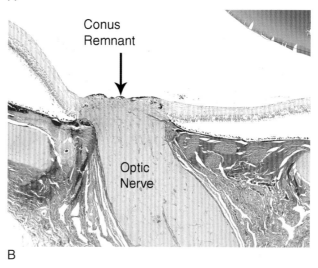

B

Figure 16.13 *Alligator mississippiensis,* American alligator. **(A)** Fundus of alligator. Optic nerve head is labeled. The tissue on the surface of the optic nerve is the vascular remnant. This represents the remnant of the conus seen in most other reptiles. **(B)** Optic nerve is labeled. The conus in this species has atrophied, and all that remains is a pigmented tuft with few if any vessels labeled. *Image by Richard Dubielzig, DVM.*

surface axis of the body so that, for camouflage, only the eyes and nostrils remain out of the water, leaving only a shadowy outline of the body. The eye is large, often reaching 20 mm in diameter in the alligator and more in other crocodilians, compared with the 24-mm normal human eye.

Compared with the cornea of primates, that of the crocodile is flatter, as one might expect with a return to water.

To compensate for the loss of the refractive power of the cornea, the lens is quite large and spherical or ellipsoidal, much like a piscine lens (Figure 16.11). The eye accommodates very little if at all, probably because of the species's nocturnal nature and the participation of other sensory modalities for prey capture.

For support, cartilage lines much of the sclera, as seen in other reptiles, fish, and birds. The upper lid contains a bony plate with the lower lid and the leading edge of the translucent nictitans containing a cartilaginous plate. The retractor bulbi muscles pull the eye beneath the horizontal plane of the skull, and with the bony plate in the upper lid, the cartilaginous lower lid, and the nictitans, the combination protects the eyeball with fortress-like security for the flying debris or from kicking limbs of prey during predatory attacks (Figures 15.5–15.7).

The lacrimal gland is found under the dorsal orbital roof; crocodilians also have a Harderian gland that empties along the undersurface of the leading edge of the nictitans. Unlike the Harderian glands in birds and reptiles, those in crocodilians produce oily tears to protect an evolutionarily terrestrial eye that had to readapt to the chronic osmotic effects of water on the cornea. The secretions of the crocodilian Harderian gland drain into the oropharynx, presumably to assist in swallowing and digestion. So the proverbial "crocodile tears" do exist.

The iris is thick and darkly pigmented. It acts as a shade, preventing any stray light from entering the globe. The anterior layer of the iris is filled with lipophores, which color the iris a light brown or cream (Figures 15.5–15.7).

Alligators, which sport the vertical slit pupil of many terrestrial reptiles, have a large eye, similar in structure to those of other vertebrate tetrapods and of birds. Just like the turtles, alligators have the horizontal strip of concentrated photoreceptors that provides a linear area of concentrated visual acuity.

But some differences exist, and one such difference is that of retinal nutrition. Some vertebrates have a direct retinal vascular supply. For example, humans have a central retinal artery; birds have a pecten to serve inner retinal metabolism (Chapter 20 and Appendix C); many fish have an analogous falciform process for the same purpose; turtles and other reptiles have a homologous conus papillaris. Alligators have none of these and thus no demonstrable inner retinal nutritive support for the alligator eye. But a possible nubbin remnant of the conus on the disc (Figure 16.13) would suggest evolutionary loss of such a structure (Appendix C).

The retina has other adaptations not commonly seen in reptiles. The retinal pigment epithelium is modified in the superior half of the posterior pole to be a tapetum lucidum. This creates the eerie blood-red glow seen at swamp water's edge at night. The tapetum allows the alligator to maximize nocturnal photons by reflecting light back to the photoreceptors after these photons have passed through the retina unabsorbed. Although the retina contains cones, it is predominantly a rod retina. Even the cones could be described as rodlike in their histologic appearance, though, because these cones may represent the evolutionary intermediary between cones and rods. Alligators have four visual pigments in what cones they have, albeit with a narrower spectrum than that of diurnal turtles. Alligators' range of color vision does not extend into the "red" as much as turtles'. The reason may relate to the more aquatic and/or nocturnal nature of the crocodilians.

Although we are taught that these lethargic-appearing crocodilians are closely related to other reptiles, the relationship is not as close as might be assumed. Perhaps their closest relatives, birds, or at least the line that would lead to birds, probably split from the crocodiles during the middle Triassic as the ancestral thecodonts split into Crocodilia and eventually radiated into Aves. Homologous neurologic design and genomics provide the primary support for this conclusion.

All crocodilians possess physiologic sensory adaptations called "integumentary sense organs." These are best seen on the lower mandible, but they are also found elsewhere on the head and, in some species, all over the body. The mandibles contain at least one (and usually more) such sensory pit per scale. The sensory pits may be seen as pigmented spots represented on each scale (Figures 15.5–15.7). They are believed to be mechanoreceptors that can detect pressure waves underwater, thus providing information on the proximity of nearby fish or other animals. The differential firing of these receptors will provide data for proximity, distance, speed, and size of the potential prey. Vision is probably not good underwater, especially in the water these creatures inhabit, but is important for other reasons.

The eye provides biological data to help us understand phylogenetic relationships. Investigators discovered that the photopigments of A. mississippiensis are more closely related to those of chickens and presumably of other birds than to those of reptiles or mammals. The crocodile eye proves to be a window to the evolutionary soul.

Archosaurian Sisters

Somewhere in the early to mid-Triassic, beasts considered nearly magical came and went. These were not archosaurs but appeared at the same time as the other principal archosaurian radiations. It must have been an alluring moment in the earth's history as these creatures emerged as aquatic giants—the stuff of legends.

Ichthyosaurs

As the sea dragons of the age, ichthyosaurs might be considered mythical, but they were all too real. Their descent remains enigmatic and controversial, but our best understanding places them in a sister group to the archosaurs (Figure 14.1). Regardless of their exact classification, they hold a significant position in the evolution of the eye. These diapsids almost certainly began with the same tools as others in that lineage: four visual pigments, double cones, superficial retinal vascularization, and other traits. One of these ichthyosaurs also possessed the largest vertebrate eye ever known—perhaps even the largest eye of any type.

The Jurassic landscape of 160 million years ago must have been a fearsome habitat with terrestrial dinosaurs dreading such creatures as Dilophosaurus among other dramatic land-based predators. The seas of that era contained a group of ferocious marine predators that were contemporaries of the dinosaurs. These highly successful creatures represented a dynasty of the Triassic, Jurassic, and Cretaceous seas for more than 180 million years—the ichthyosaurs.

Some of the larger ichthyosaurs could have had *Dilophosaurus* for lunch if the latter were foolish enough to enter the shallow seas. One of the largest ichthyosaurs, and the one with the largest known vertebrate eye, was *Temnodontosaurus.*

Larger and longer than a London double-decker bus, *Temnodontosaurus* (cutting-tooth lizard) was among the largest aquatic predators ever. Perhaps only a few others such as the *Carcharodon megalodon*, the megatooth shark of the late Tertiary, exceeded its size.

Red in tooth and fin, the ichthyosaurs, such as *Temnodotosaurus*, possessed adaptations that provided a flexible body and an undulating swimming style. With a thoroughly reptilian skull, a slender backbone, and twice the vertebrae of modern reptiles and mammals, the early ichthyosaurs could thrive in the shallow, rich continental shelves.

As the ichthyosaurs evolved, they developed heavier bodies, larger vertebrae, and other adaptations that supported the added weight. The same adaptations reduced flexibility, so the animals further adapted with a more fish-like and quick swimming style and a move to deeper waters while maintaining access to the continental shelf.

Paleontological evidence suggests that these larger, heavier animals, such as *Temnodontosaurus*, could maintain a deep oceanic dive for twenty minutes or more. This allowed access to depths of 500–1000 meters or more where little light penetrates, so ichthyosaurs also had to develop large eyes with a remarkable ability to gather light.

As some creatures evolve into these darker niches, they reach the visual "quit point" visually and simply give up on sight as the principal sensory mechanism. Creatures such as the ichthyosaurs that do not respect the quit point continue to enlarge their eyes to enormous size (Figures 16.14 and 16.15).

Although the ichthyosaurs are long extinct, their larger-than-beach-ball eyes continue to teach us much about their lifestyle and habits.

The fossilized sclerotic plates, or sclerotic ring, clearly visible in Figures 16.14 and 16.15, can help us determine the size of the eyes. These plates, seen in many other vertebrates, initially developed as a site to strengthen the eye during accommodation and were likely co-opted to serve several other functions: They helped to stabilize asymmetric globes and to support the large ichthyosaurian eyes against the extreme pressure of oceanic depths of 1000 meters or more. The bony plates were large and may have been capable of marrow production. These plates

Figure 16.14 *Ophthalmosaurus icenius*. Ichthyosaur with the largest eye-to-body-weight size. This animal lived about 150 mya, but other ichthyosaurs preceded it in the Jurassic. This animal had an eye approximately 23 cm in diameter, but the largest ichthyosaur, *Temnodontosaurus,* had an eye that was 26 cm in diameter, which is the largest eye ever known in any animal. *Image © The Natural History Museum, London.*

overlapped and reinforced the globe's anterior segment, which would have been exposed externally, creating an almost complete bony chamber. Perhaps the most prized information garnered from the fossil record includes the eye size and capabilities. The sclerotic plates allow calculating the eyes' diameter; the largest found, in a specimen of *Temnodontosaurus platyodon*, was 264 mm in diameter. Debate remains as to whether the eye of the still-living

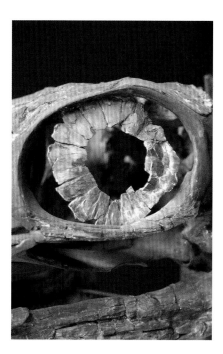

Figure 16.15 *Ophthalmosaurus icenius*, ichthysosaur. Note the individual scleral ossicles that were present in life. These bony struts permit analysis to determine the diameter of this eye. These ossicles are among the largest ossicles known. *Image © The Natural History Museum, London.*

giant squid is bigger although the largest known eye of its kind has been 250 mm. The giant squid is a very different animal than *Temnodontosaurus*, yet it probably feeds at similar, or even deeper, depths, and its eye enlarged to the same extent probably because of similar evolutionary pressures.

One other large-eyed ichthyosaur roamed the seas about the same time, with another interesting ocular distinction. Meet the *eye-lizard*.

Ophthalmosaurus icenicus

Although Temnodontosaurus *was the second- or third-largest known ichthyosaur and had the largest known vertebrate eyes ever, a related parvipelvian ichthyosaur,* Ophthalmosaurus icenicus, *was perhaps more versatile.* Ophthalmosaurus *was less than half the size of* Temnodontosaurus *and much more maneuverable, but the eyes of* Ophthalmosaurus *were only slightly smaller than those of* Temnodontosaurus. *In fact, with no contest or quibble, the eyes of* Ophthalmolsaurus, *or "eye-lizard," had the largest diameter compared to body length of any animal ever. Resembling the morphology of a modern dolphin,* Ophthalmosaurus *weighed approximately 950 kg and was only slightly longer than 4 meters, with the head comprising 20 percent of that length.*

The Ophthalmosaurus *young were born alive and were probably immediately brought to the surface by their mothers or, like today's pelagic-birthing marine mammals, they instinctually sought the surface.*

Ophthalmosaurus *had a long thin mandible that would have been an excellent tool for capturing fast, maneuverable prey, such as large fish and squid, its principal diet. This sleek ichthyosaur was probably active at night and/or was a deep-diving species, allowing it to fill a rather specialized marine niche. The ample fossil record provides understanding of the size and, to some extent, the function of the eyes.*

The fossil record reveals the eyes of Ophthalmosaurus *to have been only slightly smaller than those of* Temnodontosaurus, *up to 230 mm, but* Ophthalmolsaurus *had a much smaller body and weighed far less.*

The fossil records tell us more, such as the eye's approximate f-number (a measurement of relative light-gathering ability). The smaller the f-number is, the more light the system can absorb and use (Chapter 5). The Ophthalmosaurus *has an f-number from 0.8 to 1.1 (compared with 2.1 for a human and 1.1 for some species of owls).*

As a visual hunter, Ophthalmosaurus *and other ichthyosaurus like it successfully took the vertebrate eye to the darkest reaches of our planet and must have been a fearsome predator.*

Ichthyosaurs were a distinct group and more closely related to basal reptiles than to dinosaurs, although no representative of the lineage remains. But ichthyosaurs were not the only fantastic creatures to appear in the reptilian lineage. The Archosaurs radiated into another spectacular group that must have had many good examples of ocular physiology and morphology. These terrestrial behemoths of the earth played a key role in the earth's history—the dinosaurs.

DINOSAURS AND THEIR COMPANIONS

17

Archosaurs radiated into all manner of different lineages of reptiles, many of which have been lost to extinction. Some of these lineages spawned animals that have captured our imagination because of their enormous size and surreal qualities. Informally, they are called dinosaurs, but the dinosaurs were far from monolithic and included other radiations. There must have been at least as much diversity among the dinosaurs and related archosaurs of the Mesozoic as is seen among modern mammals. The dinosaurs would not have been as closely related to one another as modern mammals are to each other, and it is likely the visual systems exhibited greater variability than that found among mammals.

The visual witness carried through these animals and their fossil remains offer some evidence of their eyes and brains. But because, as we know, soft tissue does not fossilize, most of our knowledge is circumstantial.

Pterosaurs

The pterosaurs or pterodactyls included the pteranodon—a flying reptile popular in children's dioramas (Figure 17.1). These creatures were not true dinosaurs—rather an order of their own within the archosaurs. This order also included *Quetzalcoatlus northropi*. With a wingspan of 15.1 meters, it was the largest animal ever to fly on this earth. More than a side note to the dinosaurs, these creatures were top predators in the early Cretaceous and perhaps earlier, in the Jurassic.

Although the pterosaurs are analogous to birds, with many similar adaptations, they were not birds' direct predecessors. Many investigators believe that they were warm-blooded flying (as opposed to gliding) predators of the Triassic, Jurassic, and Cretaceous. These predators may

have been exterminated by a combination of the cosmic event that killed the dinosaurs and the isolation that specialization brings.

Pterosaurs' visual anatomy is better understood than their visual processing. Most pterosaurs had large eyes and probably good, if not excellent, vision. This would require inner retinal nutrition supplied by an elaborate structure such as the pecten, birds' conus-like structure (Appendix C). It is not known if pterosaurs had a pecten, but given the challenge of feeding on the wing, some of the species must have had excellent vision, lenticular accommodation by deformation (Appendix E), and an elaborate, thick, highly organized retina with processing at a retinal level. Oil droplets were probably present in these almost purely cone retinae. There may have been nocturnal pterosaurs, but most were probably diurnal. As in other reptiles, accommodation was done with striated muscles and was fast compared with our own accommodation.

Studies of the braincases of extinct animals show that certain brain structures were enlarged in the pterosaurs. For example, the flocculus, a structure in the brain, reached its peak in the pterosaur. The flocculus is devoted to balance and, in the pterosaurs' case, probably flight. Birds, too, have a large flocculus with many connections from the visual system. The visual abilities of most pterosaurs probably rivaled, if not exceeded, those abilities in certain predatorial birds we will find in a later Chapter. Speed on the wing requires fast and accurate visual processing. Although the pterosaurs may have appeared slow, wing speed had to be high to get off the ground. These magnificent animals would have weighed up to 65 kg and attained wing speed of up to 60 km an hour. These would have been dramatic and frightening predators.

Figure 17.1 Pteranodon. *Image © The Natural History Museum, London.*

Dinosaurs

Dinosaurs diverged from the initial archosaurian predecessors in the early Triassic, perhaps 240 mya. These include the sauropods, theropods, and others (Figure 14.1). The intimate detail of the visual witness of these animals is not known, but some information is available, much like that available for the pterosaurs.

Sauropods

Sauropods, immense vegetarians that often had long necks, were among the largest ever terrestrial animals, dwarfing mammoths. In many ways, these creatures resembled the tuatara in morphology, and they have similar ancestors. But genuine differences existed. Most of the early dinosaurs were diurnal; dinosaurs probably did not become nocturnal until later in the Mesozoic. The herbivores probably had—and needed—less acute vision than the predators. But they did gain the defensive skills of movement sense and rapid visual processing early in their evolution. Like tuataras, early dinosaurs probably had a parietal eye-pineal complex, but many later dinosaurs had no parietal eye, or at least no opening in their skull for such an eye. Different radiations of dinosaurs would have evolved different mechanisms, and some would retain the parietal eye, as some extant reptiles do.

In contrast to legend and popular culture, most dinosaurs were small diurnal animals that would exploit a rich natural environment. At the beginning of the Triassic, the climate was warm with oxygen at higher levels than today but lower than that in the Carboniferous. Reptilian herbivores would have had a sumptuous smorgasbord of luxuriant plant growth. Diurnal herbivores would have had color vision, with three, or more likely four, visual pigments and oil droplets in their photoreceptors because living descendants and their predecessors both have them. Like those of other reptiles, the herbivores' striated muscles produced fast accommodation. Because we see scleral plates and cartilaginous cups preserved along with other bony structures, we know that they were present in many dinosaurs.

Theropods

The grandparents of birds, theropods developed large, optically sophisticated, and highly efficient eyes to enable them to become top, speedy predators.

Theropods were dinosaurs and closely related to sauropods. Birds, also classified as theropods, probably sprang from the theropod lineage in the late Jurassic or perhaps earlier. Before the theropods radiated into birds, they were, for the most part, predators. They were fast and had large eyes and powerful musculature as well as distinct large jaws with an abundance of teeth. These creatures were at the top of the food chain in the Triassic, Jurassic, and much of the Cretaceous. But by the late Cretaceous, pterosaurs had probably supplanted them or, more likely, were supplemental to them. Top aerial predators, the pterosaurs did not compete directly with the terrestrial theropods—much as today's eagles and polar bears would not compete with one another. Like the diurnal animals, the pterosaurs had scleral ossicles, oil droplets, double cones, and a well-organized retina to which a conus-like structure delivered nutrition.

The nocturnal realm would have required different skills and probably different ocular morphology for herbivores, more like that of the tuatara. The predators, though, were different still. To be top predators, theropods and pterosaurs had to have good vision with a sharp fovea or two, stereopsis, and three- or four-channel (different photopigments) color vision. They needed a conus for inner retinal nutrition and thick retinae with many cells capable of aiding in visual processing.

No one knows exactly why theropods and pterosaurs became extinct, although the cause was almost certainly secondary to the meteor impact that ended the Cretaceous.

Many prey species probably wished the theropod extinction had occurred much earlier. These creatures were vicious predators with large brains that permitted tool use, ambush and group hunting, and other ruthless strategies that would assure individual as well as species survival.

The Jurassic saw the continued rise and expansion of another lineage of unusual predators that help to tell the story of the visual witness—the cephalopods. Some of these mollusks are still with us in the form of Celeoidae— octopi, cuttlefish, and squid.

CEPHALOPODS CHANGE DIRECTION

Cephalopods (Phylum Mollusca) arose in the Cambrian with an external shell and an invaginated cuplike eye without a lens. This would become the simple pinhole-camera-style eye of the *Nautilus* today and probably most or all other cephalopods of that time. By the Carboniferous up to the Jurassic, evolution made a dramatic turn within the Class Cephalopoda, leading to a subclass that appears conspicuously dissimilar to other mollusks and even other members of the Class. After the cephalopods gave rise to the coleoids, the visual witness would evolve toward its peak in the invertebrate lineage.

One of the ammonite lineages (Chapter 8) is probably a predecessor to other living cephalopods, including the octopus, squid, and cuttlefish, despite appearances to the contrary. But like all other lineages, these animals evolved and continue to do so.

Coleoids

A highly intelligent group of cephalopods, living coleoids are represented by the octopus, squid, and cuttlefish. These are all soft-bodied, multiarmed animals with no external shell.

External-shelled cephalopods positively exploded with creativity in the Silurian and Devonian, possibly related to the loss of Cambrian competitors. Initially, these animals confronted life with elegant shells. In the Ordovician these shells were straight, curved, coiled, and then spiraled, followed by all manner of gorgeous symmetrical shapes. Outside of the *Nautilus* the shelled cephalopods did not endure into present-day animals (Figures 8.8–8.10).

One lineage of cephalopods internalized or jettisoned its shell altogether for speed and agility to become the coleoids and eventually radiated into the extant squid, octopus, and cuttlefish. As a result of the process of shape-shifting, coleoid morphology is so unusual that many members may seem to be of another phylum, if not another planet. The coleoids probably radiated from an animal related to the living *Nautilus* and would have had its eye as a prototype or template (Chapter 8).

The visual witness, though, would persist, improve, and change with this radiation to the coleoids. The eye would gain not only a lens but also additional extraocular muscles to move the eye. And some would gain a cornea.

Cephalopod Lens

The early cephalopods, such as the ammonites, had eyes that probably had no lenses at all. Among the cephalopods, crystalline lenses first appeared in the coleoids, as these lenses would be required for further improvement in vision. During embryology, coleoid lenses are formed in two nearly equal segments and fused. This process is different from vertebrate lens formation, and yet the final product closely resembles a piscine lens (Chapter 10). The coleoid lens, composed of hard protein crystallins with a high index of refraction, is large compared to the eye. The coleoid lens-to-eye size generally obeys Matthiessen's ratio just as the fish eye does, despite the different embryology. This is convergent evolution at a lenticular level.

Crystalline lenses must have been important to the early coleoids because lenses would have improved light-gathering capabilities and sharpened acuity. If coleoids lost the outer shell for speed and agility, lenses would permit further visual investigation of darker and often deeper water, or even nocturnal vision as well as sharpen vision.

Extraocular Muscles

The *Nautilus* and the octopus each have seven extraocular muscles, compared with thirteen or fourteen in the squid and cuttlefish. Because bivalves and gastropods (other classes of the mollusks) do not have extraocular muscles at all, these muscles help confirm the relation of coleoids to the *Nautilus*.

Coleoid Expansion

The coleoids, which probably arose in the Carboniferous, are believed to have been rare until the Jurassic, when they escalated in number and diversity. The cause is unclear.

Perhaps a predator was lost to extinction, coleoid fecundity increased, camouflage or defense mechanisms improved, or coleoids became an artifact of preservation. But expand and diversify they did. The octupuses probably were established first, although the early phylogeny is as murky as the ink they use to confuse predators. Among other evidence, the extraocular muscle distribution would suggest that the octupuses are closer to the *Nautilus* than other coleoids, and this would point to the octupuses being among the first coleoids established.

Although the early octupuses that arose in the Jurassic are gone, the lineage has evolved, as the current species of existing octupuses were not established until the mid-Cretaceous. But, the seeds for these curious and charming creatures including the eye, though, were sewn in the Jurassic. One such octopus illustrates what the eye has become. Meet the Pacific giant octopus, the largest living octopus.

Pacific Giant Octopus, Octopus dofleini

The giant Pacific octopus (Figures 18.1–18.3) is an efficient and mostly nocturnal hunter that spends the day concealed in rocky caves in the northern Pacific. Although one specimen weighed 272 kg, most giant Pacific octupuses are in the 150-kg range, with a known maximum arm spread of 9.6 meters. It relies on sharp visual acuity honed with remarkable intelligence to subdue prey and to protect itself from its predators.

The octopus eye's similarity to a piscine eye is a good illustration of convergent evolution. But substantial and critical

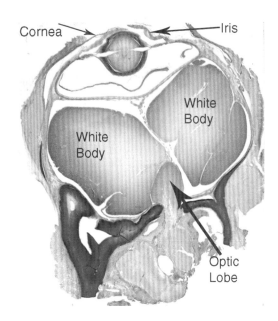

Figure 18.2 *Octopus dofleini,* Pacific giant octopus. The orbit of the octopus contains the eye, the optic lobe where the unaltered signals from the retinal cells are processed, and the white bodies. The white bodies are blood-manufacturing sites producing copper-based blood. The eye shows the lens consisting of two halves that unite during embryology. The lens is formed in two halves with the inner half developing from the same embryological tissue that will form the retina and the outer half from the same embryological tissue that will form the skin. The cornea consists of two clear halves and extends from beneath the skin folds to overlap centrally, and the iris extends from the globe itself. Thanks to Dr. Mike Murray and the Monterey Bay Aquarium for providing the specimen. *Histology by Richard Dubielzig, DVM.*

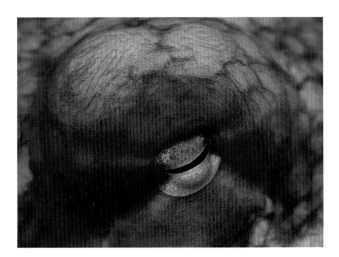

Figure 18.1 *Octopus dofleini,* Pacific giant octopus. The pupil on this and most other octupuses is horizontally oval and can close completely. The Pacific giant octopus is a continental shelf octopus and, as a result, has a thin membranous cornea. The cornea can regenerate if lost and serves mainly as a protective layer. The pelagic, or open ocean, cephalopods do not have corneas.

differences between those eyes suggest that the octopus and the fish each evolved its eye separately rather than through a common ancestor.

For example, like fish, the Pacific giant octopus has a thin, tough cornea, a pupil, a nearly spherical lens, a vitreous cavity, and photoreceptor cells lining the cavity. But the retinal anatomy of the octopus greatly differs from that of the fish. The octopus retina is everted, meaning that its photoreceptive element is facing the crystalline lens and pupil, with no interposed ganglion cells or other tissue to interfere with the image resolution (Figures 18.3 and 18.4) (Appendix A). So, based on different anatomy, no other retinal cells or blind spots exist within the eye. In all octupuses the equivalent neurologic cellular elements are immediately behind the eye in an optic lobe that contains cells that resemble those analogous cells in the vertebrate inner retina (Figure 18.2). It is as if the octopus receives the "raw" data from the photoreceptors and transfers those data to a "peripheral brain" to begin interpretation (Figures 18.2). It is an exceptional instance of convergent neurologic evolution as the optic lobe functions much like the inner retina of vertebrates.

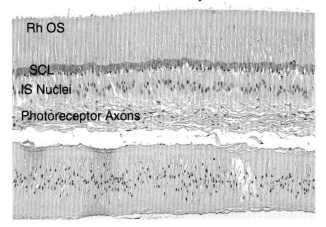

Interior of eye

Rh OS

SCL

IS Nuclei

Photoreceptor Axons

Figure 18.3 *Octopus dofleini*, Pacific giant octopus. This histologic image of the retina of the octopus eye reveals its anatomy. Rh OS, rhabdoms of the individual microvillus cells interdigitating with each other to create specific units for photon reception; SCL, supporting cell nuclei and pigment granules; IS Nuclei, layer of the inner segments of the photoreceptive cells that includes the nuclei of the photoreceptive cells. The inner segments and the outer segments are portions of the same cell. The supporting cells assist in nourishment and protection. Photoreceptor Axons, nerve trunks leading away from the photoreceptive cells. These axons will lead to and synapse in the optic lobe. Thanks to Dr. Mike Murray and the Monterey Bay Aquarium for providing the specimen. *Image by Richard Dubielzig, DVM.*

The octopus is colorblind because its retinal cells contain only a single photopigment (which is maximally sensitive at approximately 475 nm); at least two photopigments are required for color vision because a brain requires at least two visual pigments to define color—otherwise color is just a light stimulus. Furthermore, that single photoreceptive pigment is in

BOX 18.1 RHABDOMERIC ARRANGEMENT

The rhabdomeric arrangement of an octopus or other invertebrate with a camera-style eye is different from that of the arthropods (Figure 18.4). An octopus rhabdom consists of four rhabdomeres; each consists of precisely aligned and layered tubules containing the photosensitive visual pigment (Chapters 6 and 12). The rhabdomeres join one pair oriented horizontally and one pair oriented vertically so that each rhabdom has, in effect, two retinal units. Each pair of rhabdomeres (roughly a single photoreceptor equivalent) subtends a visual angle of 1°3' as compared to our own 20–30' in the fovea (our visual angle is perhaps four to five times "finer" than that of an octopus). The rhabdoms are separated from each other by pigment granules, and because of the orientation of the rhabdomeres, they can easily perceive polarized light.

a rhabdomeric arrangement that is different from that of insects (Chapter 11).

But octopuses can perceive polarized light, which may be key to their ability to change color. Instead of perceiving color, the octopus may match the polarized light they perceive.

The octopus lens differs from a fish lens in its opaque peripheral equatorial septum that divides the lens about in half (Figure 18.2). This septum extends slightly into the lens and supplies structural support but minimally restricts light entry into the eye. As in the fish eye, the spherical lens's anterior-posterior movement achieves focus and accommodation (up to 14 diopters). However, unlike fish, which can actively move their

Figure 18.4 The rhabdom consists of the combination of the finger-like projections from all the individual rhabdomeric cells that contribute these microvilli. Each individual set of microvilli from one cell is called a rhabdomere. In darkness these microvilli extend and enlarge the exposed surface area of the rhabdom. In the light a larger portion of the rhabdomeres are shielded, creating less exposure of the rhabdom to light. *Art by Tim Hengst.*

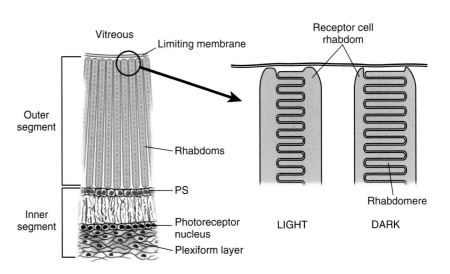

lens in only one direction, the more versatile octopus can do so in either direction. The eye's focal length is very short relative to the animal's size, so an enlarged rounded pupil can gather light well and create a bright image. The pupil's horizontal slit is unusual in a predator, and the purpose for it is not understood. When the octopus is diurnally active, the pupil will constrict to the slit or horizontal oval (Figure 18.1). This oval has been shown to limit luminance without limiting acuity, as if pulling a shade in the house on a sunny day. Horizontally concentrating longer rhabdomeres in the retina to correspond to the pupil increases the light-gathering abilities and sharpens the acuity in the horizontal meridian.

The octopus's eye movements are exquisitely controlled. The horizontal movements are similar to those of a vertebrate visual system, but the octopus's rotary, or wheel-like, "torsional" movements have a much greater range and control. Testing of torsional movements illustrates that the octopus will keep the horizontal pupil oriented to the horizontal meridian through-out marked active and passive excursions of its head. This avoids extreme retinal image motion.

Octopus eyes, on the sides of the head, often have some binocular visual field. (On the other hand, squid and cuttlefish have considerable binocularity and, presumably, stereopsis.) Connections in the octopus's brain (through a special brain tract called the ventral optic commissure) coordinate the visual system, allowing the animal to access memory units on the contralateral hemisphere and coordinate activity of all eight arms, although the exact mechanism of its "stereopsis" is not understood.

Even the cornea of the octopus is dramatically different from that of vertebrates, despite superficial similarities. The Pacific giant octopus's thin, tough cornea surely has evolved to protect the delicate outer layers of the lens from abrasive material such as sand. Only octopuses in the relatively shallow waters of the continental shelf such as the Pacific giant octopus even have a cornea; pelagic cephalopods do not. And the corneal tissue in octopuses that have one will regenerate if lost, unlike a vertebrate cornea.

The octopod cornea consists of overlapped transparent extensions of the inner edge of the upper and lower "lids." It is as if the skin extended an eyeshade from the skin edge. Although these could be called "lids," they are not true lids, as they do not contain structures to move or close them. Instead of lids that truly close, octopuses have fixed skin folds that give the appearance of lids.

The iris that surrounds and creates the pupil is an extension of the globe and, as a result, is thick and much more lidlike. Multiple layers include muscle, iridocytes, and chromatophores. The last two structures are responsible for iris colors and the ability of the octopus to change them (Figure 18.1).

About 70 percent of the genes that are known to cause the formation of the eye are the same for octopuses and humans even though the two eyes diverged from one another as much as 640 mya. Evidence suggests that that last common ancestor of the bilaterians (urbilateria) had all the machinery to make camera-style eyes (but we do not know how long before then those genes were present). This adds weight to the argument that the genetic framework for these eyes is ancient indeed.

Hunting strategies for most octopuses are complex; they include camouflage and ambush, stalking, chasing, and more sophisticated techniques. The octopus kills and begins to digest a well-armored crustacean, such as a crab, by using its tongue-like radula to drill a hole in the crab's carapace and injecting its venomous saliva. Some octopus species inject their enzymatic venom through the crab eyestalk, the weakest point in the carapace, so it allows for the most rapid entry into the body.

Most octopods can change color, but their intelligence and learning capacity are perhaps the most fascinating elements of the animals' natural history. Octopuses have drowned sharks in captivity, solved puzzles, opened jars and solved mazes. In an illustration of chicanery, one octopus was photographed climbing out of its container, crossing a lab floor to reach separately housed prey, consuming the prey, and returning to its original tank. Smart.

The coleoids' other related lineages—squid and cuttlefish—have distinct differences, interesting designs, and skill all their own. Unlike the Pacific giant octopus, at least one coleoid—the Humboldt squid—is aggressive to humans.

Humboldt Squid, Dosidicus gigas

The movie would have to be called "Beaks." Squid are generally clever, decidedly intelligent, agile, fast, and carnivorous. They have been honed by millions of years of evolution with radiation to all oceans and depths. Some of the species have grown as large as 40 kg, and at least one species has become aggressive—even to humans.

Known by fisherman as the red devils, Humboldt squid are mostly found along the western coast of Mexico and the United States (Figure 18.5). Little is known about their diurnal activities because these squid live as deep as 700 meters. Although they rise to the surface to feed at night, these 2- to 3-meter-long predators may rise only a few hundred feet. These squid may live only a year or two, so to gain such size, they must be voracious eaters. Each of the suckers on the arms of this decapod is surrounded with teeth, meant to capture, immobilize, and bring the prey to the squid's powerful central beak in the center of the mantle.

The Mexican fishermen who hunt these squid know that light will bring them to the surface, when they can be more easily caught. The fishermen also know better than to be in the water with them, as they are ferocious. They are one of the few animals, and probably the only squid, that will actively attack

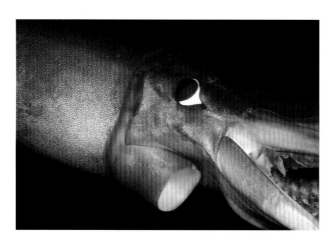

Figure 18.5 *Dosidicus gigas*, the Humboldt squid. There are no true lids, but folds of skin function to cover the eye and smooth the flow of water over the eye as the squid propels itself through the water. *Image © Peter Parks/imagequestmarine.com.*

in pods—sometimes up to 1200 individuals—and kill humans if they can.

Pack hunters, like dolphins and these squid, usually cooperate and share in the spoils, a sign of intelligence. Squid such as the Humboldt squid are highly visual and rely on good acuity for hunting and protection against predators. The eye provides clues to the evolution of the visual witness in these soft-bodied cephalopods.

The coleoids have an ocular anatomical anomaly. Most truly pelagic squid have a cornealess eye called an oegopsida eye or the eye of an oegopsid squid (Figure 18.6). Cornealess squid have skin flaps that, more or less, cover the eye when the animal

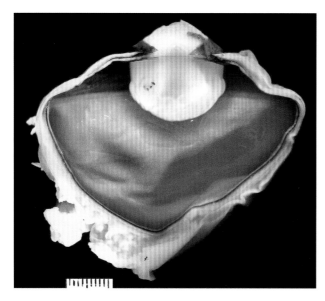

Figure 18.6 *Dosidicus gigas,* the Humboldt squid eye. This large eye measured 12 cm at the equator. *Image by Richard Dubielzig, DVM.*

swims (Figure 18.5). Often called lids, these are not true lids anatomically or embryologically (Figure 18.5).

All shallow-water coleoids that inhabit the niches of the continental shelf, such as the Pacific giant octopus, do have a cornea, thus a myopsida eye.

The squid of the open ocean, then, will have seawater bathe the surface of the lens with their oegopsida eye. This is true of the red devils discussed above but also of most other known pelagic squid.

So how do Humboldt squid and other pelagic squid protect their lens surface if they have no cornea? Although these pelagic squid do not have eyelids, they can squeeze the surrounding folds of skin to cover the eye much as a lid would close during propulsion. In addition, a small sinus flap closes the small slit that remains (Figures 18.1 and 18.5). This closure protects the lens and decreases turbulence during propulsion.

The pelagic squid were opportunistic and expanded into all niches available. More challenging niches required unusual solutions, such as the one *Histioteuthis* provided.

Histioteuthidae *species,* Histioteuthis heteropsis

The pelagic midwater region (180–600 meters) must be a frightening environment. Little light reaches this depth. Toward the deeper limits of the midwater, bioluminescent animals appear with eerie light, punctuating the darkness only by the flickers of tiny points of faint bluish light. Living at this depth requires evolutionary adaptations. Usually that adaptation involves creative sensory, morphologic, physiologic, or other mechanisms.

Histioteuthis heteropsis, *a mesopelagic squid, met the challenges of dim lighting above it and darkness below it with two dissimilar eyes: one eye is twice the diameter of the other (Figures 18.7–18.10).*

Histioteuthis *uses the larger left eye to view the 500 meters or so of ocean* above *the animal and examine that environment for prey animals. The other eye, too small to make out silhouettes or forms, looks downward for bioluminescence produced by small prey species, such as shrimp. These two eyes permit this squid to seek substantial prey between it and the sky, stars, or sun above, or the sea below.*

The eyes have much to tell us. The lens in the larger eye is yellow, whereas the lens in the smaller eye is clear. The larger, upwardly directed lens filters the near ultraviolet and deep blue wavelengths. The yellow lens represents a filter, and in optical terms is called a "high-pass filter." When Histioteuthis looks upward with this eye, any downwelling blue and UV light would be blocked from striking the retina. One of Histioteuthis's favorite meals, a species of shrimp, produces bioluminescence from small organs on the ventral surface of its body. These organs serve as camouflage; the shrimp's body would block light if not for the bioluminescence, which creates the illusion of starlight

Figure 18.7 *Histioteuthis heteropsis*. Note the large eye looking upward and the smaller eye looking downward. This is approximately the attitude and positioning of the body and eyes during propulsion. The clear structure on the larger eye on the upper surface of the animal is not a cornea but rather transparent lids. The dark spots on the undersurface of the squid are photophores. *Image © L P Madin, WHOI/CMARZ.*

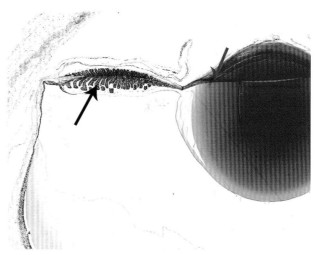

Figure 18.9 *Histioteuthis heteropsis*. This image is of the lens of the squid. Note the septum between the two different portions of the lens. This septum extends into the peripheral lens (blue arrow) but does not restrict light. The muscle attached to the lens (black arrow) moves the lens in both directions during accommodation. *Image by Richard Dubielzig, DVM.*

or sunlight. But, bioluminescence is not exactly the same wavelength as downwelling light from, for example, the moon. A high-pass filter, such as a yellow lens, would make the animal with bioluminescence stand out from downwelling light. This technique is similar to that of some fish living in deeper waters and discussed in Chapter 10.

In addition to defeating camouflage for predation, the Histioteuthis species can also produce camouflage for protection. To be successful in the midwater, an animal must be versatile and able to achieve success in two different environments. Many cephalopods, including those in the Histioteuthidae

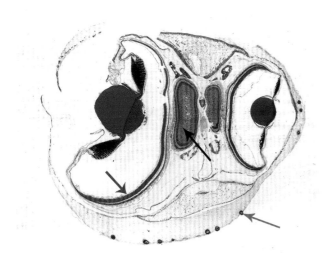

Figure 18.8 *Histioteuthis heteropsis*. A slice through the eyes across the head perpendicular to the long axis of the animal. Note the larger eye and the larger optic lobe connected to it (black arrow). Note that the retina is thicker and more concentrated in the inferior portion of the eye (red arrow). Note the photophores (green arrow). Also illustrated in more detail and discussed in Figure 18.10. *Histology by Richard Dubielzig, DVM.*

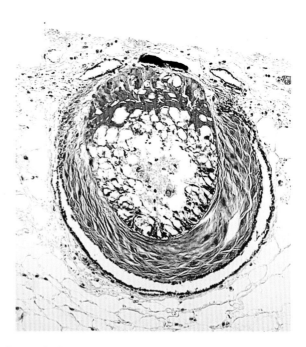

Figure 18.10 *Histioteuthis heteropsis* photophore. Photophores generate light for camouflage and, with the positioning seen in Figure 18.7, illustrate the downwelling nature of such light (see text). *Image by Richard Dubielzig, DVM.*

family, defend themselves with photophores, which surround the smaller eye on the ventral surface. These organs—also found in certain fish, krill, shrimp, lobsters and crabs, and other species—emit and release light. The bioluminescence these animals create camouflages them by producing what resembles downwelling light. Some photophores, including those of cephalopods, have crude lenses that also focus or intensify the light. And some ingenious creatures can control the amount of light and even close the photophore to stop the light.

The coleoid lineage diverged into another animal that illustrates other changes in ocular development. The squid lineage split and radiated into the cuttlefish about 100–65 mya (mid- to late Cretaceous). We explore one of the largest of these specialized mollusks, the giant cuttlefish.

The Giant Cuttlefish, Sepia apama

This octopus relative has a rectangular "W-shaped" pupil (Figure 18.11) and good vision. The pupil can close or leave only the smallest of openings in one or two spots. Presumably, the remaining slitlike pupil can function as a pinhole or a range finder, or it can act like window shades.

More than octopuses do, cuttlefish rely on vision for detection of both prey and predators. Accommodation measurements in this species of cuttlefish tell us that the resting refractive state of the cuttlefish is emmetropic (neutral) in water. Accommodation (by movement of the lens) of about five diopters occurs bilaterally and is associated with convergence at a fraction of a second before a strike.

Like certain other species, cuttlefish can detect polarization and probably can recognize light into the ultraviolet range. Recent work suggests that cuttlefish use their perception of

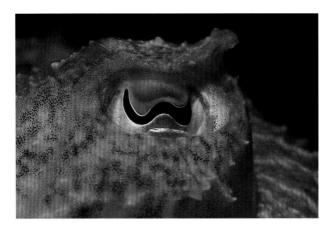

Figure 18.11 *Sepia apama* pupil. Note the "W"-shaped pupil. The small pigmented dots that surround the eye and are scattered across the skin surface are chromatophores (see text). *Image © James Brandt, MD.*

polarization to break the countershading camouflage of light-reflecting silvery fish. Like octopuses, they have but one visual pigment, so they do not see in color, and like octopuses, cuttlefish can change their own skin color to match the surround.

Cuttlefish use chromatophores, small organs or vacuoles of pigment, to perform their lightning-fast color change. Each of these organs, directly under neurologic control, consists of an elastic sacculus with granules of various colors (black, brown, red, orange, or yellow) surrounded by ten to twenty-five radial muscles. With muscular contraction or relaxation, the sacculus can express or hide true pigmentary colors, thus creating differences in both color and pattern. These animals also use iridocytes (also called iridophores) around the eyes and elsewhere to create constructive inference, thereby creating other colors, such as those of the shorter wavelengths—blues or greens.

So inability to perceive color did not stop members of this class Cephalopoda from using color for communication and camouflage.

As the lineage matured, further adaptations illustrate cooperative evolution among animals, the eye, and bacteria. At least one member has used light and color not only for camouflage but also as an instrument of attraction and illumination. Furthermore, this animal has become a bacteria farmer.

Pacific Bobtail Squid, Euprymna scolopes

The rusty glow of the bloom of a marine phytoplankton, Oxcillatoria erythraea, can be startling when it covers an entire bay. Subtler bluish bioluminescence can be seen with the gentle disturbance of the ocean surface colonized by another phytoplankton, Noctiluca. Certain pelagic animals are known to cooperate with prokaryotes to use this ethereal light for defense.

The Pacific bobtail squid, Euprymna scolopes (Figure 18.12), is born with an empty bilobed sac in the center of its mantle cavity on its ventral surface. Shortly after the squid hatches, its sac is colonized by a gram-negative rod, Vibrio fischeri, a luminous bacterial symbiont selected by the squid to create a light organ. The beating of ciliated cells on the squid's ventral surface causes seawater to pass over and into the pores of this light organ. Apparently, the ciliated epithelial cells produce mucus that can be used to "farm" the seawater of V. fischeri. As the bacteria colonize the crypts of this organ, they multiply rapidly; within twelve hours, the initial inoculum has risen to 10^6 or even 10^9 in adult squid, a concentration resulting in a thousandfold increase in the luminescence from the light organ. But this dramatic increase in luminescence is not entirely due to the bacteria, as the squid can manipulate the crypts in the light organ, changing the oxygen concentration to influence bioluminescence. As soon as initial colonization occurs,

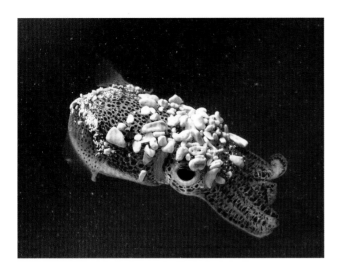

Figure 18.12 *Euprymna scolopes*, Pacific bobtail squid. Note the pebbles on her back for camouflage. *Image © Margaret McFall-Ngai, PhD.*

the ciliated epithelial cells that were key to the colonization are sloughed and entirely lost.

When V. fischeri has been established in the light organ, a repetitive cycle begins. Each morning, the squid discharges about 95 percent of the bacteria, keeping the most vigorous V. fischeri. While the squid burrows into pebbles and broken coral on the sea bottom, the bacteria multiply rapidly and are ready by nightfall.

This bilobed sac, or light organ, within the mantle of the squid is covered dorsally by the ink sac to limit stray light radiating beyond the organ. The organ itself is lined with reflectin, a newly described protein that directs the light to be emitted ventrally. Reflectins, which line the light organ, work by thin-film interference, much like the tapetal mechanisms in some animals (Chapter 10). E. scolopes has these proteins scattered across its body, including in its eye, skin, and digestive gland, the liver analogue. The nanofabrication of these proteins is unique to cephalopods and different from more common aquatic reflective tissue, such as purine or guanine.

Furthermore, a muscle-derived ventral lens with lenticular proteins similar to those of lateral eyes further concentrates and focuses the light ventrally.

E. scolopes uses this light organ to mimic the downwelling light from above. As the squid hunts for shrimp, its favorite prey, the light organ blocks the moon- or starlight above to darken the squid against predatory fish below it. When the organ replicates the downwelling light to the appropriate wavelength, the squid's predator cannot distinguish the squid from moonlight. And because some night skies are brighter than others, the squid controls its light emission with a biological diaphragm. The ink sac, which also provides ink for defense, controls the amount of light emitted.

Ventral light organs, such as this one, are common protective mechanisms in the sea and must have been successful because squid are common and widespread. Like M. niger (Chapter 10), E. scolopes may even use this submantle flashlight as a torch, illuminating its prey species from above. Perhaps the most intriguing aspect of this colonization and co-evolution of V. fischeri with the light organ of E. scolopes is this: the light organ expresses the protein from its Pax6 and rhodopsin genes, among several other genes associated with the visual transduction cascade. These are all key genes in ocular development. So the light organ is a third eye like none other. And although it may be a third eye blind, it lights the way and protects its owner. Evolution has simply found another method for establishing a lifestyle with light.

The mollusks, in the form of the cephalopods, were not the only phyla that had a direction change toward the end of Jurassic or the early Cretaceous. The reptilian clades of the time included the order Squamata. This order gave rise to animals that would lead to the modern lizards. Within that order, some of these creatures had sought life underground. They were about to emerge again during the Cretaceous.

SNAKES ARISE FROM
THE GROUND

CRETACEOUS PERIOD

145–65 MILLION YEARS AGO

19

Lepidosaurs arose in the Triassic or early Jurassic (Chapter 15) and subsequently gave rise to major successful lineages, some of which exist today. These include the tuatara, either in its own phylum or as a basal lizard, and the lizards (squamates). The order *Squamata* also includes iguanids, chameleons, geckoes, worm lizards, and eventually snakes. But snakes are an outlier, at least when it comes to the visual witness, and their peculiar eyes are a clue to snakes' beginning.

Fossorial Lizards

Burrowing lizards expanded into a fossorial (burrowing) lifestyle sometime during the Cretaceous (145–65 mya), lost their legs, and eventually radiated into snakes.

Snakes discovered that vision was not the only sensory modality or even useful underground, so the external lids fused to protect the eyes (Figure 19.1). Because burrowing snakes do not need much vision, their eyes became degenerate and lost many of their unique reptilian qualities. Eyes are physiologically and embryologically "expensive," so if they are not used, they atrophy, even very quickly. As the need for focusing declined, snake eyes abandoned their careful process of accommodation by lenticular deformation (Appendix E). Eyes also lost the need for oil droplets. Circadian rhythm was unnecessary, so the parietal eye was discarded. And the ringwulst of the lens, the scleral ossicles, and scleral cartilage to support the globe were all dumped like so much extra baggage.

Recreating an Eye

Snakes emerged from their underground lifestyle at about 95 mya, probably before they lost all of their visual mechanisms. To compete successfully with other terrestrial species, snakes had to reestablish vision and the accompanying visual mechanisms. But, once lost, some of the older reptilian morphology and physiologic mechanisms could not return, at least in the same form. To rebuild the eye, the snake sometimes had to reinvent systems that had

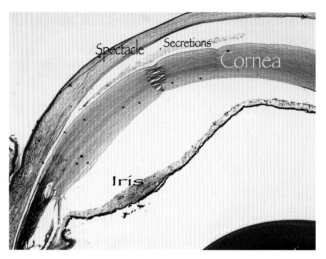

Figure 19.1 *Boa constrictor imperator.* Common boa eye. The spectacle, cornea, and iris are labeled. The lubricating, oily secretions are labeled and found between the spectacle and the cornea. These secretions are produced by the Harderian gland, which is not seen in this image. *Image by Richard Dubielzig, DVM.*

been lost. These reinventions would be different and make the eye an anomaly among the squamates.

Despite the peculiarity of the extant snake eye, its inner structure generally resembles the piscine eye: the crystalline lens is large and spherical (Figure 19.2). After all, the cornea is relatively flat beneath those fused lids. Blood vessels (called vitreal or epiretinal vessels, (Figure 19.3), which arborize across the retinal surface, supply the inner retina with nutrients and oxygen. No other reptiles or later-descended animals have this epiretinal vascularization. Snakes have fused eyelids to cover their eye, making a double layer of the cornea (Appendix D). In some ways this anatomy resembles that of fish, but this second layer has formed in an entirely different manner and is unrelated. In snakes, this second layer is called a spectacle, and here it is a tertiary spectacle, not a secondary or primary one. This provides further evidence that snakes descended from lizards and developed this spectacle to protect themselves during their underground burrowing.

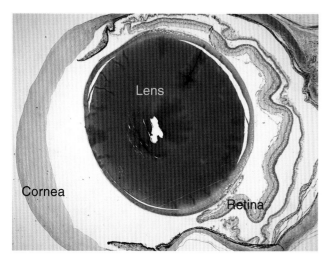

Figure 19.2 Large piscine-like lens in a pit viper. This pit viper lens is large compared to the eye size. It is spherical and resembles a fish lens. Accommodation is accomplished by movement of the lens, but the mechanism is very different from, and unrelated to, that of a fish. The retina in this image is artifactually split and detached. The spectacle has been discarded. *Image by Richard Dubielzig, DVM.*

In snakes, these lids are fused over the otherwise completely normal eye with a sulcus around the eye hidden beneath these fused lids. This inner cavity is completely lined with conjunctiva (Figure 19.1). These fused lids developed a clear scale, so the snake can see through them (Figure 19.4). A portion of this tough external skinlike covering (cuticle) is even shed during molting. This periodic molting process maintains a clear spectacle, as they do become scratched and dull. Snakes must lubricate the space between the cornea and the spectacle, so they secrete an oily substance from their Harderian gland for that purpose.

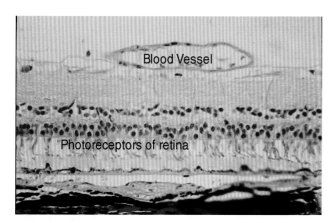

Figure 19.3 Vitreal vessels of a snake retina. The blood vessels and the retinal photoreceptors are labeled. Note that there are no vessels that radiate into the retina. *Image by Richard Dubielzig, DVM.*

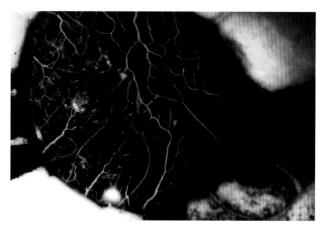

Figure 19.4 *Python regius.* Ball python spectacle. Silicone has been injected to permit photography of the blood vessels of the spectacle of this python. These vessels close after the spectacles form completely. This illustrates that the spectacles are part of the outer skin. *Image by Richard Dubielzig, DVM.*

The eyes of snakes are not particularly good at discrimination but do have good motion sense. Snake eyes accommodate by using the vitreous to push the lens forward, a unique system (Appendix E) that arose because the snakes lost their lizard-bequeathed method of lens deformation while underground.

Some species of snakes have a purely or at least primarily a cone retina, indicating that they have readjusted to diurnal life on, not in, the ground. The retina is different from that of lizards with a pure-cone retina. To assist with discrimination above ground, and probably in place of their lost oil droplets, some snakes have evolved a yellow crystalline lens, which works like sunglasses. And instead of lizards' standard double cones (Chapter 10), also left underground, snakes have their own variety of double cones, indicating that these reevolved independently of lizards'. These double cones are morphologically different—more "rod"-like—than lizards' double cones. This fits well with the theory of reemerged snakes, which simply had to reinvent an eye from the ground up.

Reemergence also required snakes to navigate the surface environment as predators, which in turn required them to reevolve the sensory component. But because snakes did not depend on vision as much as carnivorous lizards, they could explore other sensory mechanisms to integrate with vision. This is a visual witness on a thermal fringe.

Rowley's Palm Viper, Bothriechis Rowleyi, and Tree Boa, Corallus hortulanus

The sun's electromagnetic spectrum produces an energy peak in the region of the visible spectrum's shorter wavelengths. Life on earth has capitalized on this energy source for growth and sensory adaptations. But other portions of the spectrum provide useful sensory information for organisms that can recognize it. Two closely related families of snakes have independently found an elegant mechanism to use this information and integrate it into vision. Both the pit vipers (Figure 19.5) and the boas (Figure 19.6) have found the same infrared portion of the spectrum, and both employ similar neurologic mechanisms to retrieve information from it.

Rowley's palm viper is an arboreal, montane representative of Crotalinae (all pit vipers), an elite subfamily of snakes capable of radiant thermal detection and poisonous dispatch of prey (Figure 19.5). The tree boa, Corallus hortulanus, is a representative member of Boidae, a less-derived family that includes the boas, that also detects infrared but has no venom (Figure 19.6). Both families have infrared detection abilities. These are best understood and described in the pit vipers.

The nerve endings in the infrared-sensitive pits located anterior to pit vipers' eyes can detect temperature differences as subtle as 0.003C° at the surface of the receptor. These snakes can even strike based on thermal clues alone, as the pits face forward with overlapping fields, making the thermal sensitivity "stereoscopic" in nature. These ultrasensitive bolometers combine thermal isolation and extreme vascularization to create organs that are extraordinarily sensitive, matching the radiant spectrum of rodents, vipers' principal prey. These organs are bimodal receptors that absorb energy in two infrared spectrum ranges: 3–5 μm and 8–12 μm.

The infrared-sensitive neurons travel to the trigeminal ganglion (the nerve used for facial sensation) and then eventually merge with the visual input in the optic tectum—where snakes and many other vertebrates do their visual processing. The snakes are then able to "visualize" their environment mediated through an entirely different "visual" pathway. Hence, the pit vipers do not rely on a single sense to apprehend their prey.

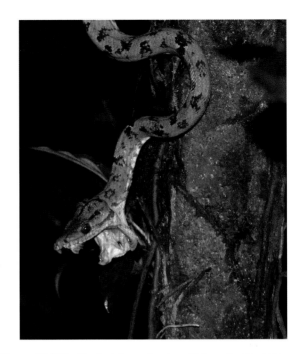

Figure 19.6 *Corallus hortulanus*, tree boa. This snake also uses thermal detection to target but will dispatch its prey by constriction not venom.

In addition to using the infrared and visible spectrum, pit vipers precisely locate their prey by integrating the chemosensory input from the vomeronasal auxiliary organ of olfaction, the organ of Jacobson, on the upper palate. The flicking, forked tongue provides the molecular input and directional clues to that organ. Even during the open-mouthed strike, thermal receptors scattered in the snake's oropharynx allow it to follow its victim.

The thermal and olfactory senses allow pit vipers to capture prey in total darkness. These senses, possibly the principal sensory mechanism on diurnal hunts, may have been retrieved from underground fossorial journey to combine with vision.

In these snakes, vision is important, but is not emphasized as it is in other reptilian (squamates) species. Unlike a few snake species, pit vipers do not have a fovea or even a macula. As in other nocturnal predators, the vertical pupil of the pit viper can constrict to an almost invisible slit to protect a rod-rich retina, especially during the day. The iris musculature is striated and fast (as in birds) to allow the creation of the slit to limit light and feed diurnally, if necessary.

These three senses, then—vision, olfaction, and thermal detection—provide a unique view of prey and extend the potential for predation from the total darkness of a subterranean burrow to full daylight.

The sensory adaptations in these serpentine lineages suggest a terrestrial awakening following a fossorial transition period. Most snakes are nocturnal but often have diurnal activity, and the eye reflects this behavior.

Figure 19.5 *Bothriechis rowleyi*, Rowley's palm viper.

For some snakes, life on or in the ground could not compare with aquatic life. As their name implies, marine snakes chose to return to the sea after successfully adapting to terrestrial life. This move required other adaptations, which can be best illustrated by the banded sea krait.

Banded Sea Krait, Laticauda colubrina

Many creatures have extraocular photoreception, even beyond the parietal eye. Most plants have photosynthesis and, hence, photoreception. But many animals have extraocular photoreception essential to their niche.

Eyes are not the only way to perceive and utilize the visible spectrum, as plants demonstrate. Most invertebrates, especially those without eyes, have some mechanism of extraocular photoreception.

Many invertebrates have novel anatomic placement of extraocular photoreceptors. For example, marine invertebrates as diverse as sea squirts, anemones, sea urchins, and annelids (worms) all have extraocular photoreception. This is probably a conserved trait from more primitive creatures, perhaps even the

Figure 19.7 *Laticauda colubrina*, sea snake.

early Archaea, the closest domain to eukaryotes and Metazoa. But, further distanced from the prokaryotes, even vertebrates such as amphibians, lizards, and snakes, including the banded krait (Figure 19.7), often have extraocular photoreception.

Most sea snakes, such as Laticauda colubrina *(Figure 19.7) and* Pelamis platurus *(a closely related, well-studied sea snake) have a thin and nonvascularized all-cone retina with a high concentration of horizontal cells and high degree of summation, indicating excellent motion perception at the expense of visual acuity. Summation means that the signals from many photoreceptor units combine into a single impulse, making the retina more sensitive but less discriminating. All sea snakes are predators with efficient neurotoxic venom. Found in the Coral Sea, the banded sea krait preys mainly on eels, sometimes on fish. Once the prey succumbs to this rear-fanged sea krait's venom, the snake swallows its meal whole.*

The sea krait also serves as prey, for sharks and sea birds such as the white-bellied sea eagle. So avoidance of predators is as important as prey capture.

An agile swimmer with a paddle-like tail for propulsion, L. colubrina and other sea snakes suffer from not knowing exactly where their tail is, so they benefit from a rear set of "eyes." Cutaneous photoreception in their tails let the snakes pull their tails under rocks or into crevices to avoid detection and predation. Thermal recognition is not part of this process, although the exact mechanism is unknown. The tails of sea snakes recognize light even underwater where the thermal difference would be insignificant.

Photonic response probably preceded life or at least coincided with it. Photoreception in the tail of sea snakes illustrates how easily evolution can add the skill of light perception and its importance (Chapter 1).

From the theropods (Chapter 17), the Cretaceous period would also see the rise of a lineage that would develop the best visual system the planet has ever known. Birds arose from the theropods, which already must have possessed excellent visual mechanisms.

THE AGE OF BIRDS—THE EYE TAKEN TO GREAT HEIGHTS

CRETACEOUS PERIOD

145–65 MILLION YEARS AGO

CENOZOIC ERA

TERTIARY PERIOD

65–2 MILLION YEARS AGO

The Mesozoic Era was the age of the reptiles, but it spawned birds—the creatures that now have the best vision on earth. The class Aves arose in the late Jurassic from one or more of the theropod dinosaurs just as the saurian medley was in full flower. Birds, which received all the tools developed by evolution to that time, refined the visual witness.

The late Jurassic was dominated by bigger-than-life, fearsome dinosaurs such as the *Tyrannosaurus* and *Brachiosaurus*. But most dinosaurs were small and furtive. Similar to our own time, many species of diverse phyla filled most available niches.

While pterosaurs (Chapter 17) plied the skies, another—completely unrelated—line of flying dinosaurs was starting to emerge. Birds—the only living descendants of dinosaurs—arose from one or more small carnivorous dinosaurs in the theropod lineage.

Birds Arise

Fossil evidence for *Archaeopteryx* shows this oldest known bird to be fully feathered and to have many other avian characteristics, such as a wishbone. Appearing about 150 million years ago, during the late Jurassic just before the Cretaceous period, this raven-sized animal was to cast a conspicuous shadow over the next 150 million years.

Although designated as the first bird, *Archaeopteryx* was probably more reptilian and cursorial than aerial in its habits. And it had to represent a long history of avian development before that fossil. Recently discovered specimens (from China) show that animals with birdlike characteristics existed well before its time. *Archaeopteryx* probably did not radiate directly into the known avian taxa anyway, as the species probably represents a sister group or out-group

to the true birds. This means that a contemporary avian-like relative alive at the time probably was responsible for the avian clade. Nevertheless, *Archaeopteryx* and other specimens suggest that the theropod dinosaurs radiated into birds, probably several times, and diversification began by the early Cretaceous.

The period warmed as its middle years approached. Flowering plants were becoming abundant, and life was brawny and rich. The seas were warmer, and some regions were more anoxic than today. If present at all, ice caps were small, and the poles contained little or no land. Some parts of the sea were replete with vertebrates (principally fish), marine reptiles, and invertebrates. Whenever it arose, this new lineage of birds would find opportunities requiring flight.

Because it would require some structural uniformity within the class, flight would somewhat restrict the morphological variation. Such characteristics as hollow, light bones, higher metabolic rates (hence endothermy or warm-bloodedness), aerodynamic shape, specialized lungs, and small compact bodies would be required to meet the biological demands of flight. These characteristics would have to be present, generally, in all flying species.

But flight had other demands and, more importantly, other implications for sight and the brain. Flight would require—or eventually lead to—speed, and speed generally requires excellent vision and faster visual processing. These requirements would have to be satisfied for birds to achieve the dominance provided by an aerial lifestyle (Figure 20.1).

Morphologically more reptilian than avian, the first fossil avian skulls represent an intermediate step between reptiles and birds. The avian brain and eyes are different from other brains and eyes in the reptilian class. These differences would be essential to conquer the aerial niche.

Figure 20.1 Evolution of birds. The evolution of birds from dinosaurs is shown in stylistic form. The illustration shows a small carnivorous theropod named *Compsognathus* (far right) leading to a more evolved theropod with feathers, named *Avimimus* (second from right). Note the larger eyes and a more birdlike head and stance. Third from the right is an artist's rendition of *Archaeopteryx*, although this bird was probably not much of a flyer. The last image (on the left) shows a more modern-appearing bird, a pigeon, with true flight. This is simplistic because *Archaeopteryx* was not in the direct lineage of birds but rather a sister group. *Image © The Natural History Museum, London. Artist John Sibbick.*

For example, the avian eye is generally (both relatively and absolutely) larger than the reptilian eye. Leuckart's ratio predicts that eye size varies directly with speed—the faster the animal, the larger the eye. So, birds have larger eyes because flight permits and demands greater speed. The large eyes of birds permitted the evolution of an unequaled quality of vision. But this enlargement would place specific physiologic and anatomic requirements on the eyes and the brain that would lead to gradual or perhaps sudden improvement in vision and visual processing. Being fierce, often fast predators, most theropod dinosaurs before these early avian species probably were highly visual animals, so some of this visual improvement probably occurred before the radiation into birds. Some evidence suggests that the theropods' visual acuity might have rivaled that of birds.

Birds entered the Cretaceous (145–65 mya) as ground-dwelling or arboreal, but probably not aerial, dinosaurs. Avian footprints suggest various species even by the early Cretaceous that had feathers for display or warmth but probably not for flight. If these early birds flew, they did so crudely, perhaps like gliding animals today. Robust flight began at least by the late Cretaceous, well before the meteorite strike of 65 mya that would end the period. The two principal characteristics that permitted survival of the avian clade probably were flight as an escape mechanism and scavenging as a feeding strategy.

Birds then flew into the Tertiary Period (65–2 mya) to become the dominant lineage. Large ground-dwelling birds evolved to become the largest predators of the early Tertiary,

and flight permitted distant colonization by smaller species. Improved visual mechanisms had to accompany flight, particularly swift flight. If you have ever watched a flycatcher or swallow master the wind for insect capture, then alight on a fine twig, especially in crepuscular light, you realize that the bird's visual acuity and processing have to be better than our own. Those advances in avian vision and flying skills had to have been present during the early Tertiary and probably much of the Cretaceous.

Many of the physiologic and anatomic changes in the eye had to evolve during the Cretaceous, if not already present by that time. The story of the avian eye is best told in its comparative anatomy and physiology because this is the crowning metazoan visual system.

Globe Morphology

In most avian species, the eyes are so big that they outweigh the brain. And, unlike our eyes, they often are individually asymmetric.

These large eyes come at a price, though. Birds have evolved eyes that fill their orbits, limiting volume of other structures, such as extraocular muscles—most birds have not retained the muscular function to move their eyes more than 1–2° from primary gaze. The bounty of extraocular muscles that originated in fish evolved into rudimentary slips of muscle in birds.

Morphologically, the basically spherical eye found in reptiles enlarged to fill the orbit in birds. As a result, some birds evolved asymmetric globes leading to three

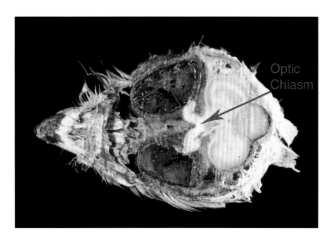

Figure 20.2 *Buteo jamaicensis*, red-tailed hawk. Slice through skull of a red-tailed hawk. Note that section occurs nearly through the center of the globes and shows the globose shape of the eye and the enormous size of the globes, especially as compared to the brain. The diameter of these eyes is approximately 23 mm, essentially equivalent to the diameter of a human eye. These eyes have collapsed anteriorly to some extent with some decrease in the anterior-posterior dimension. Generally, the globose eye has a similar anterior-posterior diameter and equatorial diameter. This section cuts directly through the optic chiasm where the optic nerves cross. This is important to stereopsis and is discussed in the text.

shapes: globose, flattened, and tubular. Only one of these shapes occurs in reptiles, or mammals for that matter, although all three shapes are seen in fish, which indicates the morphologic plasticity of the eye. But because of the variety of ecological niches occupied by birds, no single shape represents the avian eye.

Globose Globe

The avian eye began as a globose eye, inherited from birds' reptilian ancestors—specifically theropod ancestors, although some of these specializations may have occurred in these dinosaurs too. This relatively globe-shaped eye is as close to a spherical eye as is found in the Aves class. This shape—commonly seen in passerines, raptors, and some other diurnal birds—is the most common anatomical form (Figure 20.2). The eye retains the cartilaginous cup, inherited from fish through reptiles, and scleral ossicles, inherited from reptiles and possibly evolved from fish but possibly independently and convergently.

Flattened Globe

As birds evolved, flight improved, but some species chose an aquatic life. These unusual birds appeared in the late Cretaceous, a time of shallow seas around many of the continents. Holding rich fisheries, these continental shelves were a niche to be exploited, and a few birds took the bait. These birds resembled loons, cormorants, grebes, and other

diving birds. As these birds sought underwater prey, though, they would have to develop different body morphology, and a different eye.

In terrestrial birds, the cornea may account for two thirds of the focusing ability of the eye. Because the cornea is of little use as a refractive element under water, most of the aquatic birds have evolved a flattened cornea and a flattened globe—more or less like fish. The flattened globe of the loon and of other diving birds has a short anterior-posterior (AP) axis, with a flat or concave ciliary region adjacent to the cornea and scleral ossicles (Figures 20.3–20.5). The cornea, and the globe in general, are relatively flat compared to the globose eye, and the ciliary region immediately posterior to the cornea is flat, creating an asymmetrical globe. The lens, then, must do all focusing when the bird is under water.

Accommodative mechanisms to solve this problem probably appeared at about 140 mya when the diving birds first appeared in the early Cretaceous. Loons, darters, cormorants, and other diving birds have all solved this problem in a similar manner, and this may have radiated from a common ancestor.

The Australian darter illustrates the principle.

Australian Darter, Anhinga melanogaster

A diving, piscivorous bird faces special challenges that demand evolutionary creativity. A hunting vertebrate that pursues and captures prey under water must reckon with the loss of corneal refractive power. Hence, such a hunter must rely on lenticular

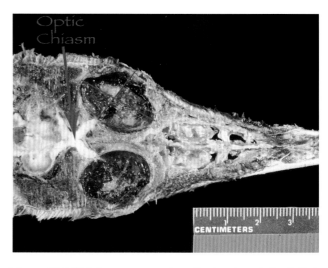

Figure 20.3 *Givia immer*, common loon. Section through the skull of a loon revealing the flattened globe with a decreased AP diameter as compared to equatorial diameter. This section through the loon skull also shows the chiasm or crossing of the optic nerves as seen with a blue arrow. The chiasm is completely crossed or decussated and is discussed in the text.

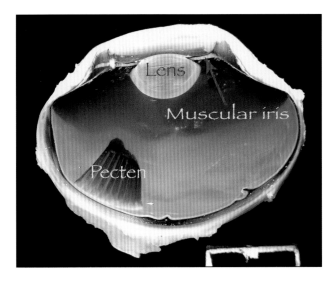

Figure 20.4 *Givia immer*, common loon, globe sectioned. Note the relatively small lens permitting easier manipulation by the muscular iris as it squeezes the ductile lens through the pupil for accommodation. *Image by Richard Dubielzig, DVM.*

Figure 20.6 *Anhinga melanogaster*, Australian darter. This bird is a powerful swimmer and will chase down fish. The bill of these birds is so sharp and powerful that darters will use their bills like daggers to impale fish. Once the fish is impaled, the darter comes to the surface and flips the fish in the air to catch and swallow it.

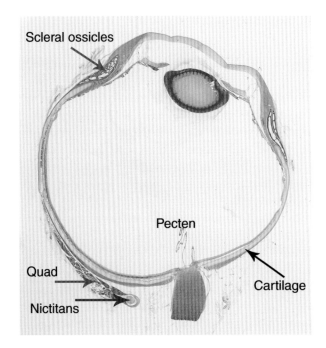

Figure 20.5 *Givia immer*, common loon. Histologic section of a loon's eye. Note scleral ossicles (blue arrow). These are bony supports for the flattened morphology of this eye. The muscles of accommodation use the bony support of these ossicles as they maintain rigidity of the sclera with contraction of the muscle. Note the pecten (labeled), which is the nutritive support to the inner retina, and note the cartilaginous cup (black arrow) that supports the globe. The quadratus muscle (Quad) becomes the sling for the tendon of the nictitans and is also labeled. At this site, the nictitans is a tendon and is pulled across the eye using the sling of the quadratus. *Image by Richard Dubielzig, DVM.*

accommodation for the entire dioptric power of the eye underwater. But birds face an additional twist when compared to predatorial fish. A bird must approximate emmetropia, or normal distance vision, when airborne as well as under water. Cormorants and their kin, such as darters, are models for the successful solution to these problems (Figures 20.6–20.8).

Perhaps the most unusual adaptation of the darter is the eye, which has evolved to solve the loss of corneal refractive power.

A review of accommodation and accommodative amplitudes helps us to understand the darter's adaptations. Crystalline lenses have inherent power measured in diopters (a unit of measurement of refractive power of a lens). In addition, the accommodative ability of the eyes of most terrestrial vertebrates to change focus from distance to near is also measured in diopters.

In the first year of life, the crystalline lens of a normal human eye has up to 20 diopters of accommodative power or focusing ability. We can change the shape of our crystalline lens for focusing at near range. As we age, we lose our accommodative abilities in an almost linear fashion, and by the age of sixty, we have but 1–2 diopters of accommodation left.

Good evidence suggests that some birds, such as the cormorant or the dipper, are capable of at least 50 diopters of accommodation and may be capable of as much as 70–80 diopters. This is accomplished by a most extraordinary mechanism.

In these birds, the iris is stiff with sturdy musculature. Accommodation involves tightening the iris sphincter muscle and contracting the ciliary muscles. The ciliary body contains three striated muscles. When accommodating, these muscle contractions force the ductile lens though the rigid pupil, creating anterior or accommodative lenticonus (Appendix E).

A   B

Figure 20.7 *Anhinga melanogaster*, Australian darter. **(A)** This bird is a superb fisher and will chase fish by using her feet, but not her wings, to propel herself. The feathers wet easily, though, because they are not inherently oily as are the feathers of other birds. Oily feathers would inhibit diving, as the bird would be more buoyant. **(B)** This bird has impaled a fish and brought it to the surface.

The pecten may contribute by filling with blood and maintaining pressure against the posterior capsule of the lens (Figure 20.9).

The lineage that includes Australian darter still has other secrets to teach us. The great cormorant, a related species, has a large Harderian gland, which is a supplementary tear/oil gland

Figure 20.8 *Anhinga melanogaster*, Australian darter. Following a hunting expedition under water, this bird must dry its wings in the sun before being able to fly.

embedded along the superior edge of the nictitans (Chapter 13). This gland secretes an oily, viscous tear to prevent osmotic dehydration related to water immersion in water or even prolonged exposure to its own—perhaps highly concentrated—tears. The lacrimal gland of cormorants and other diving birds can outperform the kidney because of its ability to excrete a highly concentrated salt solution in its tears. This mechanism allows the bird to obtain all necessary water from its fishy diet or from seawater.

The diving birds appeared on earth approximately 140 mya. The Australian darter (Figures 20.6–20.8), also known as one of the snakebirds (Anhingidae), is very closely related to boobies, gannets, and allied to pelicans. Snakebirds are strong swimmers and have long, sharp beaks with serrated edges that can grasp or spear their prey underwater and need these rapid and accurate accommodative skills. Known from at least the Eocene (56–35 mya), snakebird predecessors included several interesting birds known as plotopterids. These birds probably were the giants of the Anhingidae lineage, being as much as 25–65 percent larger than modern descendants, although we have little knowledge of their eyes.

Tubular Eye

Birds radiated into most niches where flight would prove an advantage. Twilight and the dead of night were no exceptions, and this would place special demands on the eye. The demand of maximizing photon capture would

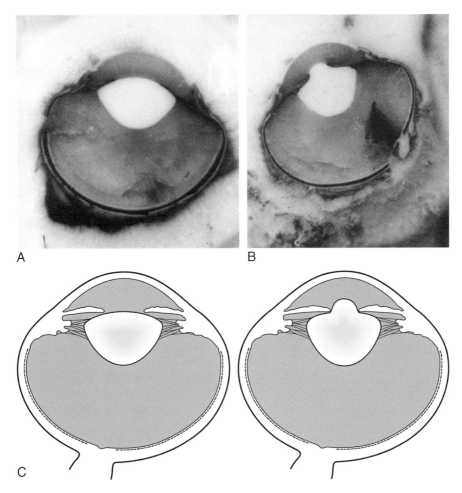

Figure 20.9 *Mergus merganser*, common merganser. Anterior or accommodative lenticonus stimulated pharmacologically. **(A)** This section of the merganser's eye is not stimulated to accommodate, and the lens is in an unaccommodated position. **(B)** This section of the merganser's eye has been stimulated to accommodate. Note the "button" of lens that has been squeezed through the pupil. This creates a "rounder" lens, at least where the light rays would strike it, and provides a stronger refractive element to compensate for the loss of the cornea as a refractive element under water. Many diving birds use this form of accommodative lenticonus. Images and insightful work provided by Jake Sivak, OD, PhD. **(C)** Artist's rendition of accommodative lenticonus illustrating this process. *Art by Tim Hengst. See accommodation panel in Figure 10.8.*

add impetus to the formation of the last and perhaps most interesting ocular shape—the tubular eye. This shape is seen mostly but not exclusively in owls.

Owls arose from a common ancestor closely related to the hawks, eagles, and vultures. Owls' passage into night (in the early to mid-Tertiary 55–60 mya in the fossil record) required larger eyes, larger lenses, and better photon capture. These requirements would bring the birth of the tubular eye. Because this morphology appeared in very different classes (birds and certain fish), the eye appears to be able to mold into the tubular shape easily on an evolutionary basis.

The most peculiar avian globe morphology, the tubular eye appears only in two distantly related orders, owls—Strigiformes, and some goatsuckers—Caprimulgiformes (Figures 20.10–20.12). This unusual tubular eye allows its owner to maximize the length of the globe relative to the size of its head. The tubular eye has the longest diameter that the orbit will permit. If the eye were spherical instead of tubular, it would be large enough to protrude outside of the orbital space. Evolution has reduced it to a tube to permit the longest diameter possible for that size orbit (Figures 20.10–20.12). To maximize the diameter of the eye, evolution has decreased the extraocular muscles to mere wisps and streamlined the globe. The retinal area is small; the anterior edge of the retina corresponds to the widest equatorial diameter. The entire eye is outsized compared with the body size, with a large lens and deep anterior chamber. This anatomy allows for sharper acuity with a larger, brighter image spread across the retina. The single fovea in each eye is steep-walled, and the retina is duplex (both rods and cones) but rod-rich. In essence, these birds have sacrificed extraocular motility, visual field, and stable structure for improved visual acuity, especially at night.

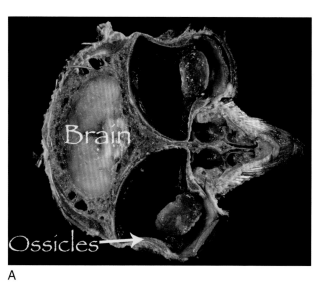

A

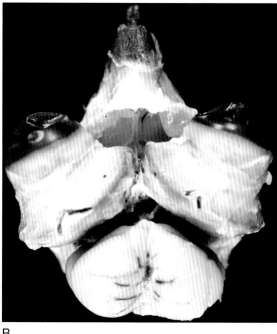

B

Figure 20.10 *Bubo virginianus*, great horned owl. **(A)** Tubular eye in a great horned owl. This section through the skull of a great horned owl shows the tubelike shape of the globe, the large lenses, and the scleral ossicles (labeled). A tubular eye performs optically like a larger eye with a diameter roughly equivalent to the anterior-posterior diameter of this eye. The ossicles help maintain the shape, as the tubular eye is less stable than a round eye. **(B)** Great horned owl's eyes and brain dissected from the skull. Note the large size of the eyes as compared to the brain. *Image by Richard Dubielzig, DVM.*

The tubular eye has other differences from most vertebrate eyes. Its unusual anatomy includes a very large anterior chamber (Appendix A), which is the space between the cornea and the iris and lens. For example, the great horned owl has approximately 3.0 ml of fluid in the anterior chamber of each eye—ten times as much as the 0.3 ml in

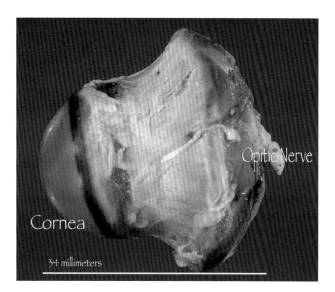

Figure 20.11 *Bubo virginianus*, great horned owl. Eye dissected from the skull. The anterior-posterior length is 34 mm. In effect, this tubular shape is optically equivalent to a circular eye that size, but without the volume to match. A normal human eye is approximately 24 mm in diameter.

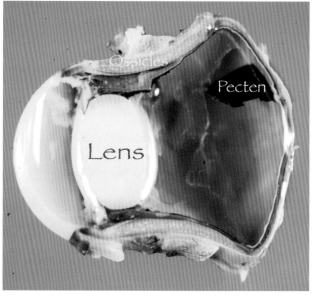

Figure 20.12 *Bubo virginianus*, great horned owl. Dissected eye is sectioned in half. Note the large sclera ossicles and the pecten discussed in the chapter. *Image by Richard Dubielzig, DVM.*

each of our eyes. The lens also is large, which helps in the collection of photons. In that same great horned owl, the eye is enormous. Its diameter is about 34 mm, compared with the 24-mm globe of a normal human (Figure 20.11), yet this owl would come up only to your knee in a standing position.

Tubular eyes get their name and character from a concave intermediate segment that elongates the globe from the front to the back. This elongation forms a tube as it joins the posterior segment. This tube makes the eye asymmetric and inherently unstable, but ten to eighteen scleral ossicles (depending on species) maintain the structural integrity. The western screech owl illustrates the principle of the scleral ossicles.

Western Screech Owl, Megascops kennicotti

As typified by the western screech owl, Megascops kennicotti *(Figures 20.13 and 20.14), the New World screech owls are the most abundant and perhaps most familiar owls in North America. They are widely distributed throughout western North America, with about forty closely related species throughout the world. These birds have tubular eyes (Figure 20.15) that fill the orbit, leaving almost no room for extraocular muscles. To compensate for these resulting rudimentary muscles, most owls can turn their heads, with great speed, nearly 270° atop a thin, lightweight spinal column. In fact, they turn their heads so quickly that myths of complete 360° cranial revolution have sprung up, adding to the lore of owls.*

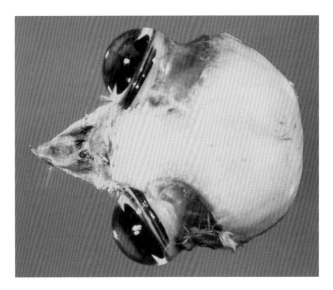

Figure 20.14 *Megascops kennicotti.* A screech owl's skull with eyes but without feathers. Note the tubular shape of the eyes, at least for the anterior half of the eye. *Image by Chris Murphy, DVM.*

The owls' eyes are frontally placed to provide as much binocular stereopsis as possible. Owls possess a stereoscopic binocular visual field of approximately 60–70°, although the total visual field is larger if one includes the monocular fields. Other avian predators such as hawks and eagles generally have only 30° of frontally placed stereoscopic binocular visual field.

As mentioned, the eyes easily outweigh the brain. In owls in particular, predation has been the principal stimulus to become

Figure 20.13 *Megascops kennicotti,* screech owl. *Image © T. Vezo/ VIREO.*

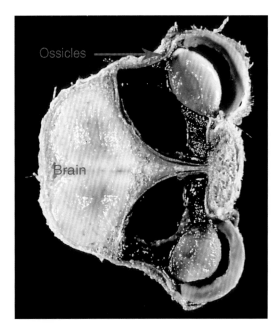

Figure 20.15 *Otus asio.* Eastern screech owl skull. Note that this much smaller bird has a very similar bony configuration to the larger great horned owl in figure 20.10. Note that the eyes together are larger than the brain.

an "eye-minded" order, one with an evolutionary concentration on visual abilities.

Orbital Size and Contents

Our globes are recessed beneath, and protected by, an orbital rim. In birds, though, much of the eye is exposed, especially on the temporal or lateral aspects, as if the eye has evolved almost beyond the orbit. The posterior aspects of the two eyes abut one another, with the eyes separated by a thin, often translucent, bony septum.

Fixed asymmetric globes present a problem to a bird because they create a limited span of the visual field. So how do birds solve it without extraocular motility or a swiveling neck?

Visual Fields

The goliath heron's (Figure 20.16) eyes sit high on the lateral aspects of its head, so it must hold down its beak to have much of a stereoscopic field. This head position provides about 10° of stereopsis directly in front of its head. To feed in the wetlands and shallows of the marshes they inhabit, these herons walk forward, keeping their heads down, to frighten or herd their victims. As the herons walk along a stream or lake, they push fish forward into the herons' stereoscopic field.

The American bittern has a very different position to maximize his visual field (Figure 20.17). This bird waits for small fish, frogs, or crustaceans to seek the shade of her body. As they seek her shade, they come within her 10° or 15° of stereoscopic binocular field directly beneath her neck and beak.

When the bittern is alarmed, its mandible goes up to permit maximal stereoscopic vision parallel to the horizon. The striations along his neck and on his breast add camouflage to the frozen posture of his alarm position.

A

B

Figure 20.17 *Botaurus lentiginosus*, American bittern. Notice the posture of the bird in an alarm or inquisitive position **(A)** and in a hunting position **(B)** as opposed to the goliath heron position in Figure 20.16.

But the title of the most interesting visual field arrangement in the bird world goes not to the American bittern but to the American woodcock.

American Woodcock, Scolopax minor

This shy upland game bird is a member of the sandpiper family. The bird is difficult, if not impossible, to see because of cryptic coloration (Figure 20.18).

Figure 20.16 *Ardea goliath*, goliath heron. Note the alarm position with the mandible down. This will maximize the binocular visual field and is similar to the feeding position.

Figure 20.18 *Scolopax minor*, American woodcock. Notice how high the eyes are on the head. *Image © J. Bruce Hallett.*

The bird's distinctive long tapered bill probes for earthworms in damp soil, in soft substrates, or in alder thickets with associated clearings and boggy fens. The bill tip may feel the vibrations of an earthworm's movement in the soil, but whatever the mechanism, the bird does not use sight to find the earthworm.

These mostly crepuscular and nocturnal birds have large eyes set high on the head (Figures 20.18 and 20.19) but probably not particularly good acuity or extraocular motility. To protect themselves from ground and aerial predators, these ground dwellers do have good motion detection and a visual field

Figure 20.19 *Scolopax minor*, American woodcock. The visual field of this bird is shown. The black field represents areas in which both eyes are used together—binocular stereopsis. The gray areas to the right and left are uniocular fields from each eye, respectively. The gray dotted fields are vertical fields representing the visual field above the bird. This visual field does not unite above the bird but leaves a blind strip directly overhead between the two individual eye fields.

documented to be among the largest in a terrestrial vertebrate (Figure 20.19) (among all vertebrates only some fish may have larger visual fields).

These birds have a binocular field with stereopsis in the frontal visual field and in the visual field behind the head! (A 5–12° binocular field in the frontodorsal quadrant begins a few degrees above the upper mandible, as shown in black in the drawing (Figure 20.19); the rear binocular field is 5°. A monocular field completes the 360° field by connecting the edges of the binocular fields (this cyclopean portion of the field is illustrated with the gray shading). The visual field also extends in a hemispheric fashion above the head nearly completely, creating a 360° hemispheric field that even extends 5° to 10° below the horizontal.

In theory, then, this bird can sit on its nest and encompass the entire visual space around it, save for the bilateral blind spots of the pecten (Appendix C). (A blind spot at the tip of the downward-pointing bill is what keeps the bird from using its sight for direct foraging.) For example, if a human had the same visual field, she could pitch a baseball to a batter at home plate, watch runners on first, second, and third base, and even watch the home run travel over the left-field fence—all without ever moving her head.

Evolutionarily, this visual arrangement makes complete sense because the bird is a prey species for various predators. The American woodcock has been documented since at least the Pleistocene (2.5 mya), illustrating the conserved nature of this adaptation. These all-encompassing visual fields probably appeared early in such birds as a result of heavy predation.

The development of the visual fields in these birds reveal that, like so many other aspects of biology, the evolution of almost any function has been niche-driven.

Eyelids

The reptilian eyelid model continues with birds. The model consists of three lids for each eye: an upper lid, a more active lower lid, and an inner lid—the nictitans. In many birds smooth muscle moves the two outer lids slowly and infrequently. In some birds the outer eyelids may close only a few times during the day and really close only during sleep. The lower lid contains a semicircular fibrous plate and covers a larger portion of the eye during blinking than the upper eyelid does. Owls and parrots tend to use their upper eyelids to blink more often than most other birds.

How lids close says as much about heritage as when they close. In reptiles, birds, and probably theropod dinosaurs, the lower lid closes to the upper lid with less movement from the upper lid. In mammals, and probably other synapsids, the lids close in the opposite direction with the upper lid being most active and closing to the lower lid. The outer lids in the synapsid line also contain striated muscle, which is much faster than smooth muscle and is capable of—and causes—more frequent movement.

Figure 20.20 *Ninox connivens*, barking owl. Note the feathers around the eyes and the resemblance to mammalian lashes.

Eyelashes are important to most terrestrial animals for protecting the ocular surface. Various classes have had different solutions, although only mammals have hair. Birds have outer eyelids with "eyelashes" made of feathers (Figures 20.20 and 20.21).

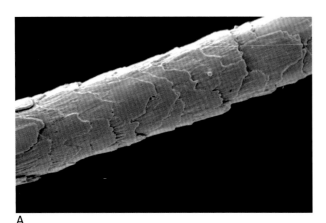

A

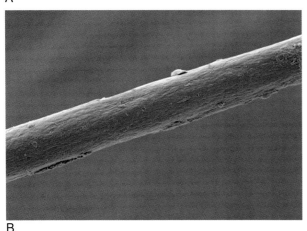

B

Figure 20.21 (A) Electron microscopic image of human lash. **(B)** Electron microscopic image of periocular filoplume of a red-tailed hawk.

Nictitans

Birds have perfected the nictitans with several innovations as compared to others with nictitating membranes (Chapters 13 and 15). With only smooth muscle in outer lids, hence lids that cannot move quickly and rarely close, birds clean and moisten the surface of their eyes with a nictitans. In birds, specialized muscles draw this thin, tough, and well-developed membrane across the globe from the nasal (or ventral-nasal) aspect of the globe nearest the beak to the temporal (or superior-temporal) aspect. Although cleansing the ocular surface is its main function, the nictitans serves other purposes such as protection.

Because the nictitans has reached its zenith in birds, theirs is fast and efficient, with a true and active blink. In fact birds' nictitans are much faster than the passive ones of mammals. In many bird species, the process works in one of two ways: the nictitans slides or flashes in front of the cornea when the eyeball is retracted into the eye socket, or the nictitans' movement is managed by a pulley system. More details are provided if desired.

BOX 20.1 HOW THE NICTITANS WORKS IN BIRDS

Retraction of the nictitans in birds occurs by special extraocular muscles including the quadratus—derived from the retractor bulbi first seen in the ancient placoderms—and the pyramidalis. Although the pyramidalis also may have come from the retractor bulbi, it probably arose from another extraocular muscle—the lateral rectus. The pyramidalis originates from the posterior sclera adjacent to the optic nerve. It loops around the optic nerve and crosses a sling formed by another muscle, the quadratus. The nictitans swings superiorly around the nerve and temporally around the globe, where it becomes tendinous and splays out into its insertion along the inner canthus. The membranous portion is often clear (Figures 20.22 and 20.23), but it is not a refractive element for any avian species, as was once thought. In lizards, crocodiles, and turtles the muscles are arranged slightly differently. In those animals the quadratus and the pyramidalis act to slide or "wink" the nictitating membrane over the cornea without significant retraction of the eye. Essentially, the retractor bulbi has been co-opted into another structure in these species.

The nictitans, then, is capable of extremely rapid sweeps across the ocular surface to clear the cornea of debris. Especially in birds, the nictitans also moistens the ocular surface with an oily or sometimes aqueous substance.

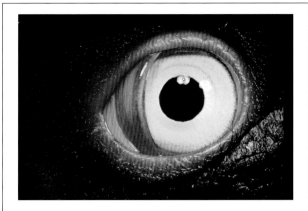

Figure 20.22 *Sarcoramphus papa*, king vulture. All birds have a nictitans used for protection and cleansing. In this image, note the blue iris can be seen through the nasal portion of the nictitans. *Image by Chris Murphy, DVM.*

A

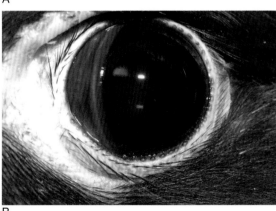

B

Figure 20.23 *Falco peregrinus*, peregrine falcon. **(A)** The head of the peregrine shows the edge of the nictitans. **(B)** The magnified image shows the nictitans being drawn across the eye. *Images by David Maggs, PhD DVM.*

The woodpeckers illustrate ocular protection by the nictitans. In the millisecond before strike, woodpeckers pull their tough and thick nictitans across the globe to hold it in place, much like a seat belt. Otherwise, the intense force could make the globes pop out of their sockets. The nictitans also protects the woodpecker against the wood fragments and splinters that might otherwise impale his eye. The woodpecker also has a Harderian gland, which is associated with the leading edge of the nictitans and is a major source of tears/oil in most birds. Diving birds such as cormorants probably exploit the oily liquid to prevent crenation, or drying out, of the membrane by the salt water.

Perhaps the most interesting birds with an oily tear film are the falcons, whose tremendous speed and acceleration pose a special challenge to keeping the ocular surface hydrated.

Peregrine Falcon, Falco peregrinus

Ferocious speed and acceleration bring their own perils, which must be either solved or avoided for success. For example, the rush of wind evaporates an aqueous tear film, especially during the spectacular stoop of a falcon approaching 300 km/hr.

Ocular surface lubrication originates not only from the Harderian gland mentioned above but, in addition, from another secretory gland—the lacrimal gland. The lacrimal gland is situated in the inferior temporal quadrant associated with the more active lower eyelid, although in many species this gland is absent or poorly developed. In falcons the Harderian gland produces a cornea-moistening viscous solution. Although the composition of these secretions is not known, they probably include mucins, polysaccharides, or hyaluronic acid rather than a more dilute tear film. This viscous solution would maintain a smooth surface but might collect debris.

The nictitans of the falcon and many similar birds has a cartilaginous-like connective-tissue fold along the leading edge of the membrane. This flangelike fold is called a marginal plait (Figure 20.24A). With each blink, it collects the tear film and any associated debris to drain through the enlarged puncta (tear drainage ducts) into the nasolacrimal system. A surface layer of "feather epithelium," believed to be unique to birds and reptiles, probably aids this corneal cleansing.

Long microvilli with club-like termini and many secondary projections from the long axis extend from the apical membrane of epithelial cells lining the bulbar surface of the third eyelid (Figure 20.24B). Those on the peregrine falcon are extremely robust and likely form a "histologic feather duster" that sweeps the cornea clean with each darting excursion of the nictitans.

Other visual problems are presented during the stoop, though, and disturbance of the tear film is just the beginning. Falcons are bifoveate (possess two fovea in each eye) with an area of concentrated photoreceptors (called an infula) between

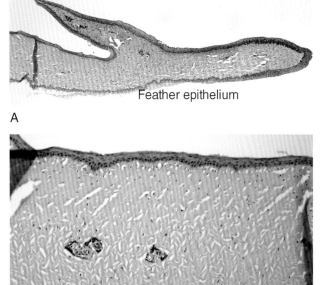

A

Feather epithelium

B

Feather epitheium

Figure 20.24 *Falco peregrinus*, peregrine falcon. **(A)** Note the feather epithelium along the edge of the nictitans. **(B)** This epithelium is in direct contact with the epithelium of the cornea and "cleans" the surface of the cornea much as a comb would clean and straighten hair. *Images by David Maggs, PhD, DVM.*

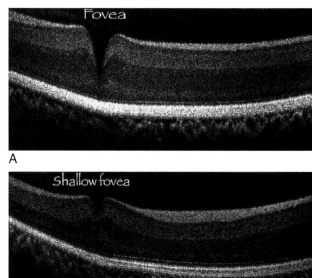

A

B

Figure 20.25 *Buteo platypterus*, broad-winged hawk. **(A)** Convexiclivate fovea (steep-walled) of broad-winged hawk taken by optical coherence tomography (OCT) by James Major, MD, PhD. OCT is an interferometry technique that permits imaging of the retina in living animals. **(B)** OCT of both foveae of the same hawk. Although this is not exactly the same appearance as would be seen in a falcon, the images would be similar. *Image © James Major, MD, PhD.*

these two foveae. How is the image acquired and successfully visualized especially during its stoop?

The nasal fovea is a deeper, steep-walled convexiclivate (Figure 20.25) and probably has better acuity. But it is the falcon's temporal fovea that is capable of simultaneous image capture for stereopsis.

The process proceeds this way: The falcon uses its deep nasal fovea to sight its avian prey from perhaps 400 meters overhead. The falcon begins its attack with a spiraling flight that allows it to keep the nasal fovea on the prey as long as possible without tilting the head sideways, which would increase aerodynamic drag. This technique does not force the bird to sacrifice acuity for stereopsis in the early phase of its hunt.

But as the bird approaches its prey, the flight pattern becomes more direct, and the image swings via the infula from the nasal to the temporal fovea, allowing the bird to maintain sharp acuity through the change in image orientation. At this point, stereopsis is achieved, and it is required as prey is now much closer and moving in three dimensions.

Members of the family Falconidae, including the peregrine falcon (Figure 20.23), have many other ocular adaptations to permit these sensational visual feats. Dense packing of the photoreceptors provides excellent discrimination, which can be measured by determining the number of cycles per degree that can be perceived. Human vision measures approximately 30 cycles per degree (cpd) for acuity of 20/20. The visual acuity of falcons has been measured to be from 75 to as high as 160 cycles per degree in full luminance combined with a surprisingly low f-number of as low as 1.4 in low light. The f-number is a measure of the light gathering capabilities of an optical instrument and is discussed in more detail in Chapter 5. Humans have an f-number of approximately 2.1, for comparison. The lower the f-number, the more sensitive the eye.

We have long recognized these fascinating visual skills and have developed a unique relationship with falcons though falconry. Furthermore, falconers have recognized their own human visual limitations and would often carry a caged shrike to alert the falconer when his bird was on the wing to return from its hunt. The shrike, himself a wing feeder, would become agitated when the falcon was returning well before the falconer could see his bird, allowing the falconer time to prepare the jesses.

A successful falcon's stoop is one of the most dramatic events in natural history, and its witness is never forgotten.

Cornea

By the time of the descent of birds, the disparate layers of the cornea had united into a single structure (Appendix D). As discussed, once tetrapods became unmistakably terrestrial, a cornea with two separate layers became less valuable and even a hindrance to its owner. The theropod dinosaurs that would later give rise to birds almost certainly had a single, united cornea, and the avian cornea would evolve further. We can best understand this evolution by a description of the various features of the avian cornea.

In most diurnal birds with a few notable exceptions, the cornea is small relative to the eye size. This disproportion probably occurred slowly, early in the Cretaceous, as the class arose and became more visually based than ever. This comparative asymmetry of the cornea as compared to the remainder of the globe is protective because there is little to no bony orbital structure surrounding the forward portions of the globe to protect it. Thus, there is less of the globe exposed if the cornea is smaller.

A smaller cornea also permits corneal accommodation, which many birds possess. That means that the muscular action can change the radius of corneal curvature to change the focus of light rays traversing the cornea. Striated musculature (like the muscles in our biceps) attaches to the peripheral cornea and can change the radius of curvature of the cornea and thus participate in accommodation, at least in many of the diurnal birds (Figure 20.26). Nocturnal birds such as owls, on the other hand, have large corneas that collect maximum light but do not participate in accommodation; owls do not depend on such fine focusing abilities. Accommodation is limited in owls as it is essentially unnecessary.

In many birds, the cornea is among the clearest known, actually clearer than our own. If the mammalian cornea were perfectly clear, it would be invisible, as it would be optically empty. Some birds, such as the Bateleur eagle, have such a clear cornea that it is barely visible even with magnification. This is simply one additional subtle strategy to improve acuity.

Iris and Pupil

Birds take the iris further in development beyond that of reptiles, but almost all birds have round pupils while reptiles have some variation in pupil shape (Chapter 15). One avian species, the blue-footed booby, even has pupillary sexual dimorphism. It is the only field sign to distinguish male from female birds—females appear to have a larger pupil, although it is just dense pigmentation around the pupil in adult females. Some birds have bright pigmentation in their irides, as do many reptiles. Some birds, such as the rock-hopper penguin, even change the color of their irides during mating season, possibly for sexual attraction. The rock-hopper penguin exhibits a red iris during times of sexual activity and a dull orange in nonmating plumage.

Accommodation

Accommodative mechanisms reach their pinnacle in birds. Birds' ability to rapidly change corneal and lens curvature produces perhaps most accurate system of accommodation of any animal on earth. It is perhaps also the fastest, with speeds up to 90 diopters of focusing ability per second, and a total accommodative range of up to 70 diopters (Appendix E).

Uvea: Choroid, Ciliary Body, and Iris

The vascular middle coat of the eye, called the uvea, includes the iris and ciliary body anteriorly and extends posteriorly as the choroid between the retina and the outer coat, the sclera. Beneath the retina in most birds, especially in the posterior pole beneath the macula, the choroid is three times thicker than our own. This diffuse layer of blood vessels provides nutrients for the photoreceptors (rods and cones) of the retina and especially the posterior pole or macula. These rods and cones are metabolically very active cells and require the constant delivery of nutrients and oxygen to function properly.

In some species, such as the eagles, the choroid also has striated muscle fibers that are thought to participate in accommodation by moving the retina forward or backward slightly for perfect focus.

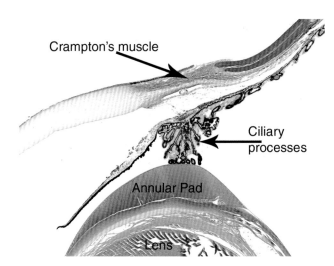

Figure 20.26 *Eurypyga helias*, sun bittern. Muscles in peripheral cornea of avian eye/histology. Note the direct contact of the ciliary body with the lens and the ringwulst or annular pad. Crampton's muscle permits corneal accommodation by traction on peripheral cornea to change radius of curvature of cornea. Ciliary processes in contact with the ringwulst of peripheral lens. *Image by Richard Dubielzig, DVM.*

Crampton's muscle

Ciliary processes

Annular Pad

Lens

No avian eye contains a true tapetum, although some of the goatsuckers, such as the nightjars, nighthawks, and related birds, have eyeshine similar to our own, and this may represent a retinal tapetum (for difference between choroidal and retinal tapetum, see Chapters 7 and 13), but opinions differ on this structure. Crocodiles have a tapetum, and early tetrapods probably had a tapetum, but the tapetum has disappeared in the reptilian clades. Thus, birds, especially as diurnal animals, would not usually need a tapetum, and hence it was lost and would not be reinvented in birds unless the goatsuckers qualify.

Retina

The radiation of stem reptiles to theropods to birds would have given the avian line a long time to develop endothermy (warm-bloodedness) and modify the retina. These animals lived in a bright and competitive world. Survival would depend largely on the best sensory mechanisms, perhaps none more valuable than vision.

Visual Processing

A thicker retina allows for much more visual processing to be done at a visual level. That would mean that these animals could command the skies with greater speed and agility than the earth had seen before or, for that matter, since.

What about the part of visual processing that occurs in the brain? It is estimated that in humans, 30 percent of the cerebral cortex is devoted to the coordination, interpretation, processing, and integration of visual signals. The percentage in birds is not known, but because the avian brain is smaller than the combined size of avian eyes by weight and volume, and because the avian retina is more complicated than our own, we can infer certain conclusions. Specifically, it appears that much visual processing goes on at the retinal level. In effect, the avian retina "digitizes" the image and sends a much more processed image to the brain than our eyes send to our brain. So in birds this requires much less brain processing and allows more rapid interpretation. A brief explanation of the complicated layered structure of the vertebrate retina shows how this is possible.

The photoreceptors (rods and cones) are the retinal elements that perceive light and begin the stimulus of the signal of an image to the brain. The photoreceptive element, closest to the choroid, is actually pointed away from the incoming light rays. Hence, the photons that will stimulate the rods and cones traverse all the retinal layers to reach them. Called an inverse retina, this is the opposite of the everse retina of the octopus (Chapter 18), whose eye has no organized structure between the lens and the "retina," only a clear jelly-like substance.

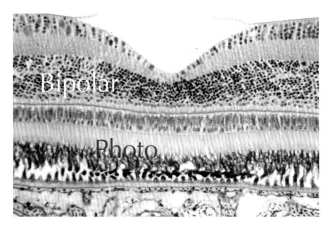

Figure 20.27 Great horned owl. The fovea of a great horned owl reveals a very complicated "middle" layer consisting of the nuclei of many bipolar, horizontal, and amacrine cells that help process or translate the signals that come from the photoreceptors. Labeling: Photo, photoreceptor cells; Bipolar, cellular layer including the bipolar, horizontal, and amacrine cells. For comparison, human retina image in Appendix A (Figures A2–A4). *Image by Richard Dubielzig, DVM.*

Stimulated by the photons, vertebrate rods and cones initiate the signal. The photoreceptors connect to the retina's middle layer, where much of the visual processing takes place (Figure 20.27). This middle layer includes the bipolar cells, which connect to the inner-layer ganglion cells, which in turn send the signal through the optic nerve to the brain.

The middle layer of the retina also contains other intermediate cells besides the bipolar cells. These are called amacrine and horizontal cells and are responsible for much of the "digitization" that takes place in the retina of any vertebrate. The avian retina contains three times as many amacrine and horizontal cells as those in the most sophisticated mammalian retina.

The benefit of these extra cells is that they encourage "talking" between the photoreceptors after the light rays containing the image have registered on them. This cross talk means that the retina can organize the image to a more sophisticated degree than our own retina can. Because the vertebrate retina is really an extension of the brain, the avian retina can organize and perfect the image at the retinal level. In birds, then, the formed image transmitted to the brain is more nearly complete and requires less visual processing at the brain level. But although this permits the bird to act on that image more quickly than we could, there is a price to pay. By acting more quickly on reflex, the bird could be more easily fooled by the image. By contrast, the primate retina will send a "cruder" or less well-processed image to the brain, and it takes primates longer to process it, but we can make more conscious decisions about how to respond to and manage it. We essentially take the raw image "upstairs" to process it.

How can we relate these skills to human terms? If you touch a hot stove with your finger, you will begin to retract your finger before the pain stimulus reaches your brain and before you sense pain. The impulse from the finger reaches the spinal cord and in a reflex arc, the spinal cord reacts without thinking and recognizes the threat. There is a reflex that will pull your finger off the stove before the pain reaches the brain. It is instinctual or "prepackaged thinking." Birds have more visual "prepackaged thinking."

The avian retina includes retinomotor responses discussed in Chapter 10. The cones can be isolated with pigment granules from the retinal pigment epithelium, much like sheathing a sword. These are organic blinders to keep the rods from being bleached by too much by bright light and permit a more rapid recovery for these cells when the bird enters a dimmer environment. In other words, these are built-in sunglasses for the retina (Appendix G).

The avian retinal adaptations are nothing short of astonishing. To begin with, certain birds see into the ultraviolet, and most species have four visual pigments—not three as do humans. This is a holdover from fish and reptiles as mentioned in Chapters 7, 10, and 15. There is some evidence that the ultraviolet pigment was lost and reestablished. This means that most bird species perceive a wider range of colors than we do. The range of sensitivity has several potential uses to birds, and there are good reasons why this sense should be retained from fish or reestablished. Flowers have an entirely different coloration when viewed with ultraviolet included in the spectrum, which may help nectar-loving birds select those special flowers for lunch. Most interestingly, fresh rodent urine radiates and radiates ultraviolet rays, providing tracking clues for raptorial species such as marsh hawks during their hunt for dinner.

In addition, birds can sense movement as slow as 15° per hour. In fact, most birds have such a sensitive movement-detection system that they can see the sun move across the horizon, and well they should. If they could not see the celestial canopy move across the horizon, they could not navigate during migration although they use other tools as well.

Some birds' movement skill is even better. To survive high speeds in a tangled environment, these birds have to sense and react to oncoming obstacles or predators. To do so, they need—and have—a rapid flicker fusion rate. Flicker fusion rate is the speed at which our retina unites images. Our retina takes individual "frames" or pictures of the world it sees, refreshes itself to process the next image, then unites the images to give us a seamless moving world. As in a motion picture, single still images must move past your eye at a certain rate for you to see it as a movie.

Some birds have a flicker fusion rate of 175 cycles per second, almost four times as fast as our own (about 48 cycles per second). This is similar to watching the movement of a bicycle wheel with a playing card between the spokes. First you see the card moving forward. As the wheel moves faster, the card appears to stand still, then to turn backward. Your eye cannot "take in" and process the pictures fast enough for them to seem to be moving forward. But a bird could process each image fast enough to see the card moving forward much longer than we can. So birds not only can distinguish more rapid movement than we can, they also can do so in much more detail. And this quick processing, especially at a retinal or subcortical level, means that the birds can make quick decisions about flight or prey capture.

Other retinal adaptations that are found in birds are not necessarily unique to them. Most birds have double cones (paired photoreceptors) scattered throughout the retina (Figure 20.28; chapter 10), which are associated with diurnal activity (Chapter 10) and probably help discrimination, among other tasks. One of the cells or both can turn off the image to create a sharper edge.

But other functions have been proposed for double cones. Aquatic birds such as gulls, terns, kittiwakes, waders, and sandpipers must deal with light reflecting off of the water. Because these birds search for food on or just below the ocean surface, their vision must penetrate beneath the reflections, glare, and irregularity of the water surface, and they must do all this with a glance. These double cones, or perhaps the oil droplets, permit many of these birds to

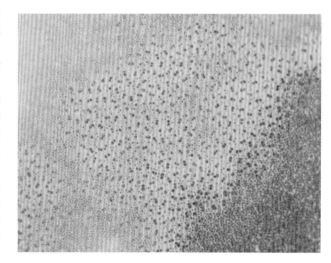

Figure 20.28 Double cones in retina of a red-tailed hawk. This retina was cut tangentially and slightly in a diagonal. Note the paired cones scattered throughout the retina. Note that these cones are not identical and hence are not twin cones but double cones. Although united, there is a principal cone and an accessory cone.

perceive polarized light, specifically to eliminate all but polarized light. In essence, these birds have polarized sunglasses.

Oil Droplets

Oil droplets (Chapters 10 and 15) represent an adaptation that birds have retained from their original theropod ancestor. The theropods probably inherited oil droplets from their reptilian ancestors.

Evolution has found specialized adaptations to suit specialized niches. All species with colored oil droplets also have different types of visual pigment. Many avian species have four such pigments, so the function of colored droplets cannot be to add color to a black-and-white world (Figure 20.29). As mentioned in Chapter 10, these beautifully colored oil droplets function as long-pass filters and will absorb photons below a specific wavelength. Current evidence suggests that oil droplets not only enhance color vision (by increasing both discrimination and the number of colors in the animals "color space") but also reduce chromatic aberration and glare.

For example, some of the ground-feeding birds have oil droplets that filter out much of the blue from the sky, creating a somewhat darker sky by increasing contrast, thus increasing their ability to see objects coming from above—such as predators.

A woodland kingfisher, the kookaburra (Figure 20.30), has five different types of oil droplets in each of the four

Figure 20.30 *Dacelo leachii,* Blue-winged kookaburra closely related to the laughing kookaburra.

types of single cones and one type of double cone. Centrally coincident with the nasal fovea, the cones become thinner, and the oil droplets lose much of their color compared to the thicker cones at the periphery (Figure 20.29). Likely, this increases color contrast in a relatively monochrome environment of the forest this bird occupies.

In addition to the yellow oil droplets in some of the photoreceptors in the inferior retina, ground feeders such as quail also have retinal photoreceptors with different-colored oil droplets above the midline. The upper retina, then, is the portion that perceives the ground beneath the birds. In that upper retina, ground feeders tend to have red oil droplets, which help discrimination of darker colors such as brown, the color of seeds. So although these birds do not have particularly good vision, they can tell nutritious seeds from ground debris.

Macular Design

AREA CENTRALIS Macular design is one of the key differences between the retina of humans and that of birds. But in both, the macula is the "sweet spot" in the retina, an area of photoreceptors whose concentration provides better acuity. This macular concentration of cells is called an area centralis. Animals with the best vision have an area of even higher photoreceptor concentration that includes a depression or pit of still more highly concentrated photoreceptors. This pit is called a fovea, the site of the eye's finest vision, which we use for reading and other close work.

The presence—or absence—of the macula is probably niche-driven. The macula really suits the needs of the birds that have one. But birds that do not require vision for

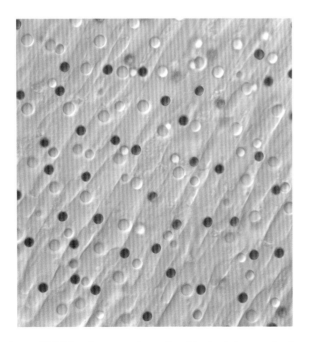

Figure 20.29 *Dacelo novaeguineae.* Laughing kookaburra retina flat mount showing oil droplets scattered in photoreceptors. *Image by Nathan Hart, PhD.*

feeding have little or no fovea or area centralis, hence little need for a macula. For example, the kiwi has poor vision because this bird relies on its sense of smell to probe the soft earth. In fact, along with vultures and tubenoses, the kiwi is one of the few birds that have a sense of smell.

Other birds whose feeding is not visually based include some species of storks. They stir up the water to get small aquatic creatures, such as crustaceans, small fish, and amphibians, to swim directly into their mandibles, which close quickly on anything that touches their beaks. But this is an inefficient mechanism, so they must rely on the shrinking of marshlands as they dry that are rich in such animals.

Ground feeders, such as the California quail, have a macula but have no real fovea. They have an area of concentration of photoreceptors but certainly no pit or fovea. Some of the ground feeders such as the guinea fowl may have an area of concentrated photoreceptors, but this is only a rudimentary fovea.

Single Fovea

Birds that are visual hunters with complicated lifestyles have a single temporal fovea. This fovea grants good visual acuity to ground-feeding birds that hunt rapidly moving prey such as lizards and snakes. The secretary bird (Figure 20.31) of the African veldt is one example. Secretary birds use their good vision to track and kill deadly snakes such as the bushmaster. (To survive their hunts, secretary birds have evolved unfeathered legs, with little for the targeted snake to bite, and resistance to snakes' venom.)

Flamingos model another interesting macula design.

Figure 20.31 *Sagittarius serpentarius.* Secretary Bird. *Image © Ed Harper.*

Infula

American (Greater) Flamingo, Phoenicopterus ruber

Flamingos, gulls, albatross, and shearwaters all have an adaptive intraocular structure known as an infula. This is a single, linear, troughlike fovea that continues in a horizontal meridian through most of the retina (Figure 20.32). In reptiles, the same concentration is called a visual strip. (Humans have it, too, but not in enough concentration to offer us any known advantage, so it may be evolutionary detritus.) In birds this horizontal band of sharp, focused visual acuity may be used for

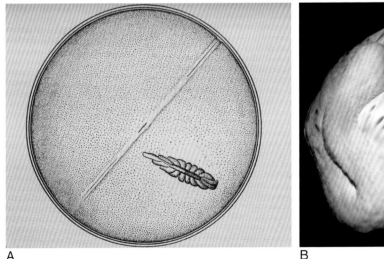
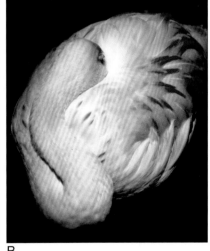

A B

Figure 20.32 (A) *Phoenicopterus roseus,* described as the common Old-World flamingo. The infulabimacular fundus as illustrated in *The Fundus Oculi of Birds,* by Casey Albert Wood with illustrations by Arthur W. Head, FZS. Chicago: The Lakeside Press, 1917, p. 60. **(B)** *Phoenicopterus ruber,* American (greater) flamingo.

protection because it would allow a flamingo to see an approaching predator such as an eagle or large hawk from a considerable distance. When a flamingo feeds, it positions its head so that the infula would be just above and parallel to the flat horizon of a shallow lake, allowing for a panoramic, almost 360° view on the horizon.

To better understand this sensory mechanism, consider this analogy. You read the newspaper by scanning lines back and forth. This bird could read the paper one line at a time without moving its eyes because the whole line would be in focus at the same time, although the linear strip is probably not concentrated enough for fine print (Figure 20.32).

Convexiclivate Fovea—Single Deep Fovea

Owls have a more temporally placed, single deep fovea, and the larger the owl is, the deeper the fovea. These foveae are so steep-walled that they probably provide refractive capabilities. The indexes of refraction of vitreous and of the retina change at that interface. Mathematically, this steep-walled (or convexiclivate) fovea could cause up to 11 percent magnification. This may have evolved to be an extra mechanism of magnification for certain birds (Figure 20.33).

Owls have a second unusual evolutionary ocular feature. Their fovea is almost completely filled with rods, rather than cones. Other birds and most other animals possess foveae nearly devoid of rods, but commanding the night skies requires sharp night vision, and that requires rods.

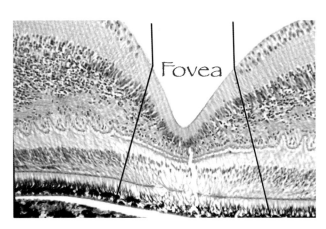

Figure 20.33 *Buteo jamaicensis*, red-tailed hawk. Convexiclivate fovea of this raptor. The steep walls of this fovea help increase image size. As parallel light rays strike the steep walls, the rays diverge slightly because the index of refraction of the retina is higher than that of the vitreous in front of the retina. This could provide as much as 7–11 percent linear magnification, depending on the angle of the interface. The fovea of a closely related bird is seen in the OCT of Figure 20.25. *Image by Richard Dubielzig, DVM.*

Birds with a single, deep fovea have excellent vision, even better than humans have. But neither see as well as birds with two foveae in each eye (Figure 20.25).

Bifoveate Birds

Bifoveate birds, such as terns, perform spectacular feats. To feed, terns hover for a bit, fold their wings behind them and drop like a stone. They pick out small fish with a high degree of success, despite the fact that their prey is underwater. To find the prey, terns use their nasal fovea. Then, to drop to the water's surface to capture the prey, terns use their two temporal foveae for stereoscopic vision. The white-fronted bee-eater, a wing feeder, is another excellent example.

White-Fronted Bee-Eater, Merops bullockoides

The vigilant bird in Figure 20.34 has chosen a perch with an unobstructed view of the surround. From there, he is able to scan the near horizon for a meal. Here the elegant high-stakes drama between predator and prey begins with the location of an erratically flying insect, often against a variably shaded background. While the bee-eater can find insects crawling on tree branches or leaves, he seems to prefer, and is exquisitely prepared for, an aerial ballet. The chase begins with uniocular tracking with the bee-eater using his nasal fovea until the range is close and requires stereopsis. Then as the bird slightly turns his head, the infula swings the image to the temporal foveae in both eyes, allowing for acute depth perception in the three-dimensional environment.

As the distance closes, the bee-eater may choose his favorite strategy to avoid the bee's stinger. The bird has learned to capture, kill, and devour truly dangerous, and even venomous insects such as wasps, without getting stung by approaching the insect on the perpendicular taking the wasp across the abdomen, often from below. He will take the insect to a favorite perch, then successively smack the insect's head against an adjacent branch, thus dispatching the insect. Then the bird scrapes the stinger from the rear of the insect, and swallows the insect.

But this remarkable acuity is only a small part of the story. The dual foveae allow these birds to establish uniocular fixation and, when ready, to slip the image along the infula to the temporal fovea. This allows for coordination with the contralateral eye, and creates stereopsis with that contralateral fovea. But much more must go into the elegance and speed of this capture. In addition to retinal processing that is more rapid than our own, the cerebellar coordination must be similarly rapid and exact, and the muscular requirements as precise as those of a ballet star (Figure 20.34).

Another species that illustrates the complex nature of the foveal design is the eagle.

stereoscopic, ray tracing suggests that stereopsis is more likely in the later stages of approach.

The eagle's eyes are large, both absolutely and relative to its body size, with a globular shape approaching our own. Many eagles have a photoreceptor concentration of more than 1 million/mm² photoreceptors in the fovea as compared to our 200,000/mm². The steep-walled nasal foveae may even afford additional linear magnification, increasing visual acuity further (Figure 20.33). In some birds, visual acuity may be as much as five times better than humans.

Combining these attributes with a clearer visual axis, increased amacrine cell concentration in the retina, and absent vascular system in front of their retina, eagles have a formidable ocular system with an "eye-mindedness" quality.

Other bifoveate birds have unusual capabilities because of specially placed foveae. The kingfisher is an excellent example.

Green and Rufous Kingfisher, Chloroceryle inda

Aquatic kingfishers have evolved special adaptations to achieve aquatic capture of prey from a hovering position. These kingfishers, as represented by the green and rufous kingfisher (Figure 20.36), and some other avian species that must look through water for their prey, have a preponderance of red oil droplets in their retinae. The droplets may help with glare or with the dispersion of light from particulate matter in the water.

As with many avian predators, kingfishers have two foveae in each eye. But this species has a real twist on the monocular and binocular foveae of other bifoveate birds. The temporal fovea is at the ora—the far periphery of the retina where the retina joins the pars plana or the base of the iris. The fovea shares the binocular field with its compatriot in the contralateral eye.

This temporal fovea is shallow and anatomically inferior. The ganglion cell density around the monocular nasal fovea is much higher than the cell density around the binocular temporal fovea, suggesting that the acuity is significantly higher at the nasal fovea. This would not be a surprise because presumably a kingfisher sights its prey with the higher-acuity nasal fovea. If the prey is aquatic, the bird hovers above it. As the bird drops toward its prey, acuity becomes less critical, but the ability to see prey movement, especially in the water, becomes paramount. The infula connecting the two foveae allows the image to swing temporally as the bird drops onto the prey. If the prey moves, the temporal fovea senses the difference in position, and the kingfisher can adjust accordingly.

But when hovering 12–15 meters above their prey, aquatic kingfishers are presented with an unusual problem. The simultaneous movement of the fish and of the hovering bird would

Figure 20.34 *Merops bullockoides*, white-fronted bee-eater. This wing feeder hunts insects and especially hymenoptera (bees, wasps, and their relatives). This magnificent bird is well equipped to hunt rapidly moving insects using two foveae. The two foveae are similar in appearance and position to those found in Figure 20.37.

The Eagle, Haliaeetus vocifer

Eagles are majestic birds, and their feeding methods often betoken their majesty. In particular, the African fish eagle (Haliaeetus vocifer) is an aerial killer, a robust fisher, and sometimes a scavenger or even a pirate of other bird's prey. Few bird species fish on the wing, but H. vocifer and the bald eagle (H. leucocephalus), its North American counterpart, are thoroughly equipped to do so.

These bifoveate birds rely on their visual abilities to facilitate their spectacular foraging methods. H. vocifer will follow his prey throughout the approach and, at the instant of strike, will redirect his gaze to keep the fish in stereoscopic view (Figure 20.35). A head-down position allows the eagle to use both temporal foveae for binocular stereopsis. The second or more nasal fovea in each eye is probably used to spot the prospective prey and perhaps to keep the prey in alignment during the initial flight approach.

The connecting infula (a linear-strip fovea connecting the nasal and temporal foveae of each eye) allows the eagle to begin stereoscopic tracking during the approach to help ensure that it will not lose the fish in the three-dimensional movements underwater. The infular strip allows for foveal quality vision as the image swings from the nasal fovea to the temporal fovea in each eye. Although it is not clear when the image becomes

Figure 20.35 *Haliaeetus vocifer*, fish eagle. This African bird is successfully approaching and catching a fish. Note in the next to the last image that the eagle has looked down to be certain that the fish remains in focus. At this point the eagle is using the shallow temporal fovea for stereopsis to be certain to know the depth and direction of the fish's movements.

create a neurologic nightmare. The solution produces a dramatic aerial dance. During the hover before the dive, kingfishers maintain a rock-steady head position relative to the movements of wing and body, permitting a steady image and depth perception. The depth of the prey varies the form of the dive. The deeper the prey, the steeper the dive angle is (with achieved speed as high as 4.5 m/sec), perhaps to reduce the effects of refraction on the apparent position of the prey. The asymmetrical lens would predict an image that can be focused on the temporal fovea when the bird is under water. Although not proven, this is a logical method for keeping up with the underwater prey that tries to escape.

Figure 20.36 *Chloroceryle inda*, green and rufous kingfisher.

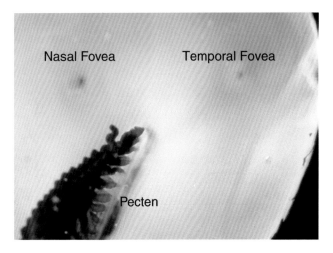

Figure 20.37 Buteo jamaicensis, red-tailed hawk. Posterior half of globe. The pecten and the two foveae are illustrated.

The position of the monocular fovea tends to be fairly constant in all bifoveate birds, but the temporal fovea is different. The two foveae in each kingfisher's eye have large angles of separation that measure up to 52°, as compared to 15° in eagles and kites. Again, we see evolution shaped by a niche.

Pecten

Birds needed and evolved more retinal cells to develop a "faster" retina with more visual processing being performed with the new skill of flight. But more retinal cells meant a thicker retina when compared to reptiles, or even humans. Birds had to somehow provide nutrients and oxygen to the high concentration of amacrine, horizontal, bipolar and ganglion cells present in the inner retina. Diffusion from the choroid would not be sufficient. Birds would need the best possible visual acuity, and intraretinal blood vessels would not be the best solution. Having to look through vessels and the oxygen-carrying pigment of the red blood cells, as primates do, would have yielded a slightly degraded image. This increased retinal thickness would have placed considerable demand on the conus that would have radiated from the reptiles and required expansion into the larger and more vascular pecten to satisfy the nutritional demands of these new cells (Appendix C). The absence of retinal blood vessels and the presence of more retinal cells solved one problem while creating another. The retina of birds needed a more robust vascular supply than the conus of reptiles would have supplied (Chapter 15).

Nutrition and oxygen delivery would require more plaits and folding of the pecten as compared to the conus,

creating more surface area of this vascular structure (Figure 20.37). The pecten is permeable to small molecules including those needed for inner retinal nutrition and oxygenation. These molecules remain pooled in the inferior vitreous next to the pecten until the bird performs a saccade, or rapid oscillatory eye moment—a shake of the eyes.

A well studied example is the Bush Stone-curlew.

Bush Stone-Curlew, Burhinus grallarius

The Bush Stone-curlew (Burhinus grallarius), a ground-dwelling omnivorous bird (Figure 20.38), is found throughout Australia and feeds principally on lizards, amphibians, and small mammals. Like other birds, the Bush Stone-curlew has no retinal vascularization, only the pecten—a collection of vessels projecting from and attached to the optic nerve head. The pecten is homologous to the central retinal artery in mammals and

Figure 20.38 *Burhinus grallarius*, Bush Stone-curlew.

Figure 20.39 *Burhinus grallarius*, Bush Stone-curlew. The pecten as seen through the pupil of a living bird. *Image by Jack Pettigrew, MD.*

Figure 20.40 *Dryocopus pileatus*, pileated woodpecker. *Image © B. Henry/VIREO.*

projects into the inferior vitreous toward the center of the eye. In some species, the pecten nearly touches the lens.

The pecten projects its vascular net into the vitreous with multiple macroscopic and microscopic pleats to increase the surface area (Figure 20.39). Although the pecten's role is nutritive, some observers hold other theories as to its purpose, from a sunshade, an azimuth, or a magnetosensor (unlikely) to an accommodation aid to a protective mechanism to warm the interior of the eye in high-flying geese and similar birds.

The process of distributing nutrients works this way: each saccade creates plumes of nutrients and oxygen rising from the inferior vitreous, like steam rising from a kettle. During these nutrient-spreading saccades, the pecten acts as an agitator to allow diffusion of oxygen to the inner retina.

These saccadic oscillations allow for inner retinal nutrition without intraretinal or epiretinal vascularization.

The pecten may play other ocular roles. For example, it may contribute to intraocular acid-base balance and elaboration of intraocular fluids, and it may mechanically agitate the vitreous during ocular movements, possibly facilitating fluid movement within the eye. Additionally, by filling with blood at a millisecond before the strike, the pecten may be responsible for maintaining a tight globe in any bird that strikes the substrate with its beak

---

Best known for striking with its beak, the woodpecker needed more than the pecten to prevent intraocular damage during drumming.

Pileated Woodpecker, Dryocopus pileatus

A canopy-dweller, woodpeckers hammer their lives away for feeding, nest construction, communication, and territorial

display. They leverage their entire body weight to increase the force of their strikes. D. pileatus (Figure 20.40) may strike the hard surface of a tree at a rate of up to twenty times a second (not a misprint) and up to 12,000 times a day, with deceleration forces of up to 1200 grams with each impact. That is equivalent to striking a wall at 16 miles an hour—face first—each time. If you spent your life intentionally battering your head against a wall in this fashion, how would you avoid headaches, concussions, or even retinal detachments?

The furtive and wary D. pileatus, as well as most of the other 300 or so known woodpecker species, has evolved several mechanisms to prevent brain and retinal damage.

To partially cushion the incessant blows, the woodpecker has a thick bony skull with relatively spongy bone, especially at the occiput (posterior aspect of the skull), and cartilage at the base of the mandible. The small subarachnoid space (space immediately around the brain) within the skull contains almost no cerebrospinal fluid, permitting tight packing of the brain. The powerful muscles that attach the mandibles to the skull contract a millisecond before strike, creating a tight but cushioned structure and distributing the force of the impact to the base and posterior aspects of the skull, thus bypassing the brain.

These birds strike in a perfect perpendicular stroke, eliminating the torsional shear force that would otherwise tear the meninges (membranes immediately around the brain) or cause concussions. The neurologic mechanisms that are responsible for these actions probably also protect against intraretinal hemorrhages and retinal detachment.

Additionally, the woodpecker is protected, at least to some extent, by its brain's size. The brain is relatively small, resulting in a small ratio of brain weight to brain surface area. Any impact would be spread out over a relatively large area, making the woodpecker's brain somewhat more resistant to concussion than a human's brain.

The same high-speed photography that has captured the woodpecker's strike mechanism has also shown that in that millisecond before strike, a thickened nictitans closes over the eye. That would protect the eye from flying debris and chips and would restrain the eyes from literally popping out of the woodpecker's head.

The woodpecker has other adaptations that may contribute to the protection from intracranial injuries. Woodpeckers possess a shock-absorbing tongue.

The tongue's construction in its complicated path from the dorsum of the maxilla to the oropharynx may be one adaptation. The tongue musculature originates on the dorsum of the maxilla, passes through the right nostril then between the eyes, divides in two, arches over the superior portion of the skull and around the occiput, passing on either side of the neck, coming forward through the lower mandible, and uniting into a single tongue muscle in the oropharyngeal cavity. Throughout this course, the musculotendinous bands encase a bone—the hyoid—which provides stability. Although one principal purpose of this bone-muscle structure is to permit protrusion of the tongue up to 4 inches beyond the tip of the bill, these musculotendinous bands also create a curious sling-like structure (Figure 20.41). This probably functions as an isometric shock absorber if it is contracted before each strike. The sling would also distribute the potential shearing forces.

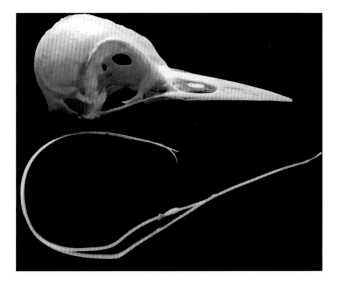

Figure 20.41 *Dryocopus pileatus*. Pileated woodpecker skull with hyoid bone. *Specimen courtesy of Jack Dumbacher and the California Academy of Sciences, San Francisco.*

To maintain a tightly compressed eye and protect the internal structures, woodpeckers enjoy a cushioned choroid with an as-yet-unknown mucopolysaccharide filling the interstices. Such a cushion could be coupled with the pecten. The pecten may cushion the strike by filling with blood and briefly elevating intraocular pressure, which would transfer to all structures within the eye to prevent damage (Figure 20.39). These two mechanisms would prevent "bouncing" of internal structures.

Neurologic Evolution

Stereopsis is important to all predatory animals but especially to aerial ones. The interpretation of three-dimensional space is important in localizing prey when both prey and predator are moving. To understand how birds achieve stereopsis, we need to understand a bit of neuroanatomy.

Mammalian stereopsis exists because of the organization of the optic nerves leading to the visual cortex. The optic nerve from each eye sends approximately one half of its fibers to connect to each side of the brain. This integration of fibers from side to side occurs at the optic chiasm. That means that fibers from each half of the retina of each eye integrate with its image partner from the other eye to help create three-dimensional space from a set of two-dimensional signals (Chapter 13).

Birds have an entirely different system for the organization of stereopsis than mammals do, suggesting that the last common ancestor of birds and mammals probably did not possess stereopsis or dealt with it differently. Before the reptilian radiation into the theropods, then, that stem reptile must have descended from the early terrestrial tetrapods to have laterally placed eyes and probably no common overlapping visual field or direct integration of signals from one side to the other.

Barn owls are a good example of how birds developed stereopsis and how important it is to them.

Barn Owls, Tyto alba

Most birds have a completely crossed chiasm, and each eye should be completely, and only, represented in the contralateral visual cortex, but such is not the case with barn owls (Figure 20.42 and 20.43). Pettigrew documented that these owls have excellent stereopsis, which the frontally placed eyes and predatorial lifestyle suggest (Figure 20.44). Pettigrew later documented a second supraoptic chiasm, which allows about 50 percent of the ipsilateral (same side) fibers to cross the brain a second time. And this allows each eye to input to each side of the brain, providing stereopsis equivalent to that in humans. The second crossing represents neurologic convergent evolution,

Figure 20.42 *Tyto alba*, barn owl.

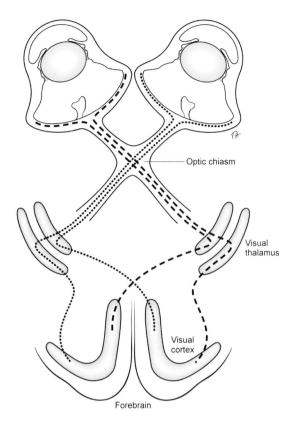

suggesting that stereopsis is especially valuable for predators and can evolve in different ways.

Vision plays a uniquely important role for the barn owl. Although it may rely on its excellent auditory system, the auditory spatial map is trained visually. Auditory nerves in the brain extend to the optic tectum to combine with impulses from the eyes. Hence, the auditory space aligns with, and is taught by, the visual system. This animal can "see" with its ears because

Figure 20.44 Stereopsis mechanisms in owls. Humans have a partially crossed chiasm so that each eye sends fibers to each half of the brain. Birds' eyes also project fibers to each side of the brain, but by an entirely different anatomical mechanism. It is an example of neurologic convergent evolution. *Art by Tim Hengst.*

Figure 20.43 *Tyto alba*, barn owl eye. Note the similar shape to the eye of the screech owl and the great horned owl. Scale is in millimeters.

these summed inputs then project into birds' equivalent of a visual cortex.

The visual system of barn owls must cope with the camouflage of cryptically colored rodents in an environment that is scotopic (nighttime with far fewer photons). To do so, this owl has other clever sensory techniques besides its auditory system.

Barn owls can make assumptions about incomplete contours, called subjective contours, to fill in the gaps. Furthermore, direct neuronal recordings have documented this "coding" of subjective contours and the process of high-order perceptual interpretation. Target rodents cannot hide in scattered weeds and foliage, as this will not successfully break up the "search image" of the owls. This added layer of sophisticated visual processing produces a highly effective nocturnal hunter.

Most owls have excellent nocturnal vision but are helpless in total darkness. But barn owls and their fellow members of Tytonidae, a small family of owls, have distinctive ear bones that permit acoustic location and keen hunting even in total darkness.

Boasting several anatomic and neurologic design features, this owl's asymmetric ears and auditory processing can detect *slight interaural time differences. That ability enables an enviable auditory sensory mechanism that provides accurate azimuth and elevation localization as well as directional motion detection to an accuracy of about 2°–3°.*

In addition to extraordinary hearing, the barn owl possesses excellent visual acuity, if only at night. The bird's predominantly rod retina restricts it to a nocturnal lifestyle. And the bird has a limited range of accommodation, probably no more than 10 diopters. Large eyes with a very large anterior chamber help to create a long anterior focal length (Figure 20.43). The 400-gram barn owl has a cornea that is 12 mm in diameter with an axial length of 18 mm. This allows the image to be spread over a wider retinal area, hence increasing image size. If photoreceptors are much smaller than 1 μm, the photonic scatter would stimulate adjacent photoreceptors or fail to admit light because of the wavelength's magnitude. The barn owl extracts maximum function from photoreceptors that approach their minimum size.

Most owls' hunting skills make the birds helpful to human agriculture. The barn owl is exceptionally so, consuming about one and a half times its weight in rodents each night during nesting season, and somewhat less at other times. The bird is not particularly fast, so it must rely on stealth to catch its prey; special adaptations to its feathers along its exposed flight surfaces allow it to be silent as it approaches dinner.

These visual and auditory predators represent an evolutionary triumph of vertebrate sensory evolution. Most of their skills were probably fully developed by the mid- to late Cretaceous, although refinements have continued through the Tertiary to the present.

The Cretaceous period fostered birds but, in addition, saw the birth of flowering plants (angiosperms). By adding food sources, angiosperms provided many new niches for vertebrates and invertebrates alike. Animalia would coevolve with these plants, and the earth would change forever.

POLLINATORS COEVOLVE

CRETACEOUS PERIOD

145–65 MILLION YEARS AGO

CENOZOIC ERA

TERTIARY PERIOD

65–2 MILLION YEARS AGO

21

The Earth in Bloom

The Cretaceous saw more expansion and diversity than just in animals. Plants, fully established by then, brought forth a dazzling innovation at about 150 mya: flowers appeared, which in turn influenced the further expansion of animals. Invertebrates, dramatically expanded in the Jurassic, helped set the stage for the birth of insect pollinators, such as bees, beetles, and flies, that coevolved with flowering plants, or angiosperms, in the Cretaceous and persisted into the Tertiary and into modern times.

A Bounty for Insects

Angiosperms stimulated great change to be sure, but many of the (mostly winged) insects that would profit from the nourishment traded for pollination were already present. They arose from a common ancestor that had been terrestrial and/or partially aquatic for millions of years. After coming ashore in the Devonian, insects radiated into various lineages, one of which would lead to the Trichopterans (caddisflies) and the lepidopterans (moths and eventually butterflies). These would prove to be important to variety within the development of visual witness. Most of these insect pollinators had apposition eyes (Chapter 6), but the other and unique forms of compound eyes appeared as well.

Although wind created mechanical pollination, flowering plants, or angiosperms, would encourage more active and accurate pollination. The flowering plants would create special niches and special problems.

Caddisfly, Trichoptera

Caddisflies (Figure 21.1) give us a window into the status of the most recent common ancestor of moths and butterflies (Lepidoptera), their sister group. Caddisflies diverged from the

Figure 21.1 *Trichoptera,* Caddisfly. This animal was caught and entombed approximately 50 million years ago in amber. *Image courtesy of and © Prof. Dr. Jes Rust, Steinmann Institut Bereich Paläontologie, Universität Bonn.*

moths an d butterflies in the early Jurassic, but both lineages dramatically expanded leading into the Cretaceous. Both also began with the superposition eye (Chapter 6), suggesting that their last common ancestor in the Jurassic must have had a similar eye.

Moths arose in the late Jurassic or early Cretaceous, but not as pollinators. Such creatures inherited a superposition eye quite suitable for nocturnal forging. Moths became detritivores (consumers of dead organic matter) or perhaps folivores (consumers of leaves). Likely they became pollinators by accident because there would have been much less nectar in the Jurassic.

The most recent common ancestor of caddisflies and moths was an ancient and early nocturnal insect that arose from an aquatic home bringing superposition eyes with it. This creature found a nocturnal, aerial world of darkness without predators and with much opportunity: plenty of vegetation and almost no competitors—at least initially. The superposition eye would be as useful for light collection on land as it was in the water.

Superposition eyes are quite versatile. Evolution would find several ways for them to adapt to daylight once established creatures had exploited the night hours. Even though the superposition eye is very sensitive, it can desensitize to permit diurnal activities while keeping its visual potential and light-gathering capabilities by various methods.

Other major lineages, such as fish, develop sunshades that permit successful navigation between very bright and very dim environments (Chapter 10). Some arthropods with superposition eyes have also developed simple pigmentary sunshades.

Lepidoptera Tobacco Hornworm Moth, Nicotiana *Species*

*The tobacco hornworm moth (*Nicotiana *sp.) is a species of hummingbird hawk moth and is cathemeral—active during both day and night. Cathemeral moths developed sunglasses that seem positively elegant. When the moth is active during the day, the pigment migrates along the column of the ommatidium to keep light from penetrating adjacent ommatidia and stimulating the rhabdom (Figures 21.2 and 21.3). Doing so allows the photons to stimulate the photosensitive pigment of that single rhabdom, which reduces the light stimulation by about 1000 times.*

The moth dark-adapts in about twenty to thirty minutes, but its sunglasses let it light-adapt in a few minutes.

Moths that prefer daylight can thank their superposition eyes for exceptional diurnal acuity, even though those

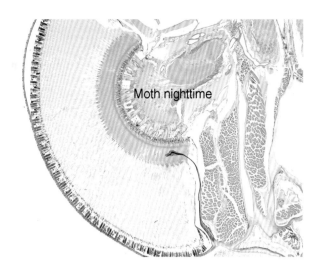

Figure 21.2 *Nicotiana* species. This tobacco hornworm moth died at night when it was fully dark-adapted. Note the pigment granules near the tips of the individual ommatidia. *Image by Richard Dubielzig, DVM.*

eyes evolved for nocturnal vision. The hummingbird hawkmoth will show us.

Hummingbird Hawkmoth, Macroglossum stellatarum

The hawkmoth (Figure 21.4) has evolved to be active in bright habitats that require complex specializations unknown in other refracting superposition eyes. And its greedy feeding habits have placed unusual demands on the eye. Flowering plants usually use nectar to attract birds or insects to help disperse their

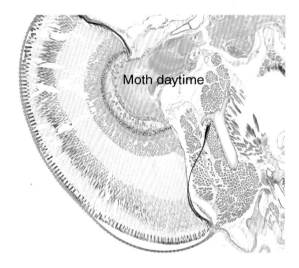

Figure 21.3 *Nicotiana* species, tobacco hornworm moth. This moth died as light adapted. Note the pigment granules spread throughout much of the length of the ommaditia. *Image by Richard Dubielzig, DVM.*

Figure 21.4 *Macroglossum stellatarum*, hummingbird hawkmoth. *Image courtesy of and © Karlheinz Baumann.*

pollen, and they must use bright colors and/or scent to advertise the nectar. Most insects, like bees, answer the ad by landing on a flower, picking up pollen as they feed and dispersing it afterward. But hummingbird hawkmoths and a few other insects do not land, so they do not live up to that part of the bargain. They feed from the air using a coiled tongue that may be three times the length of the insect to extract the nectar. Their eyes help them partake of this free lunch.

This moth's eyes possess areas of higher resolution and sensitivity that function much like the foveae of vertebrate eyes. These features violate our previous understanding of how superposition eyes function. The optical design of a classic refracting superposition eye allows the lenses of a large number of ommatidia to focus light onto each rhabdom of the retina (Chapter 6). This requires that the retinular cells be a set distance behind the lenses, that they have an empty space in between, and that the number of rhabdoms equals the number of facet lenses. But this hawkmoth eye has many more rhabdoms than facets, with the rhabdoms packed into the retina much more densely than the overlying facets would suggest.

Denser rhabdom packing creates a more detailed image everywhere it occurs—in effect creating a fovea but requiring more light. These eye areas also have the largest numbers of facets supplying light to each rhabdom. The hawkmoth's superposition eyes have thus evolved regions where the visual image is both very bright and highly resolved. This is developed to its greatest extent in the eyes' frontal regions, with which the hawkmoth closely observes the flower petals it hovers above. Looming-sensitive neurons, found in the optic lobes of the brain, help to interpret and integrate the expanding and contracting visual image of the moving flower. This allows the moth to maintain its position directly above the flower in a soft summer breeze.

The hovering flight of this agile and fast-flying member of the family Sphingidae resembles that of a hummingbird.

The flower may be wind-blown in almost any direction and in any of three dimensions, creating difficulties for the feeding moth. To keep its long proboscis in contact with the small supply of nectar at the flower's base, the moth must move deftly with the flower. The moth accomplishes this aerial positioning by sophisticated optical and neural mechanisms that result in good visual acuity and movement sensitivity.

These moths also have true trichromatic color vision. That is, they can discriminate objects on the basis of their spectral information alone, independent of intensity. To do this, the moths have at least three spectral receptor types, including an ultraviolet type. No long-wavelength pigment appears to be stimulated in the redder portion of the spectrum, which means that the trichromacy is not the same color spectrum as ours.

A related hawkmoth (Deilephila elpenor) uses the same trichromatic vision for nocturnal color vision and is one of the few animals known to have nocturnal color vision. Furthermore, both species of hawkmoth have good color constancy—the quality that allows animals to identify the true color of a lemon in different shades of illumination. But having three different visual pigments does decrease the absolute sensitivity of the eye. Still, this moth can see in full trichromatic color at night!

Other moths radiated into diurnal insects using other ocular designs and providing a different lineage. Moths radiated into butterflies—diurnal moths, really—in the early Cretaceous soon after the angiosperms appeared. These delicate animals added a one-of-a-kind twist to the compound eye, another indication that the visual witness continues to evolve, especially with new niches such as flowering plants and trees. This form of the compound eye illustrates a potential transition stage between superposition and apposition eyes. This eye also shows the plasticity of compound eyes and the ability of this eye to move between one form and another.

The Union Jack Butterfly, Delias myisis

The Union Jack butterfly (Delias myisis, Figure 21.5A) is found along the tropical northeast coast of Queensland, Australia, including the Cape York Peninsula. It has the typical languid and apparently random flight of most butterflies. But the behavior of this solitary creature belies its physiology, learning abilities, and visual mechanisms.

Of the three forms of apposition compound eyes (Chapter 6), butterflies are the only lineage that has the afocal configuration.

The afocal apposition eye's crystalline lens in the proximal tip of each ommatidial cone is a second element in an extraordinary lens system (Figure 21.5B,C). Positioned differently from that of a standard focal apposition eye, this lens permits image enlargement. In effect the lens is a mini–magnifying telescope for each ommatidium.

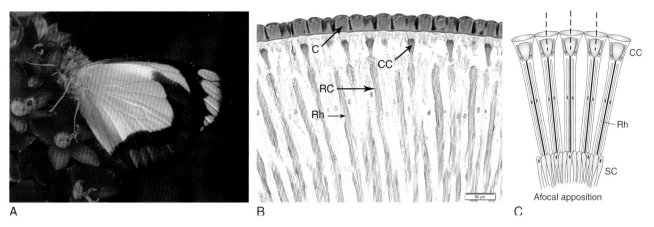

Figure 21.5 (A) *Delias myisis,* Union Jack butterflies are the only lineage with afocal apposition compound eyes. **(B)** *Danaus plexippus.* Monarch butterfly eye. Histologic section of afocal compound eye. C, corneal lens; CC, crystalline lens; Rh, rhabdom; RC, reticular cell. This unusual compound eye has a corneal lens followed by a crystalline lens. The image is focused from the corneal lens within the crystalline lens and recollimated to then strike the rhabdom. The surrounding retinular cells provide the elements that create the rhabdom. *Image by Richard Dubielzig, DVM.* **(C)** Diagrammatic illustration of the afocal compound eye seen in B. The afocal apposition eye is similar to the focal and neural superposition compound eyes. The two lenses are slightly separated, permitting the focal point of the corneal lens to be within the crystalline lens. Supporting cells (SC) surround the axons of the receptor cells as these axons travel toward the brain. *Art by Tim Hengst.*

BOX 21.1 AFOCAL APPOSITION EYE

The afocal apposition eye, found only in some Lepidoptera, is a twist on the apposition eye. Butterflies have a cornea similar to that of insects such as bees. The second refractive element, the crystalline lens, is set off a bit from the cornea, leaving a space between the cornea and the lens. The crystalline structure of this second element (following the "cornea" as mentioned above) has a focal power of 200,000 diopters. By comparison, your eye has a focal power of perhaps 65 diopters. This marvelous adaptation within the ommatidial cone, when combined with the corneal lens, creates a minute Keplerian telescope with magnification of approximately four times for each individual ommatidium. This is an afocal apposition eye because the image is focused within the second element. When that second element recollimates the light, it stimulates the rhabdom as parallel rays. This will enlarge the image somewhat and brighten it, at least in daylight. Like the focal apposition, the afocal apposition eye is restricted to diurnal activity.

But Lepidoptera have other mechanisms to improve their visual perception and photon capture. Butterflies possess colored pigments close to the rhabdom, which probably alter the spectral sensitivity of the rhabdom, much as oil droplets do in birds and reptiles. And, most Lepidoptera species have eyeshine produced by a very different variety of tapetum.

Most moths and butterflies, including the Union Jack, have interesting tapeta, the reflecting material distal to the photoreceptors. Tapeta can reflect much of the light of specific wavelengths (of one color) and are an example of a biological interference filter (Appendix H and Chapter 10). Here is how it works.

Many insects deliver oxygen to ocular tissues through a series of tubes branching directly off the trachea. This Roman-aqueduct-like system grows smaller as it approaches the internal structures of the eyes. As the system divides into smaller branches, the periodicity of the chambers directly beneath the ommatidia evolved a separation distance that is exactly one quarter of the wavelength of light. As light enters each ommatidium, striking each of the forty or so layers of these tracheal branches or tracheoles, a bit of light is reflected from each surface. Constructive interference increases and intensifies the reflection until perhaps as much as 90–100 percent (at least theoretically) of the incident light is reflected back through each photoreceptive element in the ommatidium. The light, once reflected, continues back through the photoreceptive element and exits the eye on almost the identical path it entered. So even a small flashlight will illuminate a moth or a spider with two small red (or another color of) gleaming dots (Figure 21.6).

This tapetal reflection represents convergent evolution, as vertebrate tapetal mechanisms are quite different but evolved for the same purpose.

A

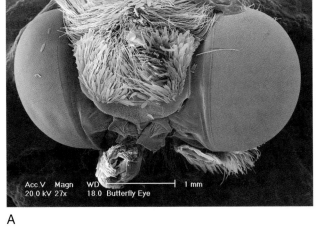

A

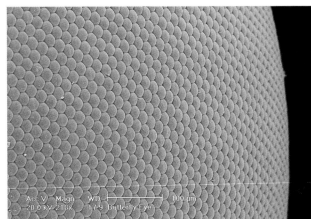

B

C

Figure 21.6 (A) Eyeshine of moth. This is produced by constructive interference of light from the tracheoles. Note the red reflex. Different colors can be produced depending on the periodicity of the tracheoles, but the reflex color in moths is usually red. **(B)** Tracheoles of moth eye. Yellow arrow points to striations of tracheoles. *Image at 100×. Image by Thomas Blankenship, PhD.*

As with most insects, butterflies have compound eyes made from ommatidia. To fit the animal's ecological needs, these optical sampling devices have evolved just as the animals have. Butterflies and some other terrestrial insects have developed several adaptations that have added to the visual abilities of the species that retain them (Figure 21.7).

Another feature to consider is related to extraocular photoreception. It does not represent a rare adaptation in itself. We have seen extraocular photoreception in horseshoe crabs, sea snakes, and other arthropods. For example, some crayfish have a caudal (tail) photoreceptive system to mediate escape. But in the Japanese yellow swallowtail butterfly (*Papilio xuthus*) and perhaps in other butterflies,

Figure 21.7 *Danaus plexippus,* monarch butterfly. **(A)** Scanning electron microscopic image of the head and compound eye. *Image by Pat Kysar.* **(B)** Single compound eye. Note multiple ommatidia. *Image by Pat Kysar.* **(C)** Very high magnification of surface of an individual ommatidium. The pebbled surface is responsible for decreasing glare. At about ½λ, these nipples are much smaller than the wavelength of light and do not interfere with the transmission of light. But this nippled surface provides external constructive interference and helps avoid reflection to permit greater absorption of light. *Image by Pat Kysar.*

BOX 21.2 OMMATIDIAL ADAPTATIONS

Ommatidia have an impressive array of modifications to satisfy the ecological needs of each species. For example, the direction of the face of each ommatidium can be determined and compared to adjacent faces of other ommatidia to see the extent of the visual field. Maps of the directions of ommatidia faces have shown a central region that has the optical axes tilted more or less toward one another. These ommatidia are parallel to or have a reduced angle from the neighboring ommatidia, creating a portion of the visual field that has greater sampling. This, then, is a foveal equivalent in an invertebrate compound eye—more detail is refracted, focused, and hence seen in that area.

Other ommatidial adaptations assist in light reflection or concentration. For example, some diurnal butterflies have a thick pebbling across the surface of the cornea. These nipple-like projections have the dimensions of ½λ in the center of the visible spectrum and create an aspheric surface, thus giving external constructive interference, avoiding reflection, and creating greater absorption of light. This adaptation is especially useful for nocturnal *Lepidoptera* (moths and some skippers) to maximize light collection, but diurnal butterflies and other insects use it for camouflage. This adaptation inspired a similar use in antireflective coating of certain commercial optics (Figure 21.7).

in the shade of a tree. Any animal that relies on color perception for food must have some degree of color constancy or it would not be able to recognize food sources in differing light. This advanced ability reveals neural processing.

Many butterflies have inborn preferences for certain colors, such as yellow and/or blue, but butterflies are capable of associative learning to help them find food, and that capability can trump innate preferences. Specifically, Japanese yellow swallowtail butterflies can be conditioned to associate a specific color with nectar after just a single presentation and can be conditioned to change that preference with just a single change in color representation.

Butterflies, then, have evolved optical devices that are clever, unique, and packaged exquisitely. Evolution has continued to tinker with the optics and the visual mechanisms among the small invertebrates.

Other insects are more important in pollination, and many of these lineages arose in the Cretaceous to seize the new niche opportunities presented by the flowering plants. The social hymenoptera are the best example.

Social Hymenoptera: Wasps, Bees, Ants, and Sawflies

Most of the social hymenoptera have apposition eyes (Chapter 6), although these have been adapted for this principal group of insect pollinators, and some extremes can be seen within the lineage. Even among the hymenoptera, evolution continued, as not all of the bees, wasps, ants, and sawflies are pollinators or even have eyes. Some of the species have become very small and parasitic—even on other insects. Meet the Fairy wasp.

Fairy Wasp, Gonatocerus ashmeadi

Size matters, even among insects. The fairy wasp is one of the smallest insects, or perhaps the smallest, with a maximum length of only 0.2 mm or 200 μm (Figures 21.8 and 21.9). This wasp's eyes are about 60 μm across (Figure 21.8).

These apposition eyes have a rather steeply curved corneal lens with a short radius of curvature, and each ommatidium has a very short length of approximately 30 μm. This may be the smallest insect so equipped (Figure 21.10). These are classic apposition eyes with an ommatidial diameter of approximately 5 μm, however, and these eyes would suffer from the same limits of diffraction as other apposition eyes. This size ommatidial facet is below the limit that diffraction would permit for most light, but the animal probably uses the shorter wavelengths such as the violet or ultraviolet range to solve this problem.

And yet, it is an accomplished parasite equipped with eyes. It is a wasp and like many other insect species, it is most favorable to man as it helps control various crop pests with its parasitic

the adaptation reaches a rare level: photoreceptors in their genitalia!

These evolved to assure successful copulation. The male butterfly knows that when these photoreceptors are no longer stimulated by light, he has correctly positioned his genitalia for fertilization.

Visual abilities in butterflies are surprisingly advanced. As with many other insects, butterflies also see into the ultraviolet, have color vision, and can even detect polarized light. At least one butterfly, the Japanese yellow swallowtail (*Papilio xuthus*), has five visual pigments, although other species may as well.

But perhaps butterflies' most impressive visual ability is their color constancy, which allows their neurologic system to draw conclusions about color (and food sources) when a change in the character of the light source changes the wavelength. For example, a flower that would appear red in broad daylight may look dark maroon on an overcast day or

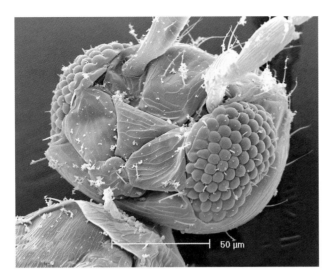

Figure 21.8 *Gonatocerus ashmeadi*, fairy wasp. Scanning electron microscopic image of the head of the fairy wasp. *Image by Brad Shibata.*

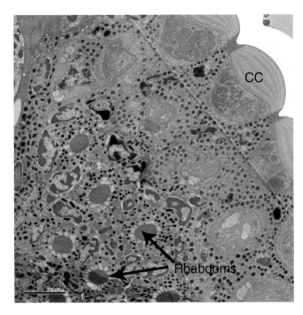

Figure 21.10 *Gonatocerus ashmeadi*, fairy wasp. Apposition eye on transmission electron microscopy. The cylinder of the crystalline cone can be seen beneath the cornea lens and is marked by CC. Two of the several scattered rhabdoms are marked. The black bar in the lower left hand corner is 5 µm. *Image by Pat Kysar.*

behavior, laying its eggs in the eggs or larvae of other harmful insects.

Still, even this tiny insect has eyes with ommatidia very much like those of much larger insects—like dragonflies, for example.

Fairy wasps date to at least the Cretaceous. The family probably arose from the wasps in the early Cretaceous, perhaps 130 mya, in coevolution with the flowering plants and in concert with beetles and other plant feeders but only to parasitize other insects.

Although bees have been the principal pollinators, with the necessary vision to recognize key signals from flowers, not all bees cooperate with flowers. But although uncooperative bees have evolved different strategies, they retain the hymenoptera visual system.

Carpenter Bee, Xylocopa micans

The carpenter bee's apposition compound eyes (Figures 21.11, and 21.12 and Chapter 6) reveal the clever ways evolution continues to solve the problem of maximizing visual acuity in small creatures. Like the creatures themselves, eyes have evolved, and the apposition eye is no exception.

Bees are descended from wasps and appeared first in the Cretaceous, as pollinators. The apposition eye evolved over the hundreds of millions of years between the Cambrian, when it first arose, and the Cretaceous. It evolved during that time for (at least) three reasons. Certainly competition was a principal driver, but the change from a watery to an aerial environment and the appearance of flowers also created new demands and new niches.

Physiologically, the apposition and other forms of compound eyes evolved much as the vertebrate eye did. Both models evolved ocular pass filters, including pigmented corneas, intracellular filters, and other unusual adaptations.

But the lateral compound eyes are not the only light-responsive structures on the carpenter bee's head. The three domelike structures on the dorsal aspect of the head of this

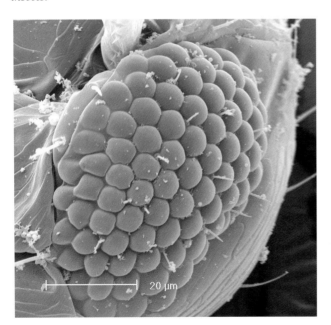

Figure 21.9 *Gonatocerus ashmeadi*, fairy wasp. Scanning electron microscopic image of compound eye. *Image by Brad Shibata.*

Figure 21.11 *Xylocopa micans,* carpenter bee. Scanning electron microscopy image. Note large eyes with individual facets, although the facets are difficult to see at this magnification. Each ommatidium of *X. micans* measures approximately 25 μm but range from 15 μm to 40 μm. *Image by Pat Kysar.*

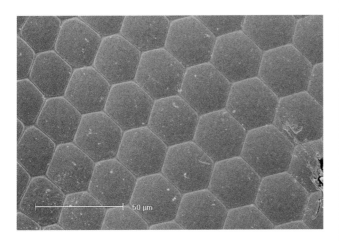

Figure 21.12 *Xylocopa micans,* carpenter bee. Scanning electron microscopy image of ommatidia facets. *Image by Pat Kysar.*

discrimination and quality of the image increase. In apposition eyes, the trade-off for this increased image quality is the loss of individual photon capture and hence overall sensitivity. Ommatidia can vary in size, ranging from 15 μm to 40 μm, and their number can vary from a few hundred to more than 25,000. As you might expect, the number and angle of units vary depending on ecological habits.

Although the individual angle has not been calculated in the carpenter bee, a closely related species has an angle that varies approximately 1° per ommatidial unit, so it sets the lower limit of resolution. These ommatidia create a mosaic of an upright image because each unit is stimulated by light of a slightly different incident angle. The carpenter bee probably cannot resolve objects smaller than those that subtend an angle of 1° or more (a human probably can resolve fifty to one hundred times this in satisfactory conditions) (Figure 21.11 and 21.12).

In the apposition compound eye, each photoreceptor cluster will have its own optical system and will provide a single spatial sample at the individual ommatidium level (Chapter 6). This provides the best resolution possible for an apposition compound eye—at or very near to the optical limits of diffraction, if the ommatidia are small enough. But an apposition eye with ommatidia that small will be less sensitive and will confine its owner to daylight.

In such an eye, pigment is housed in primary pigment cells between the individual crystalline cones. This pigment allows the insect to isolate the ommatidium optically to maintain image quality.

BOX 21.3 APPOSITION EYE RESOLUTION

To understand the resolution of an apposition eye, you must appreciate the ommatidial angle. Each ommatidium is set at a slightly different angle, and, much like a digital camera with pixels, the ommatidium adds its separate view to the combined visual field. The angle of each combined with the number of ommatidia determines the quality of the image. As the number of the ommatidia increases, and the angle of separation decreases, the

BOX 21.4 AN EVOLUTIONARY TWIST

These insects have added their own spin on these photopigments—those molecules that are light sensitive. They are bistable. This means that a photon will convert the pigment to a metarhodopsin (an intermediate thermostable molecule), which then can be converted back to rhodopsin by absorbing another photon. In short, it is a light switch that signals in both directions.

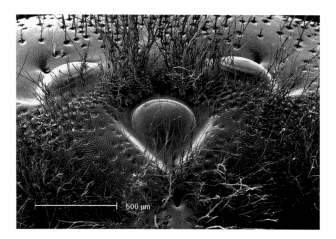

Figure 21.13 *Xylocopa micans*, carpenter bee. Three ocelli on dorsal aspect of head. *Image by Pat Kysar.*

insect (with the geniculate antenna to the left corner of the photograph) are called ocelli or "simple" eyes (Figure 21.13 and Chapter 6). They are surrounded by hairlike structures that combine to create a gravity-detection mechanism.

These ocelli can distinguish light and dark and probably polarized light. In other related species, the ocelli seem to coordinate with the ommatidia to enhance or inhibit their function.

In a related wasp, *Vespa orientalis*, investigators have confirmed that an ocellus is composed of multiple units, each called an ocellon (Figure 21.14). It is this combination organ that probably senses gravity and participates in navigation as mentioned.

The carpenter bee resembles the ubiquitous bumblebee but has distinct ecological differences. As a pollinator, the carpenter

bee is not as helpful as other hymenopteran species because it often feeds by biting through the base of the flower instead of sipping through the opening, so it is not dusted with pollen. Its vision, however, is just as important as it is to the pollinators.

The ways in which bees have coevolved with the angiosperms illustrate evolution's quirks. As creatures evolve to fill empty niches, evolution cannot revert to an older model or start from scratch; it must proceed with the tools at hand. Nocturnal bees are perhaps the best example, as they take the apposition eye to its limits.

Nocturnal Sweat Bee, Megalopta genalis

Perhaps 120 mya, close to the emergence of the angiosperms, bees arose as pollinators, despite evidence in amber that suggests the Miocene epoch (24 mya).

By the Miocene, though, bees were resolutely diurnal, with apposition compound eyes well adapted for the torrent of photons in full daylight. Even now, most bees are diurnal; their eyes do not tolerate even the limits of crepuscular light.

But there would be other niches to fill. The lineage of bees that includes *Megalotopa* (Figure 21.15) arose in the late Cretaceous. As some angiosperms became nocturnal, their pollinators coevolved into the night to help those angiosperms and the flowers of the dense rainforest to successfully reproduce.

Nocturnal angiosperms provided a beneficial niche for any pollinator that would follow them into darkness. After all, many predators and parasites attack diurnal bees, so a nocturnal life would protect bees from many of them. Moreover, nights would offer less competition for food because only some bats and moths seek pollen or nectar; nocturnal birds do not. Additionally, nocturnal angiosperms may have a higher protein and/or

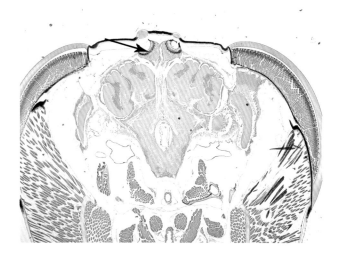

Figure 21.14 *Vespa* species. Note compound eyes laterally. The two central small eyes are ocelli. One is illustrated by the arrow. The lens of the ocellus has been artifactually detached and is above the eye. *Image by Richard Dubielzig, DVM.*

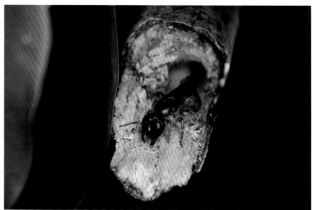

Figure 21.15 *Megalopta genalis*. This bee is at the nest, which can be seen directly behind her as a hollowed branch. *Image © Michael Pfaff.*

nectar content during anthesis (full bloom), providing richer provender to any pollinator such as a bee. But there is a catch: degraded image quality.

Megalopta genalis (Figure 21.15), a social Central American "sweat" bee, is one nocturnal bee that evolved to cope with the pressure of competition.

Because apposition compound eyes are not well suited to nocturnal activities, crepuscular or nocturnal invertebrates long ago evolved the superposition compound eye (Chapter 6). But bees were too far along the ocular evolutionary path to return to these strategies.

M. genalis, then, had to enter the night with an ill-suited eye. To permit this, evolution squeezed exceptional performance from this eye. M. genalis can find the 6-mm entrance to its home in a hollowed-out stick and forage in the rainforest understory at starlight levels. Humans would be blind to these objects at these light levels.

To accomplish this visual feat, M. genalis has unique adaptations. The diameter of the apertures of its ommatidia is nearly double the diameter of those in diurnal bees, but the number of facets is the same. This makes the eye considerably larger, thus the bee's Latin name. In addition, the retinular cells in each ommatidium make the rhabdom of the nocturnal bee four times larger and somewhat longer than that of the diurnal bee. This increased size permits the rhabdom more opportunities to capture incoming photons. As in a camera, the aperture is bigger and the shutter speed is slower, but size and speed alone are not enough to complete the task.

To help capture light, this bee has evolved at least one other neurologic mechanism: spatial summation, which means that the bee will collect light from a larger area of image that falls on the photoreceptive elements. The bee increases the receptive field by combining the input of many ommatidia through an ingenious system of neuronal wiring at the level of the connecting cells between the rhabdoms and the internal brain cells. Temporal (over time) summation also might be involved. That would mean that an image causes the stimulated cell to continue firing longer than the stimulus is present. This is like continuing to "see" an image after the flash of light that illuminated a night scene is gone because the photoreceptive elements kept firing.

These adaptations do degrade the image quality. In particular, temporal summation, if it exists, will slow the recognition of movement and blur the image. But it is better for the tropical bee, M. genalis, to see a slower and coarser world than to see nothing at all.

This bee survives in a world with light that is 8 logarithmic units dimmer than that in daytime, without being able to resort to the conventional techniques of moths, nocturnal beetles, or benthic arthropods of the ocean. So, evolution has again made the best of a bad situation.

True Flies and a New Eye

The social hymenoptera were not the only insects that would benefit from the flowering plants. True flies arose in the Triassic from an early terrestrial ancestor, then expanded somewhat and began to diverge in the Jurassic. But in the Cretaceous, as some of flies became pollinators, they exploded in diversity. Many other flies became omnivores and detritivores (eats waste and organic debris). This would require a new eye.

Diptera

Flies had new niches to fill if they could conquer the twilight of dusk and dawn without sacrificing daylight. That would require the apposition eye to take at least one more step to extend its limits and radiate into neural superposition eye.

Although most flies use the basic form of the apposition eye, some have subtle but clever specialized forms.

Stalk-Eyed Fly, Achias longividens

The Southeast-Asian stalk-eyed flies, as represented by Achias longividens (Figure 21.16A), show an ocular adaptation with astonishing visual mechanisms and visual fields. Both sexes have compound eyes at the ends of long stalks, but the males usually have longer stalks. (Some male Cyrtodiopsis whitei flies—another Borneo species—have been described as having an 8-mm body length with eyestalks that have a combined span of 20 mm. One wonders how such a creature could fly.) The males use their longer stalks to help claim and retain territory and to announce their position. Males may engage in ritualized, or even real, fights including actual "jousting" with these stalks. Competitors are driven away by the more dominant male, usually the one with the longer stalks. Size does matter!

More to our point, for any male of a species with long eyestalks, body and stalk lengths and number of ommatidia are directly related. A large male has about 2600 ommatidia per eye—not sufficient for good acuity. But the fly's visual field is probably the most complete in the animal kingdom. Without moving its head, this fly can see all of the entire visual space around it except for a few millimeters directly surrounding its body. It has an extensive binocular visual field throughout the median plane and "stereopsis" (or at least binocular field overlap) of no less than 25° (directly overhead) extending to as much as 135° in the frontal-ventral quadrant. Perhaps 70 percent of the ommatidia of each eye have a binocular partner in the opposite compound eye. In an area of the frontal and horizontal plane of the visual field, objects are viewed with more detail. We will call this area the fovea for lack of a better word, although it is not a true fovea. These "foveae" from the two compound eyes have binocular overlap. This area corresponds to one on each compound eye where the angle between the ommatidial faces is

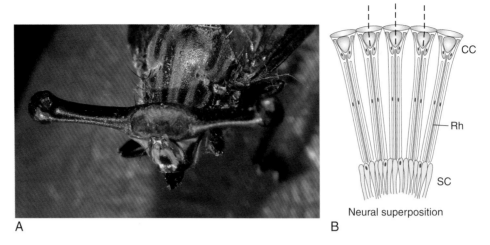

A   B

Figure 21.16 (A) *Achias longividens,* stalk-eyed fly. Ommatidia cover the tips of the stalks. **(B)** Neural superposition eye. Appears to be very similar to the focal apposition eye, but the rhabdom consists of a different arrangement of microvilli from each of the surrounding rhabdoms. In the neural superposition the individual microvilli are separated from one another, but this is not shown on this illustration. *Art by Tim Hengst.*

approximately 1 degree. For comparison, some dragonflies with a compound eye with good vision have an angle between the ommatidial faces of 0.2 degrees in an area of concentrated ommatidia.

As in many insect species, dipteran males have better acuity or different types of eyes than females. And male ornaments that influence sexual selection also are common among insects and dipterans. But ornamented eyes are uncommon.

The dipteran flies offer one more variation on apposition compound eye (Chapter 6). Most dipterans have neural super-position compound eyes (Figure 21.16B), a tweak that permits their owners to extend their feeding activity into the twilight after competitors and predators have had to go to bed.

BOX 21.5 NEURAL SUPERPOSITION EYE

The neural superposition eye (Figure 21.16B) evolved from the original focal apposition eye into a more sophisticated form. These eyes, found in the dipteran flies, represent a neurologic marvel. As in other insect eyes, each ommatidium has photoreceptive elements called rhabdomeres that compose the rhabdom. But, that is where the similarity ends. In dipteran flies, the rhabdomeres are not fused together (as they are, for example, in butterflies or bees). Seven of the eight rhabdomeres are separated at the base of the cone of each ommatidium. Each rhabdomere, then, has a slightly different image from its neighbor within the ommatidium. Seven of the eight receptor cells are "pulled" apart in the ommatidium and isolated. Two of the eight rhabdomeres overlie each other

in the center of the unit. The physical difference in the arrangement of the receptor cells and the rhabdom reaps neurological differences: as each rhabdomere sends its signal proximally and exits the ommatidium, that signal is matched with a signal from each of seven adjacent rhabdomeres in seven surrounding ommatidia. Each rhabdomere then sends a signal via its axon to the brain where, in a complicated neurological arrangement, the cells in each ommatidium combine their signals within the brain to create a single unified signal for that point in space. Although this combination process does not appreciably improve the image, it increases the photon sensitivity or light capture by about seven times without enlarging the rhabdom and sacrificing acuity. This allows the insect to be more active into twilight, effectively, adding as much as thirty minutes for feeding or reproduction. This system requires perfect alignment by these ommatidia, and indeed, that is what happens.

These dipterans push the envelope of vision and genetics with a single stalk. Evolution continues to look for creative ways to compete in even the slimmest of niches, and the extra minutes provided by this adaptation allow this dipteran fly to succeed.

Brachyceran Flies

Brachyceran flies (Figure 21.17) diverged from the dipteran lineage of flies at about 200 mya. Because both lineages

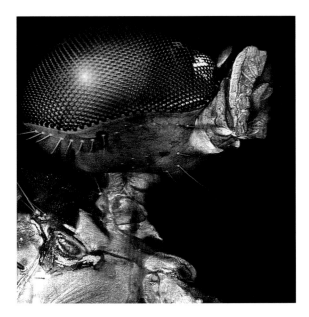

Figure 21.17 *Palaeochrysotus* Brachyceran fly in amber from the Eocene approximately 50 million years old. Note the individual ommatidia. *Image courtesy of and © Prof. Dr. Jes Rust, Steinmann Institut Bereich Paläontologie, Universität Bonn.*

A

B

Figure 21.18 (A) *Diogmites* species, robber fly. This spectacular robber fly has captured a bee. Note the large hemispherical compound eyes and note the three ocelli in the midline between the two compound eyes. Note the demarcation between the area of more concentrated ommatidia that are frontally placed. This can be seen in both eyes. *Image © Thomas Shahan, www.ThomasShahan.com.* **(B)** *Trioria interrupta*, robber fly. This particularly aggressive species of robber fly has been successful in its hunt for a dragonfly. It has been successful, in part at least, because of the advantage of its visual system, as it can see better than the dragonfly in crepuscular light. *Image © Thomas Shahan, www.ThomasShahan.com.*

have neural superposition eyes, the two groups' most recent common ancestor must have had them, too. This tells us that the neural superposition eye probably has been present for much longer. *Palaeochrysotus* (Figure 21.17) belonged to a family of long-legged brachyceran flies that were predatory and probably vicious, like other family members. Vision had to be important to this creature, so it was probably quite good.

The eyes of other extant brachyceran flies can tell us a bit about those of *Palaeochrysotus*. The neural superposition eye plays a key role in their success.

Robber Flies: Diogmites *Species and* Trioria interrupta

Voracious and predatory, these extant flies will feed on any insect they can catch, including much larger, stronger, and faster insects such as dragonflies. And robber flies can catch them in part because their vision (Figure 21.18) is similar to or in some cases better than that of the dragonflies. These robber flies are equipped with neural superposition compound eyes and three ocelli in the center of their head much like the carpenter bees discussed earlier. These compound eyes have many more ommatidia, and thus better vision, than do the stalk-eyed flies. Because these predators can extend their visible environment into the crepuscular range, and their ocelli help them to determine gravitation force, they can attack insects with more pedestrian apposition eyes that see poorly in dim light.

The arrival of plants, especially the flowering plants, heralded another food source for insects—a parasitic one, and several insect orders took advantage of it. Perhaps one of the more interesting children of the angiosperms illustrates the diversity of insects and the visual witness that accompanies them.

Male Scale Insect, Puto albicans

The Class Insecta surely found flowering plants to be evolution's manna. The variety of plant pests exploded with the appearance of these plants. Some unusual eyes appeared, too, such as those of the male scale insects.

Scale insects arose perhaps as early as the Triassic. By the early Cretaceous, when the angiosperms arrived on the scene, scale insects had radiated into a number of groups. The Putoidae family or its close ancestor in Figure 21.19 arose in the mid-Cretaceous, although this specimen was not entombed in amber until the Eocene 50 million years ago.

Scale insects are members of the superfamily Coccoidea with more than 7000 species, and this superfamily has been evolutionarily successful.

The female scale insect applies herself to a vascular plant and siphons the sap, and the life, out of it. After she becomes an adult, she has the ultimate in sedentary lifestyles: she rarely if ever moves, so she has lost her wings and often part of her antennae, legs, and eyes.

Male scale insects (Figures 21.20–21.22) have a variety of eyes suspected to have evolved from an original ancestral compound eye. Evidence suggests that an ancestral animal with a

Figure 21.20 *Puto albicans,* male scale insect. The male lives for only a few days and does not eat. When it is flying the angle of flight is not horizontal but oblique, providing a forward and backward field for these eyes. Specimen courtesy of P. Gullan BSc. *Image © Lynn Kimsey.*

compound eye gave rise to animals with fewer but larger units consisting of unified ommatidia. As these insects diverged further, these units continued to combine until only a few remaining eyes ringed the head.

Thousands of male scale insect species have an unusual distribution of eyes, with one pair where most insects have compound eyes, another pair where most insects have a mouth, and another pair placed laterally. Adaptations of these insects' eyes permit an f-number of 0.59 (Chapter 5) for increased photon capture. The rhabdomes are shorter and different from

Figure 21.19 *Putoidae* species, male scale insect. Confocal scanning electron microscopy image of *Putoidae* species caught in amber of the Eocene 50 million years ago. The image has been false colored. Note the necklace of eyes that encircle the head. *Image courtesy of and © Prof. Dr. Jes Rust, Steinmann Institut Bereich Paläontologie, Universität Bonn.*

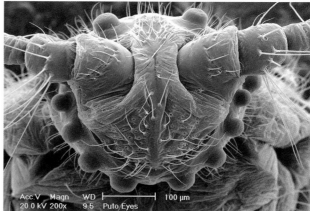

Figure 21.21 *Puto albicans,* male scale insect. Scanning electron microscopy image of this insect. Note that there are more eyes on the ventral aspect of the animal. When it is flying at an oblique angle the majority of the eyes would be facing forward with only the two on the dorsal surface facing posteriorly. Specimen courtesy of P. Gullan BSc. *Image by Pat Kysar.*

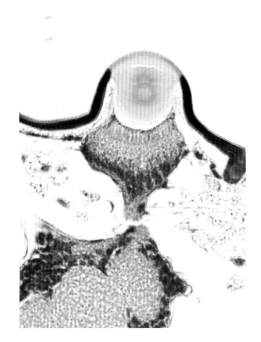

Figure 21.22 *Puto albicans*, male scale insect. Histologic image of one of the eyes. Note that the photoreceptive elements (rhabdoms) of the photoreceptive cells are facing the lens, as is found in most invertebrates. *Image by Richard Dubielzig, DVM.*

those of other insects, and the membrane structures are similar to the stacking discs of vertebrate cones' outer segments. This unique adaptation among arthropod eyes suggests that it is

original and perhaps evolved separately. In fact, the variability of these male scale insects' eyes (when compared to vertebrate eyes, for example) illustrate that eyes probably arose several times, combined into compound eyes and diverged again into individual units.

Males of some species of scale insects have rather unusual form of compound eyes, and these "extra" eyes may derive from the compound eyes of the last common ancestor, or they may have arisen de novo. In other words, these eyes probably devolved *from an ancestor with a more "united" compound eye.*

Male scale insects have supernumerary eyes that ring the head like a necklace. These are simple eyes, which are believed to be useful in guiding flight. Similarly, many researchers believe that the numerous eyes that ring the head help to stabilize the horizontal course as the male scale insect flies, holding its long axis in an oblique angle. The eyes probably help the insect to navigate toward the chemical potion emitted by the female, rather than to her body itself.

The approaching end of the Cretaceous saw the expansion and radiation of small, furtive mammals that carried eyes with individual units and visual pigments for dichromacy and sometimes for trichromacy. As birds expanded to become dominant in the Tertiary, mammals would come of age.

The story continues with the age of those early mammals—the monotremes and the marsupials.

MAMMALIA DIVERSIFIES

CRETACEOUS PERIOD

145–65 MILLION YEARS AGO

CENOZOIC ERA

EARLY TO MID-TERTIARY PERIOD

65–56 MILLION YEARS AGO

Pangaea, the last supercontinent, began separating into Laurasia and Gondwana at about 170 mya. That separation coincided with the radiation of the mammals from a synapsid ancestor, starting with the monotremes (platypus and echidnas) and followed, at about 150–125 mya, by marsupial and placental mammals. Further continental drift and competition would eventually isolate marsupials to Gondwana, specifically Antarctica, Australia, and South America. After that, with minimal competition from placental mammals, radiation continued in the latter two areas, giving rise to many different lineages of marsupials.

Mammals Thrive

At about 65 mya when a meteorite struck the Yucatan peninsula, ending the Cretaceous and snuffing nonavian dinosauria forever, the evolutionary door swung open for a synapsid clade that included all mammalian taxa (Figure 22.1). Diverse mammalian clades had existed and prospered, though, well before this cataclysm. In fact, mammals probably were so abundant then that more mammalian species have become extinct than are alive today.

Synapsids (Chapters 14 and 15) arose in the Carboniferous. This lineage would eventually evolve and radiate into the mammals, although the last common ancestor that would lead to these creatures would not appear until the mid-Triassic. These creatures would survive into the Tertiary and eventually become the dominant species. All of other synapsids besides the mammals would become extinct.

The monotremes are the most reptilian of the mammalian clades. These pivotal species deserve closer attention, as they are the closest extant relatives we have to that early synapsid line. The platypus and two closely related species

of echidnas, the only living representatives of this clade, can best tell this story.

Although the initial radiation into the monotremes occurred about 170 mya in the mid- to late-Jurassic, the eventual divergence of the platypus and echidna lineage from one another occurred about 100 million years later. The stem organism that led to the monotremes did not directly resemble either of the two living groups because evolution continued throughout the intervening years.

The extant species of the echidna and the platypus are Prototherian mammals, displaying features of mammalian specializations while retaining many whispers of our synapsid reptilian and, for that matter, amniotic beginnings. For example, the Prototherians (echidna and platypus) are the only mammals that excrete milk (the echidna's milk is pink from the iron content) from pores on the female's belly, have a cloaca, and still lay eggs. (A cloaca is cavity and opening in many fish, amphibians, reptiles, birds, and monotremes but not placental mammals, at the end of the digestive tract into which the intestinal, genital, and urinary tracts open. Hence the name monotreme or "one opening.")

Impermeable monotreme eggs represent the cleidoic egg produced by the first successful completely terrestrial tetrapod that arose at about 350 mya. In fact, that first amniotic egg may have resembled the egg laid by the present-day echidna.

Monotremes Platypus, Ornithorhynchus anatinus

The platypus is a strange animal because it sits astride a reptilian-mammalian gulf as a bit of a missing link. In fact, it is so strange that European explorers who described specimens

MAMMALIAN PHYLOGENY

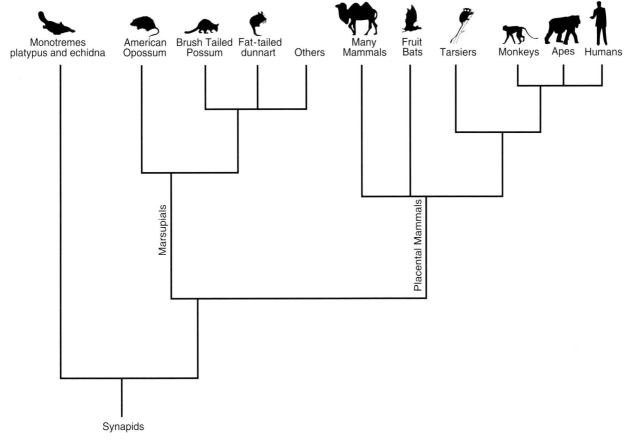

Figure 22.1 Mammalian cladogram.

were accused of attempting a hoax. But the early chroniclers did not know just how strange these venomous mammals really are.

Although the platypus enjoys an unusual sensory mechanism—electroreception—it is the eye that interests us (Figures 22.2–22.5). And the eye tells us much about the state of development when the monotremes diverged.

The story of the platypus eye is pivotal for the progress of the visual witness through the synapsid lineage. Because the platypus represents the oldest living "twig" of the mammalian tree, it can help us understand the status of the synapsid line at the time of that divergence (Figure 22.1).

With a cartilaginous cup in the sclera, a spherical shape, and rods and cones, the platypus eye resembles a reptilian eye (Figure 22.3 - 22.5). But it is the retina that is most instructive (Figure 22.4). The cones are both single and double, with some of the double cones having oil droplets that assist in discrimination and spatial tuning (Chapter 10), as in reptiles. Perhaps the most interesting story of the platypus retina, though, swirls about its genetic expression. Its cone opsins include one that provides evidence about the ancestral synapsid.

Figure 22.2 *Ornithorhynchidus anatinus*, platypus. Note the large white spot of fur just behind the bill meant to frighten predators. This is not the eye, but rather, the eye is slightly behind it and is marked with the arrow. The platypus does not use vision much below water.

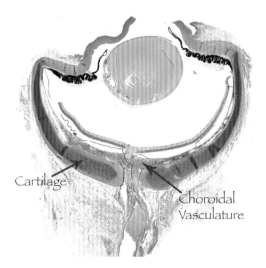

Figure 22.3 *Ornithorhynchidus anatinus*. Platypus eye. Note the choroidal vasculature which provides nutrients to the retina. The choroid vascular supply resembles the choroidal gland found in many fish. *Image by Richard Dubielzig, DVM.*

The most recent common ancestor to the monotremes and the remainder of the mammalian clade had three or all four (doubtful) of the visual pigments (opsins) bequeathed from the tetrapodian ancestors. After the monotreme lineage diverged from the main mammalian line but before the divergence into the platypus and echidna, the monotreme lineage lost one or possibly two visual pigments from the retinal cones, leaving only two opsins. Other mammals, including many of the marsupials and placentals that would arise later, incurred a similar loss. But that lost visual pigment of the ancestral monotreme was not the same one that was lost from the genome of the marsupial or placental lineage. Although monotremes, most marsupials, and most placental mammals became dichromats (having color vision from only two visual pigments), this process occurred differently in the monotremes.

The Descent of Color Vision

Here is a possible scenario of how the opsin distribution story occurred. Fish and probably the early tetrapods had four different visual pigments, each containing a different opsin with a different peak wavelength sensitivity. The sensitivities of the four opsins included one with a peak sensitivity in the long wavelengths that we will call "red"; one with a peak sensitivity in the medium wavelengths that we will call "green"; one in the short wavelengths that we will call "blue"; and one in the very short wavelengths that we will call "ultraviolet." To be sure, these visual pigments are sensitive to other wavelengths of light, but the names selected could be considered to be approximately the peak sensitivities, so they provide an easy mechanism for discussion. But the names are somewhat misleading because they imply that the opsins are sensitive only to that color and that the tuning of the opsins never changes—implications that are not true.

The synapsid lineage would eventually lead to the mammalian lineage, and it probably lost one of the four visual pigments before the divergence of the monotremes.

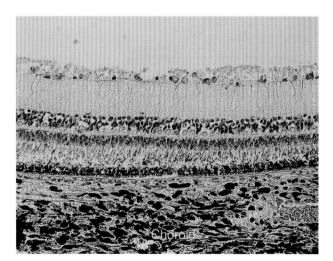

Figure 22.4 *Ornithorhynchidus anatinus*. Platypus retina. Labeled structures include the choroid with its large vascular channels and the photoreceptors labeled with Ph, and the ganglion cell layer labeled with Gang. *Image by Richard Dubielzig, DVM.*

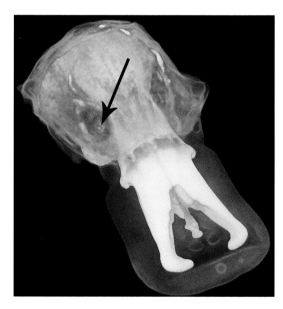

Figure 22.5 *Ornithorhynchidus anatinus*, platypus. Computerized tomographic scan with image modification to emphasize cartilage. Note the eye (*black arrow*) with scleral cartilage and lens. The bony structures can be seen in the mandible. Note the small eyes in comparison to the mandible and head. *Image by J. Anthony Seibert, PhD.*

The initial monotreme lineage, then, carried three of four visual pigments from the tetrapods, having lost the "green" visual pigment or, less likely, the "red." When the monotreme lineage diverged into platypus and echidnas, they both lost the "ultraviolet" pigment, leaving only two visual pigments.

The eyes of placental mammals evolved differently. After they diverged from the vertebrate synapsid lineage, the placental mammals lost the "blue" visual pigment but kept the "ultraviolet." (In many placental mammals, including humans, the "ultraviolet" opsin has changed somewhat to a peak at a slightly longer blue range, more toward a "blue" peak.) The remaining two visual pigments make up the complement for most mammals. But primates are among the few exceptions. Some have evolved a third visual pigment (probably by gene duplication) from the longer-wavelength pigment to have both a red and a green visual pigment for a total of three (see Chapter 23 for more detail).

After the platypus appeared, the monotreme lineage led to the divergence of the echidnas at about 55 mya. Subtle changes illustrate the evolution of that early mammalian eye into available niches at the time. And if the monotreme lineage diverged from the vertebrate lineage at about 170 mya, and the echidna diverged from the lineage after the platypus, as is currently thought, then the echidna adds evidence to the status of the last common ancestor of the monotremes and the remainder of the mammalian lineage—the marsupials or Metatherians, and placentals or Eutherians. The echidna is worth a visit.

Echidna: **Tachyglossus aculcatus** *and* **Zaglossus bruijni**

The echidnas, Tachyglossus aculcatus and Zaglossus bruijni (Figures 22.6–22.7), are less derived than the placental mammals, with eyes to match. The similarities of the two main monotreme branches (platypus and echidna), compared to the differences of the marsupial and placental mammal eyes, provide a glimmer of evidence as to status of the eye at the time of divergence of the monotremes.

The echidna retains all the ocular features (Figures 22.6–22.8) of the reptiles after the Cretaceous extinction, including a cartilaginous cup within the sclera, an avascular retina, an annular pad for the crystalline lens, and a Harderian gland. The echidna added a keratinized corneal epithelium, presumably to protect its eyes from the formic acid produced by the ants on which it preys.

Compared with placental mammals, the echidna has a more substantial tarsus in the lower lid, and the blink mechanism more resembles that of reptiles—also thought to evolutionarily protect against their prey.

Figure 22.6 *Tachyglossus aculcatus,* short-beaked echidna.

The iris sphincter of the echidna is smooth muscle and is the only intraocular musculature. This differs from the classic reptilian iris, which, like the iris of birds (and probably dinosaurs), has striated musculature. The synapsids may have come from a most recent common ancestor (such as an early tetrapod more closely related to fish and amphibians) that did not have much iris musculature. The sauropsids (lizards and their kin) went on to develop striated iris muscles and muscles of accommodation.

But the more nocturnal synapsids (mammals and their extinct kin) developed smooth muscle in their iris, perhaps because their nocturnal niche did not require vigorous pupillary responses or significant accommodation.

Another anomaly in the monotremes that provides clues about the eye at divergence is the scleral cartilage in both the echidna and the platypus but no other mammals. This suggests that the last common ancestor to the mammalian line had this structure, but it was lost by the time the marsupials and the later placental mammals appeared (Figure 22.8).

Figure 22.7 *Zaglossus bruijni,* long-beaked echidna.

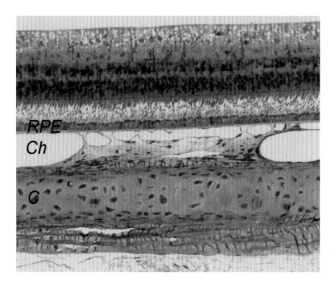

Figure 22.8 *Tachyglossus aculcatus,* short beaked echidna. Retina illustrating cartilage (C), choroid (Ch), and retinal pigment epithelium (RPE). *Image by Paul McMenamin PhD.*

The brownish fundus (the interior surface of the eye) of the echidna resembles that of some reptiles, and like reptiles, the choroid lacks a tapetum. The retina is composed of nearly all rods, which suits its crepuscular and nocturnal lifestyle, although a few cones have been found. Some oil droplets similar to those in many fish including lungfish, some reptiles, turtles, and birds have recently been discovered. Oil droplets (Chapter 10) represent an adaptation that is unlikely to have appeared twice in convergent fashion. Hence, like the lungfish and other lobefins (closely related to the presumed predecessors of the early tetrapods), the most recent common ancestor that gave rise to the sauropsids and the synapsids probably had oil droplets in its photoreceptors. This adaptation was lost by the time the placental mammals appeared, but a fragment of this morphology remains in the echidna. Like the intraocular smooth muscle, this remnant implies that earlier synapsids, such as Dimetrodon, had oil droplets in their photoreceptors and smooth muscle in their irides.

Although the echidna's retina has no fovea or area centralis, it does have a hint of a horizontal visual streak and the multilayered retinal structure typical for mammals however primitive and undifferentiated. Investigation of the visual capabilities suggests that the echidna's visual abilities resemble a rat's.

The monotreme's anatomical similarities to reptiles and birds and dissimilarities to other living marsupials and mammals suggest that the monotremes diverged from an ancient common ancestor that may not have looked much like a mammal at all.

Once the monotremes became established, the vertebrate lineage would diverge into yet another peculiar order that would perplex and amuse early explorers. This diver-

gence of marsupials and the placental mammals was not to occur for another 25–50 million years, about 150–125 mya (Figure 22.1). The next creature may demonstrate many of the features of that original synapsid, suggesting the possibility of the same lineage but not necessarily the same common ancestor that diverged into the monotremes or would give rise to the placental mammals. A good illustration of these shared features is the fat-tailed dunnart.

Fat-Tailed Dunnart, Sminthopsis crassicaudata

Small mammals arose from the original reptilian lineage (the synapsids) about 220 mya in the early to mid-Triassic as a shy, secretive creature. These early mammals probably had trichromacy at this stage. Some mammalian families would become nocturnal and lose one more visual pigment (the third visual pigment) to become dichromats, but that may not have occurred before the marsupials were well established because some Australian marsupials are trichromats although this is controversial.

So the marsupials are pivotal species in understanding the evolution of color vision. If trichromacy remains unaltered from ancient Jurassic mammalian ancestors, marsupials and other lineages retain those traces. Indeed, traces of trichromacy can be found in at least three marsupials: the honey possum, the quokka, and the fat-tailed dunnart. These species are probably closer to the ancestral stock that populated Australia when that continent broke free from Gondwana.

The fat-tailed dunnart (Figure 22.9) is a shrewlike insectivore that is arrhythmic or cathemeral (active day or night). The dunnart's vision is good for a creature this size, as would be expected for a predator, especially one that manipulates the insects it captures.

Figure 22.9 *Sminthopsis crassicaudata,* fat-tailed dunnart. *Image © Catharine Arrese, PhD.*

It has a rather pronounced horizontal visual streak and 2.3 cycles per degree resolution in the area centralis (foveal equivalent), compared with humans' 30 cycles per degree for 20/20. This visual streak, a concentration of photoreceptors and ganglion cells in the horizontal meridian, is common to animals that have a relatively unobstructed view of a horizontal horizon, such as a lion on the African veldt. In the dunnart's case, the visual streak may be appropriate retinal morphology for the vast central Australian desert.

The dunnart's retina is rod-dominant, although not to the same extent as that of a purely nocturnal animal. As in reptiles, birds, and monotremes but not placentals, some of the dunnart's cones possess transparent oil droplets. The visual system has a large binocular overlap and probably stereopsis. But the mystery is trichromacy: the marsupials that retained trichromacy have visual pigments with their peak sensitivities in the "blue," and distinct "red" range. There is an additional visual pigment that appears to be in the "green" range but is controversial. However, this extra pigment would account for trichromacy in these animals. In some marsupials, the sensitivity of the "blue" photopigment range extends into ultraviolet. What advantage would all of these visual pigments, especially sensitivity into the ultraviolet, confer on the dunnart?

Although the 20-gram adult dunnart has been known to eat small reptiles and young rodents, it is primarily insectivorous. The foliage they inhabit often camouflages insects. Most leaves reflect ultraviolet, but insects do not, so an insect will stand out to any animal capable of vision into this range.

The fat-tailed dunnart, then, represents a mammalian keystone species with color vision that extends into the ultraviolet and retains visual pigments from its synapsid ancestors.

Marsupials

The marsupials would evolve too, and as the eye became more complex, the retina would need a redesign to enable improved vision. This new order among marsupials needed more neurologic elements in their retinas. That meant thicker retinas with more cellular elements and the vascular system to supply them with nutrients. Some phyla such as fish or reptiles addressed this problem with projections of derived vessels into the central vitreous cavity (Appendix C). Mammals would need a more robust and intimate delivery system than those of fish or reptiles. Evolution responded with various methods of supplying blood. The North American opossum illustrates one mechanism (Appendix C).

North American Opossum, Didelphis virginiana, and Brush-Tailed Possum, Trichosurus vulpecula

The marsupial radiation provides us clues to the status of retinal vascularization at the time and later.

Seven marsupial orders are extant, with three represented in the New World. With the eventual connection of South America to North America through the Isthmus of Panama about 3 mya, one of the later-evolving marsupials successfully wandered north to establish a marsupium on North America.

The North American opossum (Didelphis virginiana) is its title continent's only extant marsupial (Figures 22.10–22.13). This house-cat-sized omnivore demonstrates an unusual attempt at inner retinal nutrition, suggesting that retinal vessels have evolved many times.

Much like their synapsid reptilian ancestors, most marsupials have an avascular retina. Reptiles generally have a conus, a collection of vessels protruding from the optic nerve head and feeding the inner retina. The retinal vascular system of these extinct reptilian synapsids probably resembled that of the tetrapods or fish from which they sprang.

The brush-tailed possum (Figure 22.14) belongs to the order Diprotodontia (including kangaroos and koalas) and has an avascular retina. But the retinal vasculature of D. virginiana is unusual because the arterial and venous segments of retinal vessels, including capillaries of the smallest caliber, occur in pairs (Figures 22.11 and 22.12) in a manner different from other animals with retinal vessels. This was evolution's response to the need for increased retinal nutrition after the marsupials had diverged into different orders, and the tapetum may have played a role.

At some point, nocturnality beckoned for the marsupials, especially arboreal ones. This led to other significant changes in visual function and morphology.

D. virginiana has a common solution for nocturnality—a tapetum lucidum or eyeshine (Chapters 7 and 13) in the superior half of its retina (Figure 22.13), although the brush-tailed possum does not. D. virginiana is believed to be the only

Figure 22.10 *Didelphis virginiana,* North American opossum.

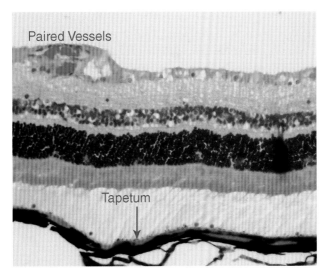

Figure 22.11 *Didelphis virginiana.* Histology of retina of North American opossum eye. Note the paired blood vessels and the retinal tapetum, both labeled. *Image by Richard Dubielzig, DVM.*

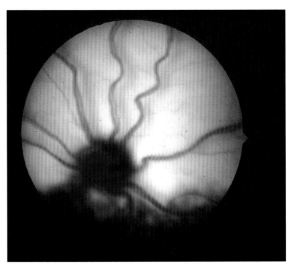

Figure 22.13 *Didelphis virginiana.* Tapetum of North American opossum. Note the paired vessels extending from the optic nerves. *Image by Nedim Buyukmihci, VDM.*

mammal to have retinal tapeta (other mammals have choroidal tapeta), although tapeta can be found in such disparate groups as certain fish and crocodiles. In tapetal regions, the pigment epithelial cells enlarge, become relatively free of the more typical melanosomes or pigment granules, and are filled with reflective, cholesterol-containing spheres. In the case of the opossum, these crystals are made of riboflavin (vitamin B2). The unusually thickened retinal pigment epithelium may interfere with diffusion from the choroidal circulation to the photoreceptors, so the deeply penetrating retinal vessels may need to compensate.

The tapetum, then, evolved as a nocturnal adaptation, but it probably also stimulated the retinal vascularization. This is convergent evolution. It is doubtful that the early synapsids had

a tapetum because diurnal animals would not have needed one, and none is found in any other marsupial.

D. virginiana has a nocturnal eye with other adaptations including a large spherical lens dedicated to gathering light. Typical for nocturnal animals, the large and nearly spherical lens is a substantial part of the eye's weight, as a larger lens assists with light capture. As an eye enlarges in an attempt to acquire more photons, the lens increases volumetrically in three dimensions, compared with the cornea, which increases only in two. This increase contributes to a disproportionately larger lens even as the cornea enlarges.

D. virginiana has a distinct area centralis (but not a true fovea), rather than the visual streak typically observed in many of the Australasian marsupials. The opossum has an almost pure rod retina with few cones, perhaps only 8000/mm2.

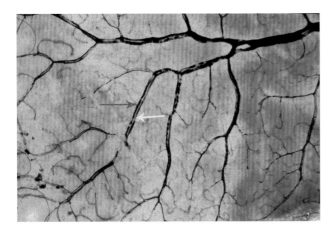

Figure 22.12 *Didelphis virginiana* retina. Note the artery (yellow arrow) and vein (red arrow) paired to the periphery of the retina. This is a very different pattern when compared to primates whose vessels are not paired so closely. *Image by Richard Dubielzig, DVM.*

Figure 22.14 *Trichosurus vulpecula,* brush-tailed possum.

They include both single and double cones with oil droplets. As in the echidna (Tachyglossus), the retinal appearance and retinal mosaic are more typically reptilian than placental mammal. But the reptilian retina is classically diurnal, not nocturnal. Rods are usually a relatively minor component in most lizards' retinas, for example, but are found to be abundant and differentiated in marsupials and monotremes. This suggests that the move to nocturnality came well after the most recent common ancestor of synapsids and sauropsids because some marsupials remain diurnal, or arrhythmic or cathemeral.

Evolution was exploring different mechanisms for retinal vascularization and light capture.

At the split of marsupials from placental mammals in the early Cretaceous, the retinae of these primitive creatures were avascular. To satisfy the oxygen-greedy neurologic mechanisms that would support the mammalian retina, vessels would have to be brought directly into the retina. The eye would require a more intimate vascular system of intraretinal vessels, and evolution would respond (Appendix C). Some placenta mammals, though, would require more than just supplementary nutrition.

THE AGE OF MAMMALS

MESOZOIC ERA

LATE CRETACEOUS PERIOD

100–65 MILLION YEARS AGO

CENOZOIC ERA

TERTIARY PERIOD

65–1.8 MILLION YEARS AGO

QUATERNARY PERIOD

1.8 MILLION YEARS AGO

23

Mammals Extend Their Dominance

The meteorite strike that brought on the end of the Cretaceous expelled millions of tons of ejecta into the stratosphere, bringing dramatic atmospheric cooling to the early Cenozoic. This era took another 40 million years for significant warming to return. But the Earth was much quicker to explode with new life as it had done so many times after a devastating cosmic event. As the strike and the resultant climate change banished niches, many other ones opened.

Birds—the only fragment of the dinosaur clades that remained—would become dominant (at least initially) in a new world with less competition.

Birds were well equipped to dominate with a variety of skills such as, of course, the ability to literally rise above the troubles on the earth, as well as supreme visual skills. These skills were probably well developed even in the Cretaceous, but they evolved to their modern form in the Tertiary. Nevertheless, birds would rule the roost only so long. This was to be the age of mammals (Figure 22.1).

Formerly small and nocturnal or isolated in southern continental landmasses, mammals began to flourish and diversify in the warmer mid-Tertiary. Mammals proved to be formidable competitors. They filled niches that might not seem to be within the mammalian purview. Bats prove that the aerial medium could indeed be a mammalian one, too.

Placental Mammals

Eastern Tube-Nosed Fruit Bat, Nyctimene robinsoni

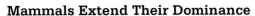

Bats arose in the late Cretaceous and diverged from one another in the Eocene (56–34 mya), but controversy remains as to the phylogeny and the timing. A likely scenario that fits the evidence would have the most recent common ancestor diverging from its lineage of small mammals in the late Cretaceous. This animal may have had good vision but not flight. When flight eventually developed, echolocation and the decline of vision followed it some families, but good vision remained with others.

When nocturnal mammalian flight began in the early Eocene night, there were few mammalian competitors for the feast of nocturnal insects and flowering plants. Highly successful and almost pancontinental, these bats have radiated into two major lineages—the crepuscular or nocturnal megachiroptera (megabats) and the strictly nocturnal microchiroptera (microbats). Although initially both groups were believed to navigate successfully without eyesight, bats are far from blind. Visual adaptations address the bats' niches.

Whereas the carnivorous microbats comprise the echolocating families, the frugivorous (fruit-eating) and nectarivorous (nectar-eating) megabats rely on olfaction and vision. In fact, in many species of megabats (also called fruit bats), the olfactory skills rival those of dogs. The tube-nosed fruit bats, such as Nyctimene robinsoni (Figure 23.1), possess such a highly developed sense of smell that they can determine the ripeness of the individual fruit to optimize their time spent foraging.

Compared with microbats, fruit bats also have better vision, which has to help them to navigate the tangle of their native tropical rainforest. The retina is composed primarily of rods, as you might expect in a crepuscular or nocturnal animal. The visual photopigments that provide fruit bats' dichromatic color vision include a short-wavelength opsin, with sensitivity that extends into the ultraviolet, and a long-wavelength opsin, with a peak sensitivity that could be best described as red. Fruit bats use their color vision to help discern colored fruit or blossoms.

Figure 23.1 *Nyctimene robinson*i, eastern tube-nosed fruit bat. This fruit bat is able to follow an odor plume with a "stereo" nose. Nocturnal vision in the fruit bats is excellent as compared to human nocturnal vision. *Image by Jack Pettigrew, MD.*

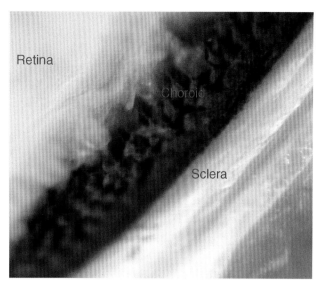

Figure 23.3 *Pteropus scapulatus*. Posterior segment of fruit bat. The retina has been peeled back to reveal the peglike appearance or the choroid. The three major layers of the inside of the eye are labeled. *Histology by William Lloyd, MD.*

Unlike the eye of some other mammals, that of the fruit bat has no retinal vasculature, and a retina that is 250 μm thick—too thick for the diffusion of nutrients from the choroid. So how do fruit bats nourish their inner retina?

The choroid consists of spikelike projections called papillae that stud the retina much like rivets uniting two pieces of metal (Figure 23.2). These projections create retinal undulations (Figures 23.3 and 23.4) called papillations. The tip of each choroidal papilla, which projects 125–150 μm into the retina, is believed to nourish the inner retina. This arrangement ensures that no portion of retina is much more than 100 μm from its blood supply. It also eliminates the need for a retinal vascular system and helps improve the image by not requiring these bats to look through their own blood supply. (This and other mammalian vascular supply systems are discussed in more detail in Appendix C.)

What does this undulating retina do to the retinal image? Apparently nothing. Or if the undulation has some effect, the bats seem adjusted to it.

Depending on light levels the nocturnal visual acuity of fruit bats ranges from 3 to 6 cycles per degree (cpd), comparable to that of cats. To compare further, our acuity in twilight or darkness drops to only 3–4 cpd, but our diurnal vision is better than that of fruit bats.

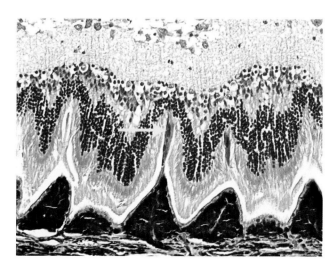

Figure 23.2 *Pteropus scapulatus*. Fruit bat retina. Note the vascular tufts that extend into the retina (*yellow arrow*). *Histology by William Lloyd, MD.*

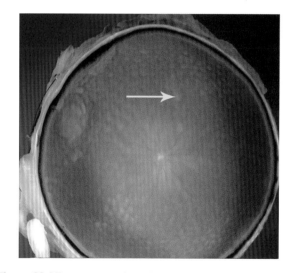

Figure 23.4 *Pteropus scapulatus*. Posterior half of the eye of a fruit bat. Note the multiple white dots from the choroidal pegs noted in Figure 23.3. Yellow arrow points to one of the many choroidal pegs that show through the overlying retina. *Histology by William Lloyd, MD.*

Even the microbats are known to use visual cues to locate larger prey, especially when echolocation would not be useful. But at about 1–2 cpd, their nocturnal vision is somewhat less than that of rodents.

The undulating retina or papillae of the fruit bats contribute to the visual acuity, not just to the eye's nutrition. These papillae increase the number of photoreceptors and perhaps depth of field and improves light-gathering ability, but it may require a different form of retinal processing or cerebral integration.

Primate Tuning of Color Vision

These bats and most other mammals are dichromats, and most mammals remain so, as the mammalian eye was well established at the beginning of the Cenozoic Era. But there would need to be improvements in acuity and color vision if mammals were to become diurnal and compete with the birds. The most notable change came in the primate line. It is best described with the lemur's story.

Black and White Ruffed Lemur, Varecia variegata variegata

Native to Madagascar, these diurnal prosimians (Figure 23.5) are masters of their arboreal domain. "Lemur" means "ghost" in Latin, and watching these magnificent animals travel through, and seemingly disappear, into the trees around them illustrates why the name fits. Lemurs vault from limb to limb, clinging to the spiniest trees with the greatest of ease, and are even able to hang by their feet and feed with their forelimbs.

These lemurs spend almost all of their lives in trees, even building nests there. In fact, they are so intricately woven into the forest tapestry that they have become the sole pollinator

Figure 23.5 *Varecia variegata variegata,* black-and-white ruffed lemur.

of the traveler's palm, with which they appear to have coevolved.

Lemurs are among the less derived primates and can help us understand primate visual development. Primates are divided into three lineages: Prosimii (lemurs, bush babies, lorises, pottos, and tarsiers), Platyrrhina (New World monkeys), and Catarrhina (Old World monkeys and hominids). The Prosimii are regarded as prosimians ("before the monkeys"), and except for the diurnal black-and-white ruffed lemur and a few other lemurs, the Prosimii are nocturnal or cathemeral.

The prosimians are viewed as perhaps preserving "primitive" or less-derived features of our early ancestors. So to compare the prosimians to the other primate suborders is to consider the particulars of our direct early evolution, although evolution would have continued in the interim since they arose.

Most authorities believe that Madagascar and India were carried away from Africa as Gondwana broke up about 180 mya during the Jurassic. At that time the greater subcontinent of India probably abutted Tanzania, Kenya, and Somalia. The new supercontinent formed by India and Madagascar existed until about 100 mya. At that point, the two islands separated, eventually sending India headlong into Asia, which created the Himalayas. Early primates (plesidadapiforms) were not in evidence until the early Paleocene Epoch (65–55 mya) but were probably present 30 million years earlier when Madagascar separated.

So Madagascar lemurs appear to occupy a very special position. They were evolving as an early primate or primate predecessor and survived on Madagascar as it broke away from India. Current evidence suggests that all of the Madagascar primates, including the bizarre aye-aye, evolved from a single ancestral species.

Independently evolving and speciating without competition from other primates, this single ancestral species and its progeny (thirty-three extant species) help the understanding of the evolution to color vision in primates.

All sighted mammals possess cones in their retinas, but color vision among mammals, even primates, is not universal. As discussed (Chapter 22), most placental mammals have lost two of the four original visual photopigments, but Old World monkeys (all catarrhines), humans, and presumably all previous hominoids possess three different visual photopigments. Lemurs are key to understanding how this difference occurred.

When primates diverged, most mammals had but two "color" visual pigments (Chapter 22). Understanding the sensitivities of the visual pigments helps the understanding of color vision. Stating the peak of sensitivity of the cone visual pigment as a wavelength is more accurate than as a specific color because these opsins are subject to genetic drift and interpretation in the brain. Additionally, the color perceptions of humans with normal color vision vary widely, making the description of color vision very complicated. But in the interest of brevity, colors are used here.

The two visual photopigments contained in the cones of most mammals include one with a spectral peak sensitivity in the short wavelengths best described as "blue" although originally it was "ultraviolet" and drifted to "blue." The description of the peak sensitivity of the second cone is complicated because it varies among species but is described here as "red." Because of gene duplication of the longer-wavelength opsin (or other more complicated genetic mechanisms), a third visual pigment has reevolved in Old World primates, such as gorillas, chimpanzees, and humans, as well as some New World primates. This visual pigment is best described as "green" or "yellow," as the wavelength is shorter than the original. It is not the same as the one found in birds, reptiles or fish, which suggests that it did not descend from the medium- or long-wavelength pigment in the synapsids of long ago. Instead, it was probably an error in duplicating the second longer-wavelength visual pigment in a common ancestor of the Old World primates.

The visual photopigments in New World monkeys are not as consistent, and as a result, they illustrate visual photopigment evolution in progress. The howler monkeys species (Alouatta seniculus and A. caraya) are the only exceptions: each animal has three (relatively) predictable visual photopigments similar to those of the Old World monkeys, and each is a trichromat. As in Old World monkeys, gene duplication at a relatively unstable locus formed the third opsin. That means that when New World and Old World monkeys separated, all were dichromats. Because the opsins in the howlers are similar, but not identical to those in the Old World monkeys, this suggests that the genetic material (called polymorphisms) was present in both Old World and New World monkeys when they separated, but these genetic tools were not active in either group.

At least three other species of New World monkeys are dichromats now but are developing a third active visual photopigment in some of their cones, suggesting that evolution is progressing toward trichromacy in these animals, too.

The females of two diurnal lemurs (closely related to the black-and-white ruffed lemur) have recently been found to have polymorphisms that would permit some individuals to have the visual photopigments required for trichromacy. This suggests that the lemurs are ready to develop three visual photopigments and hence trichromacy, but this has not happened. It helps confirm that the different lineages of primates were dichromats when they diverged. Furthermore, the lack of trichromacy among lemurs suggests that competition has not yet demanded it.

The distribution of cones is at least as important to color vision as their spectral sensitivity is. Current evidence suggests that lemurs have a relatively low concentration of cones in the macular area and are afoveate (possess no fovea). Furthermore, the prosimians (including the lemurs) appear to lack the necessary neural components to process the more complex visual signals of trichromacy. When lemurs were tested against the dichromatic New World monkeys, the diurnal prosimians proved unable to discriminate spectral variations that were quite apparent to the monkeys.

The principal difference between the two species is the higher concentration of cones in the monkeys' macular region. But trichromacy requires more than the necessary visual photopigments in the cones and cone concentration; the neurologic mechanisms to interpret and compare these signals must be in place as well. This adds further evidence to the principle that the eye (and other sensory mechanisms) drives the brain and not the reverse. Either the two must evolve in tandem, or the sensory mechanism evolves first and co-opts other neurologic machinery.

What, then, are the implications for the evolution of color vision? One explanation could be as follows. Trichromacy is a useful aid to perceptual color vision and to folivory (foliage) or a frugivorous (fruit) diet. Although the protoprimate that initiated the speciation of lemurs on Madagascar had dichromacy (as did most mammals that preceded the primates) and perhaps the genetic tools for trichromacy, this potential has not been realized, perhaps because of the lack of other primate competition. But in the New World monkeys, primate competition has selected the necessary wiring of the brain—the cone concentrations—and has begun moving toward trichromacy in many species. Alternatively, prosimians may have evolved genetic polymorphisms convergently and in parallel with the New World monkeys. The Old World monkeys had sufficient competition to select the necessary photopigments and the neural imperatives for trichromacy long ago. Hominids evolved from Old World monkeys as trichromats, suggesting that trichromacy may play a very important role in primate and human evolution. Hence, any lemur or closely allied species competing with other primates would be hard-pressed to maintain its niche and to avoid being replaced by trichromatic successors.

Lemurs are threatened by habitat destruction, but could they be threatened by trichromacy as well?

Primate visual evolution in humans and Old World primates has given biological evidence that some portions of color vision can be recaptured, albeit with different photopigments. Such recapitulation of design never occurs in exactly the same manner; that has been the case with the more recently evolved primates. After all, humans do have three visual photopigments, but only two of them resemble the four bequeathed from our Devonian tetrapod ancestor.

Other important changes occurred as the Old World monkeys returned to trichromacy. The tarsier illustrates the evolution of the primate eye and the bony and periocular muscle that was necessary to supply the larger eye and brain—especially in nocturnal animals.

Tarsier spectrum

Like the lemurs, tarsiers are prosimians. The oldest fossils of tarsiers date to about 50 mya, in the Eocene. Because the extant animals have changed little since those fossils, most authorities believe that tarsiers are close to the original primates that appeared perhaps as early as 90 mya.

Tarsiers may be the smallest primates, but their eyes (Figures 23.6 and 23.7) are so large that the weight and volume of the eyes, as a percentage of the total body weight, are the largest among all mammals. Tarsier eyes are relatively and absolutely large for several reasons. Eyes become relatively larger as an animal becomes smaller (known as Haller's ratio, Chapter 20), and, as we have seen before, nocturnal eyes often become as large as possible to gather photons. In fact, as in birds, the two eyes together weigh more than the brain (Figures 23.7 and 20.10B), and the eyes so fill the orbits that the extraocular muscles are rudimentary. These muscles move the eye very little, so the tarsier compensates by moving its head nearly 180° to maximize its visual field.

To enclose such large eyes, the orbits also had to enlarge (Figure 23.7). But there is more to this skull that hints of primate evolution.

One key step in the evolution of primates was the development of a bony postorbital septum (Figure 23.7), which few other mammals have. This has been viewed as essential to the enlargement of the brain case, while enabling the development of the jaws and their muscles for the sake of mastication and to stabilize vision during mastication. To provide enough leverage

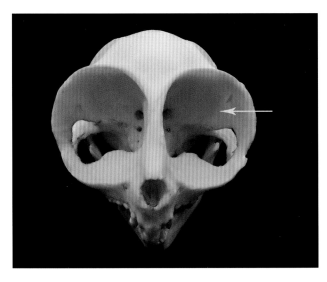

Figure 23.7 *Tarsier spectrum* skull. Note extremely large orbits. Eyes are larger than brain. Arrow points to the postorbital septum.

for the jaws, the skull had to change morphology and extend the rim of zygomatic arch (cheekbone) where the jaw muscles attach.

Concomitantly, the orbits were rotating anteriorly to face forward, providing a larger portion of the field of vision to both eyes and both sides of the brain. This dual representation provides the necessary input for the brain to establish true stereopsis—seeing in three dimensions. A fast arboreal lifestyle—swinging from tree to tree—may have demanded stereopsis. Or its presence may mean that primates were originally predators, for which stereopsis is usually essential, and the structure of the skull had to change to accommodate that lifestyle.

This Southeast Asian primate is a rare nocturnal primate. Many nocturnal primates have only one cone photopigment. For example, the owl monkey of the New World has but one cone with peak absorbance in the shorter "blue" wavelengths. Hence, this animal cannot compare this signal to any other or see in color, but it can perceive into the ultraviolet more than we can. The tarsier, though, has two cone photopigments—a middle to long wavelength and a short wavelength—so it does see in color. It also has a fovea, and this, too, is unusual for a nocturnal primate.

Cones usually serve diurnal vision, and they require photopic light levels (daylight levels). The tarsier's fovea is composed of cones, so it is the area of best, and concentrated, vision. The tarsier may have these cones and its fovea because it descended from a diurnal animal and has never lost all of its "diurnal" visual hardware. The tarsier may use its dichromacy to capture the small active insects that compose much of its diet. The animal is quite active during crepuscular hours, and the insects would still be active but not as vigilant visually in this light.

One possible use for the cones is difficult to prove. Because the longer-wavelength pigment survives in almost all mammals,

Figure 23.6 *Tarsier spectrum.* Note large eyes and very small pupils. *Image © Myron Shekelle.*

it may be necessary for circadian rhythm or some other function. Hence, the second pigment may be retained for other reasons.

As primates, tarsiers have other unusual ocular features. Among mammals, its pupil is the smallest known, probably to protect the mostly rod retina during daylight and to enable the animal to function during crepuscular or even full daylight.

With mobile ears, a 180° rotation of its head on its neck, color vision, and a fovea, this animal is well equipped to be a crepuscular insectivore capable of finding and successfully attacking large and active insects.

The tarsiers illustrate that evolution is capable of creating extremes even among mammals, and relatively quickly.

The Tertiary and Quaternary periods saw more changes than just the rise of the large mammals. Many of these resulting creatures illustrate evolution's clever approach to visual development.

PLANKTONIC SOUP EVOLVES

CENOZOIC ERA

TERTIARY PERIOD

65–1.8 MILLION YEARS AGO

QUATERNARY PERIOD

1.8 MILLION YEARS AGO TO

PRESENT

Earth's oceans are rich and full of microscopic life, and many of the larger Metazoa feed entirely on this bounty. These planktonic creatures consist of larval forms of larger animals, developing stages of vertebrates and invertebrates, and protists that are destined to remain single-celled. Surely these life forms have been changing for hundreds of millions, or billions, of years, but we know little of them because they fossilize poorly, if at all, and countless numbers have gone extinct. But many life forms remain, so we do know that most of them respond to light in some way, and some have sophisticated eyes. Although these animals represent many different phyla, some are collected in this chapter to illustrate evolution's continuing creativity.

Tiny Aquatic Arthropods

Long before the Tertiary, crustaceans developed eyes and radiated into many different species and designs. These early crustaceans were small, but evolution tends to produce ever-larger animals, at least until a global catastrophe makes large size a liability. A meteor strike, for example, will favor the smaller, more adaptable, and less specialized animal. But small animals have constraints and limitations of their own provided by physiology and physics. One such limitation is the eye.

Testing the Limits of Eye Size

Depending on the anatomy of the eye, physics will demand a certain minimum eye size and restrict the maximum as well. For example, the compound eye, used by so many arthropods, has relatively poor vision because of the size of the individual units, the ommatidia. Ommatidia cannot become much smaller than about 13 μm, their size in many insects. In smaller units, diffraction restricts and distorts the image so much that the eye would be useless. But the fairy wasp (Chapter 21) tests those limits, with ommatidia that are just 5–7 μm in diameter. However, if the insect employed a visual photopigment with peak sensitivity in the ultraviolet, these shorter wavelengths would not be disturbed by the diffraction limitations of such small ommatidia.

Insects also are limited by the quantity, not just the size, of their ommatidia. For insects using a compound eye to achieve vision comparable to that of humans would require the eye to be at least 1 meter in diameter! But an eye this large would be impossible to evolve because it would impede the species in many ways, especially from perambulation or flying. Even during the Carboniferous Period, when high oxygen levels produced some giant arthropods (griffenflies) the size of ravens, their eyes are estimated to have been no larger than 45–50 mm. With current oxygen levels, the dragonfly is probably at the upper limits of sustainable compound eye size in an insect.

The laws of physics restrict the maximum and minimum compound eye size, and these restrictions play a role in restricting the size of such organisms as well if vision is to be an important sensory mechanism. Hence, if animals become very small, novel visual mechanisms will be necessary. Plankton must face and solve these physical problems if the visual witness is to be an important sensory mechanism for them.

Copepods

Many crustaceans remain small and planktonic, as larval forms or permanent protists. These most successful of all species have evolved and are evolving unconventional approaches to their sensory systems and, in particular, their visual needs.

Copepods ("oar-footed") are an abundant and diverse subclass of crustaceans found in many aquatic habitats and near the base of the food chain, where they play an essential role. They are the primary consumers of microscopic bacteria and protists that form the base of the food pyramid. In the colder waters close to the poles, copepods' biomass is so immense that these animals sustain all manner of invertebrates and vertebrates, from worms to baleen whales. One such copepod, *Copilia quadrata*, is an excellent example of planktonic evolution and the visual perception of dimension.

Copilia quatrata

Animals with a camera-style eye must interpret a visual scene in three dimensions, although the retina has but two dimensions. Proper interpretation of a three-dimensional world with a two-dimensional retina requires sophisticated neural processing, including usually seamless assumptions about available data and perhaps binocularity. No animal possesses a three-dimensional retina, although some fish have tiered retinas that could be considered more than two-dimensional in certain situations. Other animals, such as jumping spiders (Chapter 13) and carnivorous sea snails, have a linear curved retina consisting only of a band no more than seven photoreceptors wide. But despite what is, in essence, a one-dimensional retina, the jumping spiders must have good vision with some form of three-dimensional processing because they are most successful at prey capture, and this requires a sense of depth. These creatures swing their retinae through an arc, using the principle of scanning to acquire the image (Chapter 13).

But Copilia quadrata (Figures 24.1 and 24.2) has a nondimensional retina; it is a point retina consisting of six or seven photoreceptors. In C. quadrata's copepod class, evolution has stumbled across some rare mechanisms for prey capture or mating.

C. quadrata pushes the boundaries of its epipelagic description, sometimes being found as deep as 200 meters. Except for

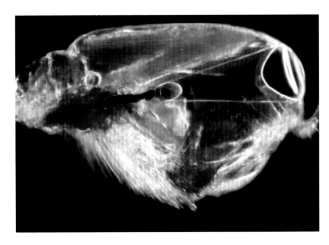

Figure 24.2 *Corycaeus* species. Note the pairing of two lenses creating a telescope. This species is related to *Copilia*. Both species have this lens system described in the text. *Image © Peter Parks/imagequestmarine.com.*

the orange pigment of the photoreceptors, this organism is transparent, except perhaps if it has recently eaten. A recent meal may be why the specimen (Figures 24.1 and 24.2) shows some pigment.

Like the jumping spider, C. quadrata scans back and forth much like the cathode ray tube that provides an image for a television screen. The creature has what looks like two eyes located anteriorly on the carapace, and it resembles other invertebrates, but the similarity with other creatures ends there. These larger external lenses are part of the carapace, and they do not move. Scanning takes place internally. The secondary lens, seen best on the sagittal image (Figure 24.2), is positioned immediately distal to the small point retina and proximal to the larger lens on the carapace.

Light initially contacts the external and larger lens with the parabolic posterior surface (Figure 24.2—Corycaeus, a closely related species, which has similar optics). The image is cast on the secondary lens, which then brings the point to focus on the receptors, much like a telescope with an eyepiece and objective lens. The eyes scan horizontally (perpendicular to the anterior-posterior axis of the animal) within an angle of 10°–15°, as measured from the distal, or first, lens. Essentially, this is a linear, one-dimensional scan. To add to the curious nature of this visual system, the eyes move in opposite directions, at a rate of between 0.5 and 10 Hz, although they do move simultaneously. A single muscle contracts to pull the yoked retinas together, assuring rapid movements medially and slow movements temporally. Consequently, stereopsis is unlikely.

Without scanning, the point retina has a field of about 3° and, in the best of circumstances, could have a linear field of 45° but probably has much less. But even with scanning, this system is one-dimensional (although arguably two-dimensional if one considers time), so how does C. quadrata measure the second

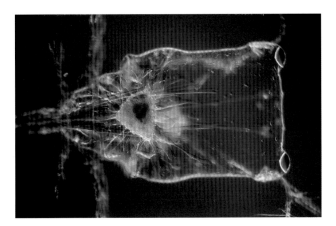

Figure 24.1 *Copilia quatrata.* Note the lenses fixed onto the anterior carapace. *Image © Peter Parks/imagequestmarine.com.*

(or third) dimension? Perhaps, as has been suggested, the animal's prey provide the animals' second dimension. C. quadrata is a ferocious predator that feeds on vertically migrating plankton it detects as they move through the horizontal scan line. If the plankton moved horizontally, they might never enter that crustacean's perceptive world. This simple but elegant perceptual system is packed into an animal that measures no more than the head of a pin (1 mm wide and 5–6 mm long, including the tail).

Certain species of pontellids, a family of copepods, have such unusual eyes that they are difficult to interpret and, as a result, they have yet to be fully understood. And yet they illustrate principles found in long-extinct predecessors. For example, *Pontella scutifer* has an ocular design that was first known among the trilobites, suggesting that such an ocular design must be an overriding principle of optics for the evolution of eyes.

Pontella scutifer

Pontella scutifer (Figures 24.3–24.5) is an epipelagic marine copepod found irregularly throughout the world's oceans, usually living well offshore. This species has three eyes, including two simple ones on the dorsal aspect of the cephalosome, or carapace equivalent (Figure 24.5). These two eyes, found in both male and female, probably have poor acuity. They are used basically to determine light, dark, and perhaps movement. These simple eyes contain a lens and few photoreceptors beneath, and

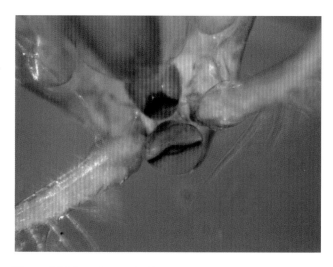

Figure 24.4 *Pontella scutifer.* Ventral view of the napulier lens. Note the paired lenses directly in front of the eye (red structure). The red eye has a single lens and an asymmetric retina.

almost in direct contact with, the lens. Despite the lens material's likely high index of refraction, the lens cannot focus an image on the photoreceptive elements. But it does not really need to.

This animal's exciting ocular feature is the third eye (Figures 24.3 and 24.4). A triplet lens system found only in the males, this is the "napulier" eye. Nauplius means larval form, which most crustaceans have. Once hatched, most crustaceans proceed though six napulier stages, each punctuated by a molt between stages. The napulier stage of many copepods has but one eye, the napulier eye. Pontella scutifer males retain this eye

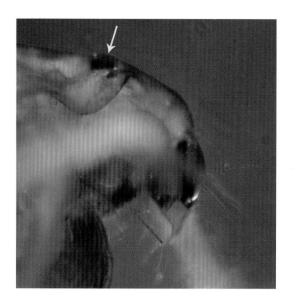

Figure 24.3 *Pontella scutifer.* Note the simple eyes on the dorsal surface (*yellow arrow*). The lens doublet can be seen anterior to the eye (*blue arrow*). This doublet is paired in such a way as to eliminate spherical aberration. The eye is simple with only one lens and six photoreceptors in an asymmetrical, center-surround pattern.

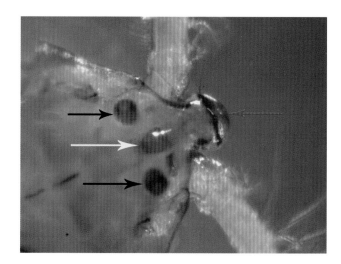

Figure 24.5 *Pontella scutifer.* A dorsal view of the head of *Pontella.* The paired black arrows show the simple eyes on the dorsum. The red arrow shows the front surface of the napulier eye with the blue arrow illustrating the junction between the two lenses in front of the eye. The yellow arrow illustrates the eye beneath and within the clear tissues of the copepod.

even after the animal develops two simple eyes (really just ocelli) on the dorsal aspect of the cephalosome.

This nauplier eye of Pontella contains three consecutive lenses with a total of only six (!) photoreceptors behind the lens system. Each lens is about 100 μm, with an index of refraction of 1.52, according to Land and Nilsson. Carotenoid pigment, appearing as red in the photographs, lines the eyecup behind the third lens. The front surface of the first lens is parabolic (Figure 24.5), which allows the triplet lens system to eliminate, or at least greatly reduce, spherical aberration. This aspheric lens system is an alternative to the graded index of refraction lens found commonly in fish (Chapter 10). By ray tracing, Land and Nilsson also show that this unique lens system creates a single point image.

The system must relate to mating because the females have a doublet rather than a triplet lens system. In addition, finding a female may be quite a challenge for the males. They form massive swarms that are almost exclusively male, which probably creates intense competition for mating because the second task must be finding the females in these swarms. The females are blue with yellow spots. The asymmetry of two of the photoreceptor elements (rhabdomes; see Chapter 6) in the male nauplier eye suggests a center-surround detection system that would respond to the female color pattern. Reflecting pigment behind the other two pairs may receive a reflected and refocused image, as in some scallops, but this is just a guess.

Other well-developed sensory mechanisms in pelagic copepods are probably more important but are likely to work cooperatively with the visual ones. The first antenna contains mechanoreceptors and chemosensory setae, which have exceptional sensitivity to movement and high frequency sound. These receptors are probably most useful in distinguishing prey from predators and in determining prospective mates.

So, although small, this copepod carnivore holds visual secrets with an eye for the ladies.

Transformation of Larval Eyes

Plankton are not a single lineage or phylum but rather many diverse drifting organisms. Many planktonic creatures are larval forms and develop into adults of different species such as jellyfish, bony fish, or even plants that sometimes look nothing like the larvae. Much transformation often occurs as the larvae become adult. When becoming an adult, one such larva changes its eye dramatically, suggesting how the adult may have evolved and illustrating how even the simplest of eyes can be most useful.

Vent Shrimp, Rimicaris exoculata

R. exoculata, *a species of shrimp, was discovered on the floor of the Atlantic Ocean at the thermal vents at 3600 meters or more*

Figure 24.6 *Rimicaris exoculata,* vent shrimp. Collection of vent shrimp with orange retinal strips visible on the dorsal surface of some. *Image courtesy of and © Cynthia Van Dover, PhD.*

below sea level where no sunlight penetrates (Figure 24.6). Most animals that live at such depths have lost their eyes or have never developed them. Eyes are of little or no use at this depth, and there is little advantage in retaining them. And yet, this shrimp does indeed have eyes, although these eyes are unlike any other. They are simply strips of exposed sensory retina without any focusing ability (Figure 24.7) or even a cup-like configuration resembling the Nautilus *eye (Chapter 8). Of what use could these eyes be to this animal?*

The thermal vents radiate heat, energy, and light in several forms. Light of a very long wavelength in the near infrared or

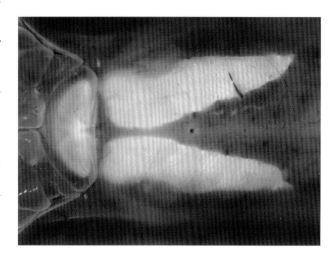

Figure 24.7 *Rimicaris exoculata,* vent shrimp. Dorsum of vent shrimp. These two linear strips are naked photoreceptor cells with photopigment that function as eyes for this adult animal. *Image courtesy of and © Edward Gaten, PhD.*

what is described as blackbody radiation and light in a range best described as green have been documented. How such light is produced is a mystery, and it is not easily visible to us, but *Rimicaris* can see it. *R. exoculata* can see these wavelengths with its photopigment with a peak response in the "greenish" range. This helps keep the animal close enough to feed on the bacteria around the vents but not so close to be cooked like scampi. Such eyes would be remarkable enough as simple strips of neurosensory naked retina running down the back of the animal, but the story does not end there.

The planktonic larvae of R. exoculata *shrimp have compound eyes and relatively normal ones at that. The larval form lives in deep water, to be sure, but not deep enough to exclude all light. When the species can use light to see, it does, but as the animal matures and becomes an adult, it loses these eyes and instead develops these naked neurosensory strips on its back and finds the vents as its new home.*

Evolution has bestowed complex compound eyes on arthropods such as related shrimp, among other phyla. R. exoculata illustrates that evolution is perfectly content to return to little more than the eyespots of Euglena *if it fits the niche of survival. Eyespots (Chapter 2) are very close to where organized sight began, and the vent shrimp's adult eye fills the animals' needs flawlessly.*

Aquatic adaptation in planktonic species, such as those described above, illustrates evolution's diverse and ingenious solutions to ocular requirements. The Cretaceous and Tertiary, though, are the periods of large mammals, and some of them have returned to an aquatic life. What adjustments would be required?

MAMMALS RETURN TO THE SEA

CENOZOIC ERA

TERTIARY PERIOD

65–1.8 MILLION YEARS AGO

QUATERNARY PERIOD

1.8 MILLION YEARS AGO

TO PRESENT

Mammals diversified and grew, especially during the latter half of the Tertiary. Even aerial niches filled with large mammalian vegans and carnivores. As mammals became more proficient at hunting, some would choose to return to an aquatic environment. This would present physiologic challenges for all systems because mammals breathe air and rely visually on an air-cornea interface. Adjustments would be required.

Aquatic Mammals

Warm blood, hair, mammary glands, middle-ear bones, and air-breathing lungs characterize mammals. A pelagic home would seem unsuitable for such an animal if not for profound physiologic adaptations that include the eye.

Cetaceans (whales, dolphins, and porpoises) began as a wolf-like creature evolved from a group of hoofed primitive mammals known as mesonychids in the Eocene about 55 mya. Probably predators, they used water either as a refuge from their own predators or possibly even as hunting grounds. Evolved to be completely free of their previous terrestrial lifestyle, these animals radiated into various diverse families, but family relationships between niches remain: hippopotamus families are a sister group to the modern whales, for example (Figure 25.1).

As these mesonychids sought a new home in an aquatic environment, they brought their eyes with them. The bottlenose dolphin is a good example of how the eyes would have to adapt to their new home.

Bottlenose Dolphins, **Tursiops truncatus**

These intelligent, social mammals rely primarily but not exclusively on echolocation. They also have excellent hearing and good vision. Their eyes are well adapted to an aquatic environment.

Although dolphins spend most of their time underwater, they also have an aerial interface when the eye is above the surface (Figure 25.2). As in most primarily aquatic animals, the cornea is flat, and the lens is large, spherical, and positioned farther forward in the globe than it is in terrestrial mammals.

Like birds, some dolphins have two foveae but only one class of functioning cone. All cones have the same visual pigment, with the peak sensitivity of the opsin shifted toward blue, although it is neither of the short-wavelength opsins— "ultraviolet" or "blue." The single opsin that remains is the longer-wavelength or "red" opsin, which has been shifted toward the shorter or "blue" wavelengths.

With only one opsin, dolphins are color-blind, but much of their world, especially for the pelagic dolphins, is monochromatic anyway. Because cones do not function in such a world, dolphins also have a lot of rods for seeing in the inky depths to which they dive. These rods have the peak of their visual pigments shifted toward the shorter wavelengths as well.

The two areas of concentrated photoreceptors in each eye provide for visual resolution that is about 9' of arc, compared with 1–2' for humans, although this depends on ideal conditions. The dolphin retina contains relatively large photoreceptors sending signals to large ganglion cells. Each area centralis contains peak concentrations of these poorly understood ganglion cells that undoubtedly are an aquatic adaptation to increase the collection of light. Consequently, this reduces discrimination. Sight beyond 50 to 100 feet even in the clearest of water is not possible or at least not of much use, but good vision within that range would be a definite benefit to a predator. Consequently, dolphins see well underwater (although not as well as a seal) and about as well as an elephant or an antelope when their eye is out of the water.

Like some reptiles and a few other mammals, the bottlenose dolphin has unusual multiple pupils. When the pupil dilates, it is circular, but it constricts into two crescent-moon-shaped

Figure 25.1 *Hippopotamus amphibius*, African hippopotamus. Weighing up to four tons, this herbivore of African rivers has evolved its eyes, nostrils, and ears to the dorsum of its skull to permit aerial sensory input while completely submerged.

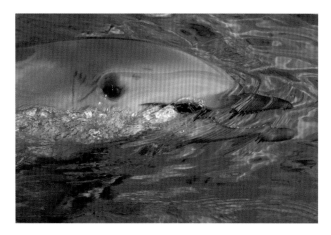

Figure 25.2 *Tursiops truncatus*, bottlenose dolphin.

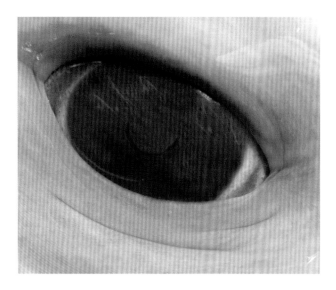

Figure 25.3 *Tursiops truncatus*. Bottlenose dolphin pupil. *Thanks to Michelle Campbell, MS, Dolphin Quest (www.dolphinquest.com) and to Bailey who posed for the image.*

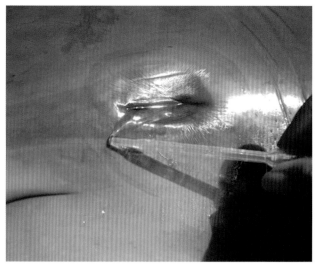

Figure 25.4 *Tursiops truncatus*, bottlenose dolphin. Note the viscous tears the consistency of jelly. The mucins in the tear film have evolved to protect the cornea from the osmotic gradients of marine waters. *Thanks to Michelle Campbell, MS, Dolphin Quest (www.dolphinquest. com) and to Bailey who posed for the image.*

pupils at the end of the horizontal slits of the original pupil (Figure 25.3). A structure called an operculum covers the center of the pupil like a window shade, leaving openings at the periphery of the slit pupil. This "two-pupil" morphology limits light flux to a retina better equipped for lower light levels. Yet these two pupils still permit functioning in the sort of bright light found above the ocean's surface. But the reason for dual pupils remains a mystery.

Dolphins face a significant problem in a constant aquatic environment that the original mammalian line never had to confront. Constant fresh and even salt water will draw electrolytes out of the corneal surface cells. This process is called crenation *when the surrounding fluid is more concentrated and* cytolysis *when it is less concentrated.*

Once the tetrapods came ashore, and a terrestrial life was assumed, any piscine protective adaptations that inhibited vision would become unnecessary or, worse, counterproductive. Consequently, the epithelium became smoother and relied on lubrication by the Harderian and lacrimal glands (Chapter 11). Once lost, protective ocular surfaces would not reappear, at least in the same molecular manner. So dolphins solve the problems of cytolysis or crenation by secreting tears. Most other mammals also secrete tears, but the dolphins' tears are very different from most. These viscous, clear-jelly-like tears (Figure 25.4) are composed of highly concentrated mucins and perhaps immunoglobulins, or the antibodies that fight infections. Dolphins' tears coat the cornea to prevent the metabolic damage from salt or fresh water. Other such tetrapods have had to adapt, too. This metabolic convergent evolution is also seen in cormorants (Chapter 20) and marine turtles (Figure 25.5).

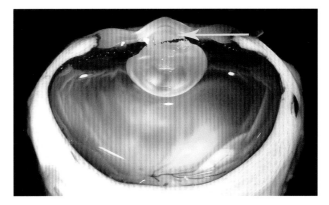

Figure 25.5 *Dermochelys coriacea,* leatherback turtle. Note the extremely viscous tears used to eliminate salt. When these large marine turtles come ashore to lay eggs, they continue to secrete tears. These tears are thick, very viscous, and probably contain large mucin molecules similar to dolphin tears, although even less is known about the ocular secretions of these animals. See Chapter 16 for further details on the leatherback turtle. *Image © http://www.arkive.org/.*

Figure 25.7 *Tursiops truncatus.* Bottlenose dolphin eye cut in a sagittal section. Note the blue-green tapetal reflections, and the accommodative lenticonus (*yellow arrow*), as the lens is squeezed through the pupil. The tapetal reflection appears bluer in this specimen than it does in life. Note the very thick posterior sclera. *Image by Richard Dubielzig, DVM.*

All of these animals have descended from clades that were originally terrestrial, with some species becoming aquatic.

Dolphins possess a choroidal tapetum that reflects photons back through the photoreceptors for a second chance at capture (Figures 25.6 and 25.7). The tapetum's (fibrosum) reflection (Chapter 10 and Chapter 13) shifts toward blue-green to be one of the few green tapeta (Figures 25.6 and 25.7). Tapeta help nocturnal terrestrial animals to perceive light at night. These reflective layers also help creatures that function at depths where few photons penetrate. Although hearing and echolocation function at depths, any small edge in vision such as a tapetum provides would be evolutionarily encouraged.

Dolphins experience life with diverse sensory abilities. They have mastered their environment, two-thirds of the surface of the earth.

Mammals returned to the sea more than once, and each time illustrates the adaptability of the eye and different adaptations. But the aquatic eye's similarities also illustrate the limitations the niche puts on existing ocular physiology. The harbor seal demonstrates how quickly and profoundly the eye can change.

Harbor Seal, Phoca vitulina

Seals and sea lions, or pinnipeds, seem to be having fun as they cavort in the water with each other or even with prey (Figure 25.8). Yet these sleek creatures are efficient marine predators. Their warm blood makes them excellent swimmers, exhibiting speed, agility, and endurance. Yet because they are air-breathing, semiterrestrials, these mammals have had to sacrifice to achieve their habitat.

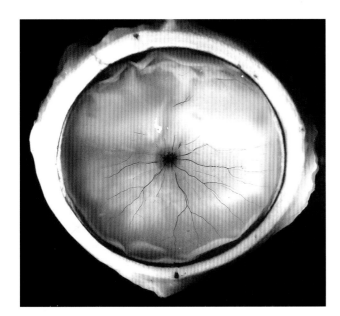

Figure 25.6 *Tursiops truncatus.* Bottlnose dolphin posterior pole. The tapetum of the dolphin reflects light in the shorter wavelengths. This fibrous tapetum reflects greenish-blue wavelengths that nearly match the peak absorption of the single opsin of the photoreceptors. The specimen appears bluer than the posterior segment appears in life because of post mortem changes. *Image by Richard Dubielzig, DVM.*

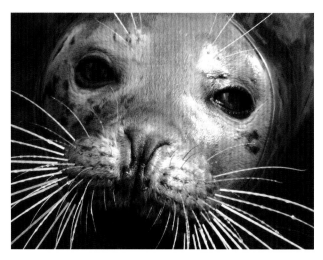

Figure 25.8 *Phoca vitulina*, harbor seal. *Image © Frederike Hanke.*

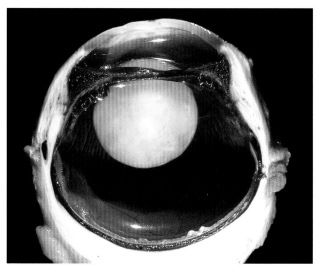

Figure 25.10 *Phoca vitulina*. Harbor seal eye. Note the round piscine-like lens and the relatively flat cornea. *Image by Richard Dubielzig, DVM.*

Seals, sea lions, and walruses—all carnivorous pinnipeds—arose from a four-legged animal resembling an otter. They probably began as a single lineage that arose fairly recently, about 35–25 mya. About the size of a medium dog, this prepinniped ancestor lived near fresh water and was semiaquatic. This descent probably began in or near the Arctic Circle when it was warmer.

Once their ancestors even partially returned to an aquatic habitat, they found a home full of fish and invertebrates. To become efficient in this world, pinnipeds evolved flippers to replace legs, feet, arms, and hands.

Vision plays an important role in this adaptation because seals and sea lions are visual predators (Figure 25.9). A seal

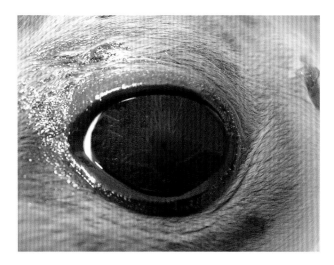

Figure 25.9 *Phoca vitulina*, harbor seal. Note the unusual inverted teardrop pupil, which has probably evolved to limit light during terrestrial basking. The seal has but one visual pigment and is color-blind. *Image © Frederike Hanke.*

spends much of its life and hunts exclusively underwater. Of necessity, then, seals have evolved a hard spherical lens much like that of a fish. The spherical piscine lens is essential to refraction for all underwater visual predators (Figure 25.10).

But once a harbor seal has hauled itself out on land the cornea again comes into play with the aerial interface, providing additional refraction. Something surprising happens, developmentally, to its cornea and those of some other seals. Seals possess a great deal of astigmatism, or steepening and flattening of the cornea in one meridian (approximately 9 diopters in the 180° meridian). The astigmatism creates a more hydrodynamic flow across the eyes during propulsion and permits a clever solution to terrestrial vision. Seals accompany high astigmatism with a vertical slit pupil in brightly lit environments (Figures 25.8 and 25.9). This slit pupil combined with—and perpendicular to—the astigmatic axis creates a pinhole effect that can improve vision in air because it minimizes the effect of the astigmatism from the cornea.

Furthermore, the harbor seal has a tapetum (cellulosum) that reflects light back through the photoreceptors and a mostly rod retina. Although these adaptations permit better vision in dim environments underwater, they would create glare and a dazzled effect on land in bright light if not for the pinhole effect from the pupil combined with the astigmatism of the cornea (Figure 25.9).

Underwater, a harbor seal relies on the spherical lens and in effect optically bypasses the corneas (Figure 25.10). This renders them emmetropic in water. Because the lens is hard and spherical, the mechanism of lens deformation does not work as well, and lens movement, as in fish, has not reevolved in this species. So seals accommodate with lens deformation,

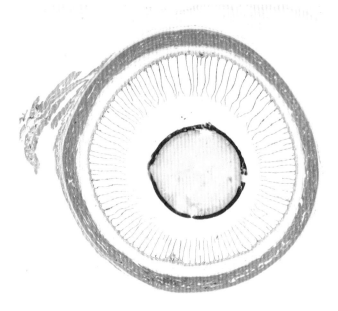

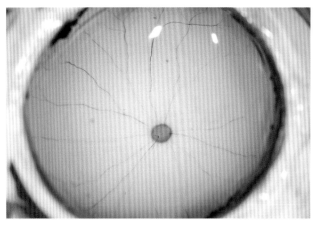

Figure 25.13 *Zalophus californianus*, California sea lion. Note greenish tapetum and retinal vascularization. *Image © Carmen M. H. Colitz, DVM, PhD, DACVO.*

Figure 25.11 *Zalophus californianus*, California sea lion. Histology: a view from inside the sea lion eye looking out. The lens in the center is yellow, and the pars plicata individual fibers reach out toward the equator of the lens. The ciliary body musculature is weak and thin. *Image by Richard Dubielzig, DVM.*

although accommodation is probably not extensive (Figures 25.11 and 25.12).

Although harbor seals are roughly the size of a human, the eyes are disproportionately large—a third longer than human eyes, with nearly twice the equatorial diameter because of the

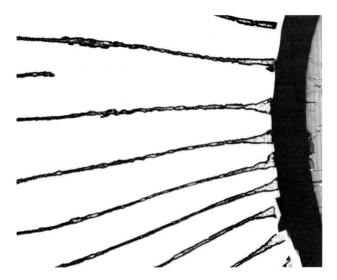

Figure 25.12 *Zalophus californianus*, California sea lion. Histology of the fibers (pars plicata) that surround and abut the lens of the eye. *Image by Richard Dubielzig, DVM.*

visual devotion to dimmer environments. The pupil dilates and becomes circular with dimmer light, and the green-reflecting tapetum (Chapter 10 and Chapter 13) supplies a second chance at photons as they are reflected (Figure 25.13). As a result, most seals have lost their second visual pigment and hence are color-blind with only a single visual pigment.

The seal lens is multifocal. Fish have this unusual adaptation so they can avoid chromatic and spherical aberration (Chapter 10), but because seals do not see in color, they should not need to adjust for chromatic aberration. The adaptation in seals probably occurs to increase the depth of focus when the pupil is dilated. This combination of morphology and physiology provides a highly sensitive eye that can see well even at the depths of 400–500 meters.

These seals are rapacious hunters. With their blubber coating and interesting physiology, they can dive deep and tolerate high levels of accumulated systemic carbon dioxide. The harbor seal is most capable of hunting for cold-water fish such as cephalopods, shrimp, cod, and flatfish, and vision is important in their capture.

The pinnipeds and the cetaceans have chosen to emphasize different sensory mechanisms with their return to an aquatic life. The cetaceans have emphasized echolocation, whereas the pinnipeds have favored vision, and the eye reflects that basic difference in the phylogeny.

The Quaternary period has seen the descent of man from a lineage of primates. Evolution brings us to a visual witness that is able to supply images to a conscious brain and to communicate and understand the sense of vision.

THE VISUAL WITNESS AND A
CONSCIOUS BRAIN
CENOZOIC ERA
QUATERNARY PERIOD
1.8 MILLION YEARS AGO
TO PRESENT

26

The vertebrate eye has evolved into an excellent instrument for many species including humans. Although not perfect, the eye does register fine detail, movement, and a wide range of light levels. Humans exhibit the most recent evolutionary step: the eye couples with a brain that not only receives and analyzes input from the eye but also can consider and communicate that input.

The Human Eye

The primate eye, as epitomized in humans, represents an achievement in mammalian descent and deserves the final vignette in our story. We can contemplate how our eye has evolved.

By the time primates appeared, the mammalian eye had changed and likely has gained and lost in the transition from the first vertebrate.

Primates have lost the retractor bulbi of many other mammals. The full complement of seven extraocular muscles was established at least as early as the placoderms of the Silurian (445–415 mya) (Chapter 9) and persists in similar anatomic distribution in many vertebrates today (Appendix B), but not in humans.

Along with the retractor bulbi, the nictitans, which served terrestrial animals so well since the Devonian, has been all but lost in the primate eye. All that remains is the semilunar fold, which is no more than an ocular appendix (Figure 26.1). The loss of the nictitans coincided with that of the retractor bulbi, so the losses are probably related.

By the time the primates arose, the mammalian retina had lost three of five opsins (basis of visual pigments) that lampreys have in their photoreceptors, and two of the four that most fish possess. Even now, living humans have nowhere near the visual pigments of some arthropods, such as the mantis shrimp with sixteen visual pigments. Primates and the placental mammals lost the brilliantly colored oil droplets that refine color vision and assist discrimination.

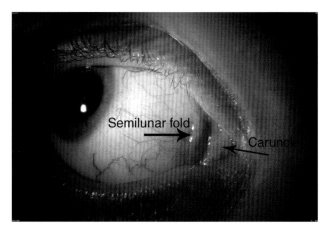

Figure 26.1 *Homo sapiens.* External image of the human eye. Note the labeled structures of the exposed external surfaces—the caruncle and the semilunar fold. The cornea can be seen to the left of the image. *Image by Ellen Redenbo.*

These oil droplets were present in the early tetrapods and still exist in monotremes, some marsupials, turtles and birds, among others.

Some innovations never occurred in the lineage that would lead to mammals because of timing. For example, the doublet lens system to thwart spherical aberration, found in the first known eye, in trilobites, was not available to the synapsid lineages because it appeared after the vertebrates and invertebrates diverged, and there was no need for it to reappear. Mammals also did not get the striated musculature that molds the lens and provides quick accommodation in reptiles and birds because it came after the sauropsids and the synapsids split. Moreover, we must look "through" much of our retina because our photoreceptors point backward instead of toward the image as they do in the octopus eye. This burden arose from the complicated manner that ciliary cells were used as photoreceptors in the vertebrate eye, and primates are stuck with it.

On the other hand, the descent of the vertebrates has provided us with an organ that has been evolving for 500 million years and does fit our niche.

The outer structures of the eye, such as the conjunctiva and cornea, have evolved to be protective and to meet the requirements of a terrestrial environment. Much of this adjustment began and occurred nearly 400 million years ago in the Devonian, long before primates appeared.

Beginning with the transition from an aquatic to a terrestrial environment, tear glands, lashes, and lids evolved and adapted to our environmental needs. Although primates have lost the retractor bulbi, they have retained the other six extraocular muscles, which are essential to our eyes' need for constant movement. These muscles can work continuously and indefinitely, without fatigue, unlike other striated muscles, and they usually work without our knowledge or consideration.

Photoreceptor evolution also continued in the new primates. After 150 million years of mostly nocturnal mammalian lifestyles, the retina would have to change if primates were to venture out into daylight, and change it did (Figure 26.2).

Some primates including humans developed a third visual pigment to create trichromacy of vision (Chapter 23). This improvement has provided those primates with the ability to perceive ripe fruit, leaves, or other nutritional morsels in its diet, an advantage that permitted it to outcompete its dichromatic confreres. Moreover, the competition to develop trichromacy and the neurologic tools to evaluate and process this input may have had a key role in the imperative for the evolution of consciousness among those primates.

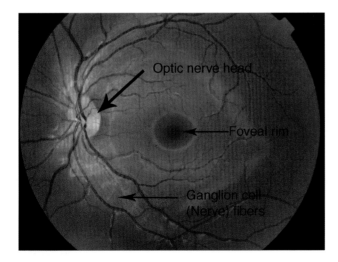

Figure 26.2 *Homo sapiens.* The posterior pole of the human eye. The optic nerve, foveal rim, and nerve fibers coming from the ganglion cells are labeled. Nerve fibers are the whitish striations coursing toward the optic nerve. *Image by Ellen Redenbo.*

Just as the eyes of other species have been doing for eons, our eyes evolved to the requirements of our evolving niche. The basic retinal and neurologic machinery had been in place for hundreds of millions of years, yet we would hone it.

The normal, healthy human eye has perhaps 110 million photoreceptor cells. These cells consist of the three visual pigments in the cones for diurnal color vision and one pigment in the rods for nocturnal vision. The photoreceptors are divided into about 100 million rods and 8 million or so cones. The primate retina is complex with many interconnecting neurons but not as complex as the retina of many birds or even some turtles. In many ways, our brain receives an inferior image from our eye, as compared with the rich, robust images received by some other animals. Humans simply do not have the best optical device in the animal world, but the brain helps us with these deficiencies.

The human retina, though, still is supremely prepared for complex visual tasks. Eighty percent of human sensory input takes place through the human retina. Much of the processing takes place in the brain, unlike in birds and reptiles, where it happens in their more complex retinas. Because more responses to birds' and reptiles' visual stimuli occur at a subcortical level, they occur more quickly. Humans, on the other hand, can use higher-level processing to understand visual stimuli. About one-third of the human brain is devoted to vision or visual processing. Because much more of our processing is consciously mediated than other animals', we are less susceptible to illusions than are most animals. For example, some camouflage works better among nonhuman animals than humans, in part for that reason. This neural machinery permits the brain to swirl and mix memory and careful analysis with the visual input. This ability probably also explains why humans do not collide with glass windows or doors as frequently as birds do.

Our eye is connected to a superlative brain that can process the gigabytes of data sent to it every instant. The brain uses this fantastic quantity of raw data to assemble moving images that pass effortlessly through our consciousness— in full color, although the perception of color exists only in our brain and not in our retina. The perception of color requires our brain's neural processing to compare the signals from at least two different color receptors—cones.

The human eye-brain combination is remarkable in more than the assemblage and interpretation of images. The human eye can function in light levels that range more than nine orders of magnitude difference (one billionfold) in the intensity of photons—from bright, full equatorial desert sunlight to dim starlight (although we may take thirty minutes to adjust to these different light levels—a process called dark adaptation). It is a truly breathtaking process.

The human eye also can perceive a lemon as yellow both in bright sunlight and in the more yellowish incandescent light of a kitchen—a process called color constancy. Although the wavelengths radiating from that lemon are very different in those two situations, we barely notice. This probably relies on the retina's ability to compare wavelengths from three different cones.

Where will this evolution lead us? As Yogi Berra is alleged to have said, "Prediction, especially about the future, is difficult." Whether he actually said it or not, the statement is a true one. Evolution is the ultimate future—and its prediction is fraught with hazard.

The Direction of the Visual Witness

Evolution is a meandering process fashioned by many factors including environment, biochemistry, external catastrophes and the niches necessary to accommodate life. In some sense, evolution begets evolution because it is not a zero-sum game. The earth teems with life, yet the impact of humans in the past several thousand years has begun the process of extinction anew, in a classic biological account of a dominant species. Specifically, an organism expands to fill every available portion of its niche, exploits it fully, then overexploits until the resource collapses—often taking other species with it. The extinction process we are undergoing will rival the Permian if it continues at our current pace. Human life will probably not disappear until most of the planet's resources have been stripped, and hence, humans will continue to be a cause of this major extinction process until our own demise.

The earth has been through this before, however. Life will continue and expand again as the biochemistry of this mature planet is ready for the continuation of life on earth. Life's resources are plentiful. More importantly, its tenacity is astonishing, and so it will persist.

Sensory systems, especially visual systems, will continue to be among the premier forces of evolution because of the advantage they afford species. Some of the ocular designs discussed in this book will undoubtedly persist. The physiology and anatomy of the eyes of birds and primates achieve superlative abilities and may continue in a similar form in new species. Evolution will probably direct the visual witness in different directions, too—perhaps directions we cannot describe or imagine. Evolution could stumble onto different visual pigments responding to longer or shorter wavelengths, if these new abilities would provide an advantage for that species. Perhaps photoreceptors will become more compact for better acuity, although some birds already possess photoreceptors that approach the wavelength of light and cannot get much smaller to create a finer "grain." In any creature, vision may not improve so much in wavelength perception or visual acuity as in better sensitivity and processing. Because the brain does much of the "heavy lifting" of vision in primates, visual processing could improve. As extinction continues apace, new species will radiate into abandoned niches, if given a chance. Surely other visual mechanisms will follow as evolution continues to tinker with available design or experiment with new modalities.

The visual witness arrives at the present through the twists and turns of random changes brought about by chance, error, and unfathomable lengths of time. As it has for 3.75 billion years, the evolution of life will continue for billions more. Countless species will continue to come and go, as they have done since life began on earth. The sensory perception of sight will accompany and, to some extent, direct the development of many animals. Evolution's witness will continue to see the light.

Appendix A

The human eye is primarily a globe consisting of three basic coats. The outer, tough coat is known as the *sclera*. It encloses the eye and becomes clear anteriorly as it becomes contiguous with the *cornea*. Inside the sclera is the second and middle vascular coat called the *uvea*. This layer is divided into three zones: (1) the *choroid* posteriorly beneath the *retina* and supplying it with nutrients; (2) the *iris* anteriorly; and (3) the *ciliary body* immediately behind the *iris*. The ciliary body is a thickening of the uvea and produces fluid for the interior of the eye. The innermost layer of the eye is the retina, which extends anteriorly but comes only to approximately the ciliary body and does not line the entire globe.

The cornea is part of the tough outer coat extending from the sclera, but its structure renders it clear. The cornea is more steeply curved than the sclera with that curvature changing at the limbus, where the clear cornea becomes the white sclera (Figure 26.1). The iris is the colored portion of the eye and is visible through the cornea in most animals. The ciliary body is within the eye and is the structure that makes intraocular fluid to maintain pressure within the eyeball. The choroid between the ciliary body and the anterior edge of the retina is an area called the *pars plana* (Figure A1).

The *lens*, also called *crystalline lens*, of the eye is clear and shaped much like a symmetrical lentil with a curved surface on the front and back. In mammals it is held in place with fine fibers from the *ciliary body* called *zonules*. The center of the eye is the largest portion of this cystic structure and is called the *vitreous cavity*. This cavity is filled with a viscous clear fluid called *vitreous*. A smaller cavity anteriorly between the inner cornea and the lens is called the *anterior chamber* and is filled with a clear fluid called the *aqueous humor*.

In mammals, reptiles, and birds, the lens can change shape in a process of deformation (Appendix E). This change in shape will result in a change in focus as the curvature of the lens is altered. This process of changing the shape of the lens (lens deformation) is called *accommodation* and permits focusing at near when accommodating and at distance

when the lens is not accommodated. Some animals, such as fish, accommodate by changing the position of the lens instead of deforming it. Some fish accommodate for distance—that is, they move the lens to focus at distance, because at rest they are already focused at near, and some fish reverse this activity to accommodate for near.

The retina consists of closely packed cells called *photoreceptors* that line up in a distinct layer with an increased concentration of these cells posteriorly in an area called the *macula* (Figure 26.2). The macula contains the highest concentration of photoreceptors and represents the area providing the best vision. In the very center of the macula, there is an area of acute vision with the maximum concentration of photoreceptors in the eye called the *fovea*. The fovea forms a small pit that permits all other cells in the retina to be drawn away from this area of critical vision (Figures A2–A4). This is the area used for the best, most critical vision—for example used for reading this sentence. Nearby, but not in the macula, is the site through which the optic nerve exits the eye. This is called the *optic disc* and has no photoreceptors on top of it, so the eye does not register any light rays that strike the optic disc. Hence, this is the blind spot of the eye (Figure 26.2).

The retina is a complex structure consisting of the aforementioned photoreceptors as the outermost layer. That is, these cells are immediately adjacent to the choroid but not the inner portion of the eye (Figures A2 and A3). Furthermore, the photoreceptive end of these cells is the outermost portion of the retina and the furthest from the inside of the eye. This means that in vertebrate eyes, the photoreceptors point *away* from the lens—they are inverted. Incoming light waves that create the image must penetrate all layers of the retina and the entire photoreceptive cell to reach the light-sensitive portion of the cell.

The retina consists of three basic cellular layers. In the outermost layer (furthest away from the incoming light), the photoreceptors receive the photons of light and turn that light energy into electrical energy. The electrical impulse then travels to a second layer of cells called the

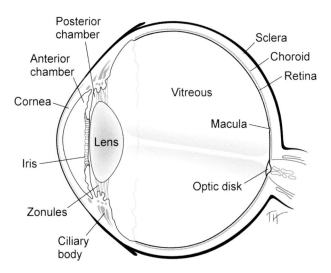

Figure A1 Human eye. The text expands on the diagram. The plan is a basic human eye. *Art by Tim Hengst.*

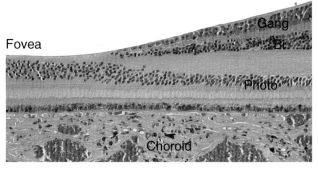

Figure A3 *Homo sapiens.* Human retina and fovea. This is a magnified view of Figure A2 showing only one half of the fovea to better illustrate the detail. Fovea, one half of the area within the retina where focusing is concentrated for best visualization. Note foveal depression is a mild slope but not steep like those of some birds and reptiles. Photo, photorecptors consisting of rods and cones; Bi, bipolar cell layer consisting of the nuclei of the bipolar cell layer; Gang, ganglion cell layer with nuclei prominently seen. *Image by Ralph C. Eagle, Jr, MD.*

bipolar cells or bipolar layer. The bipolar layer receives the impulse and transfers it to the third and final layer of cells, the ganglion cells. The ganglion cells receive the impulse and transmit out of the eye to the brain through the "tail" of the cell called an axon. The axons of the ganglion cells are long and extend into the brain to carry the signal to various brain structures (Figures A2–A4).

The bipolar layer also contains other cells including the amacrine and horizontal cells. These cells, as well as the bipolar cells, function to distribute the signal to other related portions of the retina that help interpret the signals even at the retinal level (Figures A2–A4).

The optics of the eye is complex but basically consists of the front surface of the cornea, the lens, and, in some animals, the fovea. In terrestrial animals the cornea is steep and relatively small compared to the remainder of the globe.

It creates a smooth rounded surface that will bend, or refract, any light ray that strikes it. This incoming ray is refracted because the corneal tissues slow the light ray as compared to its speed of in air, and, combined with the curvature of the cornea, will cause the light ray to change direction. This is called *refraction*. As the light ray continues into the eye, it strikes the lens of the eye. The lens refracts the light rays further and directs them to be focused on a point on the retina.

Is some species, as light is focused on the fovea, there may be further refraction at the retinal level. In certain birds and reptiles, the fovea contains very steep walls,

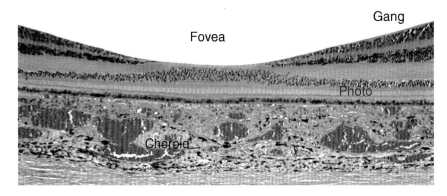

Figure A2 *Homo sapiens.* Human retina and fovea. The human retina consists of layers and is described in more detail in the text. Fovea, the sloped depression, which is the concentrated area of photoreceptors. Note that the other cells of the retina are pulled away from the fovea. Photo, photoreceptors consisting of rods and cones; Gang, ganglion cell layer with nuclei prominently seen; Choroid, the middle layer or middle coat that consists of blood vessels for nutritional support. *Image by Ralph C. Eagle, Jr, MD.*

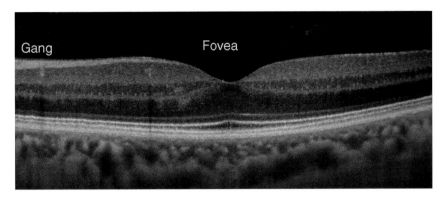

Figure A4 *Homo sapiens*. Human retina and fovea. Optical coherence tomography (OCT) image of a living human fovea. *Image courtesy of Jack Warner, PhD.*

creating a distinct pit (Figures 15.20B and 20.25). The index of refraction of the retina is greater than that of the water-like vitreous. Hence, the light rays that enter the retina from the vitreous will slow and diverge slightly. This results in further refraction of the light—but in such a way as to magnify the image. The pit of the fovea in mammals, however, is so mildly sloped that it does not create much refraction, although in other animals such as birds it can be more substantial.

In many vertebrates the iris creates a circular opening called the pupil that becomes smaller, or constricts, with light stimulus and opens, or dilates, with darkness. Some animals have slit-shaped or other oddly shaped pupils that are discussed in the text.

The vertebrate eye is in a bony protective "case" within the head called the *orbit*. The external eye beyond the cornea is covered with a mucous membrane called the *conjunctiva*, which lines the inside of the lids as well.

Many vertebrates have three "lids" with the two outer lids being similar to those of the human eye. In humans and other terrestrial vertebrates, there are lacrimal glands that produce tears beneath the upper outer lid. The location of these glands varies in terrestrial vertebrates. Many vertebrates, other than humans, possess a third eyelid called the *nictitans* (humans have the barest of remnants of it—the semilunar fold; Figure 26.1). In most animals, the *nictitans* is drawn across the eye from the nasal side to the temporal side. In many vertebrates there are accessory tear glands, called Harderian glands, that either line the nasal orbit adjacent to the *nictitans* or are housed independently on the nasal aspect of the eye. The Harderian glands produce sebaceous secretions as opposed to more aqueous secretions from the lacrimal gland. Many mammals, including primates, have a Harderian gland, but humans do not have even a remnant.

Muscle Distribution

In most vertebrates, there are three pairs of muscles positioned to be in opposition to each other to provide elevation or depression of the globe (superior and inferior rectus), horizontal adduction or abduction (medial and horizontal rectus), or clockwise and counter-clockwise torsion (superior and inferior obliques, although the rectus muscles participate in this action to some extent). Torsional movements are circular, wheel-like rotations of the eye. All movements of the eyes, including torsional ones, are innervated from the brain in such a manner as to coordinate both eyes working simultaneously.

By comparison, spiders have six extraocular muscles, many octopi have seven, and squid have more, up to twelve or thirteen, but they are arranged differently. In no way, though, are these muscles of the spider or the cephalopods homologous to those of vertebrates. Extraocular muscles in unrelated invertebrates and vertebrates display convergent evolution and emphasize the importance of such structures in all sighted species.

Extraocular muscles appear in the oldest extant vertebrates and probably were present in the earliest fish, although not in the same pattern as found in most vertebrates. It is not understood exactly how or when these muscles evolved, although some facts are known. The development of these extraocular muscles is a crucial and illustrative story in the neurologic development of movement and balance.

Why Extraocular Muscles?

Movement of the eye in vertebrates is essential to enable predators to detect and track prey. Furthermore, when any aquatic species is hunting in the water column, that predator must keep the image vertical and in the same orientation despite any movement of its own body. This requires a complex muscular system. But that is probably not the main reason the muscles evolved.

The muscular system evolved to prevent the image from fading on a set of photoreceptors. If an image persists indefinitely on a set of photoreceptors, those cells will fatigue, and the image will fade. Slight movement of the eye will prevent such fatigue. In humans and other primates all extraocular muscles participate in this fine movement, called microsaccades. The sum of the movement includes to-and-fro movement to keep the photoreceptors refreshed. Alternatively, or additionally, eye movements would stabilize the image of the prey species on the predator's retina and prevent the confusion of a moving surround. These muscles keep the eye aligned much as a gyroscope would stabilize a ship. Movement of the body would otherwise cause an image to move on the retina, leading to confusion, especially if the animal were in water. Gaze stabilization is essential for an eye to produce a usable image.

Muscle Evolution

The oldest extant vertebrates, lampreys, have six extraocular muscles, but these are not distributed in the same pattern as extraocular muscles in most other vertebrates. Lampreys have a pair of muscles, equivalent to the superior and inferior rectus in humans, to elevate or depress the eye. Lampreys have a rectus muscle to move the eye laterally, but no medial rectus. They have a retractor bulbi which inserts into the eye posteriorly and which functions somewhat like a medial rectus in some respects, but it is not precisely a medial rectus. The retractor bulbi is enigmatic, and most jawed fish do not possess one. Evolutionarily, it is possible that the retractor bulbi may have radiated into the medial rectus (used to move the eye inward toward the nose), but connections to the brain are not the same. It probably did not happen in this manner.

Medial Rectus

The medial rectus muscle opposes the lateral rectus moving the eye toward the midline of the animal, essentially, toward the nose. The nasal movement of eyes is not present, phylogenetically, until the jawed fish, such as the placoderms (Chapter 9), followed by the elasmobranchs (sharks, rays, and skates). Both of these groups are now, and have always been, predators. The medial rectus appeared as the eyes moved forward on the head, presumably to permit forward sight with binocular visualization of prey.

The placoderm lineage is believed to be the first fish with the basal distribution of seven extraocular muscles, including the medial rectus and the retractor bulbi. The bony capsule of the placoderm eye provides an understanding of the state of extraocular muscles of the jawed fish of that time.

Superior Oblique

Lampreys have a superior oblique and an inferior oblique for torsion. The superior oblique, though, is not configured in the same manner as the mammalian superior oblique. For fish this muscle is important for balance and the correct vertical positioning within the water column. Placoderms and all bony fish have a superior oblique, but it is anatomically different from that found in mammals. In fish the superior oblique originates from a site nearer to the internal rim of the orbit, although this varies from species to species. This is important for balance and correct vertical positioning within the water column.

The superior oblique muscle and its innervation from the brainstem comprise an important step in development. Placoderms and lampreys have a superior oblique muscle that originates toward the posterior orbit, but sharks are the first species, phylogenetically, to have the superior oblique originate in the superior (dorsal) *anterior* orbit as it does in all other living fish. Placoderms would not have had the same degree of ocular torsion possessed by more derived fish because of this muscular origination. Hence, they would not have had the same accuracy of balance and vertical positioning as their competitors. The placoderms were thought to have been primarily or predominantly benthic, or bottom dwellers and may not have needed the improved balance and vertical positioning offered by this change in anatomy. Placoderms, for the most part, were devastatingly effective predators, but they did not survive long. Perhaps, the more posterior origination of the superior oblique was not as favorable for balance and vertical positioning, and this difference contributed to their demise.

The elasmobranchs, however, needed the ability to remain parallel to the ocean floor below them and the sea surface above them. This change in niche likely precipitated the change in musculature. The superior oblique had changed positions, and this permitted better torsion, albeit somewhat differently from mammals. This change in position may have been a necessary improvement to help the elasmobranchs survive.

The innervation of these muscles of torsion is peculiar in all jawed vertebrates. When the head tilts obliquely and the eye is brought to the vertical position again by these muscles of torsion, there are four muscles, two muscles in each eye, involved in any one simple torsional movement. The rather complicated innervation of all four of these muscles comes from approximately the same point in the brainstem. This requires that some of the innervation cross from one side of the brainstem to the other, and some of the innervation remain on the same side of the brainstem since both eyes must tort (twist on a torsional basis). All of the other eye muscles and eye movements are innervated by nerves coming from the same side of the brain as the eye. It is not clear, evolutionarily or neurologically, why this has occurred in this manner, although it does permit close cooperation of the muscles that cause the eye to tort. Clearly, the evolution of torsional movements of the eye is important enough to require close coordination of the eyes with the vestibular system to create a gyroscopic-like control. This is the vestibulo-ocular reflex and is important to image stabilization on the retina in a three-dimensional environment.

There is another clue that might help explain this. The third semicircular canal to further assist in balance also appears at about the same time, evolutionarily, as the full complement of extraocular muscles. Lampreys have two semicircular canals, and placoderms are the first fish to have three semicircular canals. The repositioning of the superior oblique extraocular muscle and the full complement of semicircular canals provided the elasmobranchs with a predatory advantage that would permit full maneuvering in the vertical water column. This should be no surprise, as the evolution of any sensory system is usually coordinated with the evolution of its neurology.

Retractor Bulbi

As additional jawed fish arose from the vertebrate lineage after the placoderms, the retractor bulbi disappeared in most lineages but not in the lobe-finned fish. The extant lobe-finned fish, the lungfishes and the coelacanth, retain a muscle that is homologous to the retractor bulbi. It is likely that the lobe-finned fish that radiated into the tetrapods had a retractor bulbi or a homologous muscle that became the origin of the retractor bulbi muscle in the tetrapods. Although it has been though several evolutionary rearrangements over the millions of years since the placoderms, the retractor bulbi or its homologue is found in sauropsid lineages such as the reptiles and birds as well as the synapsid lineages such as the mammals.

Trochlea

In mammals there is a pulley system, called the trochlea, in the upper nasal quadrant that permits the superior oblique to begin in the *posterior* orbit yet effectively and mechanically rotate the eye from the anterior position (Figure B1).

The trochlea does not appear until the Cretaceous (145–65 mya) in monotremes, the platypus and the echidna.

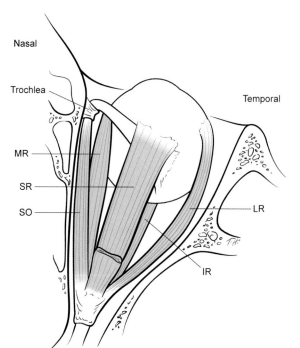

Figure B1 Human orbit as seen from above. Trochlea, trochlea; MR, medial rectus; SR, superior rectus; SO, superior oblique; IR, inferior rectus; LR, lateral rectus. The inferior oblique is not drawn and would be poorly visualized from this perspective. The muscular stub at the apex of the orbit seen in the lower medial portion of the drawing is the stub of the cut levator palpebrae superioris muscle, which raises the upper lid. *Art by Tim Hengst.*

These are the only extant representative members of the monotremes and are the "oldest" living mammals (Chapter 22). The platypus is the "older" of the two and has two slips, or two origins, of the superior oblique—one from the anterior orbit and one from the posterior orbit—as if in an intermediate stage of development. By the time of the next radiation that included the echidna and then later the marsupials, the trochlea and the superior oblique were in their current mammalian positions. The platypus is the only living mammal that has a variation from the mammalian superior oblique and trochlea pattern, and this variation is a subtle one. This is, perhaps, a transitional stage of muscular alignment.

Appendix C

Evolution of Retinal Vascularization

Early Vertebrates

Lampreys or closely related animals were among the first vertebrate species (Chapter 7) possessing eyes. It is not known exactly how this early ancestor managed retinal nutrition because vascular tissue does not fossilize, but lampreys are still with us, providing some clues, and appear not to have changed much in morphology. But they have had 500 or so million years to evolve, so their physiology is undoubtedly different from that of the original creatures of that lineage.

These most basal living vertebrates, lampreys, have nutrition supplied to their retina by diffusion from the choroid immediately underlying the retina (see the description of the human eye in Appendix A) (Figure C1). The choroid is a vascular layer and in some animals it supplies the retina by diffusion, as the retina itself has no direct vascularization. In the lamprey the rather primitive choroid consists of sinuses containing lakes of blood without the capillary structure usually seen in the choroid of more derived animals (Figure 7.6).

More Derived Fish

Elasmobranchs, sharks and rays, radiated from the lineage that had given rise to the lampreys. They have a choroid that consists of a rich blood supply with fine capillaries and embedded blood vessels that is somewhat more sophisticated than that of the lampreys. The now-extinct placoderms had a more sophisticated system, yet again. It included an embedded choroidal gland (Figure 9.1). The choroidal gland is a more organized capillary plexus once believed to be glandular but now known to be responsible for retinal nutrition as the retina became more mature and probably more demanding (Figure C1).

The choroidal gland persists into the bony fish (ray-finned and lobe-finned), but other nutritive systems for the inner retina appeared too. As the bony fish became more derived, the retina thickened as additional cellular components were added. These cells would have additional nutritive requirements, which stimulated the appearance of vascularization into the cavity of the globe, probably by attracting the vessels into the interior with cellular growth factors. In many of the teleosts these vessels within the globe first appeared in the form of a thin pigmented vascular plexus called the falciform process (Figure 10.1). This is a direct projection from choroidal vessels into the vitreous through a cleft in the entire eye wall called the fetal fissure. These vessels penetrate the eye during embryology before the forming eye can completely close this linear opening in the eye wall. Not all fish, however, have this structure. The falciform process proved to be insufficient for inner retinal nutrition as the evolving retina demanded more nutrition. The evolutionary response was the *membrana vasculosa retinae,* which is a dense preretinal vascular plexus distributed over the surface of the retina (Figure 10.17). This consists of a single vessel or network of vessels that come through the optic nerve from the neural crest and are known as hyaloidal vessels. This is in contrast to the falciform process, which is derived from the mesodermal vessels of the choroid. The *membrana vasculosa retinae* penetrates into the interior of the eye after the fetal fissure has closed, indicating that it is a more derived or mature physiologic step. The neural crest is the same tissue that spawned all of the neural components of the eye. Blood vessels from the neural crest are embryologically different from the choroidal system vessels. This hyaloidal network of vessels covers the retinal surface in some fish to supply the inner retina but does not generally send individual vessels or capillaries to penetrate the retina. This vascular system represents a new evolutionary step beyond the falciform process.

Tetrapods

The living lobe-finned fish that represent the lineage that radiated into the tetrapods have only a choroidal gland but no hyaloidal system. Extant amphibians, such as frogs, did develop a hyaloidal system, although it is not clear if the early amphibious tetrapods did so or not. So the tetrapods probably inherited a choroidal gland from their lobe-finned ancestor, and perhaps a falciform process or hyaloidal vascular system of the teleosts, although this is unknown. The extant lungfish have neither.

PHYLOGENY of RETINAL VASCULARIZATION

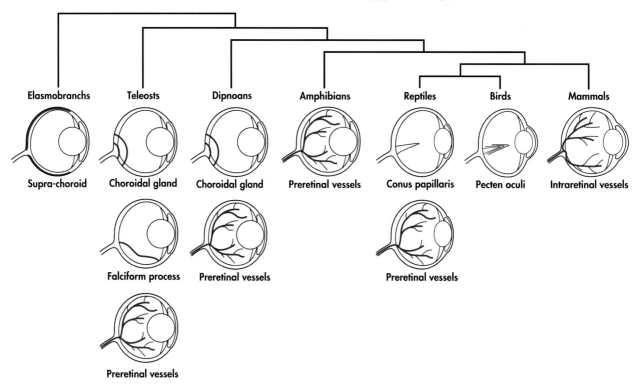

Figure C1 Cladogram of vascularization. This shows the descent of retinal vascularization. This is also discussed in Chapter 15.

The vascular morphology of a hyaloidal vascular system appears to have been passed to, or developed in, the anurans because frogs exhibit this form of preretinal vascularization as well. The hyaloidal vessels are the precursor to the central retinal artery, the principal artery providing nutrition for the inner retina in reptiles, birds, and some mammals. The central retinal artery *replaces* the hyaloid system in these animals. The hyaloidal system appears during embryology in each of these three lineages but is replaced by the central retinal artery by the time of birth.

Sauropsids

As the amphibious tetrapods evolved into the protoreptile that would later diverge into the sauropsids or "reptilian" line including turtles, crocodiles, lizards (squamates), and eventually birds, and the synapsids, which would eventually lead to the mammals, there would be further nutritional demands. As these two lineages (sauropsids and synapsids) emerged, a single core vessel coming through the optic nerve would appear in both groups. The central retinal artery or its homologous systems arose anew to replace the hyaloidal system of the early tetrapods.

As the reptiles arose, diverging into different clades, they developed a plexus of vessels, a *conus papillaris*, from the central retinal artery (Figure 15.21). But in some of the reptilian clades, there is little representation of the conus. For example, the turtles and crocodiles have only a faint pigmentation on the surface of the optic nerve head but no true intraocular vascular system (Figure 16.13). Perhaps in earlier, now-extinct species, this was present but atrophied. Nutrition in these two clades, turtles and crocodiles, still comes from the choroid by diffusion.

A major clade occurring within the squamates, the snakes, do not have a conus or other intraocular representation of the central retinal artery. Most snakes rely on the choroid for retinal nutrition with diffusion, but some snakes have a network of vessels spreading across the retinal surface much like that seen in fish. Some snakes even have a similar nutritive organ as that found in reptiles, also called a "conus." But this system is embryologically derived from mesoderm and not from neuroectoderm like the reptilian conus. In snakes this system is entirely mesodermal and is consistent with the "reevolution" of the vascular system after it was lost when the predecessors became fossorial (Figure 19.3). So despite similar morphology, the serpentine "conus" is quite different from the conus in other squamates, at least in derivation. It is unique and was incorrectly called the "conus" by initial investigators who did not understand its derivation. It is closer to that structure of fish—the falciform process––which is "squeezed" from the choroid. After reemergence from their fossorial journey, the serpents had to find a way to nourish their inner retina, and this system returned after being discarded by other lizards.

The reptilian lineage eventually radiated into the theropod dinosaurs, among other lineages, and these dinosaurs radiated into the birds. The avian retina is nourished by a vascular structure homologous to the reptilian conus—the pecten (Figures 20.5, 20.12, and 20.37). In birds, the pecten is similar to, but somewhat more derived than, the conus found in reptiles. Yet, they are similar embryologically, as both come directly from a central retinal artery. Birds do more of their visual processing in their retinae than other animals and, as a result, have a considerably thicker retina. This requires more nutrition and hence a larger pecten with more surface area through a more convoluted and pleated structure.

Synapsids

Mammals radiated from the synapsid lineage, which had separated from sauropsids approximately 325 mya. So this vascular plexus seen in the eyes of reptiles (conus) and birds (pecten) could not evolve in mammals—at least not in the same manner. It had not been present when the synapsids and the sauropsids split. A new method of retinal nutrition would have been needed for the mammals.

It is not known how those early synapsids of the Carboniferous 325 mya supplied nutrients to their retinae. But if the retinae of monotremes and most marsupials are the closest we can get to that last common ancestor, then the early synapsids had thin retinae with little or no vascularization. Alternatively, there may have been more complex vascular systems that resembled those of fish, such as the falciform fold or the epiretinal plexus, but these may have atrophied as the early mammalian lineage became nocturnal.

The monotremes (platypus and echidnas) have only their choroid for retinal nutrition. Most of the marsupials are similarly equipped. Some of the marsupials, though, do have direct vascularization of the retina with the central retinal artery in a manner much more intimate than via a conus or pecten. The North American opossum has paired retinal vessels that arise from the central retinal artery (Chapter 22), although in a pattern different from those vessels found in placental mammals. This suggests that the genetic tools for the central retinal artery were available in the lineage, and it began to appear within the early mammalian lineage. Within the placental mammals, there are several variations within the central retinal artery systems leading up to primates. As the mammalian line expanded and became established in the Eocene, nutritional demands would beckon vascularization to the retina. Placental mammals refined intraretinal vascularization, as the opossums could be considered the first extant mammals, phylogenetically, to have intraretinal vessels (Figure 22.11).

Placental Mammals

Placental mammals have varied mechanisms of retinal vascularization that fall into four basic patterns (Table C1). The first pattern is no blood vessels whatsoever, called an anangiotic retina. This requires that all retinal cells must obtain the necessary nutrients from the middle layer, or choroid. The eyes of some rodents and some ungulates are prime examples. Our shrewlike ancestors probably also relied on an anangiotic retina.

The second pattern reveals blood vessels within the retina only around the optic nerve, called a paurangiotic pattern (Figure C2). This pattern can found in mammals such as horses, elephants, and guinea pigs and might have been the first phylogenetic attempt at retinal vascularization in mammals. As the retina became more complex, the first location to require more nutrients and oxygen would be the optic nerve head.

The third pattern of vascularization of the retina is localized retinal blood vessels in specific locations, but not

TABLE *C1 Mammalian Retinal Vascularization*

Intraretinal Blood Vessel Pattern	Examples of Animals
Anangiotic	Monotremes including echidnas, platypus, and many marsupials as well as many placental mammals such as rodents
Paurangiotic	Horses, elephants, others
Merangiotic	Lagomorphs—rabbits, hares, and pikas
Holangiotic	Carnivores and primates; a few others such as the North American opossum
Special example	Fruit bats—Anangiotic retina but peglike vascularization

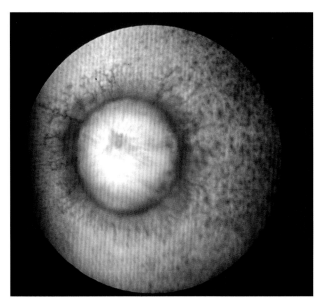

Figure C2 *Elephas maximus indicus*. Elephant fundus. Paurangiotic retinal vascular pattern. Note the vessels just seen around the optic nerve head. *Image by Nedim Buyukmihci, VMD.*

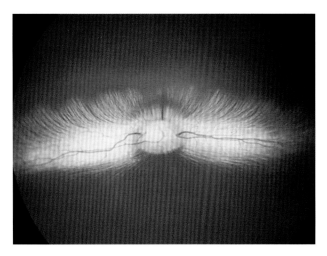

Figure C3 *Oryctolagus cuniculus*. Rabbit fundus. Merangiotic retinal vascular pattern. Note that the vessels radiate from the disc or optic nerve. The white feather-like structures are the individual fibers of nerves that have sheaths around them, called medullated nerve fibers. The sheaths only extend part way along their journey to the adjoining cells of the retina. *Image by Simon Petersen-Jones, DVetMed, PhD, Professor, Comparative Ophthalmology, Michigan State University.*

in the entire retina—called a merangiotic pattern. The rabbit exhibits this pattern (Figure C3) and shows that, as the retina became more complicated, the vessels followed the nutritional requirements.

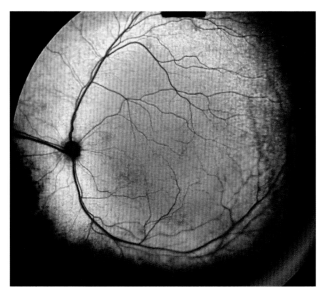

Figure C4 *Felis catus*, domestic cat. Holangiotic retinal vascular pattern. Note that the vessels go throughout the retina, although the larger vessels spare the area in the central portion of the fundus called the macula. Note the unusual color caused by the tapetal reflex. *Image by Simon Petersen-Jones, DVetMed, PhD, Professor, Comparative Ophthalmology, Michigan State University.*

The fourth classic pattern of retinal vascularization includes vessels throughout the entire retina and is called holangiotic. This is found in carnivorous mammals and primates (Figures C4 and 26.2). Certainly as visual requirements became more pressing, vascularization would have to follow. Carnivores were later additions to the mammalian clade and had a more mature vascular system to include blood vessels through the entire retina.

Primates have a central retinal artery that spreads to a capillary system throughout the retinal thickness and is the most evolved and elaborate system among the mammals.

These patterns suggest that as mammals matured and diverged, the retina became thicker, and the nutritional requirements became more demanding. The vascular system responded with increasing penetration into the retina from within the vitreous cavity.

But there is at least one notable exception within the placental mammals. Fruit bats are an exception, and their story is illustrative (Chapter 23). The clever method of supplying retinal nutrition to the inner retina in fruit bats suggests that evolution remains capable of finding new and unique solutions to perplexing anatomical problems.

Safeguarding Precious Contents

The cornea appeared initially for protection as part of an outer layer of a sac or cavity with the sclera forming around the remainder of the eye. These outer walls provided protection for the relatively fragile tissues within the eye. Once the ocular tissues were exposed to the external environment, an outer clear and tough layer would protect them and yet provide access to light. Even invertebrates have developed a cornea as protection, like a clear shell around the front of the eye. The cyclostomes (agnathans including the hagfish and lampreys) have a primitive cornea. To lump the cyclostomes into a single group, though, misses some interesting subtleties of phylogeny of the cornea.

Hagfish are craniates and not true vertebrates. As representatives of the early less-derived chordates, they have a primitive layer that is not a true cornea but is a primitive extension of the sclera much like a shell surrounding a precious seed. An extension of the sclera, this primitive cornea is separated from the overlying skin completely and moves freely beneath the skin (Figure 7.4). Although the lampreys are cyclostomes too, they are considered the most basal vertebrate, and their cornea is a bit more derived or "mature" but can be seen as two distinct layers of clear tissue (Figure 7.6). The outer layer is skin or ectoderm and covers the clear primitive cornea in a manner similar to that seen in hagfish. But in lampreys, this outer layer of ectoderm is now somewhat more intimately involved with the primary cornea and is quite clear. This second layer of skin is called the primary spectacle. The globe still moves freely beneath the primary spectacle in lampreys with only a little connection to the overlying skin.

The jawed fish developed thicker and more protective corneas, but these could not be used for refraction. In any fully aquatic animal the cornea is of no use for refraction or focusing the light rays because the index of refraction of water is virtually identical to that of a cornea. Hence, light passes directly through the cornea as if it were water. As the jawed fish developed and evolved, the eye migrated more externally, eventually to attach to the overlying ectoderm,

creating a more united single cornea. As the eye migrated more superficially, that layer of overlying ectoderm disappeared evolutionarily, and a sulcus formed around the eye like a moat. A somewhat redundant conjunctiva formed to line the surface and the sulcus. This permitted the eye to move using the redundant tissue much like pleats. This step also defines the cornea in jawed fish.

This primary cornea, though, was not enough protection, so a second layer of corneal tissue appeared in jawed fish. Embryologically, a second wave of ectodermal tissue formed over the primary cornea. These two layers of cornea came together with varying degrees of unification, and, as a consequence, this created a potential space. There was little impetus for this space to be tightly compacted in fish. In some fish it remained separated, creating what is known as the secondary spectacle. Some fish, such as the pufferfish, use this potential space for the migration of pigment though fine channels to shield their eyes from exposure to bright light—like putting on sunglasses.

Neither of these two corneal layers is particularly smooth, and it does not have to be. The contact with water eliminates any distortion from the cornea. This would create a problem for any fish that navigates the aerial world in any meaningful visual manner, however.

What do such creatures do to overcome the generally poor visual characteristics of the lackluster piscine cornea? As fish became more aerial or even terrestrial and required better distance vision, the cornea had to become clearer and more compact. That is exactly what happened as fish came ashore and developed into ancient tetrapods. These lumbering creatures are gone, but amphibians help us understand how their eyes must have developed.

Frogs are instructive and suggest how the tetrapods may have addressed the problem of two corneal layers. Tadpoles have corneas that resemble those of fish (Figure 13.12). There are two distinct layers that resemble those of fish and are not particularly smooth. During metamorphosis, the layers unite permanently, and the cornea becomes smoother and steeper (Figure 13.12). This change provided for corneal refraction and would improve image quality in an aerial environment.

Terrestrial vertebrates, then, have a united single-layered cornea, but echoes of the watery beginnings remain. In primates (far removed from the bony fish with two separate corneal layers), there is a single corneal structure including a surface epithelium, a middle stromal layer, and an inner layer of endothelium. Within the cornea, though, we can find two distinct domains in the stroma. The anterior 80 percent or so has a different arrangement of collagen and a different concentration of two major components (dermatan sulfate and keratan sulfate) of collagen as compared to the posterior 20 percent. These are the echoes of our watery past played out in our own corneae. These different domains are embryologically different and are handed down to us from those fish with two separate layers.

Not all terrestrial vertebrates have this morphologic pattern, however, and the principal exception is worth mentioning. Snakes and some lizards have fused clear lids overlying their single-layered cornea. There is a real space between the clear fused lids and the cornea, which, in snakes at least, usually contains an oily lubricant secreted by the Harderian gland. This extra layer is called the tertiary spectacle and a portion is shed when the snake sheds its skin. Once it has been shed, the snake will promptly grow another (Figure 19.1). The fused lids likely represent this lineage's response to a fossorial lifestyle prior to reemerging as terrestrial animals.

The evolving corneal morphology and physiology reveal that ocular development responds to niche requirements and that evolution must build on previous anatomy rather than recreate wholly new structures.

Appendix E

ACCOMMODATION

Accommodation, or the process of changing focus between distant objects and near ones, is accomplished by a variety of mechanisms in different groups (Figures E1 and 10.8). Variation in the mechanisms of accommodation suggests that the best method depends on the individual niche to be filled. There are at least eight basic methods of accommodation, depending on how these are classified (Figure E-1 and Table E1). The compound eyes of insects accommodate little, if at all, but they need not do so to see well for most tasks. The camera-style eye of most vertebrates, on the other hand, does accommodate if its owner is to see both distant and near objects.

Different creatures use different strategies—there is little unifying technique or style. *Geotria australis*, one of the lampreys of the southern hemisphere, has a rather large extraocular cornealis muscle, which can change the curvature of the cornea, moving it backward (or inward) to accommodate. The lamprey cornea is not in direct contact with the surrounding environment, as there is an overlying layer of ectoderm. So changing the radius of curvature of the inner layer of the cornea would change the focus of the incoming image. But this system did not persist in fish, and this is not a robust method of accommodation if it truly works. The lamprey system is unique and is not repeated in other living animals, although one other fish, the sandlance, does have corneal accommodation.

The sandlance (Chapter 10) has a striated *retractor cornealis* muscle that can quite rapidly accommodate up to 180 diopters. This works because there is a lenticule within the cornea. This method of accommodation is unique to this fish.

Cartilaginous fish such as the elasmobranchs accommodate by moving the lens forward by using an intraocular muscle called a *protractor lentis* (Figure 10.8). This muscle moves the lens away from the retina toward the cornea when the animal wishes to focus for near. Elasmobranchs, then, are neutral (emmetropic) for distance and must accommodate for near. This makes ecological sense because sharks are generally more distant predators searching the reef, or even open ocean, for larger prey species. Near vision is not required as frequently, as it is needed only during "close work" such as a kill.

Most ray-finned fish (most living bony fish) must have the capability to change focus as well, but the system is exactly the opposite of that of the elasmobranchs. Ray-finned fish have a *retractor lentis* muscle, which moves the lens *away* from the cornea and *toward* the retina (Figures 10.8 and E2). The muscle is sometimes attached to the falciform process (Appendix C). The hard spherical lens has a relatively high index of refraction, fills much of the globe, and almost always adheres to Matthiessen's ratio (Chapter 10). Optically, then, the bony fish accommodate for distance and are emmetropic (neutral for vision and in focus at rest) for near in the exact opposite manner as the elasmobranchs. This would hint that the last common ancestor of bony fish and elasmobranchs did not accommodate for either distance or near. Other explanations are possible, though, as the radiation into bony fish and elasmobranchs may have favored different mechanisms for accommodation.

Accommodation is a very "plastic" process, and different mechanisms could have evolved rather easily. Nevertheless, because the lamprey radiated from that same common ancestor or a closely related animal, the basal state was probably without accommodative mechanisms.

Lens movement, however, does not produce as accurate an image above water as other systems do, so a new system would arise once the vertebrates gained ground in the Devonian.

The advantages to a terrestrial as compared to an aquatic eye for an improved image are considerable but require that the cornea be responsible for much of the refractive ability of the eye. A steeper cornea would provide a brighter, larger image and a clearer one as well. Since the cornea has taken over the majority of refraction of the image, the lens no longer has to be so large and spherical like a marble, as it is in fish. Hence, the lens of terrestrial animals, such as reptiles, could become more tapered and oval (Figures 15.10 and 15.12), providing for different mechanisms of accommodation. Instead of movement of the lens, terrestrial and aerial vertebrates have chosen to change the shape of the lens—called lenticular deformation.

ACCOMMODATION of VERTEBRATES

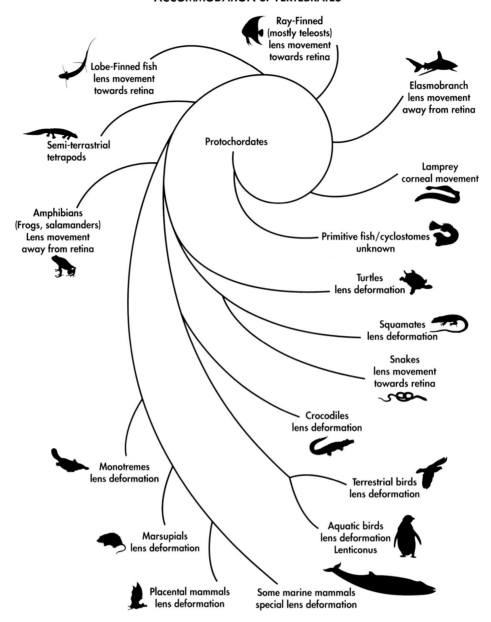

Figure E1 Descent of the mechanisms of accommodation as vertebrates radiated in different lineages.

This change to lenticular deformation does not occur in amphibians, presumably because they remain wedded to water for at least part of their lives. Interestingly, the extant amphibians accommodate by moving their lens forward, much like elasmobranchs, not posteriorly like bony fish.

It was not until the reptiles arose that lenticular deformation became key to accommodation, and then in only the more-derived reptiles. Accommodative mechanisms differ among the various reptilian lineages, and these differences provide some clue as to phylogeny.

That last common ancestor—protoreptile––that radiated into all terrestrial vertebrates diverged into the

synapsids (mammal-like reptiles) and the sauropsids (lizard-like reptiles). The eye would predict that this occurred very early in the course of reptilian development and radiation. The eyes of these two clades are different enough to suggest that the change in accommodative mechanisms occurred *after* the divergence into these two lineages. It is entirely possible that the early protoreptile retained the piscine-like method of accommodation and that the radiation into these two different lines of "reptiles" occurred just as protoreptile was shaking the mud off its hind leg as it climbed out of the ooze of being an amphibian.

Cornealis muscle—lampreys

Retractor cornealis—Sandlance

Retractor lentis—bony fish accommodates for distance

Protractor lentis—elasmobranchs-accommodates for near

Lenticular deformation several methods

    Accommodative lenticonus—turtles, birds, some marine mammals

    Ringwulst—birds, reptiles

    Smooth muscle—synapsid lineage

    Striated muscle diapsid and anapsid lineage

Lenticular deformation and lens movement—transversalis
turtles and chameleons.

Corneal curvature change—birds

Lens movement squeezing vitreous by stiff iris—snakes for near

Lenticular deformation—mammals

The synapsids radiated into Dimetrodon and other large dinosaur-like animals that ruled the earth for a period of time in the mid- and late Triassic (251–208 mya) but eventually lost that dominance and radiated into mammals, perhaps as a small furtive creature. Most mammals accommodate by lenticular deformation, as do the various clades of reptiles. But most mammals do so by using smooth muscle and fine fibers attached to the lens, whereas other reptilian clades use striated muscle, which has direct contact and applies direct pressure to the lens (Figures 10.8 and 16.6A). Striated muscle is a better choice, as is direct contact—these are faster, more accurate methods, and striated muscle can be evolved to fatigue less. Once the striated

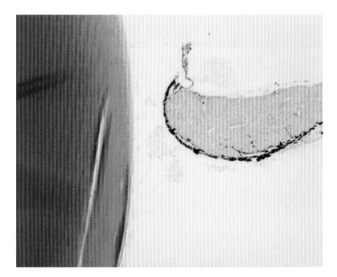

Figure E2 Retractor lentis muscle of a walleye fish. In preparation it has pulled away slightly from the lens capsule. In life it was attached to the lens to move it as described in Chapters 9 and 10. *Image by Richard Dubielzig, DVM.*

musculature had been achieved for accommodation, it was not likely that smooth muscle would ever win back its lost tasks. Hence, it is probable that the protoreptile that radiated into the synapsids did so *before* the muscles of accommodation were established because otherwise smooth muscle would never have become established.

Other explanations are possible to be sure. It is possible that mammals forsook the striated muscles during their long journey into night, because accommodation, color vision, and certain other visual functions would have been less important at night. But to switch from striated to smooth muscle for accommodation would have been an unusual "step back." It is unlikely that we will know for certain, and it is not so important to the eye, except that something like this did occur to account for the basic differences between the mammalian lineage and the various reptilian lineages.

Reptilia, however, are a diverse group, and that lineage diverged and radiated into many different classes, probably before accommodative mechanisms were fully developed. As a result, there are some peculiarities within the various groups of reptiles, as these lineages would evolve further refinements in the process of accommodation.

Lenticular deformation works better and faster when the lens is smaller and more ductile. The reptilian lens is very soft and ductile and is easily molded to become an efficient optical element. The muscles of the ciliary body are striated and fast, creating a quick response from the lens, providing quick accommodation and, hence, focus. To assist in this accommodation, some aquatic turtles and birds have a very stiff iris that is used to assist the ciliary body in accommodation. The ciliary body muscle "squeezes" the lens and forces a small button of it through the pupil. This button of lens creates anterior lenticonus, which perhaps would be better designated as accommodative lenticonus (Figure 20.9). In essence, this creates a much steeper, rounder lens capable of greater refractive abilities—much more like a piscine lens. This process is helped by a rigid equatorial beltlike structure around the lens that effectively makes the lens even smaller because this beltlike structure does not deform. This structure is called a *ringwulst* or annular pad and is enlarged and improved further in birds (Figure 20.26), but many reptiles, including the basal tuatara, have the prototype.

There are some variations in turtles and lizards as the muscular arrangement also includes a muscle called the *transversalis*. This muscle pulls the lens toward the nasal aspect of the globe for those animals that have binocular stereopsis, such as turtles and chameleons. Some turtles, such as the leatherback, though, do not have this *transversalis,* suggesting that it was lost during evolution among some turtles.

Snakes have an unusual and unique accommodative mechanism. Snakes lost the classic method of lens deformation as a mechanism for accommodation. Perhaps this occurred when they became fossorial because accommodation would be of little use underground. Whatever the reason, snakes "reinvented" accommodation when they returned to a terrestrial and above ground niche. They returned to the mechanism of moving the lens forward and backward but do so by squeezing the vitreous. Snakes have developed a stiff peripheral iris and use that musculature in the iris to apply pressure peripherally to the vitreous. This forces the lens forward though the pupil (Figure 19.2). Optically, then, snakes accommodate for near and are emmetropic for distance—just like the elasmobranchs and frogs, although there is no similarity in how this is accomplished. Snakes, too, use the process of the production of accommodative lenticonus, to some extent at least, as mentioned above for the aquatic turtles.

Birds have evolved and improved the accommodative mechanisms of reptiles and developed them in surprising ways. Birds have corneal accommodation in that they have striated muscles that can change the radius of curvature of the cornea and probably use this mechanism to fine-tune accommodation. Birds accommodate by lens deformation, and in many species there is a prominent ringwulst or annular pad encircling the peripheral lens (Figure 20.26).

Accommodation in birds is accomplished by changing the curvature of the lens, which is done by the ciliary body. The ciliary body contains two major striated muscles, which can contract in a meridional and circular fashion. These two muscles, named Brucke's and Crampton's muscles, respectively, compress the annular pad transmitting force to the more ductile portions of the lens (Figure 20.26). Crampton's muscle is more anterior and connects to the peripheral cornea, effecting corneal curvature changes when it contracts as well. These fast-acting striated muscles permit the rapid changes in accommodation required of a fast-moving animal. Such rapid accommodation requires muscular coordination to include the iris musculature. Reptiles have striated muscles in their irides, and these muscles were retained by birds. These striated muscles make a thick stiff iris, which assists with the accommodative lenticonus discussed earlier. Diving birds have perfected the use of the principle.

Mammals, in general, use a quite different mechanism of accommodation by using smooth muscles in the ciliary body to transmit or restrict tension to the lens via fine fibers. By squeezing these fibers through muscular activity of the ciliary body, these fine fibers, called zonules, relax tension on the lens for near. This seems counterintuitive, but the ciliary body musculature must contract to push closer to the lens, which relaxes these fine fibers, and this permits the lens to assume a somewhat more spherical shape and hence accommodate. This mechanism is slower and less robust than those of many reptiles and birds.

Some pinnipeds, such as certain seals and sea lions, have minimal accommodative abilities. Their lens is much more spherical than those of terrestrial mammals and more like fish lenses. Although accommodation occurs by lenticular deformation in pinnipeds, there is little of it because of the size of the lens. The ciliary body is thin, and accommodation is weak, although recently some pinnipeds have been shown to have more vigorous mechanisms and better accommodation, but little is known about them.

Appendix F

CRYSTALLINE LENS

The crystalline lens has evolved many times, and in different ways. Evolution has used enzymes, other proteins, and even minerals in the formation of the lens. Although some consistent patterns emerge, the lens components that have been selected across Animalia are diverse. These lens compounds probably were incorporated well after the photoreceptive elements, even after some other image-forming elements were in place.

Lens material rarely fossilizes, making it difficult to describe the lenses of extinct animals, yet some information is available. The early craniates definitely had lenses, as can be seen from the slightly different appearance of the anterior segment of their eyes and the manner in which the eyes fossilized, but we have no understanding of the lens composition. Trilobite lenses, though, we know to have been composed of calcite, and, as reviewed in Chapter 5, much more is known about these lenses and their optics.

Vertebrate lenses are composed of crystallins, and many of these proteins are enzymes or serve other functions within the body. There are at least eleven types of crystallins found in vertebrates with only a few being common to all. The variety suggests that different lineages recruited different proteins. Vertebrate lenses are homologous, suggesting a common ancestor, but even within the vertebrates, lenticular evolution has occurred depending on how the lens is used. Most fish lenses are rather rigid and do not mold easily but are not mineral. They are composed of protein. Lens movement, as was reviewed in Appendix E, permits accommodation in fish eyes and does not require moldable lenses. Fish lenses contain proteins that are harder than those of mammalian lenses. Hard crystalline lenses are easily to move forward and backward than soft ones, and have a relatively high index of refraction because

they must do all the refractive work of the eye. Additionally, the harder piscine lenses have a graded index of refraction from center to periphery. Many mammals have this lens feature, too, but fish lenses have a more pronounced gradation. The index of refraction of the center is quite high and declines to that of approximately water in the periphery of the lens. As the lens grows, it adds fresh living fibers at the periphery and compacts the center cells, making them denser, which accounts for the change in refraction. Such lenses are virtually free of spherical aberration, or at least have minimized it. Some fish have even added or changed proteins and as a result possess yellow or green lenses (Chapter 10).

Although frogs possess piscine-like lenses and accommodate much like sharks and rays, other truly terrestrial (as opposed to amphibious) species such as reptiles, birds, and mammals have thinner and more ductile lenses composed of different and softer crystallins. In order to accommodate by lens deformation as reptiles, birds, and mammals do, the lens proteins had to change to be softer and more ductile, and that is exactly what happened. The proteins found in some of these terrestrial vertebrates must also remain clear longer, as many of these animals have lives that extend beyond that of most fish and especially beyond those lives of invertebrates. The lens is a good example of natural selection of physiology based on the animal's niche.

In their evolution, Metazoa selected crystalline lenses much later than other principal components of ocular development, such as retinal or opsins. Different phyla cobbled together lenses from different proteins, and even minerals in the trilobites. It seems that evolution chose whatever compound was readily available and capable of being rendered transparent.

The two major classes of photoreceptive cells, ciliary and microvillus (also called rhabdomeric), appeared at least 600 mya (Chapter 3). These cells are responsible for generating a signal in response to a light stimulus.

In both classes of photoreceptors, photon capture induces structural changes in the visual pigment by stimulating the chromophore (retinal) to initiate a response. The structure and interactions of the chromophore with the opsin determine the peak wavelength that causes the configurational change. To maximize the chance of a response from any single photoreceptor, there must be many molecules of photopigment. Because photopigments are membrane bound, the more undulations of the membrane, the more photopigment can be contained in a single cell. Thus, two basic types of cells (ciliary and microvillus) have developed increased membrane surface to contain photopigment, including projections, discs, and infoldings—modified microvilli and modified cilia.

The split between ciliary and microvillus cells likely occurred before the split between the deuterostomes and the protostomes (a major embryologic split approximately 600–580 mya; Chapters 4 and 5 and Figure 4.6). Both cells went with each lineage, but each lineage dealt with them differently. The deuterostomes (mostly vertebrates) principally used the ciliary cell for photoreception and the microvillus cell for circadian rhythm, and the protostomes (invertebrates) chose the reverse. Exceptions exist, and a few animals have both cells acting as photoreceptors. Many animals have both types of cells but are using only one type as a photorecptor, at least in a visual sense. Once the division between the deuterostomes and protostomes occurred, the photoreceptors continued to evolve, and a few differences between lineages are instructive. There are several differences, but these help us understand the distribution of these cells.

Deuterostomes

Sea squirts are tunicates, and the lineage that would lead to living tunicates arose approximately 550–540 mya. These deuterostomes mature via a larval form that resembles the tadpole of a frog. This "tadpole" has an ocellus with ciliary photoreceptors that are homologous to the ciliary photoreceptors of vertebrates. This distribution suggests that the tunicate lineage and the vertebrate lineage have a common ancestor—an ancient deuterostome that had a photoreceptive organ that would permit evolution into the eyes of tunicates and vertebrates. But the ciliary photoreceptor in the ocellus was everted instead of inverted like those of the vertebrates, and the morphology of the cell was different. Further evolution would be needed to change the direction of the cell and its morphology to one that is inverted as the early vertebrates took the stage.

Agnathans, including the lineage that would lead to the lamprey, arose in the early Cambrian at approximately 530 mya. The lamprey predecessor of that time likely had an inverted photoreceptor in its retina that looked very much like all subsequent vertebrate photoreceptors. Biochemically, these ciliary photoreceptors in all members of the deuterostome lineage operate in a similar manner and were directly related, including those of the tunicates.

Ancestral lampreys probably had five cone visual pigments, although one of these was lost before living fish arrived in modern times and was probably lost in the ancestral stage. Fish, including the elasmobranch, ray-finned, and lobe-finned lineages, then began with four opsins and a basic biochemical difference in modus operandi of photoreception from the animals that would arise and use microvillus photopigments.

Primates use both classes of these cells (ciliary and microvillus), and the use of the microvillus opsins in our visual system illustrates their versatility and plasticity.

In many vertebrates, the microvillus cells are to be found in the inner retina as cells that normally receive the translated input from our photoreceptors and are used to transmit signals to the brain. These cells are the ganglion cells. Actually, only a small subset, perhaps 1–2 percent, of the ganglion cells represent these former microvillus cells. Melanopsin, the photosensitive compound found in these special microvillus-to-ganglion cells, is much closer to the opsins of invertebrates, suggesting a common origin for the ciliary and microvillus cells. Both the ciliary and microvillus cells must have evolved simultaneously or at least nearly so with the eyespots in both the protostomes and the deuterostomes or more likely the last common ancestor.

Protostomes

Platynereis, a marine annelid worm, has both microvillus cell–bound visual pigment and ciliary cell–bound visual pigment (Chapter 4). This animal illustrates the point that many animals can have both and use them in different ways. These worms are descended from the lineage of the first "worm" found in the Ediacaran, suggesting that these two different classes of photoreceptors arose and separated before bilaterians.

The sabellids (annelid worms) are interesting exceptions to this rule of protostomes having microvillus photopigments for their principal visual pigment (Chapter 4 and Figures 4.11–4.13). They are invertebrates in the protostome lineage but have modified cilia to contain their visual pigment. They also have key photobiological and neurologic differences that separate them from other invertebrates. These differences suggest that the photoreceptive cells in these animals act much like those photoreceptive cells of vertebrates.

When light strikes their photopigment, the configurational change of the photopigment initiates hyperpolarization of the membrane from a depolarized state, much as occurs in the photoreceptors in vertebrates. But in most invertebrates with rhabdomeric photopigment, configurational change initiates *de*polarization from a *hyper*polarized state. This represents a fundamental difference in ocular function and is a clue to early photoreception.

There are other animals that exhibit evolutionary changes and similarities suggesting that the microvillus and ciliary cells are derived from the same original progenitor cell 1000–600 mya.

As compared to fish, reptiles developed an enlarged brainstem, which would be a basis for further development, especially among later radiations such as birds. The vertebrate midbrain is defined as the uppermost part of the brainstem, or in the middle of the brain, as the name implies. It is involved in more sophisticated but still basic and involuntary bodily functions. Conscious activities require the upper brain or cerebrum, and evolutionarily, the expansion of this comes later, with the mammals.

The word "tectum" means covering, and in neurologic parlance, the tectum occupies the upper portion of the midbrain. In birds, reptiles, and other less-derived vertebrates, the optic tectum is within the midbrain and receives input from the optic nerves. This is the area that does the visual interpretation in many vertebrates. In birds the optic tectum reaches its peak organization and efficiency, and with good reason. Birds are the most "eye-minded" lineage. The tectum evolved from reptiles, although tectal development in the reptilian clades had been extensive during the preceding eras. We know that the avian optic tectum has reached an unrivaled degree of complexity and is larger than that in any other vertebrate class. This robust neurologic structure facilitates superb visual abilities because of superior neural control and faster visual processing. This concentrated neurologic machinery allows birds to detect more subtle changes in their environment and, in turn, permits the skills of predation and escape we see so often in the field. This degree of sophisticated machinery, though, comes at a cost to birds. They have many fewer individual limb skills as compared to primates and far less brain space devoted to such skills. Birds are unable to manipulate objects as we do as an example.

But birds may have sensory mechanisms not possessed by mammals, and these different senses may play a role in sight. Other animals such as pit vipers with infrared sense or sharks with an electromagnetic field sense (Chapter 9), illustrate that sight may integrate with other senses. Many investigators believe that some birds use the earth's magnetic field for migration. Birds possess magnetoreceptors,

although their exact location is not known. One fascinating possibility is that the photoreceptors are magnetized and function as magnetoreceptors. If true, that means that birds could *see* the magnetic field. There are other possibilities, of course, and it is difficult to understand how this would be perceived or interpreted. But similar skills exist in hammerhead sharks, as discussed in Chapter 9, although in an entirely different organ, and the degree of perception is not on this scale.

Birds have evolved large eyes, both relatively and absolutely. The larger and more complex eyes have increased local processing within the eyes, allowing for the occupation of new and more complex niches. Specifically, birds have conquered the niche of flight and predatorial flight at that. This neurologic machinery may have been present in the theropods and/or the pterosaurs, possibly in only a rudimentary form but more likely in a very sophisticated form, as some of these species were quite good at such flight as well.

There are other adaptations in the avian neurologic machinery that contribute to these improved visual abilities. In many birds there are neurons that lead from the optic tectum to the retina, called efferent axons. These cells (neurons) originate in the tectum. The area that spawns these cells reaches its peak in birds but is not exclusive to them. For example, this area (nucleus) can be found in some lizards, crocodiles, and some fish including sharks but is absent in snakes. Presumably, this portion of the brain began in some form in fish, expanded in reptiles, probably increased further in the theropods, and reached its peak in birds.

When neurons lead from the brain to points outside the brain, they are generally motor neurons that instruct, stimulate, and operate muscles. But the optic nerves of many birds contain efferent axons in what is otherwise a sensory nerve—the optic nerve. These efferent axons, though, are not motor neurons but rather terminate on the photoreceptors. Although these special efferent neurons are not completely understood, the best evidence would suggest that these cells are involved in improving discrimination, again being done at a subconscious level, as if the retina were part of the brain, which essentially it is.

These efferent fibers from the brain to the eye allow further processing of the image at the retinal level and permit the image to be transmitted to the brain in a more "formed" fashion, as discussed above. This increases the speed of interpretation of the image. Birds are all about vision and do have the best visual system on earth.

Abathochroal eye: a type of eye seen in few species of *Trilobite* in which there are few, separated lenses, each having its own cornea, and the sclera of the eye is not thick

Abyssal: pertaining to the deepest part of the ocean, generally below 4000 meters, and the creatures that live there

Accipiter: a type of hawk that generally has short wings and a long tail and is a fast agile flier

Accommodation: the adaptation of the eye to maintain focus on an object as it becomes nearer or farther away

Adenosine triphosphate (ATP): molecule formed by a plant or animal that enables the plant or animal to utilize energy obtained from food or sunlight

Adnexa: structures surrounding or attached to an organ; often participate with the organ

Agnathan: "without a jaw," generally refers to an eel-like vertebrate having no jaws

Allantois: a saclike, membranous structure that develops in the embryos of most mammals, birds and reptiles; the allantois becomes the placenta and umbilical cord

Allopatric: occurring in separate areas and not overlapping in distribution

Ammonite: coiled, chambered fossil shell of an extinct *mollusk*

Amnion: thin innermost membrane enclosing the developing embryo of higher vertebrates

Amniote: vertebrate that possesses an *amnion* during development; includes all terrestrial egg-layers

Amphioxus: small, translucent burrowing marine animal; a cephalochordate; its predecessors were early forerunners of vertebrates

Anadromous: refers to fish that migrate into fresh water from the sea to breed, lay eggs, and die

Anapsid: oldest known reptiles having no temporal openings in the skull; turtles may be the only extant anapsid, but many investigators believe that turtles are not anapsids but, rather, diapsids

Anlage: an organ or condition in its earliest stage of rudimentary development

Analogous: in biology, pertaining to structures that perform similar functions but have different evolutionary origins

Anomalocaris: extinct, large shrimplike predator whose fossils are found in the *Burgess Shale*

Arboreal: adapted for life in trees

Arborize: producing branching formations

Archaea (includes Archaeobacteria): microorganisms that resemble bacteria but have different composition of their cell outer membranes or cell walls; Archaea thrive in extreme environments such as hydrothermal vents in the deep sea; some scientists believe they are the earliest form of cellular life. Extremely old, single cell life forms that do not contain a nucleus that developed separately from bacteria

Archean eon: Time period 3750–2500 mya; includes time of first life appearing on earth

Archaeopteryx: believed to be the first bird, evolved in the *Jurassic Period*, had feathers and hollow bones but also had teeth; related to birds

Area centralis: an oval or circular area of the sensory retina, often a few millimeters temporal to the optic disc corresponding to the posterior pole of the eye in most cases; the central fovea is at its center (if a fovea is present), and the area centralis almost always contains only *retinal cones* or at least a very high density of cones in most animals

Argenta: a silvery layer on the external choroid, or within the sclera that extends onto the surface of the iris; it provides reflection and hence camouflage for fry or larval fish and provides a shimmering quality to the iris surface; differentiated from the tapetum

Aspheric: a lens that deviates slightly from a perfectly spherical shape and is relatively free from aberrations

ATP: see *Adenosine triphosphate*

Arthropod: an invertebrate having jointed limbs and a segmented body with an exoskeleton of chitin

Autotroph or autotrophic: a plant capable of synthesizing its food from simple organic substances

Bacterium: the most abundant life form on earth; single-celled organism, lacking a nucleus, which reproduces by fission or by making spores; may be free-living, saprophytic, and may cause disease in plants or animals (plural: bacteria)

Bacteriorhodopsin: a protein pigment consisting of retinal and an opsin similar to *rhodopsin* occurring in the cell membranes of archaea such as those of the genus *Halobacterium*, which photosynthesizes by converting sunlight directly into chemical energy

Bauplan: "body plan," the basis for differentiating plants and animals into the different phyla

Benthic: describes animals that live at the bottom of an ocean, a sea or a lake

Bilaterian: family of animals characterized by bilateral symmetry; includes most living animals

Blastula: early developmental stage of an animal that immediately follows the division of the fertilized egg, a wall of single cells surrounding a fluid-filled cavity, also called a blastosphere

Bolometer: extremely sensitive instrument that measures heat radiation

Bony fish: any fish that has bones rather than cartilage for structural support

Bryozoan: small invertebrate animal, member of phylum *Bryozoa*, that attach to rocks or seaweed and form mosslike colonies

Burgess shale: rock formation located in the Rocky Mountains of northern British Columbia, which contain a rich diversity of fossils of small invertebrates from the middle ***Cambrian Period***, approximately 543–490 mya

Calcite: crystalline form of calcium carbonate that is the basic constituent of limestone

Cambrian Period: time period that lasted from approximately 543 to 490 mya; characterized by proliferation of marine invertebrates

Carboniferous Period: time period that lasted from approximately 362 to 299 mya; characterized by expansion of winged insects

Catalyze: to modify or increase the rate of a chemical reaction

Catarrhine: relating to a group of primates, including Old World monkeys, apes, and humans, that are characterized by nostrils that are close together and directed to the front or downward

Cathemeral: relating to an organism that exhibits random periods of activity to seek food or socialize, showing no preference for ***diurnal*** or ***nocturnal*** activity

Cephalad: in the direction of the head

Cephalopod: member of the mollusca phylum, class *Cephalopoda*, characterized by bilateral body symmetry, a large head and eyes, and a modification of the mollusk foot into arms or tentacles

Chelonian: referring to turtles and/or tortoises

Chert: variety of silica containing microcrystalline quartz

Chicxulub crater: underwater site of meteor impact that happened 65 mya; located near the Yucutan Peninsula; results of impact thought to have caused dinosaur extinction

Chloroplast: organelle found in plant cells; responsible for photosynthesis

Chorion: membrane that surrounds an embryo; is in contact with the amniotic membrane

Chromophore: a light-reactive chemical usually bound to a protein that can absorb light wavelengths and gives color to its compound

Ciliary cell: one of the two major cells supplied with photoreceptive compounds for photoreception; relating to cilia. See Appendix G

Clade: group of organisms that share ***homologous*** features derived from a common ancestor

Cladistics: system of classification based on phylogenetic relationships as well as evolutionary history of groups of organisms

Cladogram: branching diagram used to show evolutionary relationships among organisms

Class: taxonomic category showing greater specificity than a ***Phylum*** or a Division and less specification than an ***Order***

Cleidoic egg: self-contained egg of reptiles and birds; surrounded by hard semi-impervious shell

Cloaca: cavity found in birds, fish, reptiles and some mammals through which fecal, urinary, and reproductive materials pass

Cnidarians: phylum of animals that include the corals, sea anemones, and jellyfish

Coelacanth: one of four living members of the subclass of lobe-finned fishes, *Sarcopterygii*

Color constancy: the tendency for a color to look the same under widely different viewing conditions

Compound eye: an eye seen in some invertebrates, principally arthropods, that is subdivided into multiple individual, light-receptive elements, each including a lens, a transmitting apparatus, and retinal cells

Conus: intraocular nutritive organ of certain reptiles consisting of a plexus of blood vessels originating from neural crest cells

Convergent evolution: process whereby organisms that are not closely related independently acquire similar characteristics while evolving in separate and sometimes varying ecosystems

Convexiclivate fovea: The foveal anatomy of certain fish, reptiles, and birds; it is steep walled with a well-defined pit; the fovea is the densest concentrated area of photoreceptors with the highest visual acuity

Copepod: minute marine or freshwater crustacean having an elongated body, usually with six pairs of limbs on the thorax. Often the species has a forked tail

Cotylosaur: extinct reptile from the ***Carboniferous*** and ***Permian*** periods thought to be the ancestor of all reptiles

Crepuscular: active during the time before sunrise and/or after sunset but excludes nighttime; twilight time

Cretaceous Period: time period from 145 to 65 mya; saw the sudden extinction of the Dinosaurs and the appearance of insects and flowering plants

Crypsis: ability of an organism to avoid detection

Cryogenian: time period 850–650 mya; included time of "snowball earth" in which glaciers reached nearly to the equator; controversial

Cryptochrome: molecules that are blue light sensitive and are found in plants and many animals; these molecules are important in the circadian rhythm and have one of two chromophores—flavin or pterin

Ctenophore: gelatinous marine invertebrate of the phylum *Ctenophora*; a comb jelly

Cursorial: Adapted for running

Cyanobacteria: ancient photosynthetic prokaryote containing different forms of pigment including chlorophyll; usually occurs in colonies; has been called blue-green algae, but is not algae; rather it is a bacterium

Cyclostomes: eel-like animals, including lamprey and hagfish, that have a round mouth, used for sucking behavior, with no true teeth, and no jaws; these animals are in a superphylum known as the agnatha

Cytochrome: iron-containing proteins that have an important role in cell respiration; comparison of different cytochromes helps scientists trace evolutionary relationships

Decapod: crustacean, such as a crab or lobster, that had ten appendages, each joined to a segment of the thorax

Depauperate: lacking in species variety

Descartes-Huygens correction lens: lens or lens system that corrects for spherical aberration and was devised by Rene Descartes and Christian Huygens in the seventeenth century

Detritivore: an organism, such as an earthworm, that feeds on decaying or dead organic matter

Deuterostome: animal that is defined by its embryonic development during which time the anus develops from the first opening in the embryo and the mouth develops from the second opening, which is at the opposite end of the digestive system

Devonian Period: period of time 415 to 362 mya during which the first *tetrapods*, terrestrial *arthropods*, and many fish appeared; small plants developed

Diapsid: reptile, such as a lizard or a snake, that has two pair of temporal openings in its skull

Dichromatic: possessing or exhibiting two colors

Diffraction grating: optical surface that has alternate transparent and opaque stripes, through which light (or radiation) may be passed and an interference pattern observed; patterns may be analyzed to determine the frequency of the radiation

Dimetrodon: extinct carnivorous animal of the *Permian* that had a large sail-like projection on its back; thought to be a radiation of the ancestral synapsid line that would eventually yield mammals

Dinoflagellate: single-celled organisms with nuclei (eukaryotes) of the order Dinoflagellata, usually possessing two flagella and a cellulose covering; they form the principal constituents of plankton and include bioluminescent forms and those forms that produce the red tide

Diploblast: an organism, such as a sponge, that, in its embryonic development, has two cell layers, the ectoderm and the endoderm

Domain: in biology, a supergrouping of organisms; there are three domains: *Archaea* (Archaebacteria), *bacteria* (Eubacteria), and *eukaryotes* (Eukaryota)

Double cones: a pair of connected cones found in fish, reptiles, birds, and some other groups

Echinoderm: a bilaterially symmetrical marine invertebrate, such as a sea star, a basal *deuterostome*

Ecology: the study of the relationship of biological organisms to their environments

Ediacaran: period of time 650 to 543 mya during which the first documented multicellular animals appeared; these fauna were first found in fossil beds in South Australia

Embryo: the earliest stage of development of a fertilized egg from the moment of fertilization until birth, hatching, or germination

Emmetropia: the visual state of an eye that light rays coming from a distant point (infinity) will come to focus on the retina without any device or appliance; in contrast to myopia (nearsightedness) or hyperopia (farsightedness)

Endosymbiont: an organism that lives within the body of another organism; usually implies that the combination is a single organism

Entrain: process of synchronization of one biological rhythm to another or the initiation of cycle or process

Eocene Epoch: major division of geologic time that lasted from 56 to 34 mya

Eubacteria: large group of organisms with a rigid cell wall that have no nucleus; another name for bacteria

Eukaryote: an organism whose cells contain a membrane-bound nucleus and whose DNA contains chromosomes

Euryapsid: grouping of unrelated extinct reptiles that have a single temporal opening in the skull, such as *Ichthyosaurs* and *Plesiosaurs*

Extant: current, not extinct

f-number: the ratio of the focal length of a lens to the diameter of the lens aperture; this will assist in understanding the light-gathering capabilities of a lens or an eye

Family: taxonomic division of organisms that contains fewer individuals than an *Order* but more than a *Genus*

Fetal fissure: a ventral cleft or gap in the forming embryonic optic cup; this remains as a gap in many adult fish but closes in most other vertebrates

Flagellum: organ of locomotion for a cell; resembles a whip in appearance; pl. flagella

Flavin: any of a group of water-soluble yellow pigments found in animals and plants; includes riboflavin

Folivores: organisms that eat leaves, generally foliage of trees

Foraminifera: large *Order* of marine animals that have chambered shells, many of which have minute holes in them

Fossorial: adapted for burrowing or digging

Frugivorous: fruit-eating

Gaia hypothesis: controversial postulation that all living organisms have a regulatory effect of conditions on Earth; put forward by James E. Lovelock and Lynn Margulis in 1972

Galilean telescope: type of refracting telescope having a large convex lens that produces an image that is viewed through an eyepiece that uses a concave lens

Ganglion cell: neuron whose cell body is outside the central nervous system; in the context of this text, a cell that connects the retinal cells to the brain via the optic nerve

Gastropod: Mollusk of the *Class* Gastropoda; characterized by usually having a coiled shell (or no shell), a single muscular foot, and eyes and feelers on a distinct head

Genome: an organism's genetic material

Genus: taxonomic category having fewer members than a *Family* but more than a *Species*

Gondwana: southern hemisphere supercontinent that broke up into Africa, Antarctica, Australia, India, and South America

Guanine: essential constituent of RNA and DNA

Haeckel, Ernst (1834–1919): German zoologist and evolutionist; influenced by Darwin, he attempted to develop the first genealogical tree for the animal kingdom

Hallucigenia: extinct wormlike animal found in fossils in the *Burgess Shale*

Halorhodopsin: found in Archaea, this rhodopsin is an ion pump driven by light and specific for chloride ions

Hemichordate: member of the *Phylum* Hemichordata, includes worm-shaped marine organisms

Hennig, Willi (1913–1976): German biologist who developed the concept of *cladistics*

Holochroal: a type of *trilobite* eye; these eyes have small and closely packed individual facets called ommatidia, which are round or sometimes hexagonal; there are a calcite "cornea" and a lens for focusing

Homologous: in biology, having similar structure and evolutionary origin

Hox genes: subgroup of genes that determine where limbs will grow in a developing fetus or larva

Hyalosome: lens-like structure in a dinoflagellate eye, analogous to the crystalline lens of a camera-style eye

Ichthyosaur: giant marine reptiles that resembled fish; lived during the *Mesozoic Era*

Ichthyostegid: earliest known amphibian; lived near the end of the *Devonian Period*

Iguanodon: huge, herbivorous dinosaur; lived in the early *Cretaceous Period*

Index of refraction: ratio of the speed of light in a vacuum to the speed of light in a specified medium

Inferior oblique: an extraocular muscle found in vertebrate eyes; it is responsible for extorsion of its eye and to some extent elevation; it originates on the inferior medial orbit and inserts into the posterior temporal aspect of the globe

Inferior rectus: an extraocular muscle that depresses the eye; it originates from the posterior orbit and inserts into the inferior aspect of the globe; involved with extorsion of the globe to some extent

Infula: a linear strip of concentrated photoreceptors seen in certain birds. Often seen in the form of a trough or as concentrated ganglion cells

Irides: plural of "iris"

Jurassic Period: geologic span of time from 208 to 145 mya; considered to be the Age of Reptiles

Keplerian telescope: telescope using a convex lens as an objective and a convex lens as an eyepiece

Kingdom: in Linnaean taxonomy, one of five groupings of all living organisms; these include animals, plants, fungi, **prokaryotes**, and protists

Klinotaxis: side-to-side motion of the head as an organism moves forward; a response to sensory stimuli received on either side of the head

K-T boundary: end of the **Cretaceous Period** and beginning of the **Tertiary**; coincides with the extinction of dinosaurs

Lamarckism: theory of evolution that states that organisms evolve by inheriting traits that have been acquired or altered by the use or disuse of body parts

Lateral rectus: an extraocular muscle used to move the eye laterally; it originates in the posterior orbit and inserts on the lateral aspect of the globe

Lateral gene transfer: exchange of DNA or RNA between two single-cell prokaryotes or eukaryotes, often unrelated

Laurasia: northern hemisphere supercontinent that broke up into most of the current northern hemisphere land masses

Lobe-finned fish: bony fishes with paired rounded fins; now consists of lungfish and the coelacanth

Lobopodia: an extinct group of carnivorous arthropods whose fossil records date from the early **Cambrian**; best represented in living arthropods by the **Velvet worm**

Lungfish: freshwater lobefinned fish of the subclass Dipnoi, capable of breathing air

Mach bands: an illusion established when the eye sees a wide bright area and a wide dark area with a thin gradient between them; the eye perceives a nonexistent darker band area located immediately adjacent to the gradient within the light and a nonexistent lighter band area located immediately adjacent to the gradient within the dark area

Marsupial: type of mammal whose reproduction involves the female carrying the fetus in an external pouch where the fetus nurses for several weeks or months until it emerges from the pouch

Matthiessen's ratio: a ratio of the distance from the center of the lens to the retina compared to the radius that will describe the eye; the distance from the center of the lens to the retina is almost always 2.55 times the radius of the lens

Medial rectus: an extraocular muscle used to move the eye medially or nasally; it originates in the posterior orbit and inserts on the medial aspect of the globe

Mesogleal matrix: gelatinous layer that separates the inner and outer cell layers of a jellyfish and certain other diploblasts

Mesozoic Era: geologic time period that occurred 251 to 65 mya

Metazoa: multicellular animals that have cells differentiated into tissues and organs; nearly all animals except **Protozoa**

Methanogen: ancient life forms capable of producing methane gas from decomposing organic matter

Microvillus cell: sensory cell in the eye of some invertebrates such as arthropods and mollusks; a neural sensory unit that contributes microvilli to a central unit called a rhabdom analogous to the photoreceptive portion of rods and cones in vertebrates; these individual units of a microvillus cell are called rhabdomeres; also known as rhabdomeric cell

Moa: extinct flightless ostrich-like bird endemic to New Zealand

Mollusk: chiefly marine invertebrate that has a soft unsegmented body, a mantle, and is protected by a hard shell; includes clams and land snails

Monophyletic: related to, or descended from, a single source

Monotreme: primitive, egg-laying mammal; extant animals include only two species of echidna and the platypus

Montane: pertaining to, or living in, mountains

Myopsid eye: the eye of a squid of the suborder, *Myopsina*; unique among squid in that the eye has a cornea

Nanofabrication: the making of extremely small objects or mechanisms

Nautiloid: pertaining to a mollusk in the Subclass, *Nautiloidea*, including the nautiluses

Nautilus: cephalopod mollusk that has a spiral shell and short tentacles; found in the Indian and Pacific Oceans

Nematocyst: stinging cell, most often found in the tentacles of jellyfish; used to kill or immobilize prey

Neustonic zone: area at the surface of water that is inhabited by microscopic organisms

Nictitans: the third eyelid present in many vertebrate species; it consists of a fold of conjunctiva found at the medial canthus; this reinforced membrane is moved quickly from nasal to temporal and is used to clear the ocular surface

Nictitating membrane: inner eyelid, present in many animals, that can be drawn across the eye, especially for protection from trauma; also called **nictitans**

Nodal points: an optical term that helps define the refraction or bending of light rays by a lens; lenses have two imaginary paired (conjugate) nodal points; any light ray directed at and striking the first nodal point will leave the second nodal point at the same angle to the optical axis as the light ray striking the first nodal point; since the nodal points are not overlying one another, the light ray will be bent or refracted

Ocelloid: **ocellus**-like; refers to the visual unit of single celled protists that contain subcellular elements of an eye; see **Ocellus**

Ocellus (pl. Ocelli): simple eye consisting of an eye with a single lens; technically, the human eye is a simple eye; generally, though, the term **ocellus** is used to describe the dorsal simple eyes on the head of many insects and other invertebrates; usually consists of only a few sensory cells and a single lens

Oegopsid eye: eye of a squid of the suborder *Oegopsina*; this type of eye has no cornea

Oil droplets: subcellular spheres found within the photoreceptors of some reptiles and birds; these oil droplets generally contain carotenoids

Ommatidium: single optical unit of the compound eye of insects and other arthropods; pl. ommatidia

Ontogeny: development of an individual organism from embryo to adult

Opabinia: small extinct segmented animal with an exoskeleton; had a head with five stalked eyes; lived during the Middle **Cambrian**; fossils found in the **Burgess Shale**

Optic tectum or tectum: paired structure within the brain of most animals; generally receives input from eyes and often other sensory organs and transfers inputs to other portions of the brain; often integrates with motor elements as well

Optokinetic response or nystagmus: twitching movements of the eye in response to following moving objects; following telephone poles from a moving car produces optokinetic nystagmus

Ora: edge or margin; in the eye it defines the edge of the peripheral retina

Order: taxonomic category with more organisms than a **Family**, but fewer than a **Class**

Ordovician Period: period of geologic time that lasted from 490 to 445 mya

Orthologue: see **homologous**; may be applied to similar genetic sequences or physical structures derived from the same ancestry, but not necessarily the same biochemical or physical function

Out-group: in evolutionary biology, a less-closely-related group or lineage that is used for comparison of other more closely related groups or lineages

Pakicetus: earliest known, well-preserved cetacean fossil, found in Pakistan; from the **Eocene Epoch**

Paleocene Epoch: a geologic time period extending 65–56 mya

Paleozoic Era: length of geologic time that lasted from 543 to 251 mya

Pallial eyes: simple eyes or ocelli of bilvalves usually innervated by visceral ganglia

Pangaea: ancient supercontinent that included all land masses prior to the break-up into **Laurasia** and **Gondwana**

Panspermia concept: controversial theory that life on earth originated from microorganisms from outer space

Parareptile: extinct group of **Paleozoic** reptiles

Parazoa: ancestral subkingdom of animals; organisms that have differentiated cells but do not have tissues; sponges are the only extant members of this group

Parietal eye: so-called "third eye" present in some animal species; is thought to be photoreceptive; associated with the pineal gland in regulating the animal's circadian rhythm

Parvipelvian Ichthyosaur: dinosaur group of **Ichthyosaurs** with tuna-shaped bodies that have large eyes relative to body size

Passerine: any bird of the **Order** *Passeriformes*; any perching bird

Pax6: extensively researched control gene for the development of the eyes, head, and sensory organs

Pax genes: "**pa**ired bo**x**" genes that are tissue-specific

Pecten: comb-shaped vascular structure found in the eyes of birds; thought to provide nutrition to the retina

Pelagic: living in or found in the open ocean, far from any land

Permian Period: geologic time period that lasted from 299 to 251 mya; age of reptiles

Photo-autotroph: organism that synthesizes its own food from inorganic substances by using light as an energy source; green plants

Photopic: daytime light or sunlight; vision in bright light

Photopic light conditions: conditions that allow for color vision

Photoperiodicity: the response of an organism to the length of exposure to light and dark in a twenty-four-hour period including sensitivity to a particular day length

Photopigment: light-sensitive pigment found in the rods and cones of the eye

Phototransduction: conversion of light signal by an opsin to a signal transmitted beyond the original cell

Phylogeny: evolutionary development of a particular organ or organism

Phylum: primary taxonomic division; contains more organisms than a **Class** but fewer than a **Kingdom**

Pili: hairlike structures; singular is *pilus*

Piscine: relating to fish

Placoderm: any member of an extinct class of fish from the **Silurian** and **Devonian Period**s; characterized by armor-like bony plates that covered the head and sides

Plastid: pigmented organelle found in plant cells and having specific functions, such as photosynthesis

Platyhelminths: flatworms such as planarians

Platyrrhine: descriptive term usually relating to New World monkeys; refers to their having widely separated nostrils that generally open to the side

Pleistocene Epoch: unit of geologic time that lasted from 2 million to 11,000 years ago; humans first appeared; extensive glaciation

Plesiomorphic: having features that are shared more widely than just in a particular group

Plesiosaur: large extinct marine reptile that had paddle-shaped limbs and was common during the **Mesozoic Era**

Polychaetes: marine annelid, or segmented, worms

Polyphyletic: an artificial grouping of unrelated animals, such as classifying "birds" and "mammals" as "warm-blooded"

Porifera: major division of invertebrates that includes the sponges

Potoroo: rat kangaroo of Australia

Precambrian Era: name given to the geologic time span from the formation of the earth, 4.5 billion years ago, to the appearance of the first living organisms at the beginning of the **Cambrian Period**

Priapulid: **Phylum** of mostly extinct marine worms; sixteen known species, most are small, carnivorous worms that live in muddy sediment

Prokaryotes: single-celled organisms that do not contain a nucleus; includes bacteria and **Archaea**

Prosimii: primitive primates, now comprised of only lemurs and tarsiers

Proteobacteria: major grouping of bacteria, many of which are pathogenic

Proteorhodopsin: a protein pigment consisting of retinal and an opsin similar to **rhodopsin** occurring in the cell membranes of many bacteria that photosynthesize by converting sunlight directly into chemical energy

Proterozoic Eon: length of geologic time that lasted from 2500 to 543 mya; living organisms comprised of bacteria, fungi, and other primitive multicellular organisms

Protist: any of **eukaryotic** organisms including protozoans, slime molds, and some algae; single-celled organism containing a nucleus

Protochordate: chordates, living and fossilized, that lack a skull and vertebrae such as Cephalochordata and Urochordata

Protozoa: large group of single-celled **eukaryotic** organisms, including amoebas, ciliates, flagellates, and sporozoans

Pseudotentacular eye: eyes found on or around the pseudotentacles made of folded body margins of flatworms

Pteridine, pterins: biochemical compounds consisting of a pyrazine and a pyrimidine ring; often these are colorful and active compounds

Pterosaur: extinct flying reptile of the **Jurassic** and **Cretaceous Periods**

Purine: nitrogenous compound that is one of the building blocks of DNA

Radula: toothed chitinous muscle in the mouth of all mollusks, except bivalves, used for cutting food as the food enters the mouth

Raptor: bird of prey such as a hawk, eagle, falcon, or owl

Ray-finned fish: bony fish that has fins supported by thin, bony ray-like structures; most extant fish

Retinoid: resembling or pertaining to the retina

Retinular cell: see **Microvillus cell**

Rhabdom: photosenstitive unit consisting of the **microvillus** contributions of each surrounding retinular cells providing a transparent rod-shaped structure in the center of each **ommatidium** found in compound eyes of many invertebrates including many arthropods and mollusks

Rhabdomere: the entire **microvillus** unit contributed by each retinular cell of a compound eye; if this combines with rhabdomeres from adjacent retinular cells, it becomes a rhabdom

Rhabdomeric cell: see **microvillus cell**

Rhodopsin: A photosensitive molecule consisting of an opsin and retinal; the principal compound in the rods of most vertebrates; retinal is a derivative of vitamin A

Ringwulst: annular three-dimensional structure found in the crystalline lenses of birds and reptiles; this rather rigid structure creates an effectively smaller and more mutable central lens in animals that possess it

Sabellid worm: **Polychaete** worm with no mouth organ; includes feather duster worms

Sagittal: shaped like an arrow; generally refers to the anteroposterior suture of the skull

Sagittal plane: parallel to the median plane of the body; a vertical plane through the body

Saccadic eye movements: symmetrical rapid side-to-side eye movements

Sally: sudden rush forward; a leap forward

Salp: free-swimming **tunicate**, usually found in warm seas; has a somewhat flattened, translucent body

Sanguivorous: eats or drinks blood, as do ticks, leeches, some bats

Sauropsid: classification of animals that includes turtles, lizards and snakes, crocodiles, birds, and many extinct dinosaurs

Scheiner's disc principle: an optical principle stating that light passing through two or more pinholes in an otherwise opaque disc will come to exact focus on only one point in that optical system

Schizochroal eye: one of three known designs of trilobite eyes found only in eyes of some *Phacops* species; this compound eye has larger units and each unit has a separate cornea

Scotopic: refers to vision in low-light conditions that is usually perceived exclusively by means of rod cells; dark-adapted state for perception in low-light conditions

Silurian Period: major division of geologic time that extends from 445 to 415 mya

Snell's law: the law of refraction first described by Snell, Descartes, or Harriot

Snowball Earth: controversial theory that the earth was covered in ice from 790 to 650 mya

Somnifacient: producing sleep

Species: fundamental taxonomic classification that has only one member that is capable of reproducing with another member of the same species; in nomenclature, is preceded by the **Genus** name

Spriggina: possible ancestor of the **trilobite**, found in fossil abundance in the Ediacara Hills of Australia

Smooth muscle: nonstriated muscle that is not under voluntary control, includes musculature of the gastrointestinal system and the bladder among others

Statocyst: balance organ present in some aquatic invertebrates

Stemmata: a simple eye seen in insects; these are lateral eyes, as opposed to ocelli, which are simple eyes that are dorsal and often found in or near the midline in animals that have compound eyes

Stereopsis: visual ability to have depth perception using both eyes; perception of the sensation of depth using both eyes simultaneously to receive two slightly different visual projections of the world onto each of the two separate retinae

Stigma: an eyespot

Stomata: an opening in a biological organism; as a mouth or a pore

Striated muscle: muscle tissue characterized by transverse striations of the fibers; includes skeletal, voluntary, and cardiac muscle

Stromatolite: some of the oldest recorded life forms; comprised of limestone and formed by growth of blue-green algae; responsible for producing oxygen in the primitive atmosphere enabling the start of organic life; can be seen in Shark Bay in Western Australia

Stoop: to swoop down, as is seen with a bird pursuing its prey

Structural color: color produced by interference, rather than pigments

Suborder: further taxonomic division of an **Order**

Substrate: underlying layer

Superior colliculus: important paired visual center in the brain of vertebrates; homologous to the optic tectum. In mammals and especially primates it becomes a relay station to visual cortex but retains much its importance and control

Superior oblique: an extraocular muscle found in vertebrates involved principally in intorsion but will depress the eye to some extent

Superior rectus: an extraocular muscle found in vertebrates involved with elevation of the eye and to some extent intorsion

Symbiont: organism in a **symbiotic relationship**

Symbiotic relationship: close relationship between two or more different organisms of differing **species**, which may or may not benefit one of more of the organisms but is not a parasitic relationship

Synapomorphy: a character or trait that is shared by two or more taxonomic groups and is derived through evolution from a common ancestral form

Synapsid: reptile having one temporal opening on each side of the skull

Tapetum: membranous reflective layer of the choroid layer of the eyes of certain mammals

Taxon: taxonomic category or division such as *Kingdom, Order*, etc.; plural is *taxa*

Tectum: layered structure in the brain where the nerves from the eye synapse or pass through; this site also receives other sensory input and is an important center for coordination of the muscles that move the eye and respond to ocular stimulus; this paired structure is a major portion of the brain of birds and fish; also see optic tectum

Temporal summation: the ability of a photoreceptor or photoreceptor group to retain or hold a photonic stimulus (light stimulus) over time awaiting additional photons in the same or adjacent photoreceptors to have enough stimulus to create a signal over time; used by some nocturnal animals to collect enough photons to be capable of sending a signal to the brain

Tertiary Period: unit of geologic time that lasted from 65 to 1.8 mya

Tethys Sea: ocean that existed between *Laurasia* and *Gondwana*

Tetrapod: a four-limbed vertebrate

Tetrachromatic: visual system based on four visual pigments

Therapsid: an order of *Synapsids* that includes the mammals and the their extinct predecessors

Theropod: bipedal carnivorous dinosaur that first appeared about 220 mya; predecessor of birds

Theropsid: advanced mammal-like synapsids; primitive synapsids are usually called pelycosaurs; see *Synapsid*

Transduction: process of transferring genetic material from one cell to another by a plasmid or bacteriophage

Transmembrane opsins: a light-sensitive class of proteins capable of producing a cellular signal

Trematode: any of a number of flatworms in the *Class* Trematoda that are internal or external parasites, such as a fluke

Triassic Period: unit of geologic time that lasted from 251 to 208 mya

Tribrachidium: early *Ediacaran* fossil named for its triradial symmetry

Trichromacy: possessing three different visual pigments for color vision or three different channels to produce the sensation of color vision; animals that possess trichromacy have three different visual pigments in different cones

Trilobite: well-known arthropod fossils from the *Cambrian* and *Devonian*; had complex compound eyes; exoskeleton was divided into three parts or three lobes

Triploblast: an organism that arises from fertilization of an egg that has three germ layers, the ectoderm, mesoderm, and endoderm

Trochlea: various bony or fibrous structures through or over which tendons pass or with which other structures articulate; in the eye, it is a fibrous loop in the socket through which the tendon of the superior oblique muscle passes; found only in mammals

Tuatara: one of two reptiles found in New Zealand with a *parietal eye*; also called a *sphenodon*; a reptile closely related to the lizards but in its own *Phylum*

Tunicates: chordate marine animal; includes sea squirts and *salps*

Twin cones: paired identical retinal cones found only in certain fish

Ungulate: an animal that has hooves, such as a horse or pig

Urbilateria: hypothetical single ancestor to all bilaterally symmetrical animals; the most recent common ancestor to all such animals

Uropygial gland: large gland at the base of the tail of a bird that secretes an oil that is used in preening

Velvet worm: velvety-skinned, wormlike carnivore that lives in tropical rainforests; has characteristics of *arthropods* and of *annelids*; also called an *onychophoran*

Visual streak: a line or concentration of increased photoreceptors or ganglion cells leading to the brain in any eye, but generally applied to vertebrate eyes

Wallaby: marsupial related to the kangaroo, but usually smaller and often with a colorful coat

Zygodactyl: in birds, having two toes facing forward and two toes facing back; seen in birds that climb, such as woodpeckers; in chameleons, grasping feet that look like two-finger mittens but really are bundled toes with two toes in one bundle and three toes in the other

CHAPTER 1

Alam, M., and D. Oesterhelt, *Morphology, function and isolation of halobacterial flagella.* J. Mol. Biol., 1984. 176(4): 459–475.

Allwood, A.C., et al., *Stromatolite reef from the Early Archaean era of Australia.* Nature, 2006. 441(7094): 714–718.

Arraiano, C.M., et al., *Recent advances in the expression, evolution, and dynamics of prokaryotic genomes.* J. Bacteriol., 2007. 189(17): 6093–6100.

Balashov, S.P., et al., *Xanthorhodopsin: A proton pump with a light-harvesting carotenoid antenna.* Science, 2005. 309(5743): 2061–2064.

Beronda, L.M., *Sensing the light: Photoreceptive systems and signal transduction in cyanobacteria.* Mol. Microbiol., 2007. 64(1): 16–27.

Berson, D.M., F.A. Dunn, and M. Takao, *Phototransduction by retinal ganglion cells that set the circadian clock.* Science, 2002. 295(5557): 1070–1073.

Bolhuis, H., et al., *The genome of the square archaeon* Haloquadratum walsbyi: *Life at the limits of water activity.* BMC Genomics, 2006. 7(1): 169.

Boucher, Y., and W.F. Doolittle, *Biodiversity: Something new under the sea.* Nature, 2002. 417(6884): 27–28.

Branchini, B.R., M.H. Murtiashaw, and L.A. Egan, *Synthesis of 5-nitro-2-[N-3-(4-azidophenyl)-propylamino]-benzoic acid: Photoaffinity labeling of human red blood cell ghosts with a 5-nitro-2-(3-phenylpropylamino)-benzoic acid analog.* Biochem. Biophys. Res. Commun., 1991. 176(1): 459–465.

Callaerts, P., G. Halder, and W.J. Gehring, *PAX-6 in development and evolution.* Annu. Rev. Neurosci., 1997. 20: 483–532.

Cavalier-Smith, T., *Cell evolution and Earth history: Stasis and revolution.* Phil. Trans. R. Soc. B: Biol. Sci., 2006. 361(1470): 969–1006.

Claire, L., et al., *Developmental expression of transcription factor genes in a demosponge: Insights into the origin of metazoan multicellularity.* Evol. Dev., 2006. 8(2): 150–173.

Crossman, L.C., and A. Walker, *It's hip to be square!* Nature Rev. Microbiol., 2007. 5(6): 400–401.

Di Giulio, M., *The universal ancestor and the ancestors of Archaea and Bacteria were anaerobes whereas the ancestor of the Eukarya domain was an aerobe.* J. Evol. Biol., 2007. 20(2): 543–548.

Di Giulio, M., *The origin of genes could be polyphyletic.* Gene, 2008. 426(1–2): 39–46.

Engelberg, H., and R. Schoulaker, *Sequence homologies between ribosomal and phage RNAs: A proposed molecular basis for RNA phage parasitism.* J. Mol. Biol., 1976. 106(3): 709–730.

Erren, T.C., et al., *Clockwork blue: On the evolution of non-image-forming retinal photoreceptors in marine and terrestrial vertebrates.* Naturwissenschaften, 2008. 95(4): 273–279.

Fredrickson, J.K., and T.C. Onstott, *Microbes deep inside the earth.* Sci. Am., 1996. 275(4): 68–73.

Fuhrman, J.A., M.S. Schwalbach, and U. Stingl, *Proteorhodopsins: An array of physiological roles?* Nature Rev. Microbiol., 2008. 6(6): 488–494.

Galliot, B., and D. Miller, *Origin of anterior patterning. How old is our head?* Trends Genet., 2000. 16(1): 1–5.

Gehring, W.J., and K. Ikeo, *Pax 6: Mastering eye morphogenesis and eye evolution.* Trends Genet., 1999. 15(9): 371–377.

Graham, D.M., et al., *Melanopsin ganglion cells use a membrane-associated rhabdomeric phototransduction cascade.* J. Neurophysiol., 2008. 99(5): 2522–2532.

Hadrys, T., et al., *The* Trichoplax PaxB *gene: a putative proto-PaxA/B/C gene predating the origin of nerve and sensory cells.* Mol. Biol. Evol., 2005. 22(7): 1569–1578.

Harris, W.A., *Pax-6: Where to be conserved is not conservative.* Proc. Natl. Acad. Sci. USA, 1997. 94(6): 2098–2100.

Haupts, U., J. Tittor, and D. Oesterhelt, *Closing in on bacteriorhodopsin: Progress in understanding the molecule.* Annu. Rev. Biophys. Biomol. Struct., 1999. 28: 367–399.

Horst, M.A.v.d., et al., *From primary photochemistry to biological function in the blue-light photoreceptors PYP and AppA.* Photochem. Photobiol. Sci., 2005. 4(9): 688–693.

Hoshiyama, D., N. Iwabe, and T. Miyata, *Evolution of the gene families forming the Pax/Six regulatory network: Isolation of genes from primitive animals and molecular phylogenetic analyses.* FEBS Lett., 2007. 581(8): 1639–1643.

Huber, C., and G. Wächtershäuser, *α-Hydroxy and α-amino acids under possible Hadean, volcanic origin-of-life conditions.* Science, 2006. 314(5799): 630–632.

James, F.H., et al., *Diversity among three novel groups of hyperthermophilic deep-sea* Thermococcus *species from three sites in the northeastern Pacific Ocean.* FEMS Microbiol. Ecol., 2001. 36(1): 51–60.

Jung, K.H., V.D. Trivedi, and J.L. Spudich, *Demonstration of a sensory rhodopsin in eubacteria.* Mol. Microbiol., 2003. 47(6): 1513–1522.

Kozmik, Z., *Pax genes in eye development and evolution.* Curr. Opin. Genet. Dev., 2005. 15(4): 430–438.

Lee, Y.S., and M. Krauss, *Dynamics of proton transfer in bacteriorhodopsin.* J. Am. Chem. Soc., 2004. 126(7): 2225–2230.

Levy, O., et al., *Light-responsive cryptochromes from a simple multicellular animal, the coral* Acropora millepora. Science, 2007. 318(5849): 467–470.

Lin, C., and D. Shalitin, *Cryptochrome structure and signal transduction.* Annu. Rev. Plant Biol., 2003. 54(1): 469–496.

Man, D., et al., *Diversification and spectral tuning in marine proteorhodopsins.* EMBO J., 2003. 22(8): 1725–1731.

McFadden, G.I., *Endosymbiosis and evolution of the plant cell.* Curr. Opin. Plant Biol., 1999. 2(6): 513–519.

Melyan, Z., et al., *Addition of human melanopsin renders mammalian cells photoresponsive.* Nature, 2005. 433(7027): 741–745.

Montgomery, B.L., *Sensing the light: Photoreceptive systems and signal transduction in cyanobacteria.* Mol. Microbiol., 2007. 64(1): 16–27.

Mullineaux, C.W., *How do cyanobacteria sense and respond to light?* Mol. Microbiol., 2001. 41(5): 965–971.

Nilsson, D.E., *Eye evolution: A question of genetic promiscuity.* Curr. Opin. Neurobiol., 2004. 14(4): 407–414.

Oesterhelt, D., *The structure and mechanism of the family of retinal proteins from halophilic archaea.* Curr. Opin. Struct. Biol., 1998. 8(4): 489–500.

Oparin, A.I., *The origin of life.* 2nd ed. 1953, New York: Dover Publications.

Panda, S., et al., *Melanopsin (Opn4) requirement for normal light-induced circadian phase shifting.* Science, 2002. 298(5601): 2213–2216.

Pichaud, F., and C. Desplan, *Pax genes and eye organogenesis.* Curr. Opin. Genet. Dev., 2002. 12(4): 430–434.

Plachetzki, D.C., B.M. Degnan, and T.H. Oakley, *The origins of novel protein interactions during animal opsin evolution.* PLoS One, 2007. 2(10): e1054.

Qiu, X., et al., *Induction of photosensitivity by heterologous expression of melanopsin.* Nature, 2005. 433(7027): 745–749.

Raup, D.M., and J.W. Valentine, *Multiple origins of life.* Proc. Natl. Acad. Sci. USA, 1983. 80(10): 2981–2984.

Sharma, A.K., J.L. Spudich, and W.F. Doolittle, *Microbial rhodopsins: Functional versatility and genetic mobility.* Trends Microbiol., 2006. 14(11): 463–469.

Sharma, A.K., et al., *Evolution of rhodopsin ion pumps in haloarchaea.* BMC Evol. Biol., 2007. 7: 79.

Spudich, J.L., *The multitalented microbial sensory rhodopsins.* Trends Microbiol., 2006. 14(11): 480–487.

Spudich, J.L., et al., *Retinylidene proteins: Structures and functions from archaea to humans.* Annu. Rev. Cell. Dev. Biol., 2000. 16: 365–392.

Sundstrom, V., *Femtobiology.* Annu. Rev. Phys. Chem., 2008. 59: 53–77.

Theobald, D.L., *A formal test of the theory of universal common ancestry.* Nature, 2010. 465(7295): 219–222.

Tu, D.C., et al., *Nonvisual photoreception in the chick iris.* Science, 2004. 306(5693): 129–131.

Van Gelder, R.N., *Non-visual photoreception: Sensing light without sight.* Curr. Biol., 2008. 18(1): R38–R39.

Venter, J.C., et al., *Environmental genome shotgun sequencing of the Sargasso Sea.* Science, 2004. 304(5667): 66–74.

Vorobyov, E., and J. Horst, *Getting the proto-Pax by the tail.* J. Mol. Evol., 2006. 63(2): 153–164.

Wells, J.G., *Cooperation.* Science, 1972. 176(4034): 459.

Xu, Y., T. Mori, and C.H. Johnson, *Cyanobacterial circadian clockwork: Roles of KaiA, KaiB and the kaiBC promoter in regulating KaiC.* EMBO J., 2003. 22(9): 2117–2126.

Yokoyama, S., *Molecular evolution of retinal and nonretinal opsins.* Genes Cells, 1996. 1(9): 787–794.

Chapter 2

Boucher, Y. and W.F. Doolittle, *Biodiversity: Something new under the sea.* Nature, 2002. 417(6884): 27–28.

Cavalier-Smith, T., *Cell evolution and Earth history: Stasis and revolution.* Phil. Trans. R. Soc. Lond. B: Biol. Sci., 2006. 361(1470): 969–1006.

de Duve, C., *The origin of eukaryotes: A reappraisal.* Nature Rev. Genet., 2007. 8(5): 395–403.

Di Giulio, M., *The universal ancestor and the ancestors of Archaea and Bacteria were anaerobes whereas the ancestor of the Eukarya domain was an aerobe.* J. Evol. Biol., 2007. 20(2): 543–548.

Di Giulio, M., *The origin of genes could be polyphyletic.* Gene, 2008. 426(1–2): 39–46.

Ebnet, E., et al., *Volvoxrhodopsin, a light-regulated sensory photoreceptor of the spheroidal green alga* Volvox carteri. Plant Cell, 1999. 11(8): 1473–1484.

Foster, K.W., *Eye evolution: Two eyes can be better than one.* Curr. Biol., 2009. 19(5): R208–R210.

Francis, D., *On the eyespot of the dinoflagellate,* Nematodinium. J. Exp. Biol., 1967. 47(3): 495–501.

Greuet, C., *Structure fine de l'ocelle d'*Eryihropsis pavillardi. C. R. Hebd. Seanc. Acad. Sci., 1965. 261: 4.

Hadrys, T., et al., *The Trichoplax PaxB gene: A putative proto-PaxA/B/C gene predating the origin of nerve and sensory cells.* Mol. Biol. Evol., 2005. 22(7): 1569–1578.

Horst, M.A.v.d., et al., *From primary photochemistry to biological function in the blue-light photoreceptors PYP and AppA.* Photochem. Photobiol. Sci., 2005. 4(9): 688–693.

Iseki, M., et al., *A blue-light-activated adenylyl cyclase mediates photoavoidance in Euglena gracilis.* Nature, 2002. 415(6875): 1047–1051.

Kreimer, G., *Reflective properties of different eyespot types in dinoflagellates.* Protist, 1999. 150(3): 311–323.

Panda, S., et al., *Melanopsin (Opn4) requirement for normal light-induced circadian phase shifting.* Science, 2002. 298(5601): 2213–2216.

Pennisi, E., *Evolution of developmental diversity. Evo-devo devotees eye ocular origins and more.* Science, 2002. 296(5570): 1010–1011.

Prochnik, S.E., et al., *Genomic analysis of organismal complexity in the multicellular green alga* Volvox carteri. Science, 2010. 329(5988): 223–226.

Qiu, X., et al., *Induction of photosensitivity by heterologous expression of melanopsin.* Nature, 2005. 433(7027): 745–749.

Roger, A.J., and L.A. Hug, *The origin and diversification of eukaryotes: Problems with molecular phylogenetics and molecular clock estimation.* Phil. Trans. R. Soc. Lond. B: Biol. Sci., 2006. 361(1470): 1039–1054.

Terakita, A., *The opsins.* Genome Biol., 2005. 6(3): 213.

Vesteg, M., J. Krajcovic, and L. Ebringer, *On the origin of eukaryotic cells and their endomembranes.* Riv. Biol., 2006. 99(3): 499–519.

Wells, J.G., *Cooperation.* Science, 1972. 176(4034): 459.

Chapter 3

Martin, V.J.*Photoreceptors of cnidarians.* Can. J. Zool., 2002. 80: 1703–1722.

Satterlie, RA.*Neuronal control of swimming in jellyfish: A comparative story.* Can. J. Zool., 2002. 80: 1654–1669.

Albani, A.E., et al., *Large colonial organisms with coordinated growth in oxygenated environments 2.1 Gyr ago.* Nature, 2010. 466(7302): 100–104.

Alieva, N.O., et al., *Diversity and evolution of coral fluorescent proteins.* PLoS One, 2008. 3(7): e2680.

Anthony, M.P., and P. David, *Evaluating hypotheses for the origin of eukaryotes.* BioEssays, 2007. 29(1): 74–84.

Aronova, M.Z., and T.A. Kharkevich, *[Secondary messengers in the locomotor-sensory system in primary multicellular organisms. Cytochemical study of inositol-containing compartments in receptor cell of Ctenophora and Coelenterata].* Zh. Evol. Biokhim. Fiziol., 2001. 37(4): 315–322.

Carpenter, K.S., M. Morita, and J.B. Best, *Ultrastructure of the photoreceptor of the planarian* Dugesia dorotocephala. *I. Normal eye.* Cell Tissue Res., 1974. 148(2): 143–158.

Catmull, J., et al., *Pax-6 origins—implications from the structure of two coral pax genes.* Dev. Genes Evol., 1998. 208(6): 352–356.

Chentsov, B.V., *[On the antimicrobial action of the tissues of* Ctenophora Beroe cucumis *fabr.].* Antibiotiki, 1962. 7: 900–902.

Clack, J.A., *Devonian climate change, breathing, and the origin of the tetrapod stem group.* Integr. Comp. Biol. 47(4): 13.

Claire, L., et al., *Developmental expression of transcription factor genes in a demosponge: Insights into the origin of metazoan multicellularity.* Evol. Dev., 2006. 8(2): 150–173.

Coates, M.M., *Visual ecology and functional morphology of cubozoa (Cnidaria).* Integr. Comp. Biol., 2003. 43(4): 542–548.

Coates, M., *Vision in cubozoan jellyfish,* Tripedalia cystophora, *in Biological Sciences.* 2004. Palo Alto, CA: Stanford University.

Coates, M.M., et al., *The spectral sensitivity of the lens eyes of a box jellyfish,* Tripedalia cystophora (Conant). J. Exp. Biol., 2006. 209(Pt 19): 3758–3765.

Colley, N.J., and R.K. Trench, *Selectivity in phagocytosis and persistence of symbiotic algae by the scyphistoma stage of the jellyfish* Cassiopeia xamachana. Proc. R. Soc. Lond. B: Biol. Sci., 1983. 219(1214): 61–82.

Colley, N.J., and R.K. Trench, *Cellular events in the reestablishment of a symbiosis between a marine dinoflagellate and a coelenterate.* Cell Tissue Res., 1985. 239(1): 93–103.

Collin, S.P., et al., *The evolution of early vertebrate photoreceptors.* Phil. Trans. R. Soc. Lond. B: Biol. Sci., 2009. 364(1531): 2925–2940.

Collins, A.G., and J.W. Valentine, *Defining phyla: Evolutionary pathways to metazoan body plans.* Evol. Dev., 2001. 3(6): 432–442.

Dawson, W.W., *Is the brain behind the eye? Implications of processing by the retina.* Invest. Ophthalmol. Vis. Sci., 1973. 12(6): 398–399.

Derelle, R., et al., *Homeodomain proteins belong to the ancestral molecular toolkit of eukaryotes.* Evol. Dev., 2007. 9(3): 212–219.

Ender, A., and B. Schierwater, *Placozoa are not derived cnidarians: Evidence from molecular morphology.* Mol. Biol. Evol., 2003. 20(1): 130–134.

Francis, D., *On the eyespot of the dinoflagellate, Nematodinium.* J. Exp. Biol., 1967. 47(3): 495–501.

Galliot, B., and D. Miller, *Origin of anterior patterning: How old is our head?* Trends Genet., 2000. 16(1): 1–5.

Garm, A., F. Andersson, and D.E. Nilsson, *Unique structure and optics of the lesser eyes of the box jellyfish* Tripedalia cystophora. Vision Res, 2008. 48(8): 1061–1073.

Garm, A., et al., *The lens eyes of the box jellyfish* Tripedalia cystophora *and* Chiropsalmus *sp. are slow and color-blind.* J. Comp. Physiol. A: Neuroethol. Sens. Neural. Behav. Physiol., 2007. 193(5): 547–557.

Garm, A., et al., *Rhopalia are integrated parts of the central nervous system in box jellyfish.* Cell Tissue Res., 2006. 325(2): 333–343.

Hadrys, T., et al., *The* Trichoplax *PaxB gene: A putative proto-PaxA/B/C gene predating the origin of nerve and sensory cells.* Mol. Biol. Evol., 2005. 22(7): 1569–1578.

Hartwick, R.F., *Observations on the anatomy, behaviour, reproduction and life cycle of the cubozoan* Carybdea sivickisi. Hydrobiologia, 1991. 216–217(1): 171–179.

Hernandez-Nicaise, M.L., *Ultrastructural evidence for a sensory-motor neuron in Ctenophora.* Tissue Cell, 1974. 6(1): 43–47.

Hofer, T., J.A. Sherratt, and P.K. Maini, Dictyostelium discoideum: *Cellular self-organization in an excitable biological medium.* Proc. Biol. Sci., 1995. 259(1356): 249–257.

Hoppenrath, M., et al., *Molecular phylogeny of ocelloid-bearing dinoflagellates (Warnowiaceae) as inferred from SSU and LSU rDNA sequences.* BMC Evol. Biol., 2009. 9: 116.

Korotkova, G.P., and I.V. Pylilo, *[Regenerative phenomena in Ctenophora larvae].* Vestn. Leningr. Univ. Biol., 1970. 1: 21–28.

Kozmik, Z., et al., *Role of Pax genes in eye evolution: A cnidarian PaxB gene uniting Pax2 and Pax6 functions.* Dev. Cell, 2003. 5(5): 773–785.

Kozmik, Z., et al., *Assembly of the cnidarian camera-type eye from vertebrate-like components.* Proc. Natl. Acad. Sci., 2008. 105(26): 8989–8993.

Lamb, T.D., S.P. Collin, and E.N. Pugh, Jr, *Evolution of the vertebrate eye: Opsins, photoreceptors, retina and eye cup.* Nature Rev. Neurosci., 2007. 8(12): 960–976.

Larroux, C., et al., *Developmental expression of transcription factor genes in a demosponge: Insights into the origin of metazoan multicellularity.* Evol. Dev., 2006. 8(2): 150–173.

Levy, O., et al., *Light-responsive cryptochromes from a simple multicellular animal, the coral* Acropora millepora. Science, 2007. 318(5849): 467–470.

Lichtenegger, H.C., et al., *Zinc and mechanical prowess in the jaws of* Nereis, *a marine worm.* Proc. Natl. Acad. Sci. USA, 2003. 100(16): 9144–9149.

Margulis, L., and D. Sagan, *Acquiring genomes: A theory of the origins of species.* 1st ed. 2002, New York: Basic Books.

Maria, C.R., *Genomic analyses and the origin of the eukaryotes.* Chem. Biodiversity, 2007. 4(11): 2631–2638.

Martin, V.J., *Photoreceptors of cnidarians.* Can. J. Zool., 2002. 80(10): 1703.

Martindale, M.Q., and J.Q. Henry, *Reassessing embryogenesis in the Ctenophora: the inductive role of e1 micromeres in organizing ctene row formation in the "mosaic" embryo,* Mnemiopsis leidyi. Development, 1997. 124(10): 1999–2006.

Matus, D.Q., et al., *Expression of Pax gene family members in the anthozoan cnidarian,* Nematostella vectensis. Evol. Dev., 2007. 9(1): 25–38.

Nelson, M.M., et al., *Lipids of gelatinous Antarctic zooplankton: Cnidaria and Ctenophora.* Lipids, 2000. 35(5): 551–559.

Nilsson, D.E., *Photoreceptor evolution: Ancient siblings serve different tasks.* 2005. 15(3): R94–R96.

Nilsson, D.E., et al., *Advanced optics in a jellyfish eye.* Nature, 2005. 435(7039): 201–205.

Nilsson, D.-E. and D. Arendt, *Eye evolution: The blurry beginning.* Current Biology, 2008. 18(23): R1096–R1098.

Nordstrom, K., et al., *A simple visual system without neurons in jellyfish larvae.* Proc. Biol. Sci., 2003. 270(1531): 2349–2354.

Nordström, K., et al., *A simple visual system without neurons in jellyfish larvae.* Proc. Biol. Sci., 2003. 270(1531): 2349–2354.

Pennisi, E., *Evolution of developmental diversity: Evo-devo devotees eye ocular origins and more.* Science, 2002. 296(5570): 1010–1011.

Piatigorsky, J., et al., *The cellular eye lens and crystallins of cubomedusan jellyfish.* J. Comp. Physiol. A: Neuroethol. Sens. Neural Behav. Physiol., 1989. 164(5): 577–587.

Plachetzki, D.C., B.M. Degnan, and T.H. Oakley, *The origins of novel protein interactions during animal opsin evolution.* PLoS One, 2007. 2(10): e1054.

Plachetzki, D.C., J.M. Serb, and T.H. Oakley, *New insights into the evolutionary history of photoreceptor cells.* Trends Ecol. Evol., 2005. 20(9): 465–467.

Podar, M., et al., *A molecular phylogenetic framework for the phylum Ctenophora using 18S rRNA genes.* Mol. Phylogenet. Evol., 2001. 21(2): 218–230.

Pylilo, I.V., *[Mitosis in regenerating comb row and the double-nucleated cells of Ctenophora].* Ontogenez, 1975. 6(2): 187–189.

Rompler, H., et al., *G protein-coupled time travel: Evolutionary aspects of GPCR research.* Mol. Interv., 2007. 7(1): 17–25.

Salih, A., et al., *Fluorescent pigments in corals are photoprotective.* Nature, 2000. 408(6814): 850–853.

Schierwater, B., *My favorite animal,* Trichoplax adhaerens. BioEssays, 2005. 27(12): 1294–1302.

Schierwater, B., D. de Jong, and R. DeSalle, *Placozoa and the evolution of Metazoa and intrasomatic cell differentiation.* Int. J. Biochem. Cell Biol., 2009. 41(2): 370–379.

Schwab, I.R., and A.A. Sadun, *An out-pouching of the eye?* Br. J. Ophthalmol., 2007. 91(9): 1107–1108.

Skogh, C., et al., *Bilaterally symmetrical rhopalial nervous system of the box jellyfish* Tripedalia cystophora. J. Morphol., 2006. 267(12): 1391–1405.

Srivastava, M., et al., *The* Amphimedon queenslandica *genome and the evolution of animal complexity.* Nature, 2010. 466(7307): 720–726.

Stierwald, M., et al., *The Sine oculis/Six class family of homeobox genes in jellyfish with and without eyes: Development and eye regeneration.* Dev. Biol., 2004. 274(1): 70–81.

Suga, H., et al., *Flexibly deployed Pax genes in eye development at the early evolution of animals demonstrated by studies on a hydrozoan jellyfish.* Proc. Natl. Acad. Sci. USA, 2010. 107(32): 14263–14268.

Sun, H., et al., *Isolation of Cladonema Pax-B genes and studies of the DNA-binding properties of cnidarian Pax paired domains.* Mol. Biol. Evol., 2001. 18(10): 1905–1918.

Taddei-Ferretti, C., and C. Musio, *Photobehaviour of* Hydra (Cnidaria, Hydrozoa) *and correlated mechanisms: A case of extraocular photosensitivity.* J. Photochem. Photobiol. B: Biol., 2000. 55(2–3): 88–101.

Treisman, J.E., *How to make an eye.* Development, 2004. 131(16): 3823–3827.

Warrant, E.J., and D.E. Nilsson, *Absorption of white light in photoreceptors.* Vision Res., 1998. 38(2): 195–207.

Welch, V., et al., *Optical properties of the iridescent organ of the comb-jellyfish* Beroe cucumis (Ctenophora). Phys. Rev. E Stat. Nonlin. Soft Matter Phys., 2006. 73(4 Pt 1): 041916.

Yamasu, T., and M. Yoshida, *Fine structure of complex ocelli of a cubomedusan,* Tamoya bursaria Haeckel. Cell Tissue Res., 1976. 170(3): 325–339.

Yokoyama, S., *Molecular evolution of retinal and nonretinal opsins.* Genes Cells, 1996. 1(9): 787–794.

CHAPTER 4
Aizenberg, J., and G. Hendler, *Designing efficient microlens arrays: Lessons from Nature.* J. Mater. Chem., 2004. 14(14): 2066–2072.

Aizenberg, J., et al., *Calcitic microlenses as part of the photoreceptor system in brittlestars.* Nature, 2001. 412(6849): 819–822.

Albrecht, F., and D. Adriaan, *The polychaete* Platynereis dumerilii (Annelida): A laboratory animal with spiralian cleavage, lifelong segment proliferation and a mixed benthic/pelagic life cycle. BioEssays, 2004. 26(3): 314–325.

Arendt, D., et al., *Development of pigment-cup eyes in the polychaete* Platynereis dumerilii *and evolutionary conservation of larval eyes in Bilateria.* Development, 2002. 129(5): 1143–1154.

Arendt, D., et al., *Ciliary photoreceptors with a vertebrate-type opsin in an invertebrate brain.* Science, 2004. 306(5697): 869–871.

Asano, Y., et al., *Rhodopsin-like proteins in planarian eye and auricle: Detection and functional analysis.* J. Exp. Biol., 1998. 201(9): 1263–1271.

Bellingham, J., and R. Foster, *Opsins and mammalian photoentrainment.* Cell Tissue Res., 2002. 309(1): 57–71.

Birgit, R., *Development and differentiation of the eye in* Platynereis dumerilii (Annelida, Polychaeta). J. Morphol., 1992. 212(1): 71–85.

Brandenburger, J.L., and R.M. Eakin, *Cytochemical localization of acid phosphatase in light- and dark-adapted eyes of a polychaete worm,* Nereis limnicola. Cell Tissue Res., 1985. 242(3): 623–628.

Delsuc, F., et al., *Tunicates and not cephalochordates are the closest living relatives of vertebrates.* Nature, 2006. 439(7079): 965–968.

Duque, C., et al., *Main sterols from the ophiuroids* Ophiocoma echinata, Ophiocoma wendtii, Ophioplocus januarii *and* Ophionotus victoriae. Biochem. Syst. Ecol., 1997. 25(8): 775–778.

Eakin, R.M., and J.L. Brandenberger, *Unique eye of probable evolutionary significance.* Science, 1981. 211(4487): 1189–1190.

Eakin, R.M., J.A. Westfall, and M.J. Dennis, *Fine structure of the eye of a nudibranch mollusc,* Hermissenda crassicornis. J. Cell Sci., 1967. 2(3): 349–358.

Ermak, T.H. and R.M. Eakin, *Fine structure of the cerebral and pygidial ocelli in* Chone ecaudata (Polychaeta: Sabellidae). J. Ultrastruct. Res., 1976. 54(2): 243–260.

Finnerty, J.R., et al., *Homeobox genes in the Ctenophora: Identification of paired-type and Hox homologues in the atentaculate ctenophore,* Beroe ovata. Mol. Mar. Biol. Biotechnol., 1996. 5(4): 249–258.

Fischer, A., *[On the structure and light-dark adaptation of the eyes of the polychaete* Platynereis dumerilii.] Z. Zellforsch. Mikrosk. Anat., 1963. 61: 338–353.

Fischer, A., and J. Brokelmann, *[The eye of* Platynereis dumerilii (Polychaeta): Its fine structure in ontogenetic and adaptive change]. Z. Zellforsch. Mikrosk. Anat., 1966. 71(2): 217–244.

Fong, P.P., *Lunar control of epitokal swarming in the polychaete* Platynereis bicanaliculata (Baird) from central California. Bull. Marine Sci., 1993. 52: 911–924.

Foster, K.W., *Eye evolution: Two eyes can be better than one.* Curr. Biol., 2009. 19(5): R208–210.

Fu, Y., et al., *Non-image-forming ocular photoreception in vertebrates.* Curr. Opin. Neurobiol., 2005. 15(4): 415–422.

Graham, D.M., et al., *Melanopsin ganglion cells use a membrane-associated rhabdomeric phototransduction cascade.* J. Neurophysiol., 2008. 99(5): 2522–2532.

Gregory, T., *Understanding evolutionary trees.* Evol. Educ. Outreach, 2008. 1(2): 121–137.

Hattar, S., et al., *Melanopsin-containing retinal ganglion cells: Architecture, projections, and intrinsic photosensitivity.* Science, 2002. 295(5557): 1065–1070.

Hughes, H.P., *The larval eye of the aeolid nudibranch* Trinchesia aurantia (Alder and Hancock). Z. Zellforsch. Mikrosk. Anat., 1970. 109(1): 55–63.

Inoue, T., et al., *Morphological and functional recovery of the planarian photosensing system during head regeneration.* Zool. Sci., 2004. 21(3): 275–283.

Isoldi, M.C., et al., *Rhabdomeric phototransduction initiated by the vertebrate photopigment melanopsin.* Proc. Natl. Acad. Sci. USA, 2005. 102(4): 1217–1221.

Johnsen, S., *Extraocular sensitivity to polarized light in an echinoderm.* J. Exp. Biol., 1994. 195: 281–291.

Lacalli, T., *Evolutionary biology: Light on ancient photoreceptors.* Nature, 2004. 432(7016): 454–455.

Lacalli, T.C., and L.Z. Holland, *The developing dorsal ganglion of the salp* Thalia democratica, *and the nature of the ancestral chordate brain.* Phil. Trans. R. Soc. Lond. B: Biol. Sci., 1998. 353(1378): 1943–1967.

MacRae, E.K., *The fine structure of photoreceptors in a marine flatworm.* Cell Tissue Res., 1966. 75(2): 469–484.

Mellon, D., et al., *Electrical interactions between the giant axons of a polychaete worm* (Sabella penicillus L.). J. Exp. Biol., 1980. 84(1): 119–136.

Nilsson, D.-E., *Eyes as optical alarm systems in fan worms and ark clams.* Phil. Trans. Biol. Sci., 1994. 346(1316): 195–212.

Nilsson, D.E., *Photoreceptor evolution: Ancient siblings serve different tasks.* Curr. Biol., 2005. 15(3): R94–R96.

Nilsson, D.E., R. Odselius, and R. Elofsson, *The compound eye of* Leptodora kindtii *(Cladocera). An adaptation to planktonic life.* Cell Tissue Res., 1983. 230(2): 401–410.

Osorio, D., and D.E. Nilsson, *Visual pigments: Trading noise for fast recovery.* Curr. Biol., 2004. 14(24): R1051–R1053.

Pernet, B., *Persistent ancestral feeding structures in nonfeeding annelid larvae.* Biol. Bull., 2003. 205(3): 295–307.

Pineda, D., et al., *The genetic network of prototypic planarian eye regeneration is* Pax6 *independent.* Development, 2002. 129(6): 1423–1434.

Plachetzki, D.C., J.M. Serb, and T.H. Oakley, *New insights into the evolutionary history of photoreceptor cells.* Trends. Ecol. Evol., 2005. 20(9): 465–467.

Pradillon, F., et al., *Influence of environmental conditions on early development of the hydrothermal vent polychaete* Alvinella pompejana. J. Exp. Biol., 2005. 208(8): 1551–1561.

Purschke, G., et al., *Photoreceptor cells and eyes in Annelida.* Arthropod Struct. Dev., 2006. 35(4): 211–230.

Sempere, L.F., et al., *Phylogenetic distribution of microRNAs supports the basal position of acoel flatworms and the polyphyly of Platyhelminthes.* Evol. Dev., 2007. 9(5): 409–415.

Stoll, C.J., *Peripheral and central photoreception in* Aplysia fasciata. Malacologia, 1979. 18(1–2): 459–463.

Sutton, M.D., D.E. Briggs, and D.J. Siveter, *A three-dimensionally preserved fossil polychaete worm from the Silurian of Herefordshire, England.* Proc. Biol. Sci., 2001. 268(1483): 2355–2363.

Yamasu, T., *Fine structure and function of ocelli and sagittocysts of acoel flatworms.* Hydrobiologia, 1991. 227(1): 273–282.

Yamasu, T., and M. Yoshida, *Fine structure of complex ocelli of a cubomedusan,* Tamoya bursaria *Haeckel.* Cell Tissue Res., 1976. 170(3): 325–339.

CHAPTER 5

Briggs, D.E.G., *The arthropod* Odaraia alata *Walcott, Middle Cambrian, Burgess Shale, British Columbia.* Phil. Trans. R. Soc. Lond. B: Biol. Sci., 1981. 291(1056): 541–582.

Briggs, D.E.G., and D. Collins, *The arthropod* Alalcomenaeus cambricus *Simonetta, from the Middle Cambrian Burgess Shale of British Columbia.* Palaeontology, 1999. 42(6): 953–977.

Brigitte, S., et al., *A miniscule optimized visual system in the Lower Cambrian.* Lethaia, 2009. 42(3): 265–273.

Clarkson, E., R. Levi-Setti, and G. Horvath, *The eyes of trilobites: The oldest preserved visual system.* Arthropod Struct. Dev., 2006. 35(4): 247–259.

Collins, D., *The "evolution" of Anomalocaris and its classification in the arthropod class Dinocarida (nov.) and order Radiodonta (nov.).* J. Paleontol., 1996. 70(2): 280–293.

Fischer, S., C.H. Muller, and V.B. Meyer-Rochow, *How small can small be: The compound eye of the parasitoid wasp* Trichogramma evanescens *(Westwood, 1833) (Hymenoptera, Hexapoda), an insect of 0.3- to 0.4-mm total body size.* Vis. Neurosci., 2010. 13:1–14.

Fordyce, D., and T.W. Cronin, *Trilobite vision: A comparison of schizochroal and holochroal eyes with the compound eyes of modern arthropods.* Paleobiology, 1993. 19(3): 288–303.

Fortey, R., and B. Chatterton, *A Devonian trilobite with an eyeshade.* Science, 2003. 301(5640): 1689.

Gabriel, W.N., et al., *The tardigrade* Hypsibius dujardini, *a new model for studying the evolution of development.* Developmental Biology, 2007. 312(2): 545–559.

Gál, J., G. Horváth, and E.N.K. Clarkson, *Reconstruction of the shape and optics of the lenses in the abathochroal-eyed trilobite* Neocobboldia chinlinica. Hist. Biol. Int. J. Paleobiol., 2000. 14(4): 193–204.

Gál, J., et al., *Image formation by bifocal lenses in a trilobite eye?* Vision Res, 2000. 40(7): 843–853.

Gee, H., *On being vetulicolian.* Nature, 2001. 414(6862): 407, 409.

Gould, S.J., *Wonderful life: The Burgess Shale and the nature of history.* 1st ed. 1989, New York: W.W. Norton.

Greven, H., *Comments on the eyes of tardigrades.* Arthropod Struct. Dev., 2007. 36(4): 401–407.

Horváth, G., E.N.K. Clarkson, and W. Pix, *Survey of modern counterparts of schizochroal trilobite eyes: Structural and functional similarities and differences.* Hist. Biol. Int. J. Paleobiol., 1997. 12(3): 229–263.

Levinton, J.S., *The big bang of animal evolution.* Sci. Am., 1992. 267(5): 84–91.

Levi-Setti, R., *Trilobites.* 2nd ed. 1993, Chicago: University of Chicago Press.

Mayer, G., *Structure and development of onychophoran eyes: What is the ancestral visual organ in arthropods?* Arthropod Struct. Dev., 2006. 35(4): 231–245.

Nelson, D.R., *Current Status of the Tardigrada.* Evol. Ecol. Integr. Comp. Biol., 2002. 42(3): 652–659.

Parker, A., *In the blink of an eye.* 2003, Cambridge, MA: Perseus.

Parker, A.R., *Colour in Burgess Shale animals and the effect of light on evolution in the Cambrian.* Proc. Biol. Sci., 1998. 265(1400): 967–972.

Schoenemann, B., *Cambrian view.* Palaeoworld. 15(3–4): 307–314.

Schoenemann, B., et al., *A miniscule optimized visual system in the Lower Cambrian.* Lethaia, 2009. 42(3): 265–273.

Shu, D.G., et al., *Primitive deuterostomes from the Chengjiang Lagerstatte (Lower Cambrian, China).* Nature, 2001. 414(6862): 419–424.

Shu, D.G., et al., *Head and backbone of the Early Cambrian vertebrate* Haikouichthys. Nature, 2003. 421(6922): 526–529.

Tautz, D., *Evolutionary biology: Debatable homologies.* Nature, 1998. 395(6697): 17–19.

Tessmar-Raible, K., and D. Arendt, *Emerging systems: Between vertebrates and arthropods, the Lophotrochozoa.* Curr. Opin. Genet. Dev., 2003. 13(4): 331–340.

Valentine, J.W., D. Jablonski, and D.H. Erwin, *Fossils, molecules and embryos: New perspectives on the Cambrian explosion.* Development, 1999. 126(5): 851–859.

Van Roy, P., et al., *Ordovician faunas of Burgess Shale type.* Nature, 2010. 465(7295): 215–218.

Whittington, H.B., *The enigmatic animal Opabinia regalis, Middle Cambrian, Burgess Shale, British Columbia.* Phil. Trans. R. Soc. Lond. B. 1975. 271 (910): 1–43.

Zhang, X., and D. Shu, *Soft anatomy of sunellid arthropods from the chengjiang lagerstätte, lower Cambrian of southwest China.* J. Paleontol., 2007. 81(6): 1412–1422.

CHAPTER 6

Baldwin, J.D., et al., *Molecular phylogeny and biogeography of the marine shrimp Penaeus.* Mol. Phylogenet. Evol., 1998. 10(3): 399–407.

Battelle, B.A., *The eyes of Limulus polyphemus (Xiphosura, Chelicerata) and their afferent and efferent projections.* Arthropod Struct. Dev., 2006. 35(4): 261–274.

Boles, L.C., and K.J. Lohmann, *True navigation and magnetic maps in spiny lobsters.* Nature, 2003. 421(6918): 60–63.

Callaerts, P., et al., *Pax6 and eye development in Arthropoda.* Arthropod Struct. Dev., 2006. 35(4): 379–391.

Chiou, T.H., et al., *Circular polarization vision in a stomatopod crustacean.* Curr. Biol., 2008. 18(6): 429–434.

Cronin, T., and M. Porter, *Exceptional variation on a common theme: The evolution of crustacean compound eyes.* Evol. Educ. Outreach, 2008. 1(4): 463–475.

Cronin, T.W., and C.A. King, *Spectral sensitivity of vision in the mantis shrimp, Gonodactylus oerstedii, determined using noninvasive optical Techniques.* Biol. Bull., 1989. 176(3): 308–316.

Cronin, T.W., N.J. Marshall, and R.L. Caldwell, *Spectral tuning and the visual ecology of mantis shrimps.* Phil. Trans. R. Soc. Lond. B:

 Biological Sciences, 2000. 355(1401): 1263–1267.

Cronin, T.W., et al., *Specialization of retinal function in the compound eyes of mantis shrimps.* Vision Res, 1994. 34(20): 2639–2656.

Dahl, D., and C.C. Krischer, *Evidence for a bistable photopigment contained in barnacle median photoreceptor.* Vision Res., 1976. 16(10): 1188–1190.

Doughtie, D.G., and K.R. Rao, *Ultrastructure of the eyes of the grass shrimp, Palaemonetes pugio.* Cell and Tissue Research, 1984. 238(2): 271–288.

Elofsson, R., *The frontal eyes of crustaceans.* Arthropod Struct. Dev., 2006. 35(4): 275–291.

Exner, S., and R.C. Hardie, *The physiology of the compound eyes of insects and crustaceans: A study.* 1989, Berlin, New York: Springer-Verlag.

Fernald, R.D., *The evolution of eyes.* Brain Behav. Evol., 1997. 50(4): 253–259.

Gaten, E., *Optics and phylogeny: Is there an insight? The evolution of superposition eyes in the Decapoda (Crustacea).* Contrib. Zool., 1998. 67(4): 223–236.

Gaten, E., *Apposition compound eyes of Spongicoloides koehleri (Crustacea : Spongicolidae) are derived by neoteny.* J. Marine Biol. Assoc. UK, 2007. 87(2): 483–486.

Hafner, et al., *Retinal development in the lobster Homarus americanus.* Cell Tissue Res., 2001. 305(1): 147–158.

Harzsch, S., and G. Hafner, *Evolution of eye development in arthropods: Phylogenetic aspects.* Arthropod Struct. Dev., 2006. 35(4): 319–340.

Harzsch, S., and R. Melzer, *Origin and evolution of arthropod visual systems. Introduction.* Arthropod Struct. Dev., 2006. 35(4): 209–210.

Harzsch, S., et al., *Evolution of arthropod visual systems: Development of the eyes and central visual pathways in the horseshoe crab Limulus polyphemus Linnaeus, 1758 (Chelicerata, Xiphosura).* Dev. Dynam., 2006. 235(10): 2641–2655.

Irving, T.H.K., A.G. Peele, and K.A. Nugent, *Optical metrology for analysis of lobster-eye x-ray optics.* Appl. Opt., 2003. 42(13): 2422–2430.

Kennedy, D., and M.S. Bruno, *The spectral sensitivity of crayfish and lobster vision.* J. Gen. Physiol., 1961. 44: 1089–1102.

Keskinen, E., and V.B. Meyer-Rochow, *Post-embryonic photoreceptor development and dark/light adaptation in the spittle bug Philaenus spumarius (L.) (Homoptera, Cercopidae).* Arthropod Struct. Dev., 2004. 33(4): 405–417.

Labhart, T., and C.A. Wiersma, *Habituation and inhibition in a class of visual interneurons of the rock lobster, Panulirus interruptus.* Comp. Biochem. Physiol. A Comp. Physiol., 1976. 55(3): 219–224.

Land, M.F., *Compound eyes: Old and new optical mechanisms.* Nature, 1980. 287(5784): 681–686.

Land, M.F., and D.E. Nilsson, *Observations on the compound eyes of the deep-sea ostracod Macrocypridina castanea.* J. Exp. Biol., 1990. 148(1): 221–233.

Lau, T.F.S., E. Gross, and V.B. Meyer-Rochow, *Sexual dimorphism and light/dark adaptation in the compound eyes of male and female Acentria ephemerella (Lepidoptera: Pyraloidea: Crambidae).* 2007, Universität Konstanz.

Lavery, S., et al., *Phylogenetic relationships and evolutionary history of the shrimp genus Penaeus s.l. derived from mitochondrial DNA.* Mol. Phylogenet. Evol., 2004. 31(1): 39–49.

Marshall, J., T.W. Cronin, and S. Kleinlogel, *Stomatopod eye structure and function: A review.* Arthropod Struct. Dev., 2007. 36(4): 420–448.

Marshall, J., et al., *Behavioural evidence for polarisation vision in stomatopods reveals a potential channel for communication.* Curr. Biol., 1999. 9(14): 755–758.

Marshall, J., and J. Oberwinkler, *The colourful world of the mantis shrimp.* Nature, 1999. 401(6756): 873–874.

Marshall, N.J., and M.F. Land, *Some optical features of the eyes of stomatopods.* J. Comp. Physiol. A: Neuroethol. Sens. Neural Behav. Physiol., 1993. 173(5): 565–582.

Melzer, R.R., et al., *Compound eye evolution: Highly conserved retinula and cone cell patterns indicate a common origin of the insect and crustacean ommatidium.* Naturwissenschaften, 1997. 84(12): 542–544.

Meyer-Rochow, V.B., *Larval and adult eye of the western rock lobster (Panulirus longipes).* Cell Tissue Res., 1975. 162(4): 439–457.

Muller, C.H., A. Sombke, and J. Rosenberg, *The fine structure of the eyes of some bristly millipedes (Penicillata, Diplopoda): Additional support for the homology of mandibulate ommatidia.* Arthropod Struct. Dev., 2007. 36(4): 463–476.

Nilsson, D.-E., *Three unexpected cases of refracting superposition eyes in crustaceans.* J. Comp. Physiol. A: Neuroethol. Sens. Neural Behav. Physiol., 1990. 167(1): 71–78.

Nilsson, D.E., *Evolutionary links between apposition and superposition optics in crustacean eyes.* Nature, 1983. 302(5911): 818–821.

Nilsson, D.E., *A new type of imaging optics in compound eyes.* Nature, 1988. 332(6159): 76–78.

Nilsson, D.E., *From cornea to retinal image in invertebrate eyes.* Trends Neurosci., 1990. 13(2): 55–64.

Nilsson, D.E., and R. Modlin, *A mysid shrimp carrying a pair of binoculars*. J. Exp. Biol., 1994. 189(1): 213–236.

Nilsson, D.-E., and A. Kelber, *A functional analysis of compound eye evolution*. Arthropod Struct. Dev., 2007. 36(4): 373–385.

Oakley, T.H., D.C. Plachetzki, and A.S. Rivera, *Furcation, field-splitting, and the evolutionary origins of novelty in arthropod photoreceptors*. Arthropod Struct. Dev., 2007. 36(4): 386–400.

Osorio, D., and J.P. Bacon, *A good eye for arthropod evolution*. Bioessays, 1994. 16(6): 419–424.

Parker, A.R., *Natural photonic engineers*. Materials Today, 2002. 5(9): 26–31.

Patek, S.N., et al., *Linkage mechanics and power amplification of the mantis shrimp's strike*. Journal of Experimenal Biology 2007. 3677–3688.

Quan, J., et al., *Phylogenetic relationships of 12 penaeoidea shrimp species deduced from mitochondrial DNA sequences*. Biochem. Genet., 2004. 42(9): 331–345.

Sandeman, D.C., G. Scholtz, and R.E. Sandeman, *Brain evolution in decapod crustacea*. J. Exp. Zool., 1993. 265(2): 112–133.

Smith, W.C., et al., *Opsins from the lateral eyes and ocelli of the horseshoe crab*, Limulus polyphemus. Proc. Natl. Acad. Sci. USA, 1993. 90(13): 6150–6154.

Staton, J.L., L.L. Daehler, and W.M. Brown, *Mitochondrial gene arrangement of the horseshoe crab* Limulus polyphemus *L.: Conservation of major features among arthropod classes*. Mol. Biol. Evol., 1997. 14(8): 867–874.

Stavenga, D.G., Harzsch, S. *Origin and evolution of arthropod visual systems: Introduction to Part II*. Arthropod Struct. Dev., 2007. 36: 371–372.

Toh, Y., *Diurnal changes of rhabdom structures in the compound eye of the grapsid crab*, Hemigrapsus penicillatus. J. Electron Microsc. (Tokyo), 1987. 36(4): 213–223.

Trevor, J.C., and J.B. Simon, *The phylogeny of arachnomorph arthropods and the origin of the Chelicerata*. Trans. Earth Sci., 2004. 94: 169–193.

Vogt, K., *Die Spiegeloptik des Flußkrebsauges*. J. Comp. Physiol. A: Neuroethol. Sens. Neural. Behav. Physiol., 1980. 135(1): 1–19.

Wiersma, C.A.G., and T. Yamaguchi, *The integration of visual stimuli in the rock lobster*. Vision Res., 1967. 7(3–4): 197–203, IN1–IN2.

Wiersma, C.A.G., and B. York, *Properties of the seeing fibers in the rock lobster: Field structure, habituation, attention and distraction*. Vision Res., 1972. 12(4): 627–640, IN1–IN5.

Wilkens, L.A., and J.L. Larimer, *The CNS Photoreceptor of crayfish: Morphology and synaptic activity*. J. Comp. Physiol. A: Neuroethol. Sens. Neural. Behav. Physiol., 1972. 80(4): 389–407.

Wolburg-Buchholz, K., *The superposition eye of* Cloeon dipterum: *The organization of the lamina ganglionaris*. Cell Tissue Res., 1977. 177(1): 9–28.

York, B., and C.A. Wiersma, *Visual processing in the rock lobster (crustacea)*. Prog. Neurobiol., 1975. 5(2): 127–166.

CHAPTER 7

Blair, J.E., and S.B. Hedges, *Molecular phylogeny and divergence times of deuterostome animals*. Mol. Biol. Evol., 2005. 22(11): 2275–2284.

Butler, A.B., *Sensory system evolution at the origin of craniates*. Phil. Trans. R. Soc. Lond. B: Biol. Sci., 2000. 355(1401): 1309–1313.

Cloney, R.A., *Ascidian larvae and the events of metamorphosis*. Am. Zool., 1982. 22(4): 817–826.

Collin, S.P., Farrel, A.P. McKenzie, D.J. and Colin, J.B. *Nervous and sensory systems in fish physiology*. 2007, New York, NY: Academic Press. 121–179.

Collin, S.P., et al., *Morphology and spectral absorption characteristics of retinal photoreceptors in the southern hemisphere lamprey (Geotria australis)*. Vis Neurosci. 2003 Mar-Apr;20(2):119-30.

Collin, S.P., et al. Vis Neurosci. 2004 Sep-Oct;21(5):765-73. *Vision in the southern hemisphere lamprey* Mordacia mordax: *Spatial distribution, spectral absorption characteristics, and optical sensitivity of a single class of retinal photoreceptor*. 2004.

Collin, S.P., et al., *Ancient colour vision: Multiple opsin genes in the ancestral vertebrates*. Curr. Biol., 2003. 13(22): R864–R865.

Collin, S.P., I.C. Potter, and C.R. Braekevelt, *The ocular morphology of the southern hemisphere lamprey* Geotria australis *gray, with special reference to optical specialisations and the characterisation and phylogeny of photoreceptor types*. Brain Behav. Evol., 1999. 54(2): 96–118.

Collin, S.P., and I.C. Pottert, *The ocular morphology of the southern hemisphere lamprey* Mordacia mordax *Richardson with special reference to a single class of photoreceptor and a retinal tapetum*. Brain Behav. Evol., 2000. 55(3): 120–138.

Collin, S.P. and A.E.O. Trezise, *The origins of colour vision in vertebrates*. Clin. Exp. Optom., 2004. 87(4–5): 217–223.

Cronin, T.W., and N.J. Marshall, *A retina with at least ten spectral types of photoreceptors in a mantis shrimp*. Nature, 1989. 339(6220): 137–140.

Donoghue, P.C.J., P.L. Forey, and R.J. Aldridge, *Conodont affinity and chordate phylogeny*. Biol. Rev., 2000. 75(2): 191–251.

Fernholm, B., and K. Holmberg, *The eyes in three genera of hagfish* (Eptatretus, Paramyxine *and* Myxine)—*A case of degenerative evolution*. Vision Res., 1975. 15(2): 253–259, IN1–IN4.

Gess, R.W., M.I. Coates, and B.S. Rubidge, *A lamprey from the Devonian period of South Africa*. Nature, 2006. 443(7114): 981–984.

Gill, H.S., et al., *Phylogeny of living parasitic lampreys (petromyzontiformes) based on morphological data*. Copeia, 2003. 2003(4): 687–703.

Gustafsson, O.S., Ekström, P., and Kröger, R.H., *A fibrous membrane suspends the multifocal lens in the eyes of lampreys and African lungfishes*. J. Morphol., 2010. 271(8): 980–989.

Janvier, P., *Evolutionary biology: Born-again hagfishes*. Nature, 2007. 446(7136): 622–623.

Kuo, C.-H., S. Huang, and S.-C. Lee, *Phylogeny of hagfish based on the mitochondrial 16S rRNA gene*. Mol. Phylogenet. Evol., 2003. 28(3): 448–457.

Kusunoki, T., and F. Amemiya, *Retinal projections in the hagfish*, Eptatretus burgeri. Brain Res., 1983. 262(2): 295–298.

Lacalli, T., *Evolutionary biology: Body plans and simple brains*. Nature, 2003. 424(6946): 263–264.

Lacalli, T.C., *Frontal eye circuitry, rostral sensory pathways and brain organization in* Amphioxus *larvae: Evidence from 3D reconstructions*. Phil. Trans. R. Soc. Lond. B: Biol. Sci., 1996. 351(1337): 243–263.

Lacalli, T.C., *New perspectives on the evolution of protochordate sensory and locomotory systems, and the origin of brains and heads*. Phil. Trans. R. Soc. Lond. B: Biol. Sci., 2001. 356(1414): 1565–1572.

Lacalli, T.C., *Sensory systems in* Amphioxus: *A window on the ancestral chordate condition*. Brain Behav. Evol., 2004. 64(3): 148–162.

Lacalli, T.C., and L.Z. Holland, *The developing dorsal ganglion of the salp* Thalia democratica, *and the nature of the ancestral chordate brain*. Phil. Trans. R. Soc. Lond. B: Biol. Sci., 1998. 353(1378): 1943–1967.

Ladich, F., *Communication in fishes*. 2006, Enfield, NH: Science Publishers.

Lamb, T.D., S.P. Collin, and E.N. Pugh Jr., *Evolution of the vertebrate eye: opsins, photoreceptors, retina and eye cup*. Nat Rev Neurosci, 2007. 8(12): 960–976.

Mallatt, J., and J.-y. Chen, *Fossil sister group of craniates: Predicted and found*. J. Morphol., 2003. 258(1): 1–31.

Maximov, V.V., *Environmental factors which may have led to the appearance of colour vision*. Philos Trans R Soc Lond B Biol Sci. 2000 September 29; 355(1401): 1239–1242.

Maximov, V.V., *Environmental factors which may have led to the appearance of colour vision*. Phil. Trans. R. Soc. Lond. B: Biol. Sci., 2000. 355(1401): 1239–1242.

Meyer-Rochow, V.B., *Axonal wiring and polarisation sensitivity in eye of the rock lobster*. Nature, 1975. 254(5500): 522–523.

Millar, R.H., S.R. Frederick, and Y. Maurice, *The Biology of Ascidians, Advances in Marine Biology*. 1971, London, Academic Press. 1–100.

Nicholas, D.H., *Hagfish embryos again—the end of a long drought*. BioEssays, 2007. 29(9): 833–836.

Nilsson, D.-E., and S. Pelger, *A Pessimistic Estimate of the Time Required for an Eye to Evolve*. Proc. R. Soc. Lond. B: Biol. Sci., 1994. 256(1345): 53–58.

Northcutt, R.G., *Lancelet lessons: Evaluating a phylogenetic model*. J. Comp. Neurol., 2001. 435(4): 391–393.

Ooka-Souda, S., et al., *A possible retinal information route to the circadian pacemaker through pretectal areas in the hagfish*, Eptatretus burgeri. Neurosci. Lett., 1995. 192(3): 201–204.

Shaun, P.C., and E.O.T. Ann, *The origins of colour vision in vertebrates*. Clin. Exp. Optom., 2004. 87(4–5): 217–223.

Shu, D., and S.C. Morris, *Response to Comment on "A New Species of Yunnanozoan with Implications for Deuterostome Evolution."* Science, 2003. 300(5624): 1372.

Shu, D.G., et al., *Lower Cambrian vertebrates from south China*. Nature, 1999. 402(6757): 42–46.

Shu, D.G., et al., *Head and backbone of the Early Cambrian vertebrate Haikouichthys*. Nature, 2003. 421(6922): 526–529.

Shu, D.G., S.C. Morris, and X.L. Zhang, *A Pikaia-like chordate from the Lower Cambrian of China*. Nature, 1996. 384(6605): 157–158.

Sweet, W.C., and P.C.J. Donoghue, *Conodonts: Past, present, future. Journal of Paleontology*; November 2001; v. 75; no. 6; p. 1174–1184.

Trinajstic, K., et al., *Exceptional preservation of nerve and muscle tissues in Late Devonian placoderm fish and their evolutionary implications*. Biol. Lett., 2007. 3(2): 197–200.

Vigh-Teichmann, I., et al., *Opsin-immunoreactive outer segments of photoreceptors in the eye and in the lumen of the optic nerve of the hagfish*, Myxine glutinosa. Cell Tissue Res., 1984. 238(3): 515–522.

Vorobyev, M., *Ecology and evolution of primate colour vision*. Clin. Exp. Optom., 2004. 87(4–5): 230–238.

Wicht, H., and T.C. Lacalli, *The nervous system of amphioxus: Structure, development, and evolutionary significance*. Can. J. Zool., 2005. 83: 122–150.

Xian-guang, H., et al., *New evidence on the anatomy and phylogeny of the earliest vertebrates*. Proc Biol Sci. 2002 September 22; 269(1503): 1865–1869.

Chapter 8

Barber, V.C., E.M. Evans, and M.F. Land, *The fine structure of the eye of the mollusc Pecten maximus*. Z. Zellforsch. Mikrosk. Anat., 1967. 76(3): 25–312.

Bever, M.M., and R.B. Borgens, *Eye regeneration in the mystery snail*. J. Exp. Zool., 1988. 245(1): 33–42.

Blumer, M., *Alterations of the eyes during ontogenesis in* Aporrhais pespelecani *(Mollusca, Caenogastropoda)*. Zoomorphology, 1996. 116(3): 123–131.

Cronly-Dillon, J.R., *Spectral sensitivity of the scallop* Pecten maximus. Science, 1966. 151(3708): 345–346.

Eakin, R.M., and J.L. Brandenburger, *Understanding a Snail's Eye at a Snail's Pace*. Am. Zool., 1975. 15(4): 851–863.

Eakin, R.M., and J.L. Brandenburger, *Retinal differences between light-tolerant and light-avoiding slugs (Mollusca: Pulmonata)*. J. Ultrastruct. Res., 1975. 53(3): 382–394.

Frýda, J., and R.B. Blodgett, *Two new Cirroidean genera (Vetigastropoda, Archaeogastropoda) from the Emsian (Late Early Devonian) of Alaska with notes on the early phylogeny of Cirroidea*. Journal of Paleontology, 1998. 72(2): 265–273.

Gillary, H.L., *Electrical potentials from the regenerating eye of Strombus*. J. Exp. Biol., 1983. 107(1): 293–310.

Hara, T., et al., *Rhodopsin and retinochrome in the retina of a tetrabranchiate cephalopod*, Nautilus pompilius. Zoological Science, 2009. 12(2): 195–201.

Hughes, H.P.I., *Structure and regeneration of the eyes of strombid gastropods*. Cell Tissue Res., 1976. 171(2): 259–271.

Jan-Olof, S., *Structure and optics of the eye of the hawk-wing conch, Strombus raninus (L.)*. J. Exp. Zool., 1994. 268(3): 200–207.

Kandel, E.R., *Small systems of neurons*. Sci. Am., 1979. 241(3): 66–76.

Land, M.F., *Image formation by a concave reflector in the eye of the scallop*, Pecten maximus. J. Physiol., 1965. 179(1): 138–153.

Land, M.F., *Activity in the optic nerve of* Pecten maximus *in response to changes in light intensity, and to pattern and movement in the optical environment*. J. Exp. Biol., 1966. 45(1): 83–99.

Land, M.F., *The spatial resolution of the pinhole eyes of giant clams (Tridacna maxima)*. Proc. Biol. Sci., 2003. 270(1511): 185–188.

Latiolais, J.M., et al., *A molecular phylogenetic analysis of strombid gastropod morphological diversity*. Mol. Phylogenet. Evol., 2006. 41(2): 436–444.

Michael, F.L., *Eyes with mirror optics*. J. Optics A: Pure Appl. Optics, 2000. 2(6): R44.

Nilsson, D.-E., *Eyes as optical alarm systems in fan worms and ark clams*. Phil. Trans. Biol. Sci., 1994. 346(1316): 195–212.

Seyer, J.O., D.E. Nilsson, and E. Warrant, *Spatial vision in the prosobranch gastropod* Ampularia sp. J. Exp. Biol., 1998. 201(10): 1673–1679.

Shu, D.G., et al., *Head and backbone of the Early Cambrian vertebrate Haikouichthys*. Nature, 2003. 421(6922): 526–529.

Speiser, D.I., and S. Johnsen, *Scallops visually respond to the size and speed of virtual particles*. J. Exp. Biol., 2008. 211(13): 2066–2070.

Speiser, D.I., and S. Johnsen, *Comparative morphology of the concave mirror eyes of scallops (Pectinoidea)*. Am. Malacol. Bull., 2008. 26(1–2): 27–33.

Van Roy, P., et al., *Ordovician faunas of Burgess Shale type*. Nature, 2010. 465(7295): 215–218.

Wilkens, L.A., *The visual system of the giant clam* Tridacna: *Behavioral adaptations*. Biol. Bull., 1986. 170(3): 393–408.

Wilkens, L.A., *Hyperpolarizing photoreceptors in the eyes of the giant clam* Tridacna: *Physiological evidence for both spiking and non-spiking cell types*. J. Comp. Physiol. A, 1988. 163(1): 73–84.

Zieger, M.V., and V.B. Meyer-Rochow, *Understanding the cephalic eyes of pulmonate gastropods: A review.* Am. Malacol. Bull., 2008. 26(1–2): 47–66.

Zieger, M.V., et al., *Eyes and vision in* Arion rufus *and* Deroceras agreste *(Mollusca; Gastropoda; Pulmonata): What role does photoreception play in the orientation of these terrestrial slugs?* Acta Zool., 2009. 90(2): 189–204.

Chapter 9

Anctil, M., and M.A. Ali, *Letter: Giant ganglion cells in the retina of the hammerhead shark* (Sphyrna lewini). Vision Res., 1974. 14(9): 903–904.

Anderson, P.S., and M.W. Westneat, *Feeding mechanics and bite force modelling of the skull of* Dunkleosteus terrelli, *an ancient apex predator.* Biol. Lett., 2007. 3(1): 76–79.

Basden, A.M., et al., *The most primitive osteichthyan braincase?* Nature, 2000. 403(6766): 185–188.

Block, B.A., and F.G. Carey, *Warm brain and eye temperatures in sharks.* J. Comp. Physiol. B, 1985. 156(2): 229–236.

Bodznick, D., *Elasmobranch vision: Multimodal integration in the brain.* J. Exp. Zool. Suppl., 1990. 5: 108–116.

Bozzano, A., and S.P. Collin, *Retinal ganglion cell topography in elasmobranchs.* Brain Behav. Evol., 2000. 55(4): 191–208.

Bromm, B., H. Hensel, and K. Nier, *Response of the ampullae of Lorenzini to static combined electric and thermal stimuli in Scyliorhinus canicula.* Cell. Mol. Life Sci., 1975. 31(5): 615–618.

Bullock, T.H., *Processing of ampullary input in the brain: Comparison of sensitivity and evoked responses among elasmobranch and siluriform fishes.* J. Physiol. (Paris), 1979. 75(4): 397–407.

Burrow, C.J., A.S. Jones, and G.C. Young, *X-ray microtomography of 410 million-year-old optic capsules from placoderm fishes.* Micron, 2005. 36(6): 551–557.

David, B., *Elasmobranch vision: Multimodal integration in the brain.* J. Exp. Zool., 1990. 256(S5): 108–116.

Demian, D.C., et al., *Predominance of genetic monogamy by females in a hammerhead shark,* Sphyrna tiburo: *Implications for shark conservation.* Mol. Ecol., 2004. 13(7): 1965–1974.

Donley, J.M., et al., *Thermal dependence of contractile properties of the aerobic locomotor muscle in the leopard shark and shortfin mako shark.* J. Exp. Biol., 2007. 210(7): 1194–1203.

Ebbesson, S.O., and D.L. Meyer, *The visual system of the guitar fish* (Rhinobatos productus). Cell Tissue Res., 1980. 206(2): 243–250.

Fernald, R.D. – 1988. Aquatic adaptation on eye design. In: J. Atema, R.R. Fay, A.N. Popper and W.N. Tavolga (eds.), Sensory biology of aquatic animals, pp. 435–466. Springer-Verlag, New York. Fernald, R.D., *Aquatic adaptations in fish eyes, in Sensory biology of aquatic animals.* 1988, New York: Springer-Verlag. 435–466.

Goldman, J.N., and G.B. Benedek, *The relationship between morphology and transparency in the nonswelling corneal stroma of the shark.* Invest. Ophthalmol. Vis. Sci., 1967. 6(6): 574–600.

Hart, N.S., T.J. Lisney, and S.P. Collin, *Visual communication in elasmobranchs, in Communication in fishes,* F. Ladich, et al., editors. 2006, Enfield, NH: Science Publishers. 56.

Hart, N.S., et al., *Multiple cone visual pigments and the potential for trichromatic colour vision in two species of elasmobranch.* J. Exp. Biol., 2004. 207(Pt 26): 4587–4594.

Hove, J.R., and S.A. Moss, *Effect of MS-222 on response to light and rate of metabolism of the little skate* Raja erinacea. Marine Biol., 1997. 128(4): 579–583.

Hueter, R.E., et al., *Refractive state and accommodation in the eyes of free-swimming versus restrained juvenile lemon sharks* (Negaprion brevirostris). Vision Res., 2001. 41(15): 1885–1889.

Joseph, T.E., *Ocular morphology in antarctic notothenioid fishes.* J. Morphol., 1988. 196(3): 283–306.

Kajiura, S.M., and K.N. Holland, *Electroreception in juvenile scalloped hammerhead and sandbar sharks.* J. Exp. Biol., 2002. 205(23): 3609–3621.

Knight, K., *Hammerheads' wide heads give impressive stereo view.* J. Exp. Biol., 2009. 212(24): i.

Ladich, F., *Communication in fishes.* 2006, Enfield, NH: Science Publishers.

Land, M.F., *Image formation by a concave reflector in the eye of the scallop,* Pecten maximus. J. Physiol., 1965. 179(1): 138–153.

Lisney, T.J., and S.P. Collin, *Relative eye size in elasmobranchs.* Brain Behav. Evol., 2007. 69(4): 266–279.

Maisey, J.G., and M.E. Anderson, *A primitive chondrichthyan braincase from the early Devonian of South Africa.* Journal of Vertebrate Paleontology, 2001. 21(4): 702–713.

McComb, D.M., T.C. Tricas, and S.M. Kajiura, *Enhanced visual fields in hammerhead sharks.* J. Exp. Biol., 2009. 212(24): 4010–4018.

Miles, R.S., *The Holonematidae (placoderm fishes), a review based on new specimens of holonema from the Upper Devonian of Western Australia.* Phil. Trans. R. Soc. Lond. B: Biol. Sci., 1971. 263(849): 101–234.

Naka, K., et al., *Dynamics of skate horizontal cells.* J. Gen. Physiol., 1988. 92(6): 811–831.

Nilsson, D.-E., *Eyes as optical alarm systems in fan worms and ark clams.* Phil. Trans. Biol. Sci., 1994. 346(1316): 195–212.

Reid, D.G., *A cladistic phylogeny of the genus* Littorina *(Gastropoda): Implications for evolution of reproductive strategies and for classification.* Hydrobiologia, 1990. 193(1): 1–19.

Sivak, J.G. and C.A. Luer, *Optical development of the ocular lens of an elasmobranch,* Raja elanteria. Vision Res., 1991. 31(3): 373–382.

Kajiura SM, Forni JB, Summers AP. *Olfactory morphology of carcharhinid and sphyrnid sharks: Does the cephalofoil confer a sensory advantage?* J Morphol. 2005 Jun;264(3):253-63.

Stoddart, D.M., *External nares and olfactory perception.* Cell. Mol. Life Sci., 1979. 35(11): 1456–1457.

Wilga, C.D., and P.J. Motta, *Durophagy in sharks: Feeding mechanics of the hammerhead* Sphyrna tiburo. J. Exp. Biol., 2000. 203(18): 2781–2796.

Young, G., *Early evolution of the vertebrate eye—fossil evidence.* Evol. Educ. Outreach, 2008. 1(4): 427–438.

Young, G.C., *Number and arrangement of extraocular muscles in primitive gnathostomes: Evidence from extinct placoderm fishes.* Biol. Lett., 2008. 4(1): 110–114.

Zigman, S., and P.W. Gilbert, *Lens colour in sharks.* Exp. Eye Res., 1978. 26(2): 227–231.

Chapter 10

Albensi, B.C., and J.H. Powell, *The differential optomotor response of the four-eyed fish* Anableps anableps. Perception, 1998. 27(12): 1475–1483.

Altringham, J.D., and B.A. Block, *Why do tuna maintain elevated slow muscle temperatures? Power output of muscle isolated from endothermic and ectothermic fish.* J. Exp. Biol., 1997. 200(Pt 20): 2617–2627.

Avery, J.A., and J.K. Bowmaker, *Visual pigments in the four-eyed fish,* Anableps anableps. Nature, 1982. 298(5869): 62–63.

Baylor, E.R., *Air and water vision of the Atlantic flying fish,* Cypselurus heterurus. Nature, 1967. 214(5085): 307–309.

B.A. Block, Physiology and ecology of brain and eye heaters in billfish. In: R.H. Stroud, Editor, *Planning the future of billfishes,* National Coalition Marine Conservation, New York (1990), pp. 123–136.

Borwein, B., and M.J. Hollenberg, *The photoreceptors of the "four-eyed" fish,* Anableps anableps *L.* J. Morphol., 1973. 140(4): 405–441.

Bowmaker, J.K., *The visual pigments of fish.* Prog. Retin. Eye Res., 1995. 15(1): 1–31.

Bowmaker, J.K., A. Thorpe, and R.H. Douglas, *Ultraviolet-sensitive cones in the goldfish.* Vision Res., 1991. 31(3): 349–352.

Bowmaker, J.K., and H.-J. Wagner, *Pineal organs of deep-sea fish: Photopigments and structure.* J. Exp. Biol., 2004. 207(14): 2379–2387.

Boyer, C.B., *The rainbow from myth to mathematics.* 1959, New York: T. Yoseloff.

Bridges, C.D., *Porphyropsin in retina of four-eyed fish,* Anableps anableps. Nature, 1982. 300(5890): 384.

Burnside, B., and S. Basinger, *Retinomotor pigment migration in the teleost retinal pigment epithelium. II. Cyclic-3',5'-adenosine monophosphate induction of dark-adaptive movement* in vitro. Invest. Ophthalmol. Vis. Sci., 1983. 24(1): 16–23.

Chang, C.H., C.C. Chiao, and H.Y. Yan, *The structure and possible functions of the milkfish* Chanos chanos *adipose eyelid.* J. Fish. Biol., 2009. 75(1): 87–99.

Chang, C.H., C.C. Chiao, and H.Y. Yan, *Ontogenetic changes in color vision in the milkfish* (Chanos chanos *Forsskal, 1775).* Zool. Sci., 2009. 26(5): 349–355.

Chen, N., et al., *Molecular cloning of a rhodopsin gene from salamander rods.* Invest. Ophthalmol. Vis. Sci., 1996. 37(9): 1907–1913.

Collin, S.P. and H.B. Collin, *Topographic analysis of the retinal ganglion cell layer and optic nerve in the sandlance* Limnichthyes fasciatus (Creeiidae, Perciformes). J. Comp. Neurol., 1988. 278(2): 226–241.

Collin, S.P., and H.B. Collin, *The foveal photoreceptor mosaic in the pipefish,* Corythoichthyes paxtoni (Syngnathidae, Teleostei). Histol. Histopathol., 1999. 14(2): 369–382.

Collin, S.P., R.V. Hoskins, and J.C. Partridge, *Tubular eyes of deep-sea fishes: a comparative study of retinal topography.* Brain Behav Evol, 1997. 50(6): 335–357.

Collin, S.P., R.V. Hoskins, and J.C. Partridge, *Seven retinal specializations in the tubular eye of the deep-sea pearleye,* Scopelarchus michaelsarsi: *A case study in visual optimization.* Brain Behav. Evol., 1998. 51(6): 291–314.

Collin, S.P., D.J. Lloyd, and H.J. Wagner, *Foveate vision in deep-sea teleosts: A comparison of primary visual and olfactory inputs.* Phil. Trans. R. Soc. Lond. B: Biol. Sci., 2000. 355(1401): 1315–1320.

Collin, S.P., and R.G. Northcutt, *The visual system of the Florida garfish,* Lepisosteus platyrhincus (Ginglymodi). IV. Bilateral projections and the binocular visual field. Brain Behav. Evol., 1995. 45(1): 34–53.

Colombini, I., et al., *Foraging strategy of the mudskipper* Periophthalmus sobrinus *Eggert in a Kenyan mangrove.* J. Exp. Marine Biol. Ecol., 1996. 197(2): 219–235.

Crescitelli, F., M. McFall-Ngai, and J. Horwitz, *The visual pigment sensitivity hypothesis: Further evidence from fishes of varying habitats.* J. Comp. Physiol. A, 1985. 157(3): 323–333.

Dill, L.M., *Refraction and the spitting behavior of the archerfish* (Toxotes chatareus). Behav. Ecol. Sociobiol., 1977. 2(2): 169–184.

Douglas, R.H., C.W. Mullineaux, and J.C. Partridge, *Long-wave sensitivity in deep-sea stomiid dragonfish with far-red bioluminescence: Evidence for a dietary origin of the chlorophyll-derived retinal photosensitizer of* Malacosteus niger. Phil. Trans. R. Soc. Lond. B: Biol. Sci., 2000. 355(1401): 1269–1272.

Douglas, R.H., et al., *Enhanced retinal longwave sensitivity using a chlorophyll-derived photosensitiser in* Malacosteus niger, *a deep-sea dragon fish with far red bioluminescence.* Vision Res., 1999. 39(17): 2817–2832.

Douglas, R.H., J.C. Partridge, and N.J. Marshall, *The eyes of deep-sea fish. I: Lens pigmentation, tapeta and visual pigments.* Prog. Retin. Eye Res., 1998. 17(4): 597–636.

Douglas, R.H., and A. Thorpe, *Short-wave absorbing pigments in the ocular lenses of deep-sea teleosts.* J. Marine Biol. Assoc. UK, 1992. 72(01): 93–112.

Easter, S.S., Jr., *Retinal growth in foveated teleosts: Nasotemporal asymmetry keeps the fovea in temporal retina.* J. Neurosci., 1992. 12(6): 2381–2392.

Elshoud, G.C.A., and Koomen, P. *A biomechanical analysis of spitting in archer fishes (Pisces, Perciformes, Toxidae).* Zoomorphology, 1985. 105(4): 240–252.

Evans, B.I., and R.D. Fernald, *Retinal transformation at metamorphosis in the winter flounder* (Pseudopleuronectes americanus). Vis. Neurosci., 1993. 10(6): 1055–1064.

Fang, M., et al., *Retinal twin cones or retinal double cones in fish: Misnomer or different morphological forms?* Int. J. Neurosci., 2005. 115(7): 981–987.

Fernald, R.D., *Retinal projections in the African cichlid fish,* Haplochromis burtoni. J. Comp. Neurol., 1982. 206(4): 379–389.

Fernald, R.D., *Cone mosaic in a teleost retina: No difference between light and dark adapted states.* Cell. Mol. Life Sci., 1982. 38(11): 1337–1339.

Fernald, R.D., and S.E. Wright, *Maintenance of optical quality during crystalline lens growth.* Nature, 1983. 301(5901): 618–620.

Ferry-Graham, L.A., P.C. Wainwright, and D.R. Bellwood, *Prey capture in long-jawed butterflyfishes (Chaetodontidae): The functional basis of novel feeding habits.* J. Exp. Marine Biol. Ecol., 2001. 256(2): 167–184.

Ferry-Graham, L.A., et al., *Evolution and mechanics of long jaws in butterflyfishes (family Chaetodontidae).* J. Morphol., 2001. 248(2): 120–143.

Franz-Odendaal, T.A., and B.K. Hall, *Skeletal elements within teleost eyes and a discussion of their homology.* J. Morphol., 2006. 267(11): 1326–1337.

Friedman, M., *The evolutionary origin of flatfish asymmetry.* Nature, 2008. 454(7201): 209–212.

Fritsches, K.A., R.W. Brill, and E.J. Warrant, *Warm eyes provide superior vision in swordfishes.* Curr. Biol., 2005. 15(1): 55–58.

Fritsches, K.A., and J. Marshall, *A new category of eye movements in a small fish.* Curr. Biol., 1999. 9(8): R272–273.

Fritsches, Kerstin A., Marshall, N. Justin and Warrant, Eric J. (2003) . Retinal specializations in the blue marlin: eyes designed for sensitivity to low light levels. *Marine and Freshwater Research* 54 , 333–341.

Fritsches, K.A., et al., *Colour vision in billfish.* Phil. Trans. R. Soc. Lond. B: Biol. Sci., 2000. 355(1401): 1253–1256.

Gonzalez, R.M., et al., *Membrane formations in the pineal cells of the teleost* Gambusia affinis. J. Pineal Res., 1989. 7(4): 325–332.

Graf, W., and R. Baker, *The vestibuloocular reflex of the adult flatfish. I. Oculomotor organization.* J. Neurophysiol., 1985. 54(4): 887–899.

Graf, W. and R. Baker, *The vestibuloocular reflex of the adult flat-fish. II. Vestibulooculomotor connectivity.* J. Neurophysiol., 1985. 54(4): 900–916.

Hawryshyn, C.W., *Ultraviolet polarization vision in fishes: Possible mechanisms for coding e-vector.* Phil. Trans. R. Soc. Lond. B: Biol. Sci., 2000. 355(1401): 1187–1190.

Hunt, D.M., et al., *The molecular basis for spectral tuning of rod visual pigments in deep-sea fish.* J. Exp. Biol., 2001. 204(Pt 19): 3333–3344.

Inoue, J.G., et al., *Mitogenomic evidence for the monophyly of elopomorph fishes (Teleostei) and the evolutionary origin of the leptocephalus larva.* Mol. Phylogenet. Evol., 2004. 32(1): 274–286.

Janvier, P., *Palaeontology: Squint of the fossil flatfish.* Nature, 2008. 454(7201): 169–170.

John, C.M., *Low temperature increases gain in the fish oculomotor system.* J. Neurobiol., 1984. 15(4): 295–298.

Joseph, T.E., *Morphological specialization in Antarctic fishes.* Antarctic J. US, 1981. 16(5): 146–147.

Joseph, T.E., *Ocular morphology in antarctic notothenioid fishes.* J. Morphol., 1988. 196(3): 283–306.

Justin Marshall, N., *Communication and camouflage with the same bright colours in reef fishes.* hilos Trans R Soc Lond B Biol Sci. 2000 September 29; 355(1401): 1243–1248.

Kanungo, J., S.K. Swamynathan, and J. Piatigorsky, *Abundant corneal gelsolin in Zebrafish and the four-eyed fish,* Anableps anableps: *Possible analogy with multifunctional lens crystallins.* Exp. Eye Res., 2004. 79(6): 949–956.

Kapoor, B.G., and T.J. Hara, *Sensory biology of jawed fishes: New insights.* 2001, Enfield, NH: Science Publishers.

Kröger, R., K. Fritsches, and E. Warrant, *Lens optical properties in the eyes of large marine predatory teleosts.* J. Comp. Physiol. A: Neuroethol. Sens. Neural Behav. Physiol., 2009. 195(2): 175–182.

Land, M.F., *Visual optics: The sandlance eye breaks all the rules.* Curr. Biol., 1999. 9(8): R286–288.

Land, M.F., *On the functions of double eyes in midwater animals.* Phil. Trans. R. Soc. Lond. B: Biol. Sci., 2000. 355(1401): 1147–1150.

Locket, N.A., *On the lens pad of* Benthalbella infans, *a scopelarchid deep-sea teleost.* Phil. Trans. R. Soc. Lond. B: Biol. Sci., 2000. 355(1401): 1167–1169.

Loew, E.R., and J.N. Lythgoe, *The ecology of cone pigments in teleost fishes.* Vision Res., 1978. 18(6): 715–722.

Lythgoe, J.N., *The structure and function of iridescent corneas in teleost fishes.* Proc. R. Soc. Lond. B: Biol. Sci., 1975. 188(1093): 437–457.

McCosker, J.E., et al., Cottoclinus canops, *a new genus and species of Blenny (Perciformes: Labrisomidae) from the Galápagos Islands.* Proceedings of the California Academy of Sciences, 54, no. 8. 2003, San Francisco: California Academy of Sciences.

McFall-Ngai, M., et al., *Biochemical characteristics of the pigmentation of mesopelagic fish lenses.* Biol. Bull., 1988. 175(3): 397–402.

McFall-Ngai, M.J., and J. Horwitz, *A comparative study of the thermal stability of the vertebrate eye lens: Antarctic ice fish to the desert iguana.* Exp. Eye Res., 1990. 50(6): 703–709.

Meyer, D.L., C.R. Malz, and A.G. Jadhao, *Nervus terminalis projection to the retina in the "four-eyed" fish,* Anableps anableps. Neurosci. Lett., 1996. 213(2): 87–90.

Miller, R.R., *Ecology, habits and relationships of the Middle American cuatro ojos,* Anableps dowi *(Pisces: Anablepidae).* Copeia, 1979. 1979(1): 82–91.

Montgomery, J.C., and J.A. Macdonald, *Oculomotor function at low temperature: Antarctic versus temperate fish.* J. Exp. Biol., 1985. 117(1): 181–191.

Montgomery, J.C., and A.R. McVean, *Brain function in antarctic fish: Activity of central vestibular neurons in relation to head rotation and eye movement.* J. Comp. Physiol. A: Neuroethol. Sens. Neural Behav. Physiol., 1987. 160(2): 289–293.

Munk, O., *Ocular anatomy of some deep-sea teleosts. The Carlsberg foundation's oceanographical expedition round the world 1928–30 and previous "Dana"-expeditions.* Dana-report. 1966, Copenhagen: Høst.

Nieder, J., *Amphibious behaviour and feeding ecology of the four-eyed blenny (Dialommus fuscus, Labrisomidae) in the intertidal zone of the island of Santa Cruz (Galapagos, Ecuador).* J. Fish Biol., 2001. 58(3): 755–767.

Nilsson, D.E., and S. Pelger, *A pessimistic estimate of the time required for an eye to evolve.* Proc. Biol. Sci., 1994. 256(1345): 53–58.

Owens, G.L., et al., *A fish eye out of water: Ten visual opsins in the four-eyed fish,* Anableps anableps. PLoS One, 2009. 4(6): e5970.

Pankhurst, N.W., and J.C. Montgomery, *Ontogeny of vision in the Antarctic fish* Pagothenia borchgrevinki *(Nototheniidae).* Polar Biol., 1990. 10(6): 419–422.

Partridge, J.C., S.N. Archer, and J. Vanoostrum, *Single and multiple visual pigments in deep-sea fishes.* J. Marine Biol. Assoc. UK., 1992. 72(01): 113–130.

Pearcy, W.G., S.L. Meyer, and O. Munk, *A "four-eyed" fish from the deep-sea:* Bathylychnops exilis *Cohen, 1958.* Nature, 1965. 207(5003): 1260–1262.

Pettigrew, J.D., and S.P. Collin, *Terrestrial optics in an aquatic eye: The sandlance,* Limnichthytes fasciatus (Creediidae, Teleostei). J. Comp. Physiol. A: Neuroethol. Sens. Neural Behav. Physiol., 1995. 177(4): 397–408.

Pettigrew, J.D., S.P. Collin, and M. Ott, *Convergence of specialised behaviour, eye movements and visual optics in the sandlance (Teleostei) and the chameleon (Reptilia).* Current Biology. 1999. 9(8): 421–424.

Reis, E.M.R., C.W. Slayman, and S. Verjovski-Almeida, *Heterologous expression of sarcoplasmic reticulum Ca-ATPase.* Biosci. Rep., 1996. 16(2): 107–113.

Robison, B.H., and K.R. Reisenbichler, Macropinna microstoma *and the Paradox of Its Tubular Eyes.* Copeia, 2008. 2008(4): 780–784.

Rossetto, E.S., H. Dolder, and I. Sazima, *Double cone mosaic pattern in the retina of larval and adult piranha,* Serrasalmus spilopleura. Cell. Mol. Life Sci., 1992. 48(6): 597–599.

Saidel, W.M., *Coherence in nervous system design: The visual system of* Pantodon buchholzi. Philos Trans R Soc Lond B Biol Sci. 2000 Sep 29;355(1401):1177–81.

Saidel, W.M., and A.B. Butler, *An atypical diencephalic nucleus in actinopterygian fishes: Visual connections and sporadic phylogenetic distribution.* Neurosci. Lett., 1997. 229(1): 13–16.

Saidel, W.M., and A.B. Butler, *Visual connections of the atypical diencephalic nucleus rostrolateralis in* Pantodon buchholzi *(Teleostei, Osteoglossomorpha).* Cell Tissue Res., 1997. 287(1): 91–99.

Saidel, W.M., and R.S. Fabiane, *Optomotor response of* Anableps anableps *depends on the field of view.* Vision Res., 1998. 38(13): 2001–2006.

Schwab, I.R., et al., *Evolutionary attempts at 4 eyes in vertebrates.* Trans. Am. Ophthalmol. Soc., 2001. 99: 145–56; discussion 156–7.

Schwab, I.R., et al., *Evolution of the tapetum.* Trans. Am. Ophthalmol. Soc., 2002. 100: 187–199; discussion 199–200.

Schwassmann, H.O., and L. Kruger, *Experimental analysis of the visual system of the four-eyed fish* Anableps microlepis. Vision Res., 1965. 5(6–7): 269–281, IN1.

Sivak, J.G., *Optics of the eye of the "four-eyed fish" (Anableps anableps)*. Vision Res, 1976. 16(5): 531–534.

Sivak, J.G., *The functional significance of the aphakic space of the fish eye*. Can J Zool, 1978. 56(3): 513–516.

Stein, D.L. and C.E. Bond, *Observations on the morphology, ecology, and behaviour of Bathylychnops exilis Cohen*. Journal of Fish Biology, 1985. 27(3): 215–228.

Stewart, K.W., *Observations on the morphology and optical properties of the adipose eyelid of fishes*. Journal of the Fisheries Research Board of Canada, 1962. 19(6): 1161–1162.

Swamynathan, S.K., et al., *Adaptive differences in the structure and macromolecular compositions of the air and water corneas of the "four-eyed" fish (Anableps anableps)*. FASEB J, 2003. 17(14): 1996–2005.

Szabo, T., et al., *Oculomotor system of the weakly electric fish Gnathonemus petersii*. J Comp Neurol, 1987. 264(4): 480–493.

Thorpe, A., R.H. Douglas, and R.J. Truscott, *Spectral transmission and short-wave absorbing pigments in the fish lens—I. Phylogenetic distribution and identity*. Vision Res, 1993. 33(3): 289–300.

Timmermans, P.J and P.M. Souren, *Prey catching in archer fish: the role of posture and morphology in aiming behavior*. Physiol Behav, 2004. 81(1): 101–110.

Timmermans, P.J and J.M. Vossen, *Prey catching in the archer fish: does the fish use a learned correction for refraction?* Behav Processes, 2000. 52(1): 21–34.

Timmermans, P.J.A., *Catching in the archer fish: markenship, endurance of squirting at an aerial target*. Netherlands Journal of Zoology, 2000. 50(4): 411–423.

Timmermans, P.J.A., *Prey catching in the archer fish: angles and probability of hitting an aerial target*. Behavioural Processes, 2001. 55(2): 93–105.

Ubels, J.L. and H.F. Edelhauser, *Healing of corneal epithelial wounds in marine and freshwater fish*. Current Eye Research, 1982. 2(9): 613–620.

Wagner, H.J., et al., *A novel vertebrate eye using both refractive and reflective optics*. Curr Biol, 2009. 19(2): 108–114.

Wagner, H.J., et al., *The eyes of deep-sea fish. II. Functional morphology of the retina*. Prog Retin Eye Res, 1998. 17(4): 637–685.

Warrant, E., *The eyes of deep-sea fishes and the changing nature of visual scenes with depth*. Philos Trans R Soc Lond B Biol Sci, 2000. 355(1401): 1155–1159.

Warrant, E., *Vision in the dimmest habitats on earth*. J Comp Physiol A Neuroethol Sens Neural Behav Physiol, 2004. 190(10): 765–789.

Warrant, E.J. and N.A. Locket, *Vision in the deep sea*. Biol Rev Camb Philos Soc, 2004. 79(3): 671–712.

Waxman, H.M. and J.D. McCleave, *Auto-shaping in the archer fish (Toxotes chatareus)*. Behavioral Biology, 1978. 22(4): 541–544.

Werneburg, I. and S.T. Hertwig, *Head morphology of the ricefish, Oryzias latipes (Teleostei: Beloniformes)*. J Morphol, 2009. 270(9): 1095–1106.

Yokoyama, S., et al., *Adaptive evolution of color vision of the Comoran coelacanth (Latimeria chalumnae)*. Proc Natl Acad Sci U S A, 1999. 96(11): 6279–6284.

CHAPTER 11

Ren, R, Nei, A, Prokop, J.*New early griffenfly, Sinomeganeura huangheensis from the Late Carboniferous of northern China (Meganisoptera: Meganeuridae)*. Insect Systematics &38; Evolution, 2008. 39: 223–229.

Arendt, D., et al., *Ciliary photoreceptors with a vertebrate-type opsin in an invertebrate brain*. Science, 2004. 306(5697): 869–871.

Berry, R.P., G. Stange, and E.J. Warrant, *Form vision in the insect dorsal ocelli: an anatomical and optical analysis of the dragonfly median ocellus*. Vision Res, 2007. 47(10): 1394–1409.

Briscoe, A.D. and L. Chittka, *The evolution of color vision in insects*. Annu Rev Entomol, 2001. 46: 471–510.

Chappell, R.L. and J.E. Dowling, *Neural organization of the median ocellus of the dragonfly. I. Intracellular electrical activity*. J Gen Physiol, 1972. 60(2): 121–147.

Dowling, J.E. and R.L. Chappell, *Neural organization of the median ocellus of the dragonfly. II. Synaptic structure*. J Gen Physiol, 1972. 60(2): 148–165.

Engel, M.S. and D.A. Grimaldi, *New light shed on the oldest insect*. Nature, 2004. 427(6975): 627–630.

Friedrich, M., *Ancient mechanisms of visual sense organ development based on comparison of the gene networks controlling larval eye, ocellus, and compound eye specification in Drosophila*. Arthropod Struct Dev, 2006. 35(4): 357–378.

Horridge, G.A., L. Marcelja, and R. Jahnke, *Light Guides in the Dorsal Eye of the Male Mayfly*. Proceedings of the Royal Society of London. Series B, Biological Sciences, 1982. 216(1202): 25–51.

Labhart, T. and D.E. Nilsson, *The dorsal eye of the dragonfly Sympetrum: specializations for prey detection against the blue sky*. Journal of Comparative Physiology A: Neuroethology, Sensory, Neural, and Behavioral Physiology, 1995. 176(4): 437–453.

Lamsdell, J.C. and S.J. Braddy, *Cope's Rule and Romer's theory: patterns of diversity and gigantism in eurypterids and Palaeozoic vertebrates*. Biol Lett, 2010. 6(2): 265–269.

Machida, R., *External features of embryonic development of a jumping bristletail, Pedetontus unimaculatus Machida (Insecta, Thysanura, Machilidae)*. J Morphol, 1981. 168(3): 339–355.

Mayhew, P.J., *Shifts in hexapod diversification and what Haldane could have said*. Proc Biol Sci, 2002. 269(1494): 969–974.

Meyer, E.P. and T. Labhart, *Morphological specializations of dorsal rim ommatidia in the compound eye of dragonflies and damselfies (Odonata)*. Cell Tissue Res, 1993. 272(1): 17–22.

Meyer-Rochow, V.B. and A.R. Liddle, *Structure and Function of the Eyes of Two Species of Opilionid from New Zealand Glow-worm Caves (Megalopsalis tumida: Palpatores, and Hendea myersi cavernicola: Laniatores)*. Proceedings of the Royal Society of London. Series B, Biological Sciences, 1988. 233(1272): 293–319.

Mikolajewski, D.J. and F. Johansson, *Morphological and behavioral defenses in dragonfly larvae: trait compensation and cospecialization*. Behavioral Ecology, 2004. 15(4): 614–620.

Paulus, H.F., *Phylogeny of the Myriapoda; Crustacea; Insecta: a new attempt using photoreceptor structure*. 2000. 189–208.

Rashed, A., et al., *Prey selection by dragonflies in relation to prey size and wasp-like colours and patterns*. Animal Behaviour, 2005. 70(5): 1195–1202.

Seki, T., S. Fujishita, and S. Obana, *Composition and distribution of retinal and 3-hydroxyretinal in the compound eye of the dragonfly*. Exp Biol, 1989. 48(2): 65–75.

Sherk, T.E., *Development of the compound eyes of dragonflies (odonata). I. Larval compound eyes*. Journal of Experimental Zoology, 1977. 201(3): 391–416.

Sherk, T.E., *Development of the compound eyes of dragonflies (Odonata). III. Adult compound eyes*. J Exp Zool, 1978. 203(1): 61–80.

Sherk, T.E., *Development of the compound eyes of dragonflies (Odonata). II. Development of the larval compound eyes.* J Exp Zool, 1978. 203(1): 47–60.

CHAPTER 12

Blest, A.D. and M.F. Land, *The Physiological Optics of Dinopis subrufus L. Koch: A Fish-Lens in a Spider.* Proceedings of the Royal Society of London. Series B, Biological Sciences, 1977. 196(1123): 197–222.

Blest, A.D., D.S. Williams, and L. Kao, *The posterior median eyes of the dinopid spider Menneus.* Cell and Tissue Research, 1980. 211(3): 391–403.

Clark, D.L. and C.L. Morjan, *Attracting female attention: the evolution of dimorphic courtship displays in the jumping spider Maevia inclemens (Araneae: Salticidae).* Proceedings of the Royal Society of London. Series B: Biological Sciences, 2001. 268(1484): 2461–2465.

Dacke, M., et al., *Built-in polarizers form part of a compass organ in spiders.* Nature, 1999. 401(6752): 470–473.

De Voe, R.D., *Ultraviolet and green receptors in principal eyes of jumping spiders.* J Gen Physiol, 1975. 66(2): 193–207.

Getty, R.M. and F.A. Coyle, *Observations on Prey Capture and Anti-Predator Behaviors of Ogre-Faced Spiders (Deinopis) in Southern Costa Rica (Araneae, Deinopidae).* Journal of Arachnology, 1996. 24(2): 93–100.

Harland, D.P. and R.R. Jackson, *Cues by which Portia fimbriata, an araneophagic jumping spider, distinguishes jumping-spider prey from other prey.* J Exp Biol, 2000. 203(22): 3485–3494.

Harland, D.P. and R.R. Jackson, *Influence of cues from the anterior medial eyes of virtual prey on Portia fimbriata, an araneophagic jumping spider.* J Exp Biol, 2002. 205(13): 1861–1868.

Hedin, M.C. and W.P. Maddison, *A Combined Molecular Approach to Phylogeny of the Jumping Spider Subfamily Dendryphantinae (Araneae: Salticidae).* Molecular Phylogenetics and Evolution, 2001. 18(3): 386–403.

Jackson, R.R. and D. Li, *One-encounter search-image formation by araneophagic spiders.* Animal Cognition, 2004. 7(4): 247–254.

Jackson, R.R. and S.D. Pollard, *Predatory Behavior of Jumping Spiders.* Annual Review of Entomology, 1996. 41(1): 287–308.

Jackson, R.R., S.D. Pollard, and A.M. Cerveira, *Opportunistic use of cognitive smokescreens by araneophagic jumping spiders.* Anim Cogn, 2002. 5(3): 147–157.

Jackson, R.R., et al., *Interpopulation variation in the risk-related decisions of Portia labiata, an araneophagic jumping spider (Araneae, Salticidae), during predatory sequences with spitting spiders.* Anim Cogn, 2002. 5(4): 215–223.

Koyanagi, M., et al., *Molecular evolution of arthropod color vision deduced from multiple opsin genes of jumping spiders.* J Mol Evol, 2008. 66(2): 130–137.

Land, M.F., *Movements of the Retinae of Jumping Spiders (Salticidae: Dendryphantinae) in Response to Visual Stimuli.* J Exp Biol, 1969. 51(2): 471–493.

Land, M.F., *Structure of the Retinae of the Principal Eyes of Jumping Spiders (Salticidae: Dendryphantinae) in Relation to Visual Optics.* J Exp Biol, 1969. 51(2): 443–470.

Land, M.F. and F.G. Barth, *The quality of vision in the ctenid spider Cupiennius salei.* 1992. 227–242.

Lim ML, Land MF, LI D. Sex-specific UV and fluorescence signals in jumping spiders. Science. 2007 Jan 26;315(5811): 481.

Michael, D.O., *The neural organization of the first optic ganglion of the principal eyes of jumping spiders (Salticidae).* The Journal of Comparative Neurology, 1977. 174(1): 95–117.

Nakamura, T. and S. Yamashita, *Learning and discrimination of colored papers in jumping spiders (Araneae, Salticidae).* Journal of Comparative Physiology A: Neuroethology, Sensory, Neural, and Behavioral Physiology, 2000. 186(9): 897–901.

Ortega-Escobar, J.n., *Evidence that the wold-spider Lycosa tarentula (Araneae, Lycosidae) needs visual input for path integration.* 2002. 481–486.

Peaslee, A.G. and G. Wilson, *Spectral sensitivity in jumping spiders (Araneae, Salticidae).* Journal of Comparative Physiology A: Neuroethology, Sensory, Neural, and Behavioral Physiology, 1989. 164(3): 359–363.

Prete, F.R., *Complex worlds from simpler nervous systems.* 2004, Cambridge, MA; London: MIT Press, 436 p., 16 of plates.

Rovner, J.S., *Conspecific Interactions in the Lycosid Spider Rabidosa rabida: The Roles of Different Senses.* Journal of Arachnology, 1996. 24(1): 16–23.

Schmitz, A., *Metabolic rates during rest and activity in differently tracheated spiders (Arachnida, Araneae): Pardosa lugubris (Lycosidae) and Marpissa muscosa (Salticidae).* Journal of Comparative Physiology B: Biochemical, Systemic, and Environmental Physiology, 2004. 174(7): 519–526.

Schmitz, A., *Spiders on a treadmill: influence of running activity on metabolic rates in Pardosa lugubris (Araneae, Lycosidae) and Marpissa muscosa (Araneae, Salticidae).* J Exp Biol, 2005. 208(7): 1401–1411.

Strausfeld, N.J., Weltzien, and F.G. Barth, *Two visual systems in one brain: neuropils serving the principal eyes of the spider Cupiennius salei.* J Comp Neurol, 1993. 328(1): 63–75.

Williams, D.S. and MeIntyre, *The principal eyes of a jumping spider have a telephoto component.* Nature, 1980. 288(5791): 578–580.

Yamashita, S. and H. Tateda, *Hypersensitivity in the anterior median eye of a jumping spider.* J Exp Biol, 1976. 65(3): 507–516.

Yin, C.-M., C.E. Griswold, and H.-M. Yan, *A new ogre-faced spider (Deinopis) from the gaoligong mountains, Yunnan, China (Araneae, Deinopidae).* Journal of Arachnology, 2009. 30(3): 610–612.

CHAPTER 13

Ahlberg, P.E. and Z. Johanson, *Osteolepiforms and the ancestry of tetrapods.* Nature, 1998. 395(6704): 792–794.

Bailes, H.J., et al., *Morphology, characterization, and distribution of retinal photoreceptors in the Australian lungfish Neoceratodus forsteri (Krefft, 1870).* J Comp Neurol, 2006. 494(3): 381–397.

Brooks, S.P.J., et al., *Temperature Regulation of Glucose Metabolism in Red Blood Cells of the Freeze-Tolerant Wood Frog.* Cryobiology, 1999. 39(2): 150–157.

Chen, N., et al., *Molecular cloning of a rhodopsin gene from salamander rods.* Invest Ophthalmol Vis Sci, 1996. 37(9): 1907–1913.

Clack, J.A., *Devonian climate change, breathing, and the origin of the tetrapod stem group.* Integrative and Comparative Biology.

Clack, J.A., *A new Early Carboniferous tetrapod with a melange of crown-group characters.* Nature, 1998. 394(6688): 66–69.

Conlon, J.M., et al., *Freeze tolerance in the wood frog Rana sylvatica is associated with unusual structural features in insulin but not in glucagon.* J Mol Endocrinol, 1998. 21(2): 153–159.

Costanzo, J.P., R.E. Lee, and P.H. Lortz, *Glucose concentration regulates freeze tolerance in the wood frog Rana sylvatica.* J Exp Biol, 1993. 181(1): 245–255.

Costanzo, J.P., R.E. Lee, Jr., and P.H. Lortz, *Physiological responses of freeze-tolerant and -intolerant frogs: clues to evolution of anuran freeze tolerance.* Am J Physiol Regul Integr Comp Physiol, 1993. 265(4): R721–725.

Costanzo, J.P., R.E. Lee, Jr., and M.F. Wright, *Glucose loading prevents freezing injury in rapidly cooled wood frogs.* Am J Physiol Regul Integr Comp Physiol, 1991. 261(6): R1549–1553.

Daeschler, E.B., N.H. Shubin, and F.A. Jenkins, Jr., *A Devonian tetrapod-like fish and the evolution of the tetrapod body plan.* Nature, 2006. 440(7085): 757–763.

Fanny, M. and E.H. Christine, *Control of retinal growth and axon divergence at the chiasm: lessons from Xenopus.* BioEssays, 2001. 23(4): 319–326.

Gao, K.Q. and N.H. Shubin, *Late Jurassic salamanders from northern China.* Nature, 2001. 410(6828): 574–577.

Hillis, D.M. and T.P. Wilcox, *Phylogeny of the New World true frogs (Rana).* Mol Phylogenet Evol, 2005. 34(2): 299–314.

Hisatomi, O., et al., *Primary structure of a visual pigment in bullfrog green rods.* FEBS Lett, 1999. 447(1): 44–48.

Jack, R.L., Jr., P.C. Jon, and E.L. Richard, Jr., *Freeze duration influences postfreeze survival in the frog Rana sylvatica.* The Journal of Experimental Zoology, 1998. 280(2): 197–201.

Janvier, P., *Wandering nostrils.* Nature, 2004. 432(7013): 23–24.

Joanisse, D.R. and K.B. Storey, *Oxidative damage and antioxidants in Rana sylvatica, the freeze-tolerant wood frog.* Am J Physiol Regul Integr Comp Physiol, 1996. 271(3): R545–553.

Kraig, A., *Extraocular photoreception in amphibians.* Photochemistry and Photobiology, 1976. 23(4): 275–298.

Lande, M.A. and J.A. Zadunaisky, *The Structure and Membrane Properties of the Frog Nictitans.* Invest. Ophthalmol. Vis. Sci., 1970. 9(7): 477–491.

Lee, M.R., et al., *Isolation of ice-nucleating active bacteria from the freeze-tolerant frog, Rana sylvatica.* Cryobiology, 1995. 32(4): 358–365.

Levine, R.P., J.A. Monroy, and E.L. Brainerd, *Contribution of eye retraction to swallowing performance in the northern leopard frog, Rana pipiens.* J Exp Biol, 2004. 207(Pt 8): 1361–1368.

Long, J.A., et al., *An exceptional Devonian fish from Australia sheds light on tetrapod origins.* Nature, 2006. 444(7116): 199–202.

Mann, F. and C.E. Holt, *Control of retinal growth and axon divergence at the chiasm: lessons from Xenopus.* BioEssays, 2001. 23(4): 319–326.

McClanahan, L.L., R. Ruibal, and V.H. Shoemaker, *Frogs and toads in deserts.* Sci Am, 1994. 270(3): 82–88.

McNally, J.D., C.M. Sturgeon, and K.B. Storey, *Freeze-induced expression of a novel gene, fr47, in the liver of the freeze-tolerant wood frog, Rana sylvatica.* Biochimica et Biophysica Acta (BBA) - Gene Structure and Expression, 2003. 1625(2): 183–191.

McNally, J.D., et al., *Identification and characterization of a novel freezing inducible gene, li16, in the wood frog Rana sylvatica.* FASEB J., 2002. 16(8): 902–904.

Montgomery, N.M., C. Tyler, and K.V. Fite, *Organization of retinal axons within the optic nerve, optic chiasm, and the innervation of multiple central nervous system targets Rana pipiens.* J Comp Neurol, 1998. 402(2): 222–237.

Nakagawa, S., et al., *Ephrin-B Regulates the Ipsilateral Routing of Retinal Axons at the Optic Chiasm.* Neuron, 2000. 25(3): 599–610.

O'Reilly, J.C., R.A. Nussbaum, and D. Boone, *Vertebrate with protrusible eyes.* Nature, 1996. 382(6586): 33–33.

Payne, A.P., *The Harderian gland: a tercentennial review.* J Anat, 1994. 185 (Pt 1): 1–49.

Phillips, J.B., et al., *The role of extraocular photoreceptors in newt magnetic compass orientation: parallels between light-dependent magnetoreception and polarized light detection in vertebrates.* J Exp Biol, 2001. 204(14): 2543–2552.

Reiss, J.O., *The phylogeny of amphibian metamorphosis.* Zoology, 2002. 105(2): 85–96.

Rubinsky, B., et al., *1H magnetic resonance imaging of freezing and thawing in freeze-tolerant frogs.* Am J Physiol Regul Integr Comp Physiol, 1994. 266(6): R1771–1777.

Schoch, R.R. and R.L. Carroll, *Ontogenetic evidence for the Paleozoic ancestry of salamanders.* Evol Dev, 2003. 5(3): 314–324.

Schwab, I.R., *An icy stare.* British Journal of Ophthalmology, 2005. 89(10): 1236.

Schwab, I.R., S. Collin, and H. Bailes, *Bringing the eyes along.* British Journal of Ophthalmology, 2006. 90(7): 818.

Schwab, I.R. and W. Saidel, *Look before you leap.* British Journal of Ophthalmology, 2003. 87(4): 391.

Selden, P.A., J.A. Corronca, and M.A. Hunicken, *The true identity of the supposed giant fossil spider Megarachne.* Biol Lett, 2005. 1(1): 44–48.

Skelly, D.K., *Microgeographic Countergradient Variation in the Wood Frog, Rana sylvatica.* Evolution, 2004. 58(1): 160–165.

Sordino, P., F. van der Hoeven, and D. Duboule, *Hox gene expression in teleost fins and the origin of vertebrate digits.* Nature, 1995. 375(6533): 678–681.

Storey, K.B., *Organ-specific metabolism during freezing and thawing in a freeze-tolerant frog.* Am J Physiol Regul Integr Comp Physiol, 1987. 253(2): R292–297.

Storey, K.B., *Life in a frozen state: adaptive strategies for natural freeze tolerance in amphibians and reptiles.* Am J Physiol Regul Integr Comp Physiol, 1990. 258(3): R559–568.

Storey, K.B., J. Bischof, and B. Rubinsky, *Cryomicroscopic analysis of freezing in liver of the freeze-tolerant wood frog.* Am J Physiol Regul Integr Comp Physiol, 1992. 263(1): R185–194.

Storey, K.B. and J.M. Storey, *Freeze tolerance and intolerance as strategies of winter survival in terrestrially-hibernating amphibians.* Comp Biochem Physiol A Comp Physiol, 1986. 83(4): 613–617.

Takahashi, Y., et al., *Distribution of blue-sensitive photoreceptors in amphibian retinas.* FEBS Lett, 2001. 501(2–3): 151–155.

Thomson, K.S., M. Sutton, and B. Thomas, *A larval Devonian lungfish.* Nature, 2003. 426(6968): 833–834.

Tsonis, P.A., et al., *A newt's eye view of lens regeneration.* Int J Dev Biol, 2004. 48(8–9): 975–980.

Yokoyama, S., et al., *Adaptive evolution of color vision of the Comoran coelacanth (Latimeria chalumnae).* Proc Natl Acad Sci U S A, 1999. 96(11): 6279–6284.

Zhu, M. and P.E. Ahlberg, *The origin of the internal nostril of tetrapods.* Nature, 2004. 432(7013): 94–97.

CHAPTER 14

Sbita, S.J., R.C. Morgan, and E.K. Buschbeck, *Eye and optic lobe metamorphosis in the sunburst diving beetle, Thermonectus marmoratus (Coleoptera: Dytiscidae).* Arthropod Struct Dev, 2007. 36(4): 449–462.

Warrant, E.J. and P.D. McIntyre, *Limitations to resolution in superposition eyes.* Journal of Comparative Physiology A: Neuroethology, Sensory, Neural, and Behavioral Physiology, 1990. 167(6): 785–803.

Chong, L.D., *Two Eyes in One.* Science, 2010. 329(5997): 1259.

Lall, A.B., et al., *Vision in click beetles (Coleoptera: Elateridae): pigments and spectral correspondence between visual sensitivity and species bioluminescence emission.* J Comp Physiol A Neuroethol Sens Neural Behav Physiol, 2010. 196(9): 629–638.

Chapter 15

Autumn, K., et al., *Adhesive force of a single gecko foot-hair.* Nature, 2000. 405(6787): 681–685.

Bowmaker, J.K., E.R. Loew, and M. Ott, *The cone photoreceptors and visual pigments of chameleons.* J Comp Physiol A Neuroethol Sens Neural Behav Physiol, 2005. 191(10): 925–932.

Casey Y-J Ung, A.C.B.M., *An enigmatic eye: the histology of the tuatara pineal complex.* 2004. 614–618.

Eakin, R.M., *The Third Eye.* 1973: Berkeley, University of California Press.

El Hassni, M., et al., *Localization of motoneurons innervating the extraocular muscles in the chameleon (Chamaeleo chameleon).* Anat Embryol (Berl), 2000. 201(1): 63–74.

Gioanni, H., M. Bennis, and A. Sansonetti, *Visual and vestibular reflexes that stabilize gaze in the chameleon.* Vis Neurosci, 1993. 10(5): 947–956.

Hedges, S.B. and L.L. Poling, *A molecular phylogeny of reptiles.* Science, 1999. 283(5404): 998–1001.

Herrel, A., J. Cleuren, and F. De Vree, *Prey capture in the lizard Agama stellio.* Journal of Morphology, 1995. 224(3): 313–329.

Kawamura, S. and S. Yokoyama, *Functional characterization of visual and nonvisual pigments of American chameleon (Anolis carolinensis).* Vision Res, 1998. 38(1): 37–44.

Klein, D.C., *The 2004 Aschoff/Pittendrigh lecture: Theory of the origin of the pineal gland—a tale of conflict and resolution.* J Biol Rhythms, 2004. 19(4): 264–279.

Land, M.F., *Fast-focus telephoto eye.* Nature, 1995. 373(6516): 658–659.

Loew, E.R., et al., *Visual pigments and oil droplets in diurnal lizards: a comparative study of Caribbean anoles.* J Exp Biol, 2002. 205(7): 927–938.

Marmor, M.F., et al., *Visual Insignificance of the Foveal Pit: Reassessment of Foveal Hypoplasia as Fovea Plana.* Arch Ophthalmol, 2008. 126(7): 907–913.

Meyer-Rochow, V.B., S. Wohlfahrt, and P.K. Ahnelt, *Photoreceptor cell types in the retina of the tuatara (Sphenodon punctatus) have cone characteristics.* Micron, 2005. 36(5): 423–428.

Murphy, C.J. and H.C. Howland, *On the gekko pupil and scheiner's disc.* Vision Research, 1986. 26(5): 815–817.

Ott, M., *Chameleons have independent eye movements but synchronise both eyes during saccadic prey tracking.* Exp Brain Res, 2001. 139(2): 173–179.

Ott, M. and F. Schaeffel, *A negatively powered lens in the chameleon.* Nature, 1995. 373(6516): 692–694.

Ott, M., F. Schaeffel, and W. Kirmse, *Binocular vision and accommodation in prey-catching chameleons.* Journal of Comparative Physiology A: Neuroethology, Sensory, Neural, and Behavioral Physiology, 1998. 182(3): 319–330.

Packard, G.C. and M.J. Packard, *Evolution of the Cleidoic Egg among Reptilian Antecedents of Birds.* American Zoologist, 1980. 20(2): 351–362.

Rieppel, O., *Turtle origins.* Science, 1999. 283(5404): 945–946.

Roth, L.S., et al., *The pupils and optical systems of gecko eyes.* J Vis, 2009. 9(3): 27 1–11.

Sandor, P.S., M.A. Frens, and V. Henn, *Chameleon eye position obeys Listing's law.* Vision Res, 2001. 41(17): 2245–2251.

Schmid, K.L., H.C. Howland, and M. Howland, *Focusing and accommodation in tuatara (Sphenodon punctatus).* Journal of Comparative Physiology A: Neuroethology, Sensory, Neural, and Behavioral Physiology, 1992. 170(3): 263–266.

Sivak, J.G., *Historical note: the vertebrate median eye.* Vision Res, 1974. 14(1): 137–140.

Solessio, E. and G.A. Engbretson, *Antagonistic chromatic mechanisms in photoreceptors of the parietal eye of lizards.* Nature, 1993. 364(6436): 442–445.

Su, C.Y., et al., *Parietal-eye phototransduction components and their potential evolutionary implications.* Science, 2006. 311(5767): 1617–1621.

Tansley, K., *The gecko retina.* Vision Research, 1964. 4(1–2): 33–37, IN9-IN14.

Wainwright, P.C. and A.F. Bennett, *The Mechanism of Tongue Projection in Chameleons: I. Electromyographic Tests of Functional Hypotheses.* 1992. 1–21.

Xiong, W.H., E.C. Solessio, and K.W. Yau, *An unusual cGMP pathway underlying depolarizing light response of the vertebrate parietal-eye photoreceptor.* Nat Neurosci, 1998. 1(5): 359–365.

Yau, K.-W., *34.1. Photoreception in the lizard parietal eye: Implications about ciliary-photoreceptor evolution.* Comparative Biochemistry and Physiology - Part A: Molecular & Integrative Physiology, 2007. 148(Supplement 1): S145–S145.

Chapter 16

A new genus of ichthyosaur from the Late Triassic Pardonet Formation of British Columbia: bridging the Triassic Jurassic gap. Canadian Journal of Earth Sciences, 2001. 38: 983–1002.

Autumn, K., et al., *Adhesive force of a single gecko foot-hair.* Nature, 2000. 405(6787): 681–685.

Brudenall, D.K., I.R. Schwab, and K.A. Fritsches, *Ocular morphology of the Leatherback sea turtle (Dermochelys coriacea).* Vet Ophthalmol, 2008. 11(2): 99–110.

Casey Y-J Ung, A.C.B.M., *An enigmatic eye: the histology of the tuatara pineal complex.* 2004. 614–618.

Eakin, R.M., *The Third Eye.* 1973: Berkeley, University of California Press.

Franz-Odendaal, T.A., *Intramembranous ossification of scleral ossicles in Chelydra serpentina.* Zoology (Jena), 2006. 109(1): 75–81.

Hedges, S.B. and L.L. Poling, *A molecular phylogeny of reptiles.* Science, 1999. 283(5404): 998–1001.

Humphries, S. and G.D. Ruxton, *Why did some ichthyosaurs have such large eyes?* J Exp Biol, 2002. 205(4): 439–441.

Janke, A. and U. Arnason, *The complete mitochondrial genome of Alligator mississippiensis and the separation between recent archosauria (birds and crocodiles).* Mol Biol Evol, 1997. 14(12): 1266–1272.

Loew, E.R., et al., *Visual pigments and oil droplets in diurnal lizards: a comparative study of Caribbean anoles.* J Exp Biol, 2002. 205(7): 927–938.

Loew, E.R. and V.I. Govardovskii, *Photoreceptors and visual pigments in the red-eared turtle, Trachemys scripta elegans.* Vis Neurosci, 2001. 18(5): 753–757.

Lutz, P.L. and J.A. Musick, *The biology of sea turtles.* Marine science series. 1997, Boca Raton, Fla: CRC Press. 432 p.

Marmor, M.F., et al., *Visual Insignificance of the Foveal Pit: Reassessment of Foveal Hypoplasia as Fovea Plana.* Arch Ophthalmol, 2008. 126(7): 907–913.

Motani, R., *Rulers of the Jurassic Seas.* Scientific American Special Edition, 2004. 14(2): 4.

Motani, R., B.M. Rothschild, and W. Wahl, *Large eyeballs in diving ichthyosaurs.* Nature, 1999. 402(6763): 747–747.

Murphy, C.J. and H.C. Howland, *On the gekko pupil and scheiner's disc.* Vision Research, 1986. 26(5): 815–817.

Ott, M. and F. Schaeffel, *A negatively powered lens in the chameleon.* Nature, 1995. 373(6516): 692–694.

Rieppel, O., *Turtle Origins.* Science, 1999. 283(5404): 945–946.

Rieppel, O. and R.R. Reisz, *The origin and early evolution of turtles.* Annual Review of Ecology and Systematics, 1999. 30(1): 1–22.

Roth, L.S., et al., *The pupils and optical systems of gecko eyes.* J Vis, 2009. 9(3): 27 1–11.

Sillman, A.J., S.J. Ronan, and E.R. Loew, *Histology and Microspectrophotometry of the Photoreceptors of a Crocodilian, Alligator mississippiensis.* Proceedings of the Royal Society of London. Series B: Biological Sciences, 1991. 243(1306): 93–98.

Smith, W.C., et al., *Alligator rhodopsin: sequence and biochemical properties.* Exp Eye Res, 1995. 61(5): 569–578.

Tansley, K., *The gecko retina.* Vision Research, 1964. 4(1–2): 33–37, IN9-IN14.

Wainwright, P.C. and A.F. Bennett, *The Mechanism of Tongue Projection in Chameleons: I. Electromyographic Tests of Functional Hypotheses.* 1992. 1–21.

Zimmer, C., *Jurassic Genome.* Science, 2007. 315(5817): 1358–1359.

CHAPTER 17

Chang, B.S., et al., *Recreating a functional ancestral archosaur visual pigment.* Mol Biol Evol, 2002. 19(9): 1483–1489.

Curry Rogers, K. and C.A. Forster, *The last of the dinosaur titans: a new sauropod from Madagascar.* Nature, 2001. 412(6846): 530–534.

Stevens, K.A., *Binocular Vision in Theropod Dinosaurs.* Journal of Vertebrate Paleontology, 2006. 26(2): 321–330.

Witmer, L.M., et al., *Neuroanatomy of flying reptiles and implications for flight, posture and behaviour.* Nature, 2003. 425(6961): 950–953.

CHAPTER 18

Akasaki, T., et al., *Extensive mitochondrial gene arrangements in coleoid Cephalopoda and their phylogenetic implications.* Mol Phylogenet Evol, 2006. 38(3): 648–658.

Budelmann, B.U. and J.Z. Young, *The Statocyst-Oculomotor System of Octopus vulgaris: Extraocular Eye Muscles, Eye Muscle Nerves, Statocyst Nerves and the Oculomotor Centre in the Central Nervous System.* Philosophical Transactions of the Royal Society of London. Series B, Biological Sciences, 1984. 306(1127): 159–189.

Budelmann, B.U. and J.Z. Young, *The oculomotor system of decapod cephalopods: eye muscles, eye muscle nerves, and the oculomotor neurons in the central nervous system.* Philos Trans R Soc Lond B Biol Sci, 1993. 340(1291): 93–125.

Cole, A.G. and B.K. Hall, *Cartilage differentiation in cephalopod molluscs.* Zoology, 2009. 112(1): 2–15.

Douglas, R.H., R. Williamson, and H.J. Wagner, *The pupillary response of cephalopods.* J Exp Biol, 2005. 208(Pt 2): 261–265.

Froesch, D., *On the fine structure of the Octopus iris.* Z Zellforsch Mikrosk Anat, 1973. 145(1): 119–129.

Grisley, M.S., P.R. Boyle, and L.N. Key, *Eye puncture as a route of entry for saliva during predation on crabs by the octopus Eledone cirrhosa (Lamarck).* Journal of Experimental Marine Biology and Ecology, 1996. 202(2): 225–237.

Hara, T. and R. Hara, *Retinochrome and rhodopsin in the extraocular photoreceptor of the squid, Todarodes.* J Gen Physiol, 1980. 75(1): 1–19.

Jagger, W.S. and P.J. Sands, *A wide-angle gradient index optical model of the crystalline lens and eye of the octopus.* Vision Res, 1999. 39(17): 2841–2852.

Jones, B.W. and M.K. Nishiguchi, *Counterillumination in the Hawaiian bobtail squid, Euprymna scolopes Berry (Mollusca: Cephalopoda).* Marine Biology, 2004. 144(6): 1151–1155.

Mather, J.A. and R.C. Anderson, *Personalities of Octopuses (Octopus rubescens).* Journal of Comparative Psychology, 1993. 107(3): 336–340.

Mather, J.A. and R.C. Anderson, *Exploration, Play, and Habituation in Octopuses (Octopus dofleini).* Journal of Comparative Psychology, 1999. 113(3): 333–338.

Matthew, P., *A description of the nuchal organ, a possible photoceptor, in Euprymna scolopes and other cephalopods.* Journal of Zoology, 2000. 252(2): 163–177.

McFall-Ngai, M.J., *Negotiations between animals and bacteria: the diplomacy' of the squid-vibrio symbiosis.* Comparative Biochemistry and Physiology - Part A: Molecular & Integrative Physiology, 2000. 126(4): 471–480.

Muntz, W.R.A. and J. Gwyther, *Short Communication: The Visual Acuity of Octopuses for Gratings of Different Orientations.* J Exp Biol, 1989. 142(1): 461–464.

Ogura, A., K. Ikeo, and T. Gojobori, *Comparative Analysis of Gene Expression for Convergent Evolution of Camera Eye Between Octopus and Human.* 2004. 1555–1561.

Packard, A., *Visual acuity and eye growth in octopus vulgaris.* Monitore zoologico. Italian journal of zoology, 1969. 3: 13.

Packard, A., *Cephalopods and fish: the limits of convergence.* Biological Reviews, 1972. 47(2): 241–307.

Parry, M., *A description of the nuchal organ, a possible photoreceptor, in Euprymna scolopes and other cephalopods.* Journal of Zoology, 2000. 252(2): 163–177.

Schaeffel, F., C.J. Murphy, and H.C. Howland, *Accommodation in the cuttlefish (Sepia officinalis).* J Exp Biol, 1999. 202(22): 3127–3134.

Schwab, I.R., *A well armed predator.* Br J Ophthalmol, 2003. 87(7): 812.

Shashar, N., et al., *Cuttlefish use polarization sensitivity in predation on silvery fish.* Vision Res, 2000. 40(1): 71–75.

Sivak, J.G., *Optical properties of a cephalopod eye (the short finned squid, Illex illecebrosus).* Journal of Comparative Physiology A: Neuroethology, Sensory, Neural, and Behavioral Physiology, 1982. 147(3): 323–327.

Sweeney, A.M., S.H.D. Haddock, and S. Johnsen, *Comparative visual acuity of coleoid cephalopods.* Integrative and Comparative Biology, 2007. 47(6): 808–814.

Wentworth, S.L. and W.R.A. Muntz, *Asymmetries in the sense organs and central nervous system of the squid Histioteuthis.* Journal of Zoology, 1989. 219(4): 607–619.

Wentworth, S.L. and W.R.A. Muntz, *Development of the eye and optic lobe of Octopus.* Journal of Zoology, 1992. 227(4): 673–684.

Young, J.Z., *Regularities in the retina and optic lobes of octopus in relation to form discrimination.* Nature, 1960. 186: 836–839.

Young, J.Z., *The Visual System of Octopus : (1)Regularities in the Retina and Optic Lobes of Octopus in Relation to Form Discrimination.* Nature, 1960. 186(4728): 836–839.

Young, R.E., *Function of the Dimorphic Eyes in the Midwater Squid Histioteuthis dofleini.* Pacific Sciences, 1975. 29(2): 211–218.

Young, R.E., Michael Vecchione, and Katharina M. Mangold, *Analysis of morphology to determine primary sister-taxon relationships with coleoid cephalopods.* 2008.

Young, R.E. and C.F. Roper, *Bioluminescent countershading in midwater animals: evidence from living squid.* Science, 1976. 191(4231): 1046–1048.

Young, R.E., M. Vecchione, and D.T. Donovan, *The evolution of coleoid cephalopods and their present biodiversity and ecology.* South African Journal of Marine Science, 1998. 20(1): 393–420.

CHAPTER 19
Campbell, A., *Biological infrared imaging and sensing.* Micron, 2002. 33: 211–225.

CHAPTER 20
Ault, S.J., *Electroretinograms and retinal structure of the eastern screech owl and great horned owl.* Raptor Research, 1984. 18: 62–66.

Bellhorn, M.B., R.W. Bellhorn, and D.S. Poll, *Permeability of fluorescein-labelled dextrans in fundus fluorescein angiography of rats and birds.* Experimental Eye Research, 1977. 24(6): 595–605.

Bloch, S. and C. Martinoya, *Are colour oil droplets the basis of the pigeon's chromatic space?* Vision Res, 1971. Suppl 3: 411–418.

Bohorquez Mahecha, G.A. and C. Aparecida de Oliveira, *An additional bone in the sclera of the eyes of owls and the common potoo (Nictibius griseus) and its role in the contraction of the nictitating membrane.* Acta Anat (Basel), 1998. 163(4): 201–211.

Bowmaker, J.K., *The visual pigments, oil droplets and spectral sensitivity of the pigeon.* Vision Res, 1977. 17(10): 1129–1138.

Bowmaker, J.K., et al., *Visual pigments and oil droplets from six classes of photoreceptor in the retinas of birds.* Vision Res, 1997. 37(16): 2183–2194.

Bowmaker, J.K. and G.R. Martin, *Visual pigments and colour vision in a nocturnal bird, Strix aluco (tawny owl).* Vision Res, 1978. 18(9): 1125–1130.

Bravo, H. and J.D. Pettigrew, *The distribution of neurons projecting from the retina and visual cortex to the thalamus and tectum opticum of the barn owl, Tyto alba, and the burrowing owl, Speotyto cunicularia.* J Comp Neurol, 1981. 199(3): 419–441.

Broxmeyer, H.E., et al., *Involvement of Interleukin (IL) 8 receptor in negative regulation of myeloid progenitor cells* in vivo: *evidence from mice lacking the murine IL-8 receptor homologue.* J Exp Med, 1996. 184(5): 1825–1832.

Chen, D.M., J.S. Collins, and T.H. Goldsmith, *The ultraviolet receptor of bird retinas.* Science, 1984. 225(4659): 337–340.

Chiappe, L.M., *Glorified dinosaurs : the origin and early evolution of birds.* 2007, Sydney, Australia Hoboken, N.J.: University of South Wales Press; John Wiley & Sons. ix, 263 p.

Cooper, G., F. Grieser, and S. Biggs, *Butyl Acrylate/Vinyl Acetate Copolymer Latex Synthesis Using Ultrasound As an Initiator.* J Colloid Interface Sci, 1996. 184(1): 52–63.

Davies, M.N.O. and P.R. Green, *Head-Bobbing During Walking, Running and Flying: Relative Motion Perception in the Pigeon.* J Exp Biol, 1988. 138(1): 71–91.

Edwards, L., *Sharp-eyed robins can see magnetic fields.* physorg. com, 2010.

Emmerton, J., *Pattern discrimination in the near-ultraviolet by pigeons.* Percept Psychophys, 1983. 34(6): 555–559.

Fite, K.V. and S. Rosenfield-Wessels, *A comparative study of deep avian foveas.* Brain Behav Evol, 1975. 12(1–2): 97–115.

Flannery, M.C., *Looking at fossile in new ways.* American Biology Teacher, 2005. 67(1): 5.

Galifret, Y., et al., *Centrifugal control in the visual system of the pigeon.* Vision Res, 1971. Suppl 3: 185–200.

Gibson, L.J., *Woodpecker pecking: how woodpeckers avoid brain injury.* Journal of Zoology, 2006. 270(3): 462–465.

Goldsmith, T.H., *Hummingbirds see near ultraviolet light.* Science, 1980. 207(4432): 786–788.

Gordon, D., *Letter: Woodpeckers, gannets, and head injury.* Lancet, 1976. 1(7963): 801–802.

Hackett, S.J., et al., *A phylogenomic study of birds reveals their evolutionary history.* Science, 2008. 320(5884): 1763–1768.

Hall, M.I., *The anatomical relationships between the avian eye, orbit and sclerotic ring: implications for inferring activity patterns in extinct birds.* J Anat, 2008. 212(6): 781–794.

Hart, N.S., *Variations in cone photoreceptor abundance and the visual ecology of birds.* J Comp Physiol [A], 2001. 187(9): 685–697.

Hart, N.S., *The visual ecology of avian photoreceptors.* Prog Retin Eye Res, 2001. 20(5): 675–703.

Hodos, W., et al., *Normative data for pigeon vision.* Vision Res, 1985. 25(10): 1525–1527.

Howland, H.C., M. Howland, and K.L. Schmid, *Focusing and accommodation in the brown kiwi (Apteryx australis).* Journal of Comparative Physiology A: Neuroethology, Sensory, Neural, and Behavioral Physiology, 1992. 170(6): 687–689.

Howland, H.C. and J.G. Sivak, *Penguin vision in air and water.* Vision Res, 1984. 24(12): 1905–1909.

Hu, D., et al., *A pre-Archaeopteryx troodontid theropod from China with long feathers on the metatarsus.* Nature, 2009. 461(7264): 640–643.

Hunt, D.M., et al., *Evolution and spectral tuning of visual pigments in birds and mammals.* Philosophical Transactions of the Royal Society B: Biological Sciences, 2009. 364(1531): 2941–2955.

Ji, Q., et al., *The distribution of integumentary structures in a feathered dinosaur.* Nature, 2001. 410(6832): 1084–1088.

Johnston, M.C., et al., *Origins of avian ocular and periocular tissues.* Exp Eye Res, 1979. 29(1): 27–43.

Jones, M.P., K.E. Pierce Jr, and D. Ward, *Avian Vision: A Review of Form and Function with Special Consideration to Birds of Prey.* Journal of Exotic Pet Medicine, 2007. 16(2): 69–87.

Katzir, G. and H.C. Howland, *Corneal power and underwater accommodation in great cormorants (Phalacrocorax carbo sinensis).* J Exp Biol, 2003. 206(5): 833–841.

Katzir, G. and G.R. Martin, *Visual fields in herons (Ardeidae)— panoramic vision beneath the bill.* Naturwissenschaften, 1994. 81(4): 182–184.

Kelber, A. and U. Henique, *Trichromatic colour vision in the hummingbird hawkmoth, Macroglossum stellatarum L.* Journal of Comparative Physiology A: Neuroethology, Sensory, Neural, and Behavioral Physiology, 1999. 184(5): 535–541.

Kevan, P.G., L. Chittka, and A.G. Dyer, *Limits to the salience of ultraviolet: lessons from colour vision in bees and birds.* J Exp Biol, 2001. 204(Pt 14): 2571–2580.

King-Smith, P.E., *Absorption spectra and function of the coloured oil drops in the pigeon retina.* Vision Res, 1969. 9(11): 1391–1399.

Kolmer, W., *Über das Auge des Eisvogels (Alcedo attis attis).* Pflügers Archiv European Journal of Physiology, 1924. 204(1): 266–274.

Kram, Y.A., S. Mantey, and J.C. Corbo, *Avian Cone Photoreceptors Tile the Retina as Five Independent, Self-Organizing Mosaics.* PLoS One, 2010. 5(2): e8992.

Kram, Y.A., S. Mantey, and J.C. Corbo, *Avian cone photoreceptors tile the retina as five independent, self-organizing mosaics.* PLoS One, 2010. 5(2): e8992.

Kurochkin, E.N., et al., *A fossil brain from the Cretaceous of European Russia and avian sensory evolution.* Biol Lett, 2007. 3(3): 309–313.

Levy, B. and J.G. Sivak, *Mechanisms of accommodation in the bird eye.* Journal of Comparative Physiology A: Neuroethology, Sensory, Neural, and Behavioral Physiology, 1980. 137(3): 267–272.

Macko, K.A. and W. Hodos, *Near point of accommodation in pigeons.* Vision Res, 1985. 25(10): 1529–1530.

Marshall, A.J., *Biology and comparative physiology of birds.* 1960, New York, : Academic Press. 2 v.

Martin, G.R., *Visual fields in woodcocks Scolopax rusticola (Scolopacidae; Charadriiformes).* Journal of Comparative Physiology A: Neuroethology, Sensory, Neural, and Behavioral Physiology, 1994. 174(6): 787–793.

Martin, G.R., et al., *Vision in Birds, in The Senses: A Comprehensive Reference.* 2008, Academic Press: New York. 25–52.

May, P.R., et al., *Woodpecker drilling behavior. An endorsement of the rotational theory of impact brain injury.* Arch Neurol, 1979. 36(6): 370–373.

May, P.R., et al., *Woodpeckers and head injury.* Lancet, 1976. 1(7957): 454–455.

Meyer-Rochow, V.B. and J. Gál, *Dimensional limits for arthropod eyes with superposition optics.* Vision Research, 2004. 44(19): 2213–2223.

Moller, A., et al., *Retinal cryptochrome in a migratory passerine bird: a possible transducer for the avian magnetic compass.* Naturwissenschaften, 2004. 91(12): 585–588.

Muller, B., et al., *Bat eyes have ultraviolet-sensitive cone photoreceptors.* PLoS One, 2009. 4(7): e6390.

Murphy, C.J. and H.C. Howland, *Owl eyes: Accommodation, corneal curvature and refractive state.* Journal of Comparative Physiology A: Neuroethology, Sensory, Neural, and Behavioral Physiology, 1983. 151(3): 277–284.

Nieder, A. and H. Wagner, *Perception and neuronal coding of subjective contours in the owl.* Nat Neurosci, 1999. 2(7): 660–663.

Pettigrew, J.D., *Binocular Visual Processing in the Owl's Telencephalon.* Proceedings of the Royal Society of London. Series B. Biological Sciences, 1979. 204(1157): 435–454.

Pettigrew, J.D. and M. Konishi, *Neurons selective for orientation and binocular disparity in the visual Wulst of the barn owl (Tyto alba).* Science, 1976. 193(4254): 675–678.

Pettigrew, J.D., J. Wallman, and C.F. Wildsoet, *Saccadic oscillations facilitate ocular perfusion from the avian pecten.* Nature, 1990. 343(6256): 362–363.

Prum, R.O., *Palaeontology: Dinosaurs take to the air.* Nature, 2003. 421(6921): 323–324.

Reymond, L., *Spatial visual acuity of the eagle Aquila audax: a behavioural, optical and anatomical investigation.* Vision Res, 1985. 25(10): 1477–1491.

Ritz, T., S. Adem, and K. Schulten, *A Model for Photoreceptor-Based Magnetoreception in Birds.* Biophysical Journal, 2000. 78(2): 707–718.

Ritz, T., et al., *Resonance effects indicate a radical-pair mechanism for avian magnetic compass.* Nature, 2004. 429(6988): 177–180.

Ritz, T., et al., *Magnetic Compass of Birds Is Based on a Molecule with Optimal Directional Sensitivity.* 2009. 96(8): 3451–3457.

Roberts, N.W., et al., *A biological quarter-wave retarder with excellent achromaticity in the visible wavelength region.* Nat Photon, 2009. 3(11): 641–644.

Ruggeri M et al., *Retinal structure of birds of prey revealed by ultra-high resolution spectral-domain optical coherence tomography.* Invest Ophthalmol Vis Sci., 2010. 51(11): 5789-95.

Schwab, I.R., *Cure for a headache.* British Journal of Ophthalmology, 2002. 86(8): 843.

Shaw, N.A., *The neurophysiology of concussion.* Prog Neurobiol, 2002. 67(4): 281–344.

Sivak, J.G., *Avian mechanisms for vision in air and water.* Trends in Neurosciences, 1980. 3(12): 314–317.

Tiemeier, O.W., *The os opticus of birds.* Journal of Morphology, 1950. 86(1): 25–46.

Tucker, V.A., *The deep fovea, sideways vision and spiral flight paths in raptors.* J Exp Biol, 2000. 203(Pt 24): 3745–3754.

Vincent, J.F.V., Sahinkaya, M. N., O'Shea, W., *A woodpecker hammer.* Proceedings of the Institution of Mechanical Engineers, Part C: . Journal of Mechanical Engineering Science, 2007. 221(10): 1141–1147.

Vorobyev, M., *Coloured oil droplets enhance colour discrimination.* 2003. 1255–1261.

Witmer, L.M., *Palaeontology: Feathered dinosaurs in a tangle.* Nature, 2009. 461(7264): 601–602.

Wood, C.A., *The fundus oculi of birds, especially as viewed by the ophthalmoscope; a study in the comparative anatomy and physiology.* 1917, Chicago, : The Lakeside Press. 3 ., 5–180 p.

Wygnanski-Jaffe, T., et al., *Protective ocular mechanisms in woodpeckers.* Eye, 2005. 21(1): 83–89.

Xu, X., et al., *Four-winged dinosaurs from China.* Nature, 2003. 421(6921): 335–340.

Young, S.R. and G.R. Martin, *Optics of retinal oil droplets: a model of light collection and polarization detection in the avian retina.* Vision Res, 1984. 24(2): 129–137.

Chapter 21

Altner, I. and D. Burkhardt, *Fine structure of the ommatidia and the occurrence of rhabdomeric twist in the dorsal eye of male Bibio marci (Diptera, Nematocera, Bibionidae).* Cell Tissue Res, 1981. 215(3): 607–623.

Arikawa, K., *Hindsight of Butterflies.* BioScience, 2001. 51(3): 219–225.

Arikawa, K., K. Inokuma, and E. Eguchi, *Pentachromatic visual system in a butterfly.* Naturwissenschaften, 1987. 74(6): 297–298.

Arikawa, K. and D. Stavenga, *Random array of colour filters in the eyes of butterflies.* J Exp Biol, 1997. 200(Pt 19): 2501–2506.

Arikawa, K., D. Suyama, and T. Fujii, *Hindsight by genitalia: photo-guided copulation in butterflies.* Journal of Comparative Physiology A: Neuroethology, Sensory, Neural, and Behavioral Physiology, 1997. 180(4): 295–299.

Awata, H., M. Wakakuwa, and K. Arikawa, *Evolution of color vision in pierid butterflies: blue opsin duplication, ommatidial heterogeneity and eye regionalization in Colias erate.* J Comp Physiol A Neuroethol Sens Neural Behav Physiol, 2009. 195(4): 401–408.

Barlow, H.B., *The Size of Ommatidia in Apposition Eyes.* J Exp Biol, 1952. 29(4): 667–674.

Beardsley, J.W., *A New Fossil Scale Insect (Homoptera Coccoidea) From Canadian Amber.* Psyche, 1969(3): 270–279.

Bernhard, C.G. and D. Ottoson, *Quantitative Studies on Pigment Migration and Light Sensitivity in the Compound Eye at Different Light Intensities.* J Gen Physiol, 1964. 47: 465–478.

Briscoe, A.D., *Reconstructing the ancestral butterfly eye: focus on the opsins.* J Exp Biol, 2008. 211(Pt 11): 1805–1813.

Briscoe, A.D. and L. Chittka, *The evolution of color vision in insects*. Annu Rev Entomol, 2001. 46(1): 471–510.

Burkhardt, D., I. Motte, and K. Lunau, *Signalling fitness: larger males sire more offspring. Studies of the stalk-eyed fly Cyrtodiopsis whitei (Diopsidae, Diptera)*. Journal of Comparative Physiology A: Neuroethology, Sensory, Neural, and Behavioral Physiology, 1994. 174(1): 61–64.

Buschbeck, E. and M. Hauser, *The visual system of male scale insects*. Naturwissenschaften, 2009. 96(3): 365–374.

Cook, L.G., P.J. Gullan, and H.E. Trueman, *A preliminary phylogeny of the scale insects (Hemiptera: Sternorrhyncha: Coccoidea) based on nuclear small-subunit ribosomal DNA*. Mol Phylogenet Evol, 2002. 25(1): 43–52.

Duelli, P., *An insect retina without microvilli in the male scale insect, Eriococcus sp. (eriococcidae, homoptera)*. Cell Tissue Res, 1978. 187(3): 417–427.

Eguchi, E., *Retinular fine structure in compound eyes of diurnal and nocturnal sphingid moths*. Cell Tissue Res, 1982. 223(1): 29–42.

Erren, T., et al., *Clockwork blue: on the evolution of non-image-forming retinal photoreceptors in marine and terrestrial vertebrates*. Naturwissenschaften, 2008. 95(4): 273–279.

Fischer, S., C.H. Muller, and V.B. Meyer-Rochow, *How small can small be: The compound eye of the parasitoid wasp Trichogramma evanescens (Westwood, 1833) (Hymenoptera, Hexapoda), an insect of 0.3- to 0.4-mm total body size*. Vis Neurosci, 2010: 1–14.

Gilbert, C., *Visual determinants of escape in tiger beetle larvae (Cicindelidae)*. Journal of Insect Behavior, 1989. 2(4): 557–574.

Greiner, B., et al., *Neural organisation in the first optic ganglion of the nocturnal bee Megalopta genalis*. Cell Tissue Res, 2004. 318(2): 429–437.

Grimaldi, D.A. and M.S. Engel, *Evolution of the insects*. 2005, Cambridge [U.K.]; New York: Cambridge University Press.xv, 755 p.

Gullan, P.J. and M. Kosztarab, *Adaptations in scale insects*. Annu Rev Entomol, 1997. 42: 23–50.

HODGSON, C. and I. FOLDI, *A review of the Margarodidae sensu Morrison (Hemiptera: Coccoidea) and some related taxa based on the morphology of adult males*. Vol. 1. 2006: Magnolia Press. 250.

Holmes, R.S., D.W. Cooper, and J.L. Vandeberg, *Marsupial and monotreme lactate dehydrogenase isozymes: phylogeny, ontogeny, and homology with eutherian mammals*. J Exp Zool, 1973. 184(1): 127–148.

Hoonkanen, A., Meyer-Rochow, V. B., *The eye of the parthenogenetic and minute moth Ectoedemia argyropeza (Zeller) (Lepidoptera: Nepticulidae)*. European Journal of Entomology, 2009. 160(4): 619–629.

Kelber, A., *Innate preferences for flower features in the hawkmoth Macroglossum stellatarum*. J Exp Biol, 1997. 200(4): 827–836.

Kelber, A., A. Balkenius, and E.J. Warrant, *Scotopic colour vision in nocturnal hawkmoths*. Nature, 2002. 419(6910): 922–925.

Kern, R. and D. Varju, *Visual position stabilization in the hummingbird hawk moth, Macroglossum stellatarum L. I. Behavioural analysis*. J Comp Physiol A, 1998. 182(2): 225–237.

Kjer, K.M., R.J. Blahnik, and R.W. Holzenthal, *Phylogeny of caddisflies (Insecta, Trichoptera)*. Zoologica Scripta, 2002. 31(1): 83–91.

Koshitaka, H., et al., *Tetrachromacy in a butterfly that has eight varieties of spectral receptors*. Proc Biol Sci, 2008. 275(1637): 947–954.

Land, M.F., *Visual acuity in insects*. Annu Rev Entomol, 1997. 42(1): 147–177.

Land, M.F. and R.D. Fernald, *The Evolution of Eyes*. 1992. 1–29.

Marcos, S. and R. Navarro, *Determination of the foveal cone spacing by ocular speckle interferometry: limiting factors and acuity predictions*. J Opt Soc Am A Opt Image Sci Vis, 1997. 14(4): 731–740.

Nilsson, D.E., M.F. Land, and J. Howard, *Afocal apposition optics in butterfly eyes*. Nature, 1984. 312(5994): 561–563.

Obara, Y., H. Koshitaka, and K. Arikawa, *Better mate in the shade: enhancement of male mating behaviour in the cabbage butterfly, Pieris rapae crucivora, in a UV-rich environment*. J Exp Biol, 2008. 211(Pt 23): 3698–3702.

Post, C.T.J. and T.H. Goldsmith, *Pigment migration and light-adaptation in the eye of the moth, Galleria mellonella*. Biol Bull, 1965. 128(3): 473–487.

Qiu, X., et al., *Ommatidial heterogeneity in the compound eye of the male small white butterfly, Pieris rapae crucivora*. Cell Tissue Res, 2002. 307(3): 371–379.

Rosenzweig, E., et al., *Micromorphology of the dorsal ocelli of the Oriental hornet*. J Gravit Physiol, 1998. 5(1): P113–114.

Shaw, S.R., *The Photoreceptor Axon Projection and Its Evolution in the Neural Superposition Eyes of Some Primitive Brachyceran Diptera (Part 2 of 2)*. Brain, Behavior and Evolution, 1990. 35(2): 116–125.

Stavenga, D.G., *Colour in the eyes of insects*. J Comp Physiol A Neuroethol Sens Neural Behav Physiol, 2002. 188(5): 337–348.

Stavenga, D.G., *Reflections on colourful ommatidia of butterfly eyes*. J Exp Biol, 2002. 205(Pt 8): 1077–1085.

Stavenga, D.G., *Partial coherence and other optical delicacies of lepidopteran superposition eyes*. J Exp Biol, 2006. 209(Pt 10): 1904–1913.

Stavenga, D.G. and K. Arikawa, *Evolution of color and vision of butterflies*. Arthropod Struct Dev, 2006. 35(4): 307–318.

Stavenga, D.G., et al., *Light on the moth-eye corneal nipple array of butterflies*. Proc Biol Sci, 2006. 273(1587): 661–667.

Stavenga, D.G., et al., *Retinal regionalization and heterogeneity of butterfly eyes*. Naturwissenschaften, 2001. 88(11): 477–481.

Takeuchi, Y., K. Arikawa, and M. Kinoshita, *Color discrimination at the spatial resolution limit in a swallowtail butterfly, Papilio xuthus*. J Exp Biol, 2006. 209(Pt 15): 2873–2879.

Varela, F.G., *The vertebrate and the (insect) compound eye in evolutionary perspective*. Vision Res, 1971. Suppl 3: 201–209.

Warrant, E., K. Bartsch, and G.N. C, *Physiological optics in the hummingbird hawkmoth: a compound eye without ommatidia*. J Exp Biol, 1999. 202 (Pt 5): 497–511.

Warrant, E.J., et al., *Nocturnal vision and landmark orientation in a tropical halictid bee*. Curr Biol, 2004. 14(15): 1309–1318.

Warrant, E.J., et al., *Ocellar optics in nocturnal and diurnal bees and wasps*. Arthropod Struct Dev, 2006. 35(4): 293–305.

Wilkinson, G.S., D.C. Presgraves, and L. Crymes, *Male eye span in stalk-eyed flies indicates genetic quality by meiotic drive suppression*. Nature, 1998. 391(6664): 276–279.

Yack, J.E., et al., *The eyes of Macrosoma sp. (Lepidoptera: Hedyloidea): a nocturnal butterfly with superposition optics*. Arthropod Struct Dev, 2007. 36(1): 11–22.

CHAPTER 22

Arrese, C., *Pupillary mobility in four species of marsupials with differing lifestyles*. Journal of Zoology, 2002. 256(2): 191–197.

Arrese, C., et al., *Retinal Structure and Visual Acuity in a Polyprotodont Marsupial, the Fat-Tailed Dunnart (Sminthopsis crassicaudata)*. Brain, Behavior and Evolution, 1999. 53(3): 111–126.

Arrese, C.A., et al., *Trichromacy in Australian marsupials*. Curr Biol, 2002. 12(8): 657–660.

Arrese, C.A., et al., *Topographies of retinal cone photoreceptors in two Australian marsupials.* Vis Neurosci, 2003. 20(3): 307–311.

Asher, R.J., I. Horovitz, and M.R. Sanchez-Villagra, *First combined cladistic analysis of marsupial mammal interrelationships.* Mol Phylogenet Evol, 2004. 33(1): 240–250.

Bininda-Emonds, O.R.P., et al., *The delayed rise of present-day mammals.* Nature, 2007. 446(7135): 507–512.

Cardillo, M., et al., *A species-level phylogenetic supertree of marsupials.* Journal of Zoology, 2004. 264(1): 11–31.

Chiappe, L.M. and S. Bertelli, *Palaeontology: skull morphology of giant terror birds.* Nature, 2006. 443(7114): 929.

Cowing, J.A., et al., *The molecular mechanism for the spectral shifts between vertebrate ultraviolet- and violet-sensitive cone visual pigments.* Biochem J, 2002. 367(Pt 1): 129–135.

Davies, W.L., et al., *Visual pigments of the platypus: a novel route to mammalian colour vision.* Curr Biol, 2007. 17(5): R161–163.

Grutzner, F. and J.A. Graves, *A platypus' eye view of the mammalian genome.* Curr Opin Genet Dev, 2004. 14(6): 642–649.

Hemmi, J.M. and U. Grunert, *Distribution of photoreceptor types in the retina of a marsupial, the tammar wallaby (Macropus eugenii).* Vis Neurosci, 1999. 16(2): 291–302.

Hunt, D.M., et al., *The rod opsin pigments from two marsupial species, the South American bare-tailed woolly opossum and the Australian fat-tailed dunnart.* Gene, 2003. 323: 157–162.

Hunt, D.M., et al., *Evolution and spectral tuning of visual pigments in birds and mammals.* Philosophical Transactions of the Royal Society B: Biological Sciences, 2009. 364(1531): 2941–2955.

Jacobs, G.H., et al., *Opsin gene and photopigment polymorphism in a prosimian primate.* Vision Res, 2002. 42(1): 11–18.

Luo, Z.X., *Transformation and diversification in early mammal evolution.* Nature, 2007. 450(7172): 1011–1019.

Luo, Z.X., et al., *An Early Cretaceous tribosphenic mammal and metatherian evolution.* Science, 2003. 302(5652): 1934–1940.

Manger, P.R. and J.D. Pettigrew, *Electroreception and the Feeding Behaviour of Platypus (Ornithorhynchus anatinus: Monotremata: Mammalia).* Philosophical Transactions: Biological Sciences, 1995. 347(1322): 359–381.

McMenamin, P.G. and W.J. Krause, *Morphological observations on the unique paired capillaries of the opossum retina.* Cell Tissue Res, 1993. 271(3): 461–468.

Messer, M., et al., *Evolution of the Monotremes: Phylogenetic Relationship to Marsupials and Eutherians, and Estimation of Divergence Dates Based on α-Lactalbumin Amino Acid Sequences.* Journal of Mammalian Evolution, 1998. 5(1): 95–105.

Nicol, S.C., et al., *The echidna manifests typical characteristics of rapid eye movement sleep.* Neurosci Lett, 2000. 283(1): 49–52.

Pettigrew, J.D., P.R. Manger, and S.L. Fine, *The sensory world of the platypus.* Philos Trans R Soc Lond B Biol Sci, 1998. 353(1372): 1199–1210.

Proske, U. and E. Gregory, *Electrolocation in the platypus—some speculations.* Comp Biochem Physiol A Mol Integr Physiol, 2003. 136(4): 821–825.

Rowe, T., et al., *The oldest platypus and its bearing on divergence timing of the platypus and echidna clades.* Proc Natl Acad Sci U S A, 2008. 105(4): 1238–1242.

Shi, Y. and S. Yokoyama, *Molecular analysis of the evolutionary significance of ultraviolet vision in vertebrates.* Proc Natl Acad Sci U S A, 2003. 100(14): 8308–8313.

van Rheede, T., et al., *The platypus is in its place: nuclear genes and indels confirm the sister group relation of monotremes and Therians.* Mol Biol Evol, 2006. 23(3): 587–597.

Wakefield, M.J., et al., *Cone visual pigments of monotremes: filling the phylogenetic gap.* Vis Neurosci, 2008. 25(3): 257–264.

Wroe, S., *Killer kangaroos and other murderous marsupials.* Sci Am, 1999. 280(5): 68–74.

Young, H.M. and J.D. Pettigrew, *Cone photoreceptors lacking oil droplets in the retina of the echidna, Tachyglossus aculeatus (Monotremata).* Vis Neurosci, 1991. 6(5): 409–420.

Young, H.M. and D.I. Vaney, *The retinae of Prototherian mammals possess neuronal types that are characteristic of non-mammalian retinae.* Vis Neurosci, 1990. 5(1): 61–66.

CHAPTER 23

Blakeslee, B. and G.H. Jacobs, *Color vision in the ring-tailed lemur (Lemur catta).* Brain Behav Evol, 1985. 26(3–4): 154–166.

Bowmaker, J.K., et al., *Photosensitive and photostable pigments in the retinae of Old World monkeys.* J Exp Biol, 1991. 156: 1–19.

Brudenall, D.K., et al., *Optimized architecture for nutrition in the avascular retina of Megachiroptera.* Anat Histol Embryol, 2007. 36(5): 382–388.

Carrie C. Veilleux, D.A.B., *Opsin gene polymorphism predicts trichromacy in a cathemeral lemur.* 2009. 86–90.

Collin, S.P. and A.E.O. Trezise, *The origins of colour vision in vertebrates.* Clinical and Experimental Optometry, 2004. 87(4–5): 217–223.

Dulai, K.S., et al., *Sequence divergence, polymorphism and evolution of the middle-wave and long-wave visual pigment genes of great apes and old world monkeys.* Vision Res, 1994. 34(19): 2483–2491.

Dyer, M.A., et al., *Developmental sources of conservation and variation in the evolution of the primate eye.* Proc Natl Acad Sci U S A, 2009. 106(22): 8963–8968.

Gerald H. Jacobs, J.N., *Color Vision: How Our Eyes Reflect Primate Evolution.* Scientific American, 2009.

Hunt, D.M., et al., *Sequence and Evolution of the Blue Cone Pigment Gene in Old and New World Primates.* Genomics, 1995. 27(3): 535–538.

Hunt, D.M., et al., *Molecular evolution of trichromacy in primates.* Vision Res, 1998. 38(21): 3299–3306.

Jacobs, G.H., *The distribution and nature of colour vision among the mammals.* Biol Rev Camb Philos Soc, 1993. 68(3): 413–471.

Jacobs, G.H., *Primate photopigments and primate color vision.* Proc Natl Acad Sci U S A, 1996. 93(2): 577–581.

Jacobs, G.H., *A perspective on color vision in platyrrhine monkeys.* Vision Res, 1998. 38(21): 3307–3313.

Jacobs, G.H., *Evolution of colour vision in mammals.* Philos Trans R Soc Lond B Biol Sci, 2009. 364(1531): 2957–2967.

Jacobs, G.H. and J.F. Deegan, *Photopigments and colour vision in New World monkeys from the family Atelidae.* 2001. 695–702.

Jacobs, G.H. and J.F. Deegan, 2nd, *Diurnality and cone photopigment polymorphism in strepsirrhines: examination of linkage in Lemur catta.* Am J Phys Anthropol, 2003. 122(1): 66–72.

Jacobs, G.H., J.F. Deegan, 2nd, and J. Neitz, *Photopigment basis for dichromatic color vision in cows, goats, and sheep.* Vis Neurosci, 1998. 15(3): 581–584.

Jacobs, G.H., et al., *Opsin gene and photopigment polymorphism in a prosimian primate.* Vision Res, 2002. 42(1): 11–18.

Jacobs, G.H., et al., *Trichromatic colour vision in New World monkeys.* Nature, 1996. 382(6587): 156–158.

Jacobs, G.H., M. Neitz, and J. Neitz, *Mutations in S-cone pigment genes and the absence of colour vision in two species of nocturnal primate.* Proc Biol Sci, 1996. 263(1371): 705–710.

Jacobs, G.H. and M.P. Rowe, *Evolution of vertebrate colour vision.* Clinical and Experimental Optometry, 2004. 87(4–5): 206–216.

Jones, K.E., et al., *A phylogenetic supertree of the bats (Mammalia: Chiroptera).* Biol Rev Camb Philos Soc, 2002. 77(2): 223–259.

Kraft, T.W., J. Neitz, and M. Neitz, *Spectra of human L cones.* Vision Res, 1998. 38(23): 3663–3670.

Luo, Z.X., *Transformation and diversification in early mammal evolution.* Nature, 2007. 450(7172): 1011–1019.

Mollon, J.D., *"Tho' she kneel'd in that place where they grew. . ."* *The uses and origins of primate colour vision.* J Exp Biol, 1989. 146(1): 21–38.

Mollon, J.D. and J.K. Bowmaker, *The spatial arrangement of cones in the primate fovea.* Nature, 1992. 360(6405): 677–679.

Mollon, J.D., J.K. Bowmaker, and G.H. Jacobs, *Variations of colour vision in a New World primate can be explained by polymorphism of retinal photopigments.* Proc R Soc Lond B Biol Sci, 1984. 222(1228): 373–399.

Nakashige, M., A.L. Smith, and D.S. Strait, *Biomechanics of the macaque postorbital septum investigated using finite element analysis: implications for anthropoid evolution.* J Anat, 2011. 218(1): 142–150.

Nei, M., J. Zhang, and S. Yokoyama, *Color vision of ancestral organisms of higher primates.* 1997. 611–618.

Neitz, J., T. Geist, and G.H. Jacobs, *Color vision in the dog.* Vis Neurosci, 1989. 3(2): 119–125.

Neitz, J. and G.H. Jacobs, *Polymorphism of the long-wavelength cone in normal human colour vision.* Nature, 1986. 323(6089): 623–625.

Nikaido, M., et al., *Monophyletic origin of the order chiroptera and its phylogenetic position among mammalia, as inferred from the complete sequence of the mitochondrial DNA of a Japanese megabat, the Ryukyu flying fox (Pteropus dasymallus).* J Mol Evol, 2000. 51(4): 318–328.

Pirie, A., *Crystals of riboflavin making up the tapetum lucidum in the eye of a lemur.* Nature, 1959. 183(4666): 985–986.

Rehorek, S.J. and T.D. Smith, *The primate Harderian gland: Does it really exist?* Ann Anat, 2006. 188(4): 319–327.

Ross, C.F. and W.L. Hylander, In vivo *and in vitro* bone strain in the owl monkey circumorbital region and the function of the postorbital septum. Am J Phys Anthropol, 1996. 101(2): 183–215.

Ross, C.F. and E.C. Kirk, *Evolution of eye size and shape in primates.* J Hum Evol, 2007. 52(3): 294–313.

Shi, Y. and S. Yokoyama, *Molecular analysis of the evolutionary significance of ultraviolet vision in vertebrates.* Proc Natl Acad Sci U S A, 2003. 100(14): 8308–8313.

Shyue, S.K., et al., *Adaptive evolution of color vision genes in higher primates.* Science, 1995. 269(5228): 1265–1267.

Springer, M.S., et al., *Placental mammal diversification and the Cretaceous Tertiary boundary.* 2003. 1056–1061.

Travis, D.S., J.K. Bowmaker, and J.D. Mollon, *Polymorphism of visual pigments in a callitrichid monkey.* Vision Res, 1988. 28(4): 481–490.

Wang, D., et al., *Molecular Evolution of Bat ColorVision Genes.* 2004. 295–302.

Wang, D., et al., *Molecular evolution of bat color vision genes.* Mol Biol Evol, 2004. 21(2): 295–302.

Whitmore, A.V. and J.K. Bowmaker, *Differences in the Temporal Properties of Human Longwave- and Middlewave-sensitive Cones.* European Journal of Neuroscience, 1995. 7(6): 1420–1423.

Wikler, K.C. and Rakic, *Distribution of photoreceptor subtypes in the retina of diurnal and nocturnal primates.* J Neurosci, 1990. 10(10): 3390–3401.

CHAPTER 24

Johnsen, S., *Transparent animals.* Sci Am, 2000. 282(2): 80–89.

Land, M.F. and D.-E. Nilsson, *Animal eyes.* Oxford animal biology series. 2002, Oxford; New York: Oxford University Press. xii, 221 p., 4 of plates.

Wolken, J.J., *Photobehavior of marine invertebrates: extraocular photoreception.* Comp Biochem Physiol C, 1988. 91(1): 145–149.

Lakin, RC, Jinks, RN, Battelle BA, et al: *Retinal Anatomy of Chorocaris chacei, a Deep-Sea Hydrothermal Vent Shrimp from the Mid-Atlantic Ridge.* J Comp Neuro, 1997: 385: 503–514.

Van Dover, CL, Szuts EZ, Chamberlain, SC Cann JR: *A novel eye in "eyeless" shrimp from hydrothermal vents of the Mid-Atlantic Ridge.* Nature1989; 337: 458–460.

Pelli, DG, Chamberlain SC: *The visibility of 350°black-body radiation by the shrimpRimicaris exoculata and man.* Nature 1989; 337: 460–462.

CHAPTER 25

Colitz, C.M., et al., *Characterization of progressive keratitis in Otariids.* Vet Ophthalmol, 2010. 13 Suppl: 47–53.

Dawson, W.W., et al., *Static and kinetic properties of the dolphin pupil.* Am J Physiol, 1979. 237(5): R301–305.

Dawson, W.W., J.P. Schroeder, and J.F. Dawson, *The ocular fundus of two cetaceans.* Marine Mammal Science, 1987. 3(1): 1–13.

Fasick, J.I., et al., *The visual pigments of the bottlenose dolphin (Tursiops truncatus).* Vis Neurosci, 1998. 15(4): 643–651.

Fasick, J.I. and P.R. Robinson, *Spectral-tuning mechanisms of marine mammal rhodopsins and correlations with foraging depth.* Vis Neurosci, 2000. 17(5): 781–788.

Hanke, F., et al., *Basic mechanisms in pinniped vision.* Experimental Brain Research, 2009. 199(3): 299–311.

Hanke, F.D., et al., *Corneal topography, refractive state, and accommodation in harbor seals (Phoca vitulina).* Vision Res, 2006. 46(6–7): 837–847.

Hanke, F.D., et al., *Multifocal lenses in a monochromat: the harbour seal.* J Exp Biol, 2008. 211(Pt 20): 3315–3322.

Herman, L.M., et al., *Bottle-nosed dolphin: double-slit pupil yields equivalent aerial and underwater diurnal acuity.* Science, 1975. 189(4203): 650–652.

Mass, A.M., *A high-resolution area in the retinal ganglion cell layer of the Steller's sea lion (Eumetopias jubatus): a topographic study.* Dokl Biol Sci, 2004. 396: 187–190.

Mass, A.M. and A.Y. Supin, *Adaptive features of aquatic mammals' eye.* Anat Rec (Hoboken), 2007. 290(6): 701–715.

Miller, S.N., C.M. Colitz, and R.R. Dubielzig, *Anatomy of the California sea lion globe.* Vet Ophthalmol, 2010. 13 Suppl: 63–71.

Peichl, L., G. Behrmann, and R.H. Kroger, *For whales and seals the ocean is not blue: a visual pigment loss in marine mammals.* Eur J Neurosci, 2001. 13(8): 1520–1528.

Pepper, R.L. and J.V. Simmons, Jr., *In-air visual acuity of the bottlenose dolphin.* Exp Neurol, 1973. 41(2): 271–276.

Rybczynski, N., M.R. Dawson, and R.H. Tedford, *A semi-aquatic Arctic mammalian carnivore from the Miocene epoch and origin of Pinnipedia.* Nature, 2009. 458(7241): 1021–1024.

Sivak, J.G., et al., *The eye of the hooded seal, Cystophora cristata, in air and water.* J Comp Physiol A, 1989. 165(6): 771–777.

Spurr-Michaud, S., Argueso, and I. Gipson, *Assay of mucins in human tear fluid.* Exp Eye Res, 2007. 84(5): 939–950.

Welsch, U., et al., *Microscopic anatomy of the eye of the deep-diving Antarctic Weddell seal (Leptonychotes weddellii).* J Morphol, 2001. 248(2): 165–174.

CHAPTER 26

Gamlin, P.D., et al., *Human and macaque pupil responses driven by melanopsin-containing retinal ganglion cells.* Vision Res, 2007. 47(7): 946–954.

Mollon, J.D., B.C. Regan, and J.K. Bowmaker, *What is the function of the cone-rich rim of the retina?* Eye (Lond), 1998. 12 (Pt 3b): 548–552.

Sekaran, S., et al., *Melanopsin-dependent photoreception provides earliest light detection in the mammalian retina.* Curr Biol, 2005. 15(12): 1099–1107.

GENERAL

The eye as a replicating and diverging, modular developmental unit. Trends in Ecology and Evolution, 2003. 18: 623–627.

Ankel-Simons, F. and D.T. Rasmussen, *Diurnality, nocturnality, and the evolution of primate visual systems.* Am J Phys Anthropol, 2008. Suppl 47: 100–117.

Archer, S.N., *Adaptive mechanisms in the ecology of vision.* 1998: Kluwer Academic Publishers.

Arendt, D., *Evolution of eyes and photoreceptor cell types.* Int J Dev Biol, 2003. 47(7–8): 563–571.

Berner, R.A., J.M. VandenBrooks, and P.D. Ward, *Oxygen and Evolution.* Science, 2007. 316(5824): 557–558.

Busettini, C., G.S. Masson, and F.A. Miles, *A role for stereoscopic depth cues in the rapid visual stabilization of the eyes.* Nature, 1996. 380(6572): 342–345.

Carroll, S.B., *Endless forms most beautiful : the new science of evo devo and the making of the animal kingdom.* 1st ed. 2005, New York: W.W. Norton & Co. xi, 350 p., [16] of plates.

Collin, S.P., et al., *The evolution of early vertebrate photoreceptors.* Philos Trans R Soc Lond B Biol Sci, 2009. 364(1531): 2925–2940.

Cvekl, A., et al., *Regulation of gene expression by Pax6 in ocular cells: a case of tissue-preferred expression of crystallins in lens.* Int J Dev Biol, 2004. 48(8–9): 829–844.

Darwin, C., On the origin of species by means of natural selection, or, The preservation of favoured races in the struggle for life. 1859, London: John Murray. ix, [1], 502 p., [1] folded leaf of plates.

Dawkins, R., *Evolutionary biology. The eye in a twinkling.* Nature, 1994. 368(6473): 690–691.

Dawkins, R., *Climbing mount improbable.* 1996, New York: Norton. xii, 340 p.

Dawkins, R., *The ancestor's tale : a pilgrimage to the dawn of life.* 2004, London: Weidenfeld & Nicolson Illustrated. 528 p.

De Duve, C., *Life evolving : molecules, mind, and meaning.* 2002, New York: Oxford University Press. xv, 341 p.

Donoghue, M.J., et al., *The Importance of Fossils in Phylogeny Reconstruction.* Annual Review of Ecology and Systematics, 1989. 20(1): 431–460.

Doolittle, W.F., *Uprooting the tree of life.* Sci Am, 2000. 282(2): 90–95.

Dunn, C.W., et al., *Broad phylogenomic sampling improves resolution of the animal tree of life.* Nature, 2008. 452(7188): 745–749.

Fernald, R.D., *The evolution of eyes.* Brain Behav Evol, 1997. 50(4): 253–259.

Fernald, R.D., *Evolving eyes.* Int J Dev Biol, 2004. 48(8–9): 701–705.

Fernald, R.D., *Casting a genetic light on the evolution of eyes.* Science, 2006. 313(5795): 1914–1918.

Fortey, R.A., *Life, an unauthorised biography : a natural history of the first four thousand million years of life on earth.* 1997, London: HarperCollins Publishers. xiv, 398 p., [32] of plates.

Fuhrman, J., *Genome sequences from the sea.* Nature, 2003. 424(6952): 1001–1002.

Gee, H., *In search of deep time : beyond the fossil record to a new history of life.* 1999, New York: Free Press. 267 p.

Gehring, W.J., *Historical perspective on the development and evolution of eyes and photoreceptors.* Int J Dev Biol, 2004. 48(8–9): 707–717.

Gehring, W.J., *New perspectives on eye development and the evolution of eyes and photoreceptors.* J Hered, 2005. 96(3): 171–184.

Gehring, W.J. and K. Ikeo, *Pax 6: mastering eye morphogenesis and eye evolution.* Trends Genet, 1999. 15(9): 371–377.

Goldsmith, T.H., *Optimization, constraint, and history in the evolution of eyes.* Q Rev Biol, 1990. 65(3): 281–322.

Goldsmith, T.H., *Optimization, constraint, and history in the evolution of eyes.* Q Rev Biol, 1990. 65(3): 281–322.

Gould, S.J., *The evolution of life on the earth.* Sci Am, 1994. 271(4): 84–91.

Gregory, T., *Understanding Evolutionary Trees.* Evolution: Education and Outreach, 2008. 1(2): 121–137.

Halder, G., Callaerts, and W.J. Gehring, *New perspectives on eye evolution.* Curr Opin Genet Dev, 1995. 5(5): 602–609.

Harvey, P.H. and C.J. Godfray, *Evolution: A Horn for an Eye.* Science, 2001. 291(5508): 1505–1506.

Herring, P.J. and Marine Biological Association of the United Kingdom., *Light and life in the sea : a volume arising from the Symposium on Light and Life in the Sea.* 1990, Cambridge [England]; New York, NY, USA: Cambridge University Press. 357 p.

Hunt, D.M., et al., *Vision in the ultraviolet.* Cell Mol Life Sci, 2001. 58(11): 1583–1598.

Hutchinson, G.E., et al., *Fossils, Early Life, and Atmospheric History: Discussion.* Proceedings of the National Academy of Sciences of the United States of America, 1965. 53(6): 1213–1215.

Jacobs, G.H., *Comparative color vision.* Academic Press series in cognition and perception. 1981, New York: Academic Press. viii, 209 p.

Jonasova, K. and Z. Kozmik, *Eye evolution: lens and cornea as an upgrade of animal visual system.* Semin Cell Dev Biol, 2008. 19(2): 71–81.

Knoll, A.H., *Life on a young planet : the first three billion years of evolution on Earth.* 2003, Princeton, N.J.: Princeton University Press. x, 277 p.

Lamb, T.D., S.P. Collin, and E.N. Pugh, Jr., *Evolution of the vertebrate eye: opsins, photoreceptors, retina and eye cup.* Nat Rev Neurosci, 2007. 8(12): 960–976.

Land, M.F., *The optics of animal eyes.* Contemporary Physics, 1988. 29(5): 435–455.

Land, M.F., *Biological optics: deep reflections.* Curr Biol, 2009. 19(2): R78–80.

Land, M.F. and R.D. Fernald, *The Evolution of Eyes.* Annual Review of Neuroscience, 1992. 15(1): 1–29.

Land, M.F. and D.-E. Nilsson, *Animal eyes.* Oxford animal biology series. 2002, Oxford; New York: Oxford University Press. xii, 221 p., 4 of plates.

Luo, D.-G., T. Xue, and K.-W. Yau, *How vision begins: An odyssey.* Proceedings of the National Academy of Sciences, 2008. 105(29): 9855–9862.

Marshall, N.J. and M.F. Land, *Some optical features of the eyes of stomatopods.* Journal of Comparative Physiology A: Neuroethology, Sensory, Neural, and Behavioral Physiology, 1993. 173(5): 565–582.

Murphy, C.J. and H.C. Howland, *The optics of comparative ophthalmoscopy.* Vision Res, 1987. 27(4): 599–607.

Nilsson, D.-E., *The evolution of eyes and visually guided behaviour.* Philosophical Transactions of the Royal Society B: Biological Sciences, 2009. 364(1531): 2833–2847.

Nilsson, D.E., *Eye evolution: a question of genetic promiscuity.* Curr Opin Neurobiol, 2004. 14(4): 407–414.

Nilsson, D.E. and S. Pelger, *A pessimistic estimate of the time required for an eye to evolve.* Proc Biol Sci, 1994. 256(1345): 53–58.

Nilsson, D.E. and E.J. Warrant, *Visual discrimination: Seeing the third quality of light.* Curr Biol, 1999. 9(14): R535–537.

Nilsson, D.-E. and D. Arendt, *Eye Evolution: The Blurry Beginning.* Current biology : CB, 2008. 18(23): R1096–R1098.

Plachetzki, T.H.O.a.D.C., *Encyclopedia of Eye*, in*The Evolution of Opsins*. 2010, Elsevier LTD. 82–88.

Queiroz, A.D., *Do Image-Forming Eyes Promote Evolutionary Diversification?* Evolution, 1999. 53(6): 1654–1664.

Rowe, M., *Inferring the retinal anatomy and visual capacities of extinct vertebrates.* Palaeontologia Electronica, 2000. 3(1): 3–43.

Vorobyev, M., *Coloured oil droplets enhance colour discrimination.* Proc Biol Sci, 2003. 270(1521): 1255–1261.

Walls, G.L., *The vertebrate eye and its adaptive radiation.* Cranbrook institute of science Bulletin. 1942, Bloomfield Hills, Mich: Cranbrook Institute of Science. xiv p., 1 l., 785 p.

Yokoyama, S., *Molecular evolution of vertebrate visual pigments.* Prog Retin Eye Res, 2000. 19(4): 385–419.

Yoshida, T., et al., *Rapid evolution drives ecological dynamics in a predator-prey system.* Nature, 2003. 424(6946): 303–306.

Zeki, S., *Comparative Color Vision. By G. H. Jacobs. Pp. 209. (Academic Press,* London, *1982.)* Experimental Physiology, 1983. 68(4): 747.

Zuker, C.S., *On the evolution of eyes: would you like it simple or compound?* Science, 1994. 265(5173): 742–743.

APPENDIX

Beisel, K.W., et al., *Development and evolution of the vestibular sensory apparatus of the mammalian ear.* J Vestib Res, 2005. 15(5–6): 225–241.

Bemis, W.E. and R.G. Northcutt, *Innervation of the basicranial muscle of Latimeria chalumnae.* Environmental Biology of Fishes, 1991. 32(1): 147–158.

Carey, J. and N. Amin, *Evolutionary changes in the cochlea and labyrinth: Solving the problem of sound transmission to the balance organs of the inner ear.* Anat Rec A Discov Mol Cell Evol Biol, 2006. 288(4): 482–489.

Fritzsch, B., *Evolution of the vestibulo-ocular system.* Otolaryngol Head Neck Surg, 1998. 119(3): 182–192.

Fritzsch, B. and R.G. Northcutt, *Origin and migration of trochlear, oculomotor and abducent motor neurons in Petromyzon marinus L.* Brain Res Dev Brain Res, 1993. 74(1): 122–126.

Fritzsch, B., et al., *Organization of the six motor nuclei innervating the ocular muscles in lamprey.* J Comp Neurol, 1990. 294(4): 491–506.

Gehring, W.J., *Historical perspective on the development and evolution of eyes and photoreceptors.* Int J Dev Biol, 2004. 48(8–9): 707–717.

Gillum, W., *Mechanisms of accommodation in vertebrates.* Ophthalmic Semin, 1976. 1(3): 253–286.

Graf, W., *Motion detection in physical space and its peripheral and central representation.* Ann N Y Acad Sci, 1988. 545: 154–169.

Graf, W. and W.J. Brunken, *Elasmobranch oculomotor organization: anatomical and theoretical aspects of the phylogenetic development of vestibulo-oculomotor connectivity.* J Comp Neurol, 1984. 227(4): 569–581.

Guyton, D., *Ocular torsion: Sensorimotor principles.* Graefe's Archive for Clinical and Experimental Ophthalmology, 1988. 226(3): 241–245.

Howland, H.C., et al., *Restricted range of ocular accommodation in barn owls (Aves:Tytonidae).* Journal of Comparative Physiology A: Neuroethology, Sensory, Neural, and Behavioral Physiology, 1991. 168(3): 299–303.

Maisey, J.G., *Remarks on the inner ear of elasmobranchs and its interpretation from skeletal labyrinth morphology.* J Morphol, 2001. 250(3): 236–264.

Martin, G.R., *How do birds accommodate?* Nature, 1987. 328(6129): 383.

Mazan, S., et al., *Otx1 gene-controlled morphogenesis of the horizontal semicircular canal and the origin of the gnathostome characteristics.* Evol Dev, 2000. 2(4): 186–193.

Mittelstaedt, H., *Interaction of eye-, head-, and trunk-bound information in spatial perception and control.* J Vestib Res, 1997. 7(4): 283–302.

Neal, H.V., *The history of the eye muscles.* Journal of Morphology, 1918. 30(2): 433–453.

Ott, M., *Visual accommodation in vertebrates: mechanisms, physiological response and stimuli.* J Comp Physiol A Neuroethol Sens Neural Behav Physiol, 2006. 192(2): 97–111.

Robinson, D.A., *The use of matrices in analyzing the three-dimensional behavior of the vestibulo-ocular reflex.* Biol Cybern, 1982. 46(1): 53–66.

Simpson, J.I. and W. Graf, *Eye-muscle geometry and compensatory eye movements in lateral-eyed and frontal-eyed animals.* Ann N Y Acad Sci, 1981. 374: 20–30.

Sivak, J.G., *Accommodation in vertebrates: a contemporary survey.* Curr Top Eye Res, 1980. 3: 281–330.

Sivak, J.G., T. Hildebrand, and C. Lebert, *Magnitude and rate of accommodation in diving and nondiving birds.* Vision Research, 1985. 25(7): 925–933.

Trinajstic, K., et al., *Exceptional preservation of nerve and muscle tissues in Late Devonian placoderm fish and their evolutionary implications.* Biol Lett, 2007. 3(2): 197–200.

Young, G.C., *Number and arrangement of extraocular muscles in primitive gnathostomes: evidence from extinct placoderm fishes.* Biol Lett, 2008. 4(1): 110–114.

Note: Page numbers followed by "*b*", "*f*", or "*t*" indicate boxes, figures, or tables, respectively.

Permian period, 133–37
Silurian period, 77–84
pallial eyes, of scallop, 70, 71f
Pandalus platyceros. See spot prawns
Pantodon buchholzi. See African
    butterfly fish
parabolic superposition eye, 51, 51t, 61b,
    61f, 62f
    of hermit crab, 61, 61f
Parazoa, 6
parietal eye
    hagfish and, 64
    of placoderms, 77
    primates and, 64
    of reptiles, 150–51, 150f
    of tuatara, 149–50, 149f, 150f
Parker, Andrew, 22
Parson's chameleon, 139, 139f, 143–45,
    143f, 144f, 145f
Parvancornia, 39, 39f
Pax6 gene
    development of, 29
    eye and, 17, 22
PaxB gene, 29
Pax genes, 10
    in eye development, 29–30
    in metazoans, 22
    in *Trichoplax adhaerens*, 17
pecten, of birds, 197–98, 197f
pelycosaurs, 131
peregrine falcon, 187–88, 187f, 188f
Peridinium foliaceum, 14
Periophthalmus argentilineatus. See
    silverstripe mudskipper
Permian period, 133–37
    extinction during, 136–38
    invertebrates of, 133–37
Petromyzon marinus, 66f
Phacopina, 41b
Phacops rana milleri, 39–42, 40b, 40f,
    41f, 42f
Phoca vitulina. See harbor seal
Phocops rana crassituberculata, 39–42, 40b,
    40f, 41f, 42f
Phoenicopterus ruber. See American
    flamingo
photo-autotrophs, 8
photolyases, 9
photoperiodicity, cryptochromes and, 9
photophores, *Histioteuthis heteropsis* and,
    169–70
photopigment, signal creation by, 34b
photoreception. *See also* ciliary cells;
        rhabdomeric cells
    of corals, 17
    of cyanobacteria, 7–8
    evolution of, 21
photoreceptors, 260–61
    of birds, 190
    in building an eye, 22–23
    of cuatro ojos, 90–91

of dolphins, 235
early types of, 25, 26f, 27f
evolution of, 241
of giant clams, 73
of human eye, 243, 244f
of piscine eye, 96
of scallop, 71f, 72, 72f
of silverstripe mudskipper, 95
photosynthesis, of cyanobacteria, 7–8
Phyllomedusa bicolor. See frogs
pigmentation
    in barreleye, 101–2, 101f
    for chromatic aberration, 101
pigment granules
    in building an eye, 22
    of conchs, 75, 76f
Pikaia gracilens, 63
pileated woodpecker, 198–99, 198f
pineal gland
    hagfish and, 64
    of placoderms, 77
    primates and, 64
pinhole camera, 23, 23t
pinhole eyes, of giant clams, 73, 73f
piscine camouflage, 107–8
piscine eye
    anatomy of, 85
    argenta of, 97
    development of, 77–84
    EOMs, 92–96
    habitat expansion and, 100–107
    lens of, 87–92
        accommodation, 88–89, 88f, 89f
        crystalline, 87–88
        Matthiessen's ratio, 89–91
    maturity of, 85–108
    neurology and optics of, 97–100, 98f
    octopus eye compared to, 165, 165f
    outer coats of, 85–87
        cornea, 85–87, 86f
        sclera, 85, 86f
    photoreceptor design of, 96
    retinal vascularization of, 96–97, 97f
    retina of, 96
    retinomotor pigment movement
        of, 97
placental mammals, 224–26
    retinal vascularization in, 251–52,
        251t, 252f
placoderms, 77–81, 77f, 78f, 79f, 80f
    Dunkleosteus, 79–80, 80f
    retinal vascularization in, 249
planktonic soup evolution, 230–34
Platyhelminthes, 30, 30f
Platynereis, 30–31, 31f, 32f
Platynereis bicaniculata, 31, 31f
platypus, 216–18, 217f, 218f
polarized light
    mantis shrimp and, 55, 55f
    octopus and, 166
    wolf spider and, 115

pollinators, 202–7, 206f
Polyclad flatworms. *See Platyhelminthes*
Pomacanthus semicirculatus. See koran
    angelfish
Pontella scutifer, 232–33, 232f
Portia fimbriata. See jumping spiders
postorbital septum, of *Tarsier spectrum*,
    226, 226f
predation, evolution and, 43
primary spectacle, 65f, 66–67
primates, 226–29
    black and white ruffed lemur,
        226–27, 226f
    color vision tuning of, 226–27
    humans, 240–42, 240f, 241f
    parietal eye and, 64
    pineal gland and, 64
    retinal vascularization in, 252, 252f
    Tarsier spectrum, 226–27, 226f
proboscis, of *Opabinia*, 44
prokaryotes, 5–8
    Archaea, 6, 6f
    bacteria, 6, 6f
    eukaryotes relationship to, 6
    first life, 5–6, 6f, 7f
    first witness, 6–8
proteorhodopsin, in prokaryotes, 8
Proterozoic eon, 12–16
protists, 6, 6f, 12
    brain and eye in, 21–22
    Erythropsidium, 14, 14f
    Euglena gracilis, 13, 13f
    Metazoa development from, 15–16
    nucleation of, 12–13
    subcellular eyes of, 13–14, 14f
    Warnowia, 14, 14f
protochordates, 63
proton pump, retinal, 8
protostomes, 29, 29f, 45b, 261
protractor lentis muscle, in
        elasmobranchs, 82, 255
pseudobranch, 79
Pseudoceros dimidiatus, 30, 30f
Pteropus scapulatus, 225, 225f
pterosaurs, 161–62, 162f
Ptilosarcus gurneyi, 25, 26f
pupil
    of birds, 189
    of conchs, 75, 76f
    of crocodilians, 158
    of cubozoan jellyfish, 20
    of dolphins, 235–36, 236f
    of human eye, 245
    of koran angelfish, 87–88, 87f, 88f
    of octopus, 165, 166f
Puto albicans. See male scale insect
pyramidalis muscle
    in nictitans operation, 186–87b, 187f
    in reptilian eye, 141–42
Pyrophorus phosphorescens. See lantern
    click beetle